FUNDAMENTALS
OF MULTICOMPONENT
DISTILLATION

McGraw-Hill Chemical Engineering Series

BUILDING THE LITERATURE OF A PROFESSION

Fifteen prominent chemical engineers first met in New York more than 50 years ago to plan a continuing literature for their rapidly growing profession. From industry came such pioneer practitioners as Leo H. Baekeland, Arthur D. Little, Charles L. Reese, John V. N. Dorr, M. C. Whitaker, and R. S. McBride. From the universities came such eminent educators as William H. Walker, Alfred H. White, D. D. Jackson, J. H. James, Warren K. Lewis, and Harry A. Curtis. H. C. Parmelee, then editor of *Chemical and Metallurgical Engineering*, served as chairman and was joined subsequently by S. D. Kirkpatrick as consulting editor.

After several meetings, this committee submitted its report to the McGraw-Hill Book Company in September 1925. In the report were detailed specifications for a correlated series of more than a dozen texts and reference books which have since become the McGraw-Hill Series in Chemical Engineering and which became the cornerstone of the chemical engineering curriculum.

From this beginning there has evolved a series of texts surpassing by far the scope and longevity envisioned by the founding Editorial Board. The McGraw-Hill Series in Chemical Engineering stands as a unique historical record of the development of chemical engineering education and practice. In the series one finds the milestones of the subject's evolution: industrial chemistry, stoichiometry, unit operations and processes, thermodynamics, kinetics, and transfer operations.

Chemical engineering is a dynamic profession, and its literature continues to evolve. McGraw-Hill and its consulting editors remain committed to a publishing policy that will serve, and indeed lead, the needs of the chemical engineering profession during the years to come.

THE SERIES

Bailey and Ollis: *Biochemical Engineering Fundamentals*
Bennett and Myers: *Momentum, Heat, and Mass Transfer*
Beveridge and Schechter: *Optimization: Theory and Practice*
Carberry: *Chemical and Catalytic Reaction Engineering*
Churchill: *The Interpretation and Use of Rate Data—The Rate Concept*
Clarke and Davidson: *Manual for Process Engineering Calculations*
Coughanowr and Koppel: *Process Systems Analysis and Control*
Danckwerts: *Gas Liquid Reactions*
Finlayson: *Nonlinear Analysis in Chemical Engineering*
Gates, Katzer, and Schuit: *Chemistry of Catalytic Processes*
Harriott: *Process Control*
Holland: *Fundamentals of Multicomponent Distillation*
Johnson: *Automatic Process Control*
Johnstone and Thring: *Pilot Plants, Models, and Scale-up Methods in Chemical Engineering*
Katz, Cornell, Kobayashi, Poettmann, Vary, Elenbaas, and Weinaug: *Handbook of Natural Gas Engineering*
King: *Separation Processes*
Klinzing: *Gas-Solid Transport*
Knudsen and Katz: *Fluid Dynamics and Heat Transfer*
Lapidus: *Digital Computation for Chemical Engineers*
Luyben: *Process Modeling, Simulation, and Control for Chemical Engineers*
McCabe and Smith, J. C.: *Unit Operations of Chemical Engineering*
Mickley, Sherwood, and Reed: *Applied Mathematics in Chemical Engineering*
Nelson: *Petroleum Refinery Engineering*
Perry and Chilton (Editors): *Chemical Engineers' Handbook*
Peters: *Elementary Chemical Engineering*
Peters and Timmerhaus: *Plant Design and Economics for Chemical Engineers*
Ray: *Advanced Process Control*
Reed and Gubbins: *Applied Statistical Mechanics*
Reid, Prausnitz, and Sherwood: *The Properties of Gases and Liquids*
Resnick: *Process Analysis and Design for Chemical Engineers*
Satterfield: *Heterogeneous Catalysis in Practice*
Sherwood, Pigford, and Wilke: *Mass Transfer*
Slattery: *Momentum, Energy, and Mass Transfer in Continua*
Smith, B. D.: *Design of Equilibrium Stage Processes*
Smith, J. M.: *Chemical Engineering Kinetics*
Smith, J. M., and Van Ness: *Introduction to Chemical Engineering Thermodynamics*
Thompson and Ceckler: *Introduction to Chemical Engineering*
Treybal: *Mass Transfer Operations*
Van Winkle: *Distillation*
Volk: *Applied Statistics for Engineers*
Walas: *Reaction Kinetics for Chemical Engineers*
Wei, Russell, and Swartzlander: *The Structure of the Chemical Processing Industries*
Whitwell and Toner: *Conservation of Mass and Energy*

FUNDAMENTALS
OF MULTICOMPONENT
DISTILLATION

Charles D. Holland

Texas A&M University

McGraw-Hill Book Company

New York   St. Louis   San Francisco   Auckland   Bogotá   Hamburg
Johannesburg   London   Madrid   Mexico   Montreal   New Delhi
Panama   Paris   São Paulo   Singapore   Sydney   Tokyo   Toronto

To My Wife
Eleanore

This book was set in Times Roman.
The editor was Diane D. Heiberg.
The production supervisor was Leroy A. Young.
Fairfield Graphics was printer and binder.

FUNDAMENTALS OF MULTICOMPONENT DISTILLATION

34567890 FGFG 898765432

Library of Congress Cataloging in Publication Data

Holland, Charles Donald.
    Fundamentals of multicomponent distillation.

    (McGraw-Hill chemical engineering series)
    Includes bibliographical references and index.
    1. Distillation. I. Title.
TP156.D5H64   660.2'8425   80-29238
ISBN 0-07-029567-0

CONTENTS

Chapter 15   Selected Topics in Matrix Operations and Numerical Methods for Solving Multivariable Problems   563

Appendixes

Indexes   619

PREFACE

This book constitutes an in-depth treatment of the subject of multicomponent distillation. It begins with first principles and goes to the frontiers of the subject. Each topic is introduced in an elementary and fundamental manner which makes the book suitable for the undergraduate student, the graduate student, and the practicing engineer. The subject matter is presented in the order of increasing difficulty and complexity.

The gap between the treatment of binary and multicomponent mixtures is closed in Chap. 1. This chapter is initiated by presenting the fundamental relationships and techniques needed for making bubble-point and dew-point calculations, and it is concluded by the presentation of techniques for solving a variety of special types of problems such as the separation of a multicomponent mixture by a single-stage flash process and the separation of a multicomponent mixture by use of multiple stages at the operating condition of total reflux.

In Chaps. 2 through 5, the theta methods and variations of the Newton-Raphson method are applied to all types of single columns and systems of columns in the service of separating both ideal and nonideal solutions. Applications of the techniques presented in Chaps. 2 through 5 to systems of azeotropic and extractive distillation columns are presented in Chap. 6. An extension of these same techniques as required for the solution of problems involving energy exchange between recycle streams is presented in Chap. 7. Special types of separations wherein the distillation process is accompanied by chemical reactions are treated in Chap. 8.

In Chap. 9, all of the techniques developed in Chaps. 1 through 8 are brought to bear in the design and operation of conventional and complex distillation columns. To complete the in-depth treatment of multicomponent distillation, the special topics of total reflux, minimum reflux, design of valve and sieve trays, plate efficiencies, design of packed columns, thermodynamic relationships, and selected numerical methods are presented in Chaps. 10 through 15. A Solutions Manual may be obtained (without cost) by Faculty members by writing directly to me or to McGraw-Hill.

Since the writings of any author are influenced by all that he has met, I am deeply indebted to all of my teachers, fellow faculty members, students, and the many past and present workers in the field of distillation. In particular, I wish to thank Professors P. T. Eubank and D. T. Hanson of the Department of Chemical Engineering as well as Professors Emeriti K. C. Klipple and H. A. Luther of the Department of Mathematics for their many helpful suggestions. The direction and advice generously offered by N. J. Tetlow and Ron McDaniel of Dow Chemical Company is appreciated immensely. I also want to thank Dr. Tetlow for his contributions to Chap. 14 (Thermodynamic Relationships for Multicomponent Mixtures) as coauthor. The helpful suggestions provided by W. L. Bolles of Monsanto are very much appreciated. For the support of the research (upon which much of this book is based) by David L. Rooke, Donald A. Rikard, and Holmes H. McClure (all of Dow Chemical Company), I am most thankful. Also, support of research provided by the Texas Engineering Experiment Station and the Center for Energy and Mineral Resources is appreciated. To both former and present graduate students (Najeh S. Al-Haj-Ali, G. W. Bentzen, Andy Feng, S. E. Gallun, Alejandro Gomez M., J. R. Haas, F. E. Hess, Alicia Izarraraz, P. E. Mommessin, and G. P. Pendon) who participated in this research, I salute you for your many contributions and I shall always be indebted to you. Finally, I want to pay special tribute to my Staff Assistant, Mrs. Wanda Greer, who helped make this book possible through her loyal service and effective assistance over the past eight years in the discharge of my administrative responsibilities.

Charles D. Holland

INTRODUCTION TO
THE FUNDAMENTALS OF DISTILLATION

In this chapter, the fundamental principles and relationships involved in making multicomponent distillation calculations are developed from first principles. To enhance the visualization of the application of the fundamental principles to this separation process, a variety of special cases are considered which include the determination of bubble-point and dew-point temperatures, single-stage flash separations, multiple-stage separation of binary mixtures, and multiple-stage separation of multicomponent mixtures at the operating conditions of total reflux.

The general objective of distillation is the separation of compounds that have different vapor pressures at any given temperature. The word *distillation* as used herein refers to the physical separation of a mixture into two or more fractions that have different boiling points.

If a liquid mixture of two volatile materials is heated, the vapor that comes off will have a higher concentration of the lower boiling material than the liquid from which it was evolved. Conversely, if a warm vapor is cooled, the higher boiling material has a tendency to condense in a greater proportion than the lower boiling material. The early distillers of alcohol for beverages applied these fundamental principles. Although distillation was known and practiced in antiquity and a commercial still had been developed by Coffey in 1832, the theory of distillation was not studied until the work of Sorel[14] in 1893. Other early workers were Lord Rayleigh[11] and Lewis.[8] Present-day technology has permitted the large-scale separation by distillation of ethylbenzene and *p*-xylene, which have only a 3.9°F difference in boiling points (Ref. 1).

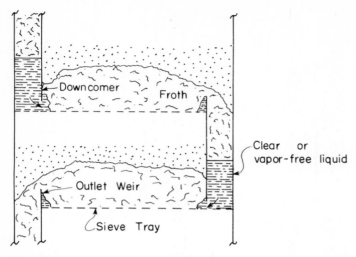

Figure 1-1 Interior of a column equipped with sieve trays.

A distillation column consists of a space for contacting vapor and liquid streams for the purpose of effecting mass transfer between the two phases. Although the contacting of two phases is generally effected by a series of plates (or trays), packed columns are becoming more widely used as discussed in Chap. 13. However, in the development of the fundamentals of the various calculational procedures in this and subsequent chapters, it is supposed that the column is equipped with plates.

In normal operation, there is a certain amount of liquid on each plate, and some arrangement is made for ascending vapors to pass through the liquid and make contact with it. The descending liquid flows down from the plate above through a downcomer, across the next plate, and then over a *weir* and into another downcomer to the next lower plate as shown in Fig. 1-1. For many years, *bubble caps* were used for contacting the vapor with the liquid. A variety of designs of bubble caps are shown in Fig. 1-2. These contacting devices promote the production of small bubbles of vapor with relatively large surface areas.

Over the past 20 years, most of the bubble-cap trays have been replaced by other types of contacting devices. New columns are usually equipped with either *valve trays* (see Fig. 1-3) or *sieve trays* (see Fig. 1-1), sometimes called *perforated trays*. In valve trays, the valve opens wider as the vapor velocity increases and closes as the vapor velocity decreases. This feature of opening and closing allows the valve to remain immersed in liquid and thereby preserve a liquid seal over wide ranges of liquid and vapor flow rates.

Distillation columns have been built as high as 338 feet. Diameters as large as 50 feet have been used. Operating pressures for distillation columns have been reported which range from 15 mmHg to 500 lb/in² abs. A typical commercial installation is shown in Fig. 1-4.

Figure 1-2 Various types of bubble caps used in distillation columns. (*By courtesy of Glitsch, Inc.*)

Figure 1-3 Portion of a Glitsch V-1 ballast tray. (*By courtesy of Glitsch, Inc.*)

Figure 1-4 Typical view of distillation columns at the Gulf refinery at Alliance, Louisiana. (*By courtesy Gulf Oil Corporation and Glitsch, Inc.*)

As indicated in Fig. 1-5, the overhead vapor V_2, upon leaving the top plate, enters the condenser where it is either partially or totally condensed. The liquid formed is collected in an accumulator from which the liquid stream L_1 (called *reflux*) and the top product stream D (called the *distillate*) are withdrawn. When the overhead vapor V_2 is totally condensed to the liquid state and the distillate D is withdrawn as a liquid, the condenser is called a *total condenser*. If V_2 is partially condensed to the liquid state to provide the reflux L_1 and the distillate D is withdrawn as a vapor, the condenser is called a *partial condenser*. The

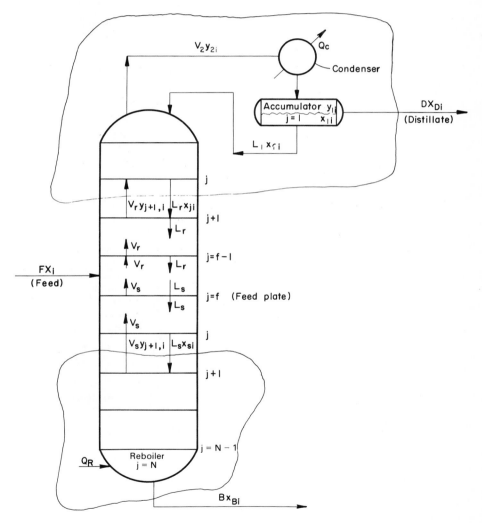

Figure 1-5 Sketch of a conventional column in which the total-flow rates are constant within the rectifying and stripping sections.

amount of liquid reflux is commonly expressed in terms of the *reflux ratio*, L_1/D. Although the internal liquid-to-vapor ratio L/V is sometimes referred to as the internal reflux ratio, the term reflux ratio will be reserved herein to mean L_1/D.

The liquid that leaves the bottom plate of the column enters the reboiler, where it is partially vaporized. The vapor produced is allowed to flow back up through the column, and the liquid is withdrawn from the reboiler and called the *bottoms* or *bottom product B*. In practice, the reboiler is generally located externally from the column.

1-1 FUNDAMENTAL PRINCIPLES INVOLVED IN DISTILLATION

To compute the composition of the top product D and the bottom product B which may be expected by use of a given distillation column operated at a given set of conditions, it is necessary to obtain a solution to equations of the following types:

1. Equilibrium relationships
2. Component-material balances
3. Total-material balances
4. Energy balances

Consider first the subject of equilibrium relationships.

Physical Equilibrium

A two-phase mixture is said to be in physical equilibrium if the following conditions are satisfied (Ref. 3).

1. The temperature T^V of the vapor phase is equal to the temperature T^L of the liquid phase.
2. The total pressure P^V throughout the vapor phase is equal to the total pressure P^L throughout the liquid phase. $\qquad\qquad$ (1-1)
3. The tendency of each component to escape from the liquid phase to the vapor phase is exactly equal to its tendency to escape from the vapor phase to the liquid phase.

In the following analysis it is supposed that a state of equilibrium exists, $T^V = T^L = T$, $P^V = P^L = P$, and the escaping tendencies are equal.

Now consider the special case where the third condition may be represented by Raoult's law

$$Py_i = P_i x_i \qquad\qquad (1\text{-}2)$$

where x_i and y_i are the mole fractions of component i in the liquid and vapor phases, respectively, and P_i is the vapor pressure of pure component i at the temperature T of the system.

The separation of a binary mixture by distillation may be represented in two-dimensional space while n-dimensional space is required to represent the separation of a multicomponent mixture ($i > 2$). The graphical method proposed by McCabe and Thiele[9] for the solution of problems involving binary mixtures is presented in a subsequent section. The McCabe-Thiele method makes use of an equilibrium curve which may be obtained from the "boiling-point diagram."

Construction and Interpretation of the Boiling-Point Diagram for Binary Mixtures

When a state of equilibrium exists between a vapor and a liquid phase composed of two components A and B, the system is described by the following set of independent equations

$$\left. \begin{array}{c} Py_A = P_A x_A \\ Py_B = P_B x_B \\ y_A + y_B = 1 \\ x_A + x_B = 1 \end{array} \right\}$$

Equilibrium relationships (1-3)

where it is understood that Raoult's law is obeyed. Since the vapor pressures P_A and P_B depend upon T alone, Eq. (1-3) consists of four equations in six unknowns. Thus, to obtain a solution to this set of equations, two variables must be fixed. [Observe that this result is in agreement with the Gibbs phase rule: $\mathscr{P} + \mathscr{V} = c + 2$. For the above case, the number of phases $\mathscr{P} = 2$, the number of components $c = 2$, and thus the number of degrees of freedom $\mathscr{V} = 2$, that is, the number of variables which must be fixed is equal to 2.] In the construction of the boiling-point diagram for a binary mixture, the total pressure P is fixed and a solution is obtained for each of several temperatures lying between the boiling-point temperature T_A of pure A and the boiling-point temperature T_B of pure B at the total pressure P. That is, when $T = T_A$, $P_A = P$ and when $T = T_B$, $P_B = P$.

The solution of the set of equations [Eq. (1-3)] for x_A in terms of P_A, P_B, and P is effected as follows. Addition of the first two equations followed by the elimination of the sum of the y's by use of the third expression yields

$$P = P_A x_A + P_B x_B \tag{1-4}$$

Elimination of x_B by use of the fourth equation of the set given by Eq. (1-3) followed by rearrangement of the result so obtained yields

$$x_A = \frac{P - P_B}{P_A - P_B} \tag{1-5}$$

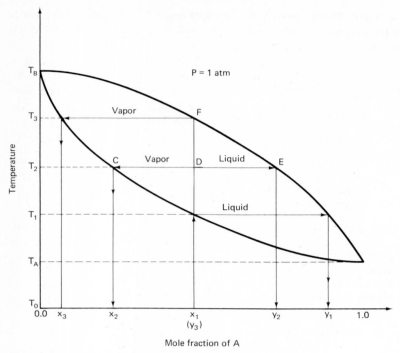

Figure 1-6 The boiling-point diagram.

From the definition of a mole fraction ($0 \le x_A \le 1$), Eq. (1-5) has a meaningful solution at a given P for every T lying between the boiling-point temperatures T_A and T_B of pure A and pure B, respectively. After x_A has been computed by use of Eq. (1-5) at the specified P and T, the corresponding value of y_A which is in equilibrium with the value of x_A so obtained is computed by use of the first expression of Eq. (1-3), namely,

$$y_A = \left(\frac{P_A}{P}\right) x_A \tag{1-6}$$

By plotting T versus x_A and T versus y_A, the lower and upper curves, respectively, of Fig. 1-6 are typical of those obtained when component A is more volatile than B. Component A is said to be more volatile than component B, if for all T in the closed interval $T_A \le T \le T_B$, the vapor pressure of A is greater than the vapor pressure of B, that is, $P_A > P_B$. The horizontal lines such as $\overline{CE}$ that join equilibrium pairs (x, y), computed at a given T and P by use of Eqs. (1-5) and (1-6), are commonly called *tie lines*.

> **Example 1-1** (Taken from Ref. 6 by courtesy *Instrument Society of America*).
> By use of the following vapor pressures for benzene and toluene [taken from *The Chemical Engineer's Handbook*, 2d ed., J. H. Perry (ed.) McGraw-

Hill, New York, 1941], compute the three equilibrium pairs (x, y) on a boiling-point diagram which correspond to the temperatures $T = 80.02°C$, $T = 100°C$, and $T = 110.4°C$. The total pressure is fixed at $P = 1$ atm. *Given:*

Temperature (°C)	P_A (benzene) (mmHg)	P_B (toluene) (mmHg)
80.02	760	300.0
84.0	852	333.0
88.0	957	379.5
92.0	1078	432.0
96.0	1204	492.5
100.0	1344	559.0
104.0	1495	625.5
108.0	1659	704.5†
110.4	1748	760.0

† In the more recent editions, the vapor pressure of 704.5 mm for toluene at 108°C is inaccurately listed as 740.5 or 741 mm.

SOLUTION At $T = 80.02°C$, $P_A = 760$, $P_B = 300$, and $P = 760$. Then Eq. (1-5) gives

$$x_A = \frac{P - P_B}{P_A - P_B} = \frac{760 - 300}{760 - 300} = 1$$

Thus

$$y_A = \frac{P_A}{P} x_A = 1$$

Therefore at the temperature $T = 80.02°C$, the curves T versus x_A and T versus y_A coincide at $(1, 80.02)$.

At $T = 110.4°C$, $P_A = 1748$, $P_B = 760$, and $P = 760$. Then by Eq. (1-5)

$$x_A = \frac{760 - 760}{1748 - 760} = 0$$

and thus

$$y_A = \left(\frac{1748}{760}\right)(0) = 0$$

Hence, the curves T versus x_A and T versus y_A again coincide at the point $(0, 110.4)$.

At any temperature between T_A and T_B, say $T = 100°C$, the calculations are carried out as follows

$$x_A = \frac{760 - 559}{1344 - 559} = \frac{201}{785} = 0.256$$

and

$$y_A = \left(\frac{1344}{760}\right)(0.256) = 0.453$$

These results give the point $(0.256, 100)$ on the T versus x_A curve and the point $(0.453, 100)$ on the T versus y_A curve. Other points on these curves for temperatures lying between T_A and T_B are located in the same manner.

A boiling-point diagram is a most convenient aid in the visualization of phase behavior. For definiteness, suppose P is fixed at 1 atm. Consider first the case of the liquid mixture of A and B at a temperature T_0, at a pressure of 1 atm, and with the composition $x_A = x_1$, $x_B = 1 - x_1$. As indicated by Fig. 1-6, such a mixture is in the single-phase region. Suppose the pressure is held fixed at 1 atm throughout the course of the following changes. First, suppose the mixture is heated to the temperature T_1. At this temperature, the first evidence of a vapor phase, a "bubble of vapor," may be observed. The temperature T_1 is called the *bubble-point* temperature of a liquid with the composition x_1. The mole fraction of A in the vapor in equilibrium with this liquid is seen to be y_1. As the mixture is heated from T_1 to T_2, vaporization continues. Since A has a greater escaping tendency than B, the liquid becomes leaner in A $(x_2 < x_1)$. The relative amounts of A and B vaporized also depend on their relative amounts in the liquid phase. As the liquid phase becomes richer in B, the vapor phase also becomes richer in B $(y_2 < y_1)$. Point D (the intersection of the horizontal line passing through T_2 and the vertical line passing through x_1) is seen to lie in the two-phase region. As outlined in Prob. 1-22, it can be shown that the ratio of the moles of vapor to the moles of liquid formed from a feed of composition x_1 at T_2 is equal to the ratio of $\overline{CD}/\overline{DE}$. Also, note that all initial liquid mixtures (at the temperature T_0) with the mole fraction of A lying between x_2 and y_2 will have the same equilibrium composition (x_2, y_2) at the temperature T_2 and pressure $P = 1$ atm. If the particular mixture $x_A = x_1$ at T_0 is heated until point F is reached, the equilibrium mixture (x_3, y_3) at T_3 is obtained. The temperature T_3 is called the *dew-point* temperature. At F, the last point in the two-phase region, all of the liquid is vaporized with the exception of, say, one drop. Thus, the dew-point temperature is seen to be that temperature at which the first drop of liquid is formed when a vapor with the composition $y_3 = x_1$ is cooled from a temperature greater than its dew-point temperature to its dew-point temperature, T_3.

Generalized Equilibrium Relationships

Unfortunately, the phase behavior of many mixtures is not adequately described by Raoult's law. A more precise statement of the third condition of Eq. (1-1) is that the partial molar-free energy of each component in the vapor phase is equal to its partial molar-free energy in the liquid phase (see Chap. 14). From this condition the following alternative but equivalent statement may be deduced

$$\hat{f}_i^V = \hat{f}_i^L \tag{1-7}$$

where $\hat{f}_i^V$ and $\hat{f}_i^L$ are the fugacities of component i in the vapor and liquid phases, respectively, evaluated at the compositions of the respective phases and at the P and T of the system. Equation (1-7) may be restated in the following equivalent form

$$\gamma_i^V f_i^V y_i = \gamma_i^L f_i^L x_i \tag{1-8}$$

where f_i^L, f_i^V = fugacities of pure component i in the liquid and vapor states, respectively, evaluated at the total pressure P and temperature T of the system

x_i, y_i = mole fractions of component i in the liquid and vapor phases, respectively

γ_i^L, γ_i^V = activity coefficients of component i in the liquid and vapor phases, respectively. $\gamma_i^L = \gamma_i^L(P, T, x_1, \ldots, x_c)$; $\gamma_i^V = \gamma_i^V(P, T, y_1, \ldots, y_c)$

If it may be assumed that the vapor forms an ideal solution, then $\gamma_i^V = 1$ for each i, and Eq. (1-8) reduces to

$$y_i = \gamma_i^L K_i x_i \tag{1-9}$$

where $K_i = f_i^L/f_i^V$, the ideal solution K value. The expression given by Eq. (1-9) is recognized as one form of Henry's law. If the liquid phase also forms an ideal solution ($\gamma_i^L = 1$ for all i), then Eq. (1-9) reduces to

$$y_i = K_i x_i \tag{1-10}$$

In some of the literature, the activity coefficients γ_i^V and γ_i^L are absorbed in K_i, that is, the product $\gamma_i^L K_i/\gamma_i^V$ is called K_i and an equation of the form of Eq. (1-10) is obtained which is applicable to systems described by Eq. (1-8).

If the effect of total pressure on the liquid fugacity is negligible in the neighborhood of the vapor pressure of pure component i, then

$$f_i^L\bigg|_{P,\,T} \cong f_i^L\bigg|_{P_i,\,T} = f_i^V\bigg|_{P_i,\,T} \tag{1-11}$$

where P_i is the vapor pressure of pure component i at the temperature T. If in addition to the assumptions required to obtain Eqs. (1-10) and (1-11), one also assumes that the vapor phase obeys the perfect gas law ($Pv = RT$), then Eq. (1-10) reduces to Raoult's law, Eq. (1-2).

Determination of the Bubble-Point and Dew-Point Temperatures of Multicomponent Mixtures

In the interest of simplicity, the equilibrium relationship given by Eq. (1-10) is used in the following developments. The state of equilibrium for a two-phase (vapor and liquid) system is described by the following equations where any number of components c are distributed between the two phases

Equilibrium relationships
$$\left\{ \begin{array}{l} y_i = K_i x_i \\[2mm] \sum_{i=1}^{c} y_i = 1 \qquad (i = 1, 2, \ldots, c) \\[2mm] \sum_{i=1}^{c} x_i = 1 \end{array} \right. \qquad (1\text{-}12)$$

Since K_i is a function of the total pressure P and the temperature T $[K_i = K_i(P, T)]$, it is evident that the expressions represented by Eq. (1-12) consists of $c + 2$ equations in $2c + 2$ unknowns. Thus, to obtain a solution to these equations, c variables must be fixed.

For the particular case where $c - 1$ values of x_i and the total pressure P are fixed, the temperature T required to satisfy these equations is called the *bubble-point temperature*. The cth mole fraction may be found by use of the $(c - 1)$ fixed values of x_i and the last expression given by Eq. (1-12). When the first expression is summed over all components and the sum of the y_i's eliminated by use of the second expression given by Eq. (1-12), the following result is obtained

$$\sum_{i=1}^{c} K_i x_i = 1 \qquad (1\text{-}13)$$

Equation (1-13) consists of one equation in one unknown, the temperature. The form of the implicit function $K_i(T)$ generally requires that the solution of Eq. (1-13) for the bubble-point temperature be effected by a trial-and-error procedure. Of the many numerical methods for solving such a problem, only Newton's method[2, 5] is presented. In the application of this method, it is convenient to restate Eq. (1-13) in functional form as follows

$$f(T) = \sum_{i=1}^{c} K_i x_i - 1 \qquad (1\text{-}14)$$

Thus, the bubble-point temperature is that T that makes $f(T) = 0$. In the solution of this problem by use of Newton's method, an expression for $f'(T)$ is needed. Term-by-term differentiation of Eq. (1-14) yields

$$f'(T) = \sum_{i=1}^{c} x_i \frac{dK_i}{dT} \qquad (1\text{-}15)$$

Newton's method is initiated by the selection of an assumed value for T, say T_n. Then the values of $f(T_n)$ and $f'(T_n)$ are determined. The improved value of T, denoted by T_{n+1} is found by application of Newton's formula (see Prob. 1-6)

$$T_{n+1} = T_n - \frac{f(T_n)}{f'(T_n)} \qquad (1\text{-}16)$$

The value so obtained for T_{n+1} becomes the assumed value for the next trial. This procedure is repeated until $|f(t)|$ is less than some small preassigned positive number ε. Observe that when the T has been found that makes $f(T) = 0$, each term $K_i x_i$ of the summation in Eq. (1-14) is equal to y_i. In illustrative Example 1-2 as well as those which follow, synthetic functions for the K values and the enthalpies were selected in order to keep the arithmetic simple.

Example 1-2 (a) If for a three-component mixture, the following information is available, compute the bubble-point temperature at the specified pressure of $P = 1$ atm by use of Newton's method. Take the first assumed value of T_n to be equal to $100°F$.

(b) Find the composition of the vapor in equilibrium with the liquid

	$K_i = C_i \exp(-E_i/T)$, T in °R		
Component	C_i	E_i	x_i
1	$4 \times 10^3/P$†	4.6447×10^3	1/3
2	$8 \times 10^3/P$	4.6447×10^3	1/3
3	$12 \times 10^3/P$	4.6447×10^3	1/3

† P is in atm.

SOLUTION (a) The formula for $f'(T)$ which is needed in Newton's method is obtained as follows

$$\frac{df(T)}{dT} = \sum_{i=1}^{c} x_i \frac{dK_i}{dT} = \sum_{i=1}^{c} x_i (C_i e^{-E_i/T}) \left(\frac{E_i}{T^2}\right)$$

$$= \sum_{i=1}^{c} (K_i x_i)(E_i/T^2)$$

Trial 1. Assume $T = 100°F = 560°R$

Component	K_i @ 560°R and 1 atm	$K_i x_i$	E_i/T^2	$(K_i x_i)(E_i/T^2)$
1	1	1/3	0.0148	0.00493
2	2	2/3	0.0148	0.00986
3	3	3/3	0.0148	0.0148
		6/3 = 2		0.0296

Thus

$$f(560) = \sum_{i=1}^{c} K_i x_i - 1 = 2 - 1 = 1$$

$$f'(560) = \sum_{i=1}^{c} x_i \frac{dK_i}{dT} = 0.0296$$

Then

$$T_2 = T_1 - \frac{f(T_1)}{f'(T_1)} = 560 - \frac{1}{0.0296} = 526°\text{R}$$

Trial 2. Assume $T = 526°\text{R}$

Component	K_i @ 526°R and 1 atm	$K_i x_i$	E_i/T^2	$(K_i x_i)(E_i/T^2)$
1	0.585	0.195	0.0168	0.00328
2	1.170	0.390	0.0168	0.00655
3	1.755	0.585	0.0168	0.00993
		1.170		0.0197

Thus

$$f(526) = 1.17 - 1 = 0.17$$
$$f'(526) = 0.0197$$

Then

$$T_3 = 526 - \frac{(0.17)}{(0.0197)} = 517.4°\text{R}$$

Trial 3. Assume $T = 517.4°\text{R}$ and repeat the steps shown above. The results so obtained are as follows

$$f(517.4) = 1.0102 - 1 = 0.0102$$
$$f'(517.4) = 0.0175$$

Then

$$T_4 = 517.4 - \frac{(0.0102)}{(0.0175)} = 516.8°\text{R}$$

Trial 4. Assume $T = 516.8°\text{R}$

$$f(516.8) = 0.999 - 1 = -0.001$$

which is within the desired accuracy of the calculations. Thus, the bubble-point temperature is $516.8°\text{R} = 56.8°\text{F}$.

(b) At equilibrium $y_i = K_i x_i$. Thus

Component	K_i @ 516.8°R 1 atm	$y_i = K_i x_i$
1	0.5	0.167
2	1.0	0.133
3	1.5	0.500

When the $\{y_i\}$ and P are fixed rather than the $\{x_i\}$ and P, the solution temperature of the expressions given by Eq. (1-12) is called the *dew-point temperature*. By rearranging the first expression of Eq. (1-12) to the form $x_i = y_i/K_i$ and carrying out steps analogous to those described above, the dew-point function $F(T)$ is obtained

$$F(T) = \sum_{i=1}^{c} \frac{y_i}{K_i} - 1 \tag{1-17}$$

The dew-point temperature is that T that makes $F(T) = 0$. In this case

$$F'(T) = -\sum_{i=1}^{c} \frac{y_i}{K_i^2} \frac{dK_i}{dT} \tag{1-18}$$

When the T is found that makes $F(T) = 0$, each term y_i/K_i of the summation in Eq. (1-17) is equal to x_i.

Now observe that if after the bubble-point temperature has been determined for a given set of x_i's, the set of y_i's so obtained are used to determine the dew-point temperature at the same pressure, it will be found that these two temperatures are equal. For a binary mixture, this result is displayed graphically in Fig. 1-6. For example, a bubble-point temperature calculation on the basis of the $\{x_{1i}\}$ yields the bubble-point temperature T_1 and the composition of the vapor $\{y_{1i}\}$. Then a dew-point temperature on the set $\{y_{1i}\}$ yields the dew-point temperature T_1 and the original set of x_{1i}'s.

Use of the K_b Method for the Determination of Bubble-Point and Dew-Point Temperatures

Robinson and Gilliland[12] pointed out that if the relative values of the K_i's or P_i's are independent of temperature, the expressions given by Eq. (1-12) may be rearranged in a manner such that trial-and-error calculations are avoided in the determination of the bubble-point and dew-point temperatures. The ratio K_i/K_b is called the relative volatility α_i of component i with respect to component b, that is,

$$\alpha_i = \frac{K_i}{K_b} \tag{1-19}$$

where K_i and K_b are evaluated at the same temperature and pressure. Component b may or may not be a member of the given mixture under consideration.

When the $\{x_i\}$ and the pressure P are given and it is desired to determine the bubble-point temperature, the formula needed may be developed by first rewriting the first expression of Eq. (1-12) as follows

$$y_i = \left(\frac{K_i}{K_b}\right)K_b x_i = \alpha_i K_b x_i \tag{1-20}$$

Summation of the members of Eq. (1-20) over all components i, followed by rearrangement yields

$$K_b = \frac{1}{\sum\limits_{i=1}^{c} \alpha_i x_i} \qquad (1\text{-}21)$$

Since the α_i's are independent of temperature, they may be computed by use of the values of K_i and K_b evaluated at any arbitrary value of T and at the specified pressure. After K_b has been evaluated by use of Eq. (1-21), the desired bubble-point temperature is found from the known relationship between K_b and T.

If the y_i's are known instead of the x_i's, then the desired formula for the determination of the dew-point temperature is found by first rearranging Eq. (1-20) to the following form

$$K_b x_i = \frac{y_i}{\alpha_i}$$

and then summing over all components to obtain

$$K_b = \sum_{i=1}^{c} \frac{y_i}{\alpha_i} \qquad (1\text{-}22)$$

This equation is used to determine the dew-point temperature in a manner analogous to that described for the determination of the bubble-point temperature by use of Eq. (1-21).

Many families of compounds are characterized by the fact that their vapor pressures may be approximated by the Clausius-Clapeyron equation, and by the fact that their latent heats of vaporization are approximately equal. The logarithm of the vapor pressures of the members of such families of compounds fall on parallel lines when plotted against the reciprocal of the absolute temperature. For any two members i and b of such a mixture, it is readily shown that α_i is independent of temperature.

Although there exists many systems whose α_i's are very nearly constant and Eqs. (1-21) and (1-22) are applicable for the determination of the bubble-point and dew-point temperatures, respectively, the greatest use of these relationships lies in their application in the iterative procedures for solving multicomponent distillation problems as described in Chap. 2.

Example 1-3 Use the K_b method to solve Example 1-2.

SOLUTION Since K_b may be selected arbitrarily, take $K_b = K_1$. Assume $T = 100°F = 560°R$.

Component	X_i	K_i @ 560°R 1 atm	$\alpha_i = \dfrac{K_i}{K_1}$	$\alpha_i x_i$
1	1/3	1	1	1/3
2	1/3	2	2	2/3
3	1/3	3	3	$\dfrac{3/3}{6/3 = 2}$

Thus

$$K_b = \frac{1}{\displaystyle\sum_{i=1}^{c} \alpha_i x_i} = \frac{1}{2} = K_1$$

Since $K_1 = C_1 e^{-E_1/T}$, it follows that

$$T = \frac{E_1}{\ln C_1/K_b} = \frac{4.6447 \times 10^3}{\ln 8 \times 10^3} = 516.8°R$$

1-2 SEPARATION OF MULTICOMPONENT MIXTURES BY USE OF ONE EQUILIBRIUM STAGE

Each of the separation processes considered in Secs. 1-2 and 1-3 consist of special cases of the general separation problem in which a multicomponent mixture is to be separated into two or more parts through the use of any number of plates.

The boiling-point diagram (Fig. 1-6) is useful for the visualization of the necessary conditions required for a flash to occur. Suppose that feed to be flashed has the composition $X_i = x_{1i}$ ($x_{1,A}$ and $x_{1,B}$), and further suppose that this liquid mixture at the temperature T_0 and the pressure $P = 1$ atm is to be flashed by raising the temperature to the specified flash temperature $T_F = T_2$ at the specified flash pressure $P = 1$ atm. First observe that the bubble-point temperature of the feed T_{BP} at $P = 1$ atm is T_1. The dew-point temperature, T_{DP}, of the feed at the pressure $P = 1$ atm is seen to be T_3. Then it is obvious from Fig. 1-6 that a necessary condition for a flash to occur at the specified pressure is that

$$T_{BP} < T_F < T_{DP} \tag{1-23}$$

In practice, the flash process is generally carried out by reducing the pressure on the feed stream rather than by heating the feed at constant pressure as described above.

To determine whether the feed will flash at a given T_F and P, the above inequality may be used by determining the bubble-point and dew-point temperatures of the feed at the specified pressure P. In the determination of the bubble-point temperature of the feed at the specified P of the flash, the $\{x_i\}$ in Eq. (1-14) are replaced by the $\{X_i\}$ of the feed, and in the determination of the dew-point temperature at the specified pressure, the $\{y_i\}$ in Eq. (1-17) are replaced by the $\{X_i\}$. Alternatively, the inequality given by Eq. (1-23) is satisfied if at the specified T_F and P

$$f(T_F) > 0$$
$$F(T_F) > 0 \tag{1-24}$$

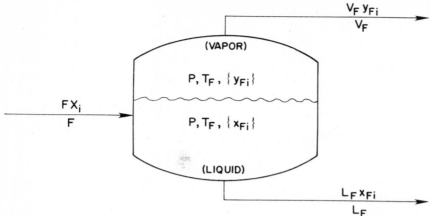

Figure 1-7 The flash process.

where

$$f(T_F) = \sum_{i=1}^{c} K_{Fi} X_i - 1$$

$$F(T_F) = \sum_{i=1}^{c} \frac{X_i}{K_{Fi}} - 1 \tag{1-25}$$

The symbol K_{Fi} represents the K value of component i evaluated at T_F and P.

The two types of flash calculations which are commonly made are generally referred to as *isothermal* and *adiabatic* flashes.

Isothermal Flash Process

The name "isothermal flash" is commonly given to the single-stage separation process shown in Fig. 1-7 for which the flash temperature T_F and pressure P are specified as well as the total flow rate F and composition $\{X_i\}$ of the feed. The name "isothermal flash" originated, no doubt, from the fact that the temperature of the contents of the flash drum as well as the vapor and liquid streams formed by the flash is fixed at T_F. The flash temperature T_F is not necessarily equal to the feed temperature prior to its flashing.

For the set of specifications stated above, the problem is to find the total flow rates V_F and L_F and the respective compositions $\{y_{Fi}\}$ and $\{x_{Fi}\}$ of the vapor and liquid streams formed by the flash process.

In addition to the $c + 2$ equations required to describe the state of equilibrium between the vapor and liquid phases [see Eq. (1-12)], c additional component-material balances which enclose the flash chamber are required to describe the isothermal flash process. Thus, the independent equations required to describe this flash process are as follows

$$\begin{cases} y_{Fi} = K_{Fi}x_{Fi} & (i = 1, 2, \ldots, c) \\ \\ \sum_{i=1}^{c} y_{Fi} = 1 \\ \\ \sum_{i=1}^{c} x_{Fi} = 1 \end{cases}$$

Equilibrium relationships $\qquad\qquad\qquad\qquad\qquad$ (1-26)

Material balances $\qquad FX_i = V_F y_{Fi} + L_F x_{Fi} \qquad (i = 1, 2, \ldots, c)$

Equation (1-26) is seen to represent $2c + 2$ equations in $2c + 2$ unknowns $[V_F, L_F, \{y_{Fi}\}, \{x_{Fi}\}]$.

This system of nonlinear equations is readily reduced to one equation in one unknown (say V_F) in the following manner. First observe that the total material balance expression (a dependent equation) may be obtained by summing each member of the third expression of Eq. (1-26) over all components to give

$$F \sum_{i=1}^{c} X_i = V_F \sum_{i=1}^{c} y_{Fi} + L_F \sum_{i=1}^{c} x_{Fi} \qquad \text{or} \qquad F = V_F + L_F \qquad (1\text{-}27)$$

The relationships given by Eq. (1-26) may be reduced to one equation in one unknown in a variety of ways, and a variety of forms of the flash function may be obtained. One form of the flash function is developed below and a different form is developed in Chap. 4 in the formulation of multiple-stage problems. Elimination of the y_{Fi}'s from the last expression given by Eq. (1-26) by use of the first expression, followed by rearrangement, yields

$$x_{Fi} = \frac{X_i}{L_F/F + V_F K_{Fi}/F} \qquad (1\text{-}28)$$

Elimination of L_F from Eq. (1-28) by use of Eq. (1-27) yields

$$x_{Fi} = \frac{X_i}{1 - \Psi(1 - K_{Fi})} \qquad (1\text{-}29)$$

where

$$\Psi = \frac{V_F}{F}$$

When each side of Eq. (1-29) is summed over all components i and the result so obtained is restated in functional form, the following expression of the flash function is obtained

$$P(\Psi) = \sum_{i=1}^{c} \frac{X_i}{[1 - \Psi(1 - K_{Fi})]} - 1 \qquad (1\text{-}30)$$

and

$$P'(\Psi) = \sum_{i=1}^{c} \frac{X_i(1 - K_{Fi})}{[1 - \Psi(1 - K_{Fi})]^2} \qquad (1\text{-}31)$$

A graph of the branch of the function $P(\Psi)$ which contains the positive root is presented in Fig. 1-8. An examination of this curve shows that Newton's method always converges to the desired root when $\Psi = 1$ is taken to be the first assumed value of the root. After the positive root that makes $P(\Psi) = 0$ has been found, both V_F and L_F may be calculated by use of the fact that $\Psi = V_F/F$ and the total material balance given by Eq. (1-27). Also, a comparison of Eqs. (1-29) and (1-30) shows that each term in the summation of $P(\Psi) = 0$ is one of the solution values of x_{Fi}. After the solution set of x_{Fi}'s has been computed, the corresponding set of y_{Fi}'s is found by using the first expression of Eq. (1-26), $y_{Fi} = K_{Fi} x_{Fi}$.

Example 1-4 (Taken from Ref. 6 by courtesy *Instrument Society of America*.) It is proposed to flash the following feed at a specified temperature $T_F = 100°F$ and a pressure $P = 1$ atm.

Component	K_i	X_i
1	$K_i = \dfrac{10^{-2}T\dagger}{3P\dagger}$	1/3
2	$K_2 = \dfrac{2 \times 10^{-2}T}{P}$	1/3
3	$K_3 = \dfrac{7 \times 10^{-2}T}{2P}$	1/3

† T is in °F and P is in atm.

If the feed rate to the flash drum is $F = 100$ mol/h, compute the vapor and liquid rates V_F and L_F leaving the flash as well as the respective mole fractions $\{y_{Fi}\}$ and $\{x_{Fi}\}$ of these streams.

SOLUTION First, the specified value of T_F will be checked to determine whether or not it lies between the bubble-point and dew-point temperatures of the feed.

Component	K_{Fi} @ $P = 1$ $T_F = 100°F$	X_i	$K_{Fi}X_i$	$\dfrac{X_i}{K_{Fi}}$
1	1/3	1/3	1/9	1
2	2	1/3	2/3	1/6
3	7/2	1/3	7/6	2/21
			1.94	1.262

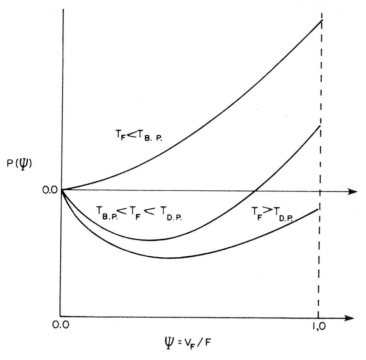

Figure 1-8 Graphical representation of the flash function $P(\Psi)$.

Thus

$$f(T_F) = \sum_{i=1}^{c} K_{Fi}X_i - 1 = 1.94 - 1 = 0.94 > 0$$

$$F(T_F) = \sum_{i=1}^{c} \frac{X_i}{K_{Fi}} - 1 = 1.262 - 1 = 0.262 > 0$$

and thus $T_{BP} < 100 < T_{DP}$.

Trial 1. Assume $\Psi = 1$.

Component	K_i	$1 - K_{Fi}$	$\Psi(1 - K_{Fi})$	$1 - \Psi(1 - K_{Fi})$
1	1/3	2/3	2/3	1/3
2	2	−1	−1	2
3	7/2	−5/2	−5/2	7/2

Components	$\dfrac{X_i}{1 - \Psi(1 - K_{Fi})}$	$\dfrac{X_i}{[1 - \Psi(1 - K_{Fi})]^2}$	$\dfrac{X_i(1 - K_i)}{[1 - (1 - K_{Fi})]^2}$
1	1.0000	3.0000	2.0000
2	0.1667	0.0833	−0.0833
3	0.0952	0.0272	−0.0680
	1.2619		1.8487

$$P(1) = 1.2619 - 1 = 0.2619$$

$$P'(1) = 1.8487$$

$$\Psi_2 = 1 - \left(\frac{0.2619}{1.8487}\right) = 1 - 0.1417 = 0.8583$$

Trial 2. Assume $\Psi = 0.8583$ and repeat the steps shown in the first trial. The results so obtained are as follows

$$P(0.8583) = 1.0651 - 1 = 0.0651$$

$$P'(0.8583) = 1.0358$$

$$\Psi_3 = 0.8583 - \left(\frac{0.0651}{1.0358}\right) = 0.7955$$

Continuation of this procedure gives the solution value of $\Psi = 0.787$. Thus, $V_F = 78.7$, $L_F = 21.3$, and the solution sets $\{x_{Fi}\}$ and $\{y_{Fi}\}$ are as follows

Compo-nent	$1 - K_{Fi}$	$\Psi(1 - K_{Fi})$	$1 - \Psi(1 - K_{Fi})$	$x_{Fi} = \dfrac{X_i}{1 - \Psi(1 - K_{Fi})}$	$y_{Fi} = K_{Fi} x_{Fi}$
1	0.667	0.525	0.475	0.701	0.234
2	−1.000	−0.787	1.787	0.187	0.374
3	−2.500	−1.968	2.968	0.112	0.392

Up to this point no mention has been made of the manner of satisfying the energy requirement of the flash. The specification of T_F implies that the feed either possesses precisely the correct amount of energy for the flash to occur at T_F at the specified P or that energy is to be added or withdrawn at the flash drum as required. It is common practice to adjust the heat content of the feed before it reaches the flash drum such that the flash occurs adiabatically; that is, the heat Q added at the flash drum is equal to zero.

After the solution $[V_F, L_F, \{y_{Fi}\}, \{x_{Fi}\}]$ has been found for a given isothermal flash problem, the heat content H that the feed must possess in order for the flash to occur adiabatically [$Q = 0$ at the flash drum] may be found by use of the enthalpy balance which encloses the entire process

$$FH = V_F H_F + L_F h_F \tag{1-32}$$

When the vapor V_F and liquid L_F form ideal solutions, the enthalpies H_F and h_F of the vapor and liquid streams, respectively, may be computed as follows

$$H_F = \sum_{i=1}^{c} H_{Fi} y_{Fi} \quad \text{and} \quad h_F = \sum_{i=1}^{c} h_{Fi} x_{Fi} \tag{1-33}$$

The above procedure may also be used in the solution of adiabatic flash problems as described below.

Example 1-5 (Holland,[6] by courtesy *Instrument Society of America.*) On the basis of the solution to Example 1-4, compute the enthalpy H which the feed must possess in order for the flash to occur adiabatically.

Given:

Component	h_i (Btu/lb mol)	H_i (Btu/lb mol)
1	$h_1 = 10,000 + 30T$†	$H_1 = 17,000 + 30T$†
2	$h_2 = 8,000 + 20T$	$H_2 = 13,000 + 20T$
3	$h_3 = 500 + T$	$H_3 = 800 + T$

† T is in °F.

SOLUTION Calculation of the enthalpy H of the feed:

Component	x_{Fi}	y_{Fi}	h_{Fi} @ $T_F = 100°F$	$h_{Fi} x_{Fi}$	H_{Fi} @ $T_F = 100°F$	$H_{Fi} y_{Fi}$
1	0.701	0.234	13,000	9,113	20,000	4,680
2	0.187	0.374	10,000	1,870	15,000	5,610
3	0.112	0.392	600	67	900	353
				11,050		10,643

Then $h_F = 11,050$ Btu/lb mol and $H_F = 10,643$ Btu/lb mol, and thus

$$H = \frac{V_F H_F}{F} + \frac{L_F h_F}{F} = (0.787)(10,643) + (0.213)(11,050)$$

$$= 10,730 \text{ Btu/lb mol}$$

Adiabatic Flash Process

The term *adiabatic flash* is used to describe the problem wherein the following specifications are made: P, $Q = 0$ [no heat is added at the flash drum], H, $\{X_i\}$, and F. In this case there are $2c + 3$ unknowns $[T_F, V_F, L_F, \{y_{Fi}\}, \{x_{Fi}\}]$. The independent equations are also $2c + 3$ in number, the $2c + 2$ given by Eq. (1-26) plus the enthalpy balance given by Eq. (1-32), that is,

$$
\begin{aligned}
& y_{Fi} = K_{Fi} x_{Fi} && (i = 1, 2, \ldots, c) \\
\text{Equilibrium} \quad & \sum_{i=1}^{c} y_{Fi} = 1 && \text{(1-34)} \\
\text{relationships} \quad & \sum_{i=1}^{c} x_{Fi} = 1 &&
\end{aligned}
$$

Material balances: $\quad FX_i = V_F y_{Fi} + L_F x_{Fi} \quad (i = 1, 2, \ldots, c)$

Enthalpy balance: $\quad FH = V_F H_F + L_F h_F$

One relatively simple method for solving an adiabatic flash problem consists of the repeated use of the procedure described above whereby an H_n is computed for each assumed T_{Fn} where n denotes the trial number. The problem then reduces to finding a T_{Fn} such that the resulting H_n is equal to the specified value H; that is, it is desired to find the T_{Fn} such that $\delta(T_{Fn}) = 0$, where

$$\delta(T_{Fn}) = \delta_n = H_n - H \qquad (1\text{-}35)$$

One numerical method for solving such a problem is called interpolation *regula falsi* (see Probs. 1-7 and 1-8). This method consists of the linear interpolation between the most recent pair of points (T_{Fn}, δ_n) and $(T_{F,n+1}, \delta_{n+1})$ by use of the following formula

$$T_{F,n+2} = \frac{T_{F,n+1}\,\delta_n - T_{Fn}\,\delta_{n+1}}{\delta_n - \delta_{n+1}} \qquad (1\text{-}36)$$

To initiate this interpolation procedure, it is necessary to evaluate δ for each of two assumed temperatures T_{F1} and T_{F2}. Then Eq. (1-36) is applied to obtain T_{F3}. After δ_3 has been obtained, the new temperature T_{F4} is found by interpolation between the points (T_{F2}, δ_2) and (T_{F3}, δ_3). When $|\delta|$ has been reduced to a value less than some arbitrarily small, preassigned positive number, the desired solution is said to have been obtained.

It should be pointed out that the equations required to describe the adiabatic flash are of precisely the same form as those required to describe the separation process which occurs on the plate of a distillation column in the process of separating a multicomponent mixture.

Other methods for solving the adiabatic flash problems are presented in Chaps. 4 and 5. The method presented in Chap. 5 is recommended for the solution of problems involving highly nonideal solutions.

1-3 MULTIPLE-STAGE SEPARATION OF BINARY MIXTURES

Although all of the separation problems involving binary mixtures may be solved by use of the general methods presented in subsequent chapters for multi-component mixtures, it is, nevertheless, rewarding to consider the special case of the separation of binary mixtures because this separation may be represented graphically in two-dimensional space. Many of the concepts of distillation may be illustrated by the graphical method of design proposed by McCabe and Thiele.[9]

The McCabe-Thiele Method

In the description of this process, the following symbols are used in addition to those explained above. The mole fraction of the most volatile component in the feed is represented by X, in the distillate by X_D, and in the bottoms by x_B. The subscript j is used as the counting integer for the number of the stages. Since the distillate is withdrawn from the accumulator ($j = 1$) and the bottoms is withdrawn from the reboiler ($j = N$), the mole fractions in the distillate and bottoms have double representation; that is, $X_{Di} = x_{1i}$ (for a column having a total condenser) and $x_{Bi} = x_{Ni}$. For the case where the column has a partial condenser (D is withdrawn as a vapor), $X_{Di} = y_{1i}$.

The *rectifying section* consists of the partial or total condenser and all plates down to the feed plate. The *stripping section* consists of the feed plate and all plates below it including the reboiler. When the total molar flow rates do not vary from plate to plate within each section of the column, they are denoted by V_r (vapor) and L_r (liquid), in the rectifying section and by V_s and L_s in the stripping section. The feed rate F, distillate rate D, bottoms rate B, and reflux rate L_1 are all expressed on a molar basis.

The design method of McCabe and Thiele[9] is best described by solving the following numerical example.

> **Example 1-6** It is desired to find the minimum number of perfect plates required to separate an equal molar mixture of benzene and toluene into a distillate product containing 96 percent benzene ($X_D = 0.96$) and a bottom product containing no more than 5 percent benzene ($x_B = 0.05$) at the following operating conditions: (1) the column pressure is 1 atm, and a total condenser is to be used (D is a liquid), (2) the thermal condition of the feed is such that the rate L_s at which liquid leaves the feed plate is given by $L_s = L_r + 0.6F$, and (3) a reflux ratio $L_1/D = 2.2$ is to be employed. The equilibrium sets $\{x_A, y_A\}$ of benzene used to construct the equilibrium curve shown in Fig. 1-9 were found by solving Prob. 1-1.

This set of specifications fixes the system; that is, the number of independent equations that describe the system is equal to the number of unknowns. Before solving this problem, the equations needed are developed. First, the equilibrium

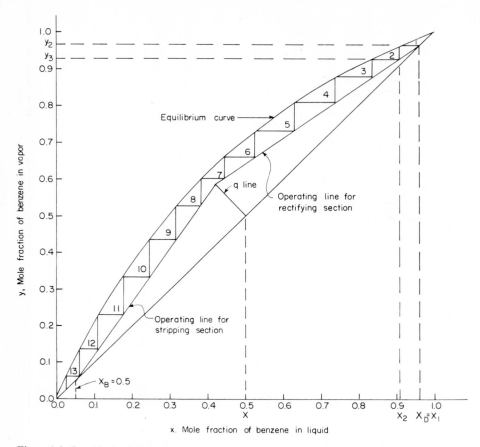

Figure 1-9 Graphical solution of Example 1-6 by the McCabe-Thiele method.

pairs $\{x, y\}$ satisfying the equilibrium relationship $y = Kx$ may be read from a boiling-point diagram (see part (a) of Prob. 1-1) and plotted in the form of y versus x to give the equilibrium curve; see Fig. 1-9. Observe that the equilibrium pairs $\{x, y\}$ are those mole fractions connected by the tie lines of the boiling-point diagram; see Fig. 1-6.

A component-material balance enclosing the top of the column and plate j (see Fig. 1-5) is given by

$$y_{j+1} = \left(\frac{L_r}{V_r}\right)x_j + \frac{DX_D}{V_r} \tag{1-37}$$

Similarly, for the stripping section, the component-material balance (see Fig. 1-5) is given by

$$y_{j+1} = \left(\frac{L_s}{V_s}\right)x_j - \frac{Bx_B}{V_s} \tag{1-38}$$

The component-material balance enclosing the entire column is given by

$$FX = DX_D + Bx_B \tag{1-39}$$

The total molar flow rates within each section of the column are related by the following defining equation for q, namely

$$L_s = L_r + qF \tag{1-40}$$

By means of a total material balance enclosing plates $f - 1$ and f, it is readily shown through the use of Eq. (1-40) that

$$V_r - V_s = (1 - q)F \tag{1-41}$$

By means of energy balances, it can be shown that q is approximately equal to the heat required to vaporize one mole of feed divided by the latent heat of vaporization of the feed (see Prob. 1-22).

Since Eqs. (1-37) and (1-38) are straight lines, they intersect at some point (x_I, y_I), provided of course they are not parallel. When the point of intersection is substituted into Eqs. (1-37) and (1-38) and L_r, V_r, L_s, V_s, x_B, and X_D are eliminated by use of Eqs. (1-37) through (1-41), the following equation for the q *line* is obtained

$$y_I = -\left(\frac{q}{1-q}\right)x_I + \left(\frac{1}{1-q}\right)X \tag{1-42}$$

SOLUTION OF EXAMPLE 1-6 With the aid of the above equations, the number of plates required to effect the specified separation may be determined. To plot the operating line [Eq. (1-37)] for the rectifying section, the y intercept (DX_D/V_r) is computed in the following manner. Since $V_r = L_r + D$, and $L_1 = L_r$, it follows that

$$\frac{DX_D}{V_r} = \frac{X_D}{(L_r/D) + 1} = \frac{0.96}{3.2} = 0.3$$

Since $y_2 = X_D$ (for a total condenser), the point (y_2, X_D) lies on the $45°$ diagonal. The y intercept and the point (y_2, X_D) locate the operating line for the rectifying section as shown in Fig. 1-9.

When $x_I = X$ is substituted in Eq. (1-42), the result $y_I = X$ is obtained, and hence the q line passes through point (X, X) which in this case is the point $(0.5, 0.5)$. Since $q = 0.6$, the y intercept of the q line [Eq. (1-42)] is computed as follows

$$\frac{X}{1-q} = \frac{0.5}{(1-0.6)} = 1.25$$

Since the operating line for the stripping section [Eq. (1-38)] passes through the point $(x_B, x_B) = (0.05, 0.05)$ and the intersection of the q line with the operating line for the rectifying section, it may be constructed by connecting these two points as shown in Fig. 1-9.

The number of perfect plates required to effect the specified separation may be determined graphically as indicated in Fig. 1-11. It is readily confirmed that the construction shown in Fig. 1-11 gives the desired solution. Since $y_2 = X_D = x_1$ (for a total condenser) and since y_2 is in equilibrium with x_2, the desired value of x_2 is determined by the point of intersection of line 1 and the equilibrium curve as shown in Fig. 1-9. Line 1 also represents plate 1. When x_2 is substituted into Eq. (1-37), the value of y_3 is obtained. Since (x_2, y_3) lies on the operating line for the rectifying section, this point is located by passing a vertical line through (x_2, y_2). The ordinate y_3 obtained is displayed graphically in Fig. 1-9. When the first opportunity to change operating lines is taken, the minimum number of total plates needed to effect the specified separation at the specified operating conditions is obtained. When the feed is introduced on stage number 8, a total of 14 stages are required, 12 plates plus the reboiler and a total condenser (see Fig. 1-9).

It should be noted that if the operating line for the rectifying section is used indefinitely instead of changing to the operating line for the stripping section, the specified value of $x_B = 0.05$ can never be attained even though infinitely many plates are employed.

Minimum Reflux Ratio

As the specified value of the reflux ratio (L_1/D) is decreased, the intersection of the two operating lines moves closer to the equilibrium curve and the minimum number of plates required to effect the specified separation $(x_B = 0.05, X_D = 0.96)$ increases. On the other hand, as L_1/D is decreased, the condenser and reboiler duties decrease. The *minimum reflux ratio* is the smallest one which can be used to effect the specified separation. This reflux ratio requires infinitely many plates in each section as demonstrated in Fig. 1-10. It should be noted that for this case, the plates at and adjacent to the feed plate have the same composition. (In the case of multicomponent systems, these limiting conditions do not necessarily occur at and adjacent to the feed plate as discussed in Chap. 11). From the standpoint of construction costs, this reflux ratio is unacceptable because infinitely many plates are required, which demands a column of infinite height.

Total Reflux

At total reflux, the operating lines coincide with the 45° line. This gives the smallest number of plates needed to effect the separation. As pointed out by Robinson and Gilliland,[12] two physical interpretations of total reflux are possible. From a laboratory or plant operational point of view, total reflux is attained by introducing an appropriate quantity of feed to the column and then

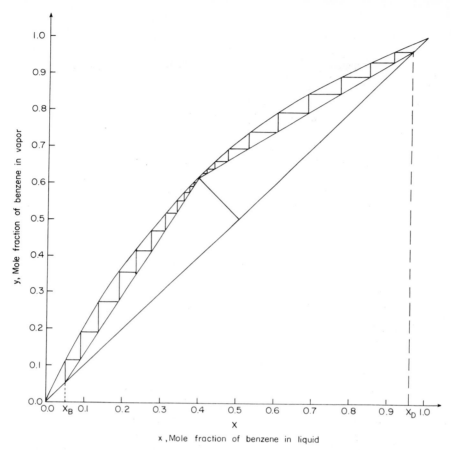

Figure 1-10 At the minimum reflux ratio (L_1/D), infinitely many plates are required to effect the specified separation (X_D, x_B).

operating so that $F = D = B = 0$. From the standpoint of design, total reflux can be thought of as a column of infinite diameter operating at infinitely large vapor and liquid rates, and with a feed that enters at a finite rate F and with distillate and bottoms that leave at the rates D and B, where $F = D + B$. Thus, infinite condenser and reboiler duties are required as well as a column having an infinitely large diameter. At total reflux, six plates, a total condenser, and a reboiler are required to effect the specified separation as shown in Fig. 1-11. A graph of the total costs per year versus the reflux rate L_1 at a fixed set of specifications is shown in Fig. 1-12 for reflux rates over the range from minimum to total reflux.

The set of equations required to describe a distillation column in the process of separating a binary mixture is merely an extension of the sets stated previously for the boiling-point diagram [Eq. (1-3)], bubble-point and dew-point temperatures [Eq. 1-12)], and the flash process [Eq. (1-26)]. The complete set of

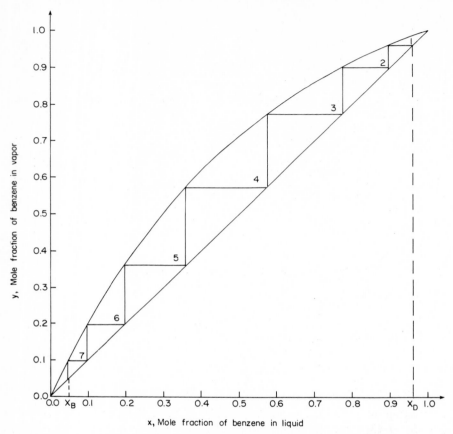

Figure 1-11 Determination of the total number of plates required to effect the specified separation at total reflux.

equations solved above by the McCabe-Thiele method are as follows

$$
\text{Equilibrium relationships}
\begin{cases}
y_{ji} = K_{ji} x_{ji} & \left(\begin{array}{l} i = 1, 2 \\ j = 1, 2, \ldots, N \end{array}\right) \\[2ex]
\displaystyle\sum_{i=1}^{2} y_{ji} = 1 & (j = 1, 2, \ldots, N) \\[2ex]
\displaystyle\sum_{i=1}^{2} x_{ji} = 1 & (j = 1, 2, \ldots, N)
\end{cases}
$$

$$
\text{Material balances}
\begin{cases}
V_r y_{j+1,\,i} = L_r x_{ji} + DX_{Di} & \left(\begin{array}{l} i = 1, 2 \\ j = 1, 2, \ldots, f-1 \end{array}\right) \\[2ex]
V_s y_{j+1,\,i} = L_s x_{ji} - Bx_{Bi} & \left(\begin{array}{l} i = 1, 2 \\ j = f, f+1, \ldots, N-1 \end{array}\right) \\[2ex]
FX_i = DX_{Di} + Bx_{Bi} & (i = 1, 2)
\end{cases}
$$

(1-43)

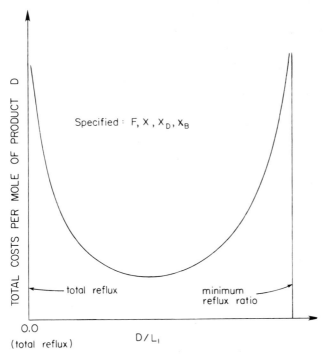

Figure 1-12 Total costs (capital plus operating costs) per mole of product D (or B) for a specified separation.

The counting integer j for stage number takes on only integral values. Examination of Eq. (1-43) shows that it consists of $6N$ independent equations. This result could have been obtained as follows. Since a single-equilibrium stage [Eq. (1-26)] is represented by $2c + 2$ independent equations and since the column represented by Eq. (1-43) has N equilibrium stages [the condenser $j = 1$, plates $j = 2$, 3, ..., $N - 1$, and the reboiler $j = N$], then one would expect to obtain $(2c + 2)N$ independent equations for a distillation column. Thus, a column in the service of separating a binary mixture is represented by $6N$ equations. Also, in the McCabe-Thiele method as presented above, it is assumed that the behavior on the feed plate may be represented by model 1, Fig. 1-13.

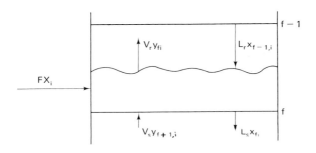

Figure 1-13 Model 1. Assumed in the McCabe-Thiele method.

For the case where the total flow rates V_j and L_j vary throughout each section of the column, these flow rates may be determined by solving the enthalpy balances simultaneously with the above set of equations. For binary mixtures, the desired solution may be found by use of either graphical methods (Refs. 10, 13) or the numerical methods proposed in subsequent chapters for the solution of problems involving the separation of multicomponent mixtures.

1-4 SEPARATION OF MULTICOMPONENT MIXTURES AT TOTAL REFLUX

The topic of total reflux is considered briefly in this chapter for the purpose of developing the well-known Fenske equation (Ref. 4) which is needed in Chaps. 2 and 3. A more general treatment of the subject area of total reflux is presented in Chap. 7.

Development of the Fenske Equation[4]

From the standpoint of design, the most useful definition of total reflux consists of the one in which the total flow rates $[L_j \ (j = 1, 2, \ldots, N - 1), V_j \ (j = 2, \ldots, N)]$ are unbounded while the feed and product rates are finite. More precisely

$$\lim_{V_{j+1} \to \infty} \frac{L_j}{V_{j+1}} = 1 - \lim_{V_{j+1} \to \infty} \frac{D}{V_{j+1}} = 1$$

and

$$F = D + B \tag{1-44}$$

where F, D, and B are all nonzero, finite, and positive. The corresponding component-material balances are given by

$$y_{j+1,\,i} = x_{ji} \Big| \lim_{V_{j+1} \to \infty} \frac{L_j}{V_{j+1}} \Big| + X_{Di} \Big| \lim_{V_{j+1} \to \infty} \frac{D}{V_{j+1}} \Big| = x_{ji} \tag{1-45}$$

As a consequence of the results given by Eqs. (1-44) and (1-45) it follows that the component-material balance is given by

$$y_{j+1,\,i} = x_{ji} \tag{1-46}$$

By repeated use of Eq. (1-46) and the equilibrium relationship

$$y_{ji} = K_{ji} x_{ji} \tag{1-47}$$

the Fenske equation[4]

$$\frac{b_i}{d_i} = \frac{B/D}{\prod\limits_{j=2}^{N} K_{ji}} \tag{1-48}$$

[for a column having a total condenser] is obtained. An abbreviated development of this equation follows. The component-material balance enclosing the condenser-accumulator section (stage 1) is given by

$$y_{2i} = x_{1i} \tag{1-49}$$

for component i, and for the case of a total condenser

$$y_{2i} = X_{Di} \tag{1-50}$$

The equilibrium relationship for stage 2 is

$$y_{2i} = K_{2i} x_{2i} \tag{1-51}$$

Elimination of y_{2i} from Eqs. (1-50) and (1-51) gives

$$x_{2i} = \frac{X_{Di}}{K_{2i}} \tag{1-52}$$

For stage 3, the component-material balance and equilibrium relationship for component i are as follows

$$y_{3i} = x_{2i} \tag{1-53}$$

$$y_{3i} = K_{3i} x_{3i} \tag{1-54}$$

Elimination of x_{2i} and y_{3i} from Eqs. (1-52) through (1-54) gives

$$x_{3i} = \frac{X_{Di}}{K_{2i} K_{3i}} \tag{1-55}$$

Continuation of this procedure for stages $j = 4$ through N (the reboiler) yields

$$x_{Ni} = \frac{X_{Di}}{K_{2i} K_{3i} \cdots K_{N-1, i} K_{Ni}} \tag{1-56}$$

Since $x_{Ni} = x_{Bi}$, it is evident that Eq. (1-48) is obtained upon multiplication of both sides of Eq. (1-56) by B/D. The steps of this derivation (which consist of the alternative use of material balances and equilibrium relationships) are seen to be the same as those involved in the graphical solution for a binary mixture (see Fig. 1-11).

An alternative form of Eq. (1-48) which reduces to an exact solution when the relative volatilities are constant is obtained as follows. First, state Eq. (1-48) for the base component b, then divide the members of Eq. (1-48) by the corresponding members for component b and rearrange the result so obtained to give

$$\frac{b_i}{d_i} = \frac{b_b/d_b}{\alpha_{2i} \alpha_{3i} \cdots \alpha_{N-1, i} \alpha_{Ni}} \tag{1-57}$$

where

$$\alpha_{ji} = K_{ji}/K_{jb} \qquad d_i = DX_{Di} \qquad \text{and} \qquad b_i = Bx_{Bi}$$

If the α_{ji}'s are independent of temperature, then Eq. (1-57) reduces to

$$\frac{b_i}{d_i} = \frac{b_b}{d_b} \alpha_i^{-(N-1)} \tag{1-58}$$

For the case of a partial condenser $(y_{1i} = X_{Di})$, the appropriate expression for b_i/d_i is obtained by replacing the exponent $(N-1)$ in the above expression by the exponent N; that is, the partial condenser counts as an additional equilibrium stage.

At a fixed number of stages N, the set of b_i/d_i's relative to b_b/d_b may be computed for a given system by use of Eq. (1-58). Then for any specified value of b_b/d_b, the corresponding set of d_i's and D may be computed by use of Eqs. (1-60) and (1-61), respectively. These formulas are obtained by first solving the component-material balance

$$Fx_i = d_i + b_i = d_i[1 + b_i/d_i] \tag{1-59}$$

for d_i

$$d_i = \frac{Fx_i}{1 + b_i/d_i} \tag{1-60}$$

and then summing over all components

$$D = \sum_{i=1}^{c} \frac{Fx_i}{1 + b_i/d_i} \tag{1-61}$$

In summary, Eq. (1-58) may be used to compute the best possible separation (the lightest possible distillate and heaviest bottoms) which may be achieved with a fixed number of plates at the limiting condition of total reflux; provided, of course, that the α_i's are constant throughout the column. At this limiting condition of total reflux, the column diameter as well as the reboiler and condenser duties become infinite. Equation (1-58) may be used to compute the composition of the distillate and bottoms of a column operating at total reflux as may be demonstrated by solving Prob. 1-24. Examination of Eq. (1-58) shows that at a given N, a set of b_i/d_i's may be found for every specified value of b_b/d_b. Furthermore, the natural logarithm of b_i/d_i versus the natural logarithm of α_i plots as a straight line for each choice of the intercept b_b/d_b. This characteristic of the Fenske equation forms the basis of the proof of the proposition that the θ method constitutes an exact solution of certain problems involving columns at total reflux as demonstrated in the next chapter.

Determination of Bubble-Point and Dew-Point Temperatures of Mixtures Containing Inert Gases and Inert Liquids

In many industrial applications, the mixtures contain inert gases and liquids. Modifications needed to make bubble-point, dew-point and flash calculations are presented in this section. An "inert gas" component is one which appears in the gas phase alone because it is insoluble in the liquid phase. An "inert liquid" component is one which appears in the liquid phase alone. The inert liquid is miscible in the liquid phase and does not exhibit a detectable vapor pressure. More precisely, an inert gas (denoted by the subscript L) is defined as a component for which $1/K_L = 0$ and $x_L = 0$ while an inert liquid (denoted by the subscript H) is defined as a component for which $K_H = 0$ and $y_H = 0$.

Components which appear in both phases are called "volatile" components. Volatile components are defined more precisely as those components having K values which are nonzero, finite, and positive. The bubble-point and dew-point functions are given in Prob. 1-11.

When a feed which contains both inert gases and an inert liquid is flashed at a specified temperature and pressure, the expression given by Eq. (1-30) for the flash function reduces to the expression given in Prob. 1-17.

NOTATION

b_i	molar flow rate of component i in the bottoms
B	total molar flow rate of bottoms
c	total number of components
d_i	molar flow rate of component i in the distillate
D	total molar flow rate of the distillate
$\hat{f}_i^L, \hat{f}_i^V$	fugacities of components i in the liquid and vapor phases (composed of any number of components), respectively; evaluated at the total pressure and temperature of the two-phase system and at the compositions of the respective phases, atm
f_i^L, f_i^V	fugacities of pure component i in the liquid and vapor phases, respectively; evaluated at the total pressure and temperature of the two-phase system, atm
$f(T)$	bubble-point function; defined by Eq. (1-14)
$F(T)$	dew-point function; defined by Eq. (1-17)
F	total molar flow rate of the feed
h_{Fi}, H_{Fi}	enthalpies of pure component i; evaluated at the temperature T_F and pressure P of the flash, Btu/lb mol
h_j	$\displaystyle\sum_{i=1}^{c} h_{ji} x_{ji}$, for an ideal solution; evaluated at the temperature T_j, pressure and composition of the liquid leaving the jth plate
H_j	$\displaystyle\sum_{i=1}^{c} H_{ji} y_{ji}$, for an ideal solution; evaluated at the temperature T_j, pressure, and composition of the vapor leaving the jth plate
H	enthalpy per mole of feed, regardless of state
K_{ji}	equilibrium vaporization constant; evaluated at the temperature and pressure of the liquid leaving the jth stage
L_j	total flow rate of liquid leaving any stage of j ($j = 1, 2, \ldots, N - 1, N$), mol/h

L_r total-flow rate of liquid leaving stages $j = 1, 2, \ldots, f - 2, f - 1$

L_s total-flow rate of liquid leaving stages $j = f, f + 1, \ldots, N - 1, N$

N total number of stages

P_i vapor pressure of component i, atm

P total pressure, atm

$P(\Psi)$ flash function; defined by Eq. (1-30)

q a factor related to the thermal condition of the feed; defined by Eq. (1-40)

Q_C condenser duty, Btu/h

Q_R reboiler duty, Btu/h

T temperature. T_{BP} = bubble-point temperature, and T_{DP} = dew-point temperature

T_F flash temperature

V_j total-flow rate of vapor leaving the jth stage ($j = 2, 3, \ldots, N$), mol/h

V_r total-flow rate of the vapor leaving stages $j = 2, 3, \ldots, f - 1, f$, mol/h

V_s total-flow rate of vapor leaving stages $j = f + 1, f + 2, \ldots, N - 1, N$, mol/h

x_{Fi} mole fraction of component i in the liquid leaving a flash process

x_{ji} mole fraction of component i in the liquid leaving the jth stage

x_{Bi} mole fraction of component i in the bottoms

x_I abscissa of the point of intersection of the operating lines for a binary mixture

X_i total mole fraction of component i in the feed (regardless of state)

X_{Di} total mole fraction of component i in the distillate (regardless of state)

y_{ji} mole fraction of component i in the vapor leaving plate j

y_I ordinate of the point of intersection of the operating lines for a binary mixture

Greek Letters

α_{ji} relative volatility, $\alpha_{ji} = K_{ji}/K_{jb}$

δ function of T_F; defined by Eq. (1-35)

γ_i^L, γ_i^V activity coefficients for component i in the liquid and vapor phases, respectively

Ψ moles of vapor formed per mole of feed by the flash process; $\Psi = V_F/F$

Subscripts

f feed plate

F variables associated with a partially vaporized feed

H inert liquid components

i component number, $i = 1, 2, \ldots, c$

j stage number; $j = 1$ for the accumulator; for the top plate $j = 2$, for the feed plate $j = f$, for the bottom plate $j = N$, and for the reboiler $j = N$; that is, $j + 1, 2, \ldots, f, \ldots, N - 1, N$, or ($j = 1, 2, \ldots, N$)

k trial number

L inert gas components

n trial number
N total number of stages
r rectifying section
s stripping section

Superscripts

L liquid phase
V vapor phase

Mathematical Symbols

$\exp(x)$ e^x

$\displaystyle\sum_{i=1}^{c} x_i$ sum over all values x_i, $i = 1, 2, \ldots, c$

$\{x_j\}$ set of all values x_j belonging to the particular set under consideration

$\displaystyle\prod_{j=1}^{c}(1+x_j)$ product of the factors $(1+x_j)$ from $j = 1$ through $j = c$

PROBLEMS

1-1 (a) Calculate the equilibrium pairs $\{x, y\}$ for each of the temperatures given in Example 1-1.

(b) From the plot obtained in part (a), construct the equilibrium curve; that is, construct a graph in which the y's of the equilibrium pairs $\{x, y\}$ are plotted on the ordinate and the corresponding x's are plotted on the abscissa.

1-2 (a) Repeat Example 1-2 for the case where the first assumed value of T_n is taken to be equal to 40°F.

Answer: 56.8°F.

(b) After the bubble-point temperature has been determined, compute the corresponding values of y_i's which are in equilibrium with the x_i's.

Answer: $y_1 = 1/6$, $y_2 = 1/3$, $y_3 = 1/2$.

1-3 Repeat Prob. 1-2 where the following vapor compositions are known instead of the liquid compositions. In this case determine the dew-point temperature at a specified total pressure of $P = 1$ atm. For the first trial, assume $T = 560°R$. Use the K values given in Example 1-2.

Given: $y_1 = 1/6$, $y_2 = 1/3$, and $y_3 = 1/2$. After the dew-point temperature has been determined, compute the corresponding x_i's which are in equilibrium with the y_i's.

Answer: $T_{DP} = 56.8°F$, $x_1 = 1/3$, $x_2 = 1/3$, $x_3 = 1/3$.

1-4 If the α_i's are not independent of temperature, then the temperature found at the end of the first trial by use of the K_b method depends upon the component selected as the base component as well as the temperature assumed to evaluate the α_i's. These facts are illustrated by solving the following problems.

Component	$K_i = C_i \exp(-E_i/T)$, T is in °R		
	C_i	E_i	x_i
1	$4.0 \times 10^3/P$†	4.6×10^3	$1/3$
2	$6.0 \times 10^3/P$	4.7×10^3	$1/3$
3	$12.0 \times 10^3/P$	4.8×10^3	$1/3$

† P is in atm.

(a) Find the correct bubble-point temperature by use of the Newton-Raphson method at a specified total pressure of 1 atm. Assume an initial temperature of 520°R.

Answer: 531.52°R.

(b) Make one trial by use of the K_b method. Evaluate the α_i's at an assumed temperature of 100°F and $P = 1$ atm. Take component 1 as the base component.

Answer: 530.7°R.

(c) Repeat part (b) for the case where component 2 is selected as the base component.

Answer: 531.3°R.

(d) Repeat part (b) for the case where the α_i's are evaluated at an assumed temperature of 50°F rather than 100°F.

Answer: 532°R.

1-5 Any iterative process in which the calculated values of the variables are used without alteration to make the next trial calculation is referred to herein as *direct iteration*. This calculational procedure is known by a variety of names such as *iteration, successive iteration,* and *successive substitution*.

(a) Suppose that it is required to find the x which satisfies the following equation

$$x - \tfrac{1}{2}x - 2 = 0$$

Although $x = 4$ is seen to be the correct answer, solve this problem by direct iteration by rearranging it to the following form

$$x_{k+1} = \tfrac{1}{2}x_k + 2$$

where the subscript k denotes the trial number. To initiate the procedure, let $x_k = 1$. Then $x_{k+1} = 5/2$. For the next trial take $x_k = 5/2$. Continue the process and show that the calculational procedure converges to the solution value $x = 4$. Prepare a graph of x_{k+1} versus x_k. Use the 45° line to transfer the calculated values of x from the ordinate to the abscissa.

(b) Repeat part (a) for the case where the original equation is solved as follows

$$x_{k+1} = 2x_k - 4$$

Begin with $x_k = 3$ and also with $x_k = 5$.

(c) Let the right-hand sides of the expressions given in parts (a) and (b) be denoted by $f(x)$. Then describe the convergence characteristics of a function for which

$$\left|\frac{df(x)}{dx}\right| < 1$$

and a function for which

$$\left|\frac{df(x)}{dx}\right| > 1$$

1-6 (a) Show that the formula for Newton's method given by Eq. (1-16) may be obtained by equating the derivative of the function $f(x)$ at x_k, denoted by $f'(x_k)$, to the slope of the tangent line at x_k, namely,

$$\frac{f(x_k) - 0}{x_k - x_{k+1}}$$

where x_{k+1} is the point of intersection of the tangent line and the x axis.

(b) Use Newton's method to find the positive x that makes $f(x) = 0$

$$f(x) = x^2 - 4x - 4$$

Begin with $x = 3$. Prepare a graph of $f(x)$ versus x, and on this graph show the path of the calculational procedure. [*Hint:* Newton's method may be regarded as a linear extension of the function from the point $(x_k, f(x_k))$ to the point $(x_{k+1}, 0)$. The slope of this line is $f'(x_k)$].

1-7 For any two assumed values of x denoted by x_k and x_{k+1}, the corresponding values of the function, $f(x_k)$ and $f(x_{k+1})$ are readily computed. Let these two points $(x_k, f(x_k))$ and $(x_{k+1}, f(x_{k+1}))$

be connected by a straight line. The equation of this line may be represented by the equation

$$\frac{f(x) - f(x_k)}{x - x_k} = \frac{f(x_k) - f(x_{k+1})}{x_k - x_{k+1}} \tag{A}$$

Extend this straight line until it intersects that x axis at the point $(x_{k+2}, 0)$. [This procedure amounts to the assumption that the function is linear over the range of the extrapolation.] Show that at $x = x_{k+2}$ and $f(x) = 0$, Eq. (A) may be solved to give the interpolation *regula falsi* formula

$$x_{k+2} = \frac{x_k f(x_{k+1}) - x_{k+1} f(x_k)}{f(x_{k+1}) - f(x_k)}$$

1-8 (*a*) Repeat Prob. 1-6 for the case where the positive value of x is found by interpolation *regula falsi* (see Prob. 1-7). Take the first assumed value of x to be $x_k = 2$ and $x_{k+1} = 6.5$. After the first trial, find the new assumed value of x by applying the method of interpolation *regula falsi* to the most recent pair of x values and their corresponding functional values.

(*b*) Produce a graph of $f(x)$ versus x, and on this graph indicate the path followed by the method of interpolation *regula falsi* in the solution of this problem.

1-9 Solve Example 1-2 by use of the method of interpolation *regula falsi*. Take $T_k = 580°\text{R}$ and $T_{k+1} = 680°\text{R}$.

1-10 (*a*) For mixtures which contain all volatile components, show that after K_b has been determined by use of the bubble-point formula [Eq. (1-21)], the y_i's which are in eqilibrium with x_i's are given by

$$y_i = \frac{\alpha_i x_i}{\sum\limits_{i=1}^{c} \alpha_i x_i}$$

(*b*) Similarly, show that after K_b has been found by use of the dew-point formula [Eq. (1-22)], the following expression may be used to compute the x_i's which are in equilibrium with the y_i's

$$x_i = \frac{y_i/\alpha_i}{\sum\limits_{i=1}^{c} y_i/\alpha_i}$$

1-11 Consider a mixture which contains both inert liquid components H and inert vapors L. For an inert vapor, $1/K_L = 0$, and for an inert liquid, $K_H = 0$. For convenience, suppose that component 1 of a given mixture is an inert liquid, components 2 through $c - 1$ are volatile, and component c is an inert gas. The equilibrium relationships for this mixture are as follows

$$y_H = y_1 = K_1 x_1 = K_H x_H$$

$$y_2 = K_2 x_2$$

$$\vdots \qquad \vdots$$

$$y_{c-1} = K_{c-1} x_{c-1}$$

$$y_L = y_c = K_c x_c = K_L x_L$$

$$x_H + x_2 + x_3 + \cdots + x_{c-1} + x_L = 1$$

$$y_H + y_2 + y_3 + \cdots + y_{c-1} + y_L = 1$$

(*a*) Show that the bubble-point function is given by

$$f(T) = \sum_{\substack{i=1 \\ i \ne H, L}}^{c} K_i x_i - (1 - y_L)$$

(b) Show that the dew-point function is given by

$$F(T) = \sum_{\substack{i=1 \\ i \ne H, L}}^{c} \frac{y_i}{K_i} - (1 - x_H)$$

1-12 For the case where inert gases L and inert liquids H are present in addition to volatile components $(i = 2, 3, \ldots, c - 2, c - 1)$, show that the following formulas for K_b apply instead of those given by Eqs. (1-21) and (1-22)

$$K_b = \frac{1 - y_L}{\sum\limits_{i=2}^{c-1} \alpha_i x_i} \qquad K_b = \frac{\sum\limits_{i=2}^{c-1} y_i / \alpha_i}{1 - x_H}$$

1-13 Find the bubble-point temperature of the following mixture.

Component	x_i	K_i	Other specifications
1	1/4	0.0	$P = 1$ atm
2	1/4	0.01 T/P†	$y_5 = 0.2$
3	1/4	0.02 T/P	
4	1/4	0.03 T/P	
5	0	$1/K_5 = 0$	

† T is in °F and P is in atm.

Answer: 53.33°F.

1-14 Find the dew-point temperature of the following mixture.

Component	y_i	K_i	Other specifications
1	......	0.0	$P = 1$ atm
2	0.1333	0.01 T/P†	$x_H = 0.25$
3	0.2667	0.02 T/P	
4	0.4000	0.03 T/P	
5	0.2000	$1/K_5 = 0$	

† T is in °F and P is in atm.

Answer: 53.33°F.

1-15 Show that by elimination of the x_{Fi} from the component-material balance $FX_i = V_F y_{Fi} + L_F x_{Fi}$ by use of the equilibrium relationship $y_{Fi} = K_{Fi} x_{Fi}$, the following form of the flash function may be obtained

$$p(\psi) = \sum_{i=1}^{c} \frac{X_i}{1 - \psi[1 - (1/K_{Fi})]} - 1$$

where $\psi = L_F/F$.

1-16 A feed having the composition given in Example 1-4 and the enthalpies given in Example 1-5 is to be flashed adiabatically at a pressure of 1 atm. The enthalpy of the feed is 10,730 Btu/lb mol. Make one complete trial by use of the procedure described in the text [see Eqs. (1-35) and (1-36)]. For the first trial take $T_{F1} = 95°F$ and $T_{F2} = 105°F$.

1-17 For the case where the feed to be flashed contains both inert gases ($K_{FL} \to \infty$) and inert liquids ($K_{FH} = 0$), show that the flash function given by Eq. (1-30) takes the following form

$$P(\Psi) = \sum_{\substack{i=1 \\ i \neq L}}^{c} \frac{X_i}{[1 - \Psi(1 - K_{Fi})]} - 1$$

1-18 Repeat Example 1-4 for the case where 25 moles per hour of an inert liquid component are added to the given feed of 100 moles per hour.

(a) Show that the maximum value of $\Psi = V_F/F$ which is physically possible is equal to 0.8.

(b) Find the solution to this problem by use of Newton's method.

1-19 Find the minimum reflux ratio (L_1/D) required to effect the following separation of an equimolar mixture of A and B. The feed is 50 percent vaporized, and the specified values of the product mole fractions for component A are as follows

$$X_D = 0.95$$

$$x_B = 0.06$$

Given the following equilibrium data for component A.

x	y
0.1	0.43
0.2	0.63
0.3	0.79
0.5	0.905
0.9	0.995

Answer: $(L_1/D)_{min} = 0.418$

1-20 Find the minimum total number of stages required to effect the separation of Prob. 1-19 for the following conditions

1. Column pressure = 1 atm.
2. Total condenser.
3. Use an operating reflux ratio L_1/D equal to twice the minimum reflux ratio found in Prob. 1-19.

Answer: Four plates plus the reboiler. The feed plate is the stage 3.

1-21 For the case where the distillate is withdrawn as a vapor, the partial condenser represents an additional separation stage. In this case

$$X_D = y_1 = K_1 x_1$$

where y_1 is the mole fraction of the vapor in equilibrium with the liquid reflux having the mole fraction x_1 in the accumulator. The material balance enclosing the condenser-accumulator section is represented by

$$y_2 = \left(\frac{L_r}{V_r}\right)x_1 + \left(\frac{D}{V_r}\right)X_D$$

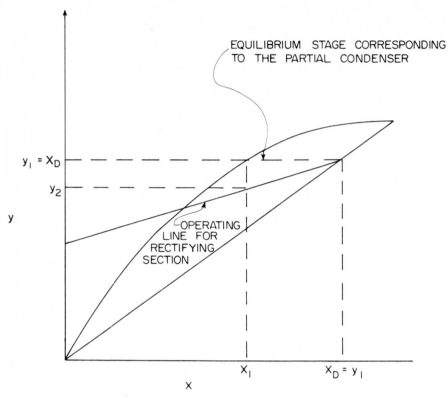

Figure P1-21 Graphical representation for a partial condenser.

The graphical representation on a McCabe-Thiele diagram is given in Fig. P1-21.

Repeat Prob. 1-20 for the case where a partial condenser rather than a total condenser is used and the distillate is withdrawn as a vapor with the vapor composition $X_D = 0.95$ rather than as a liquid with this same composition.

Answer: Three plates plus the reboiler plus the partial condenser. The feed plate is stage 2.

1-22 (*a*) On the basis of the following assumptions, show that q is approximately equal to the heat required to vaporize one mole of feed divided by the latent heat of vaporization of the feed.

1. Assume model 1 of Fig. 1-13 for the behavior of the feed plate.
2. Assume $H_r = H_s$, $h_r = h_s$, where these are the vapor and liquid enthalpies which appear in the enthalpy balance enclosing the feed plate and any number of plates above or below it.
   $V_s H_s + FH + L_r h_r - V_r H_r - L_s h_s = 0$
   The total heat content of the feed regardless of state is denoted by H.
3. The latent heat of vaporization of the feed is equal to $H_r - h_r = \lambda_r$.

(*b*) Use the boiling-point diagram (Fig. 1-6) and the component-material balance $FX = V_F y_F + L_F x_F$ to show that the ratio of the moles of vapor V_F to the moles of liquid L_F formed from the flash of the feed F at the flash temperature T_F is given

$$\frac{V_F}{L_F} = \frac{X - x_F}{y_F - X}$$

where X is the mole fraction of a given component in the feed and x_F and y_F are the mole fractions in the liquid and vapor formed by the flash.

1-23 Show that for a column possessing a partial condenser, the Fenske equation becomes

$$\frac{b_i}{d_i} = \frac{b_b}{d_b} \alpha_i^{-N}$$

1-24 A given column has two plates, a reboiler, a partial condenser, and a feed F of 100 moles per hour. The composition of the feed and the relative volatilities which are independent of temperature are as follows:

Component	X_i	α_i
1	1/3	1
2	1/3	2
3	1/3	3

Find the distillate rates D which must be employed for the above column in order to achieve the following separations of the base component (component 1)

$$(a) \quad b_b/d_b = 16$$

$$(b) \quad b_b/d_b = 8$$

at total reflux. Also find the set of b_i/d_i's at each D.

Answer:

(a)	Component	b_i/d_i
	1	16.0000
	2	1.0000
	3	0.1975

$$D = 46.4632$$

(b)	Component	b_i/d_i
	1	8.0000
	2	0.5000
	3	0.09875

$$D = 56.2629$$

1-25 (a) By use of the logarithmic form of Eq. (1-58) and a variety of choices of b_b/d_b (or D), show that for a given mixture and a fixed value for $(N - 1)$, a family of parallel straight lines is obtained.

(b) Show that the ratio of each b_i/d_i found in part (b) of Prob. 1-24 to the corresponding value found in part (a) is equal to 0.5.

REFERENCES

1. E. V. Anderson, R. Brown, and C. E. Bolton: "Styrene-crude Oil to Polymer," *Ind. Eng. Chem.*, **52**: 550 (1960).
2. Brice Carnahan, H. A. Luther, and J. O. Wilkes: *Applied Numerical Methods*, John Wiley & Sons, Inc., New York, 1964.
3. Kenneth Denbigh: *The Principles of Chemical Equilibrium*, Cambridge University Press, New York, 1955.

4. M. R. Fenske: "Fractionation of Straight-Run Pennsylvania Gasoline," *Ind. Eng. Chem.*, **24**:482 (1932).

5. C. D. Holland: *Multicomponent Distillation*, Prentice-Hall, Inc., Englewood Cliffs, N.J., 1963.

6. ———: "Introduction to the Fundamentals of Distillation," *Proceedings of the Fourth Annual Education Symposium of the ISA*, Apr. 5–7, 1972, Wilmington, Delaware.

7. ——— and J. D. Lindsay: "Distillation," *Encyclopedia Chemical Technology*, 2d ed., John Wiley & Sons, Inc., New York, vol. 7, pp. 204–208, 1965.

8. W. K. Lewis: "Theory of Fractional Distillation," *J. Ind. Chem.*, **1**:522 (1909).

9. W. L. McCabe and E. W. Thiele: "Graphical Design of Fractionating Columns," *Ind. Eng. Chem.*, **17**:605 (1925).

10. Marcel Ponchon: "Graphical Study of Fractional Distillation," *Tech. Moderne*, **13**:20 (1921).

11. Lord Rayleigh (J. Strutt): "On the Distillation of Binary Mixtures," *Phil. Mag.*, **4**:527 (1902).

12. C. S. Robinson and E. R. Gilliland: *Elements of Fractional Distillation*, 4th ed., McGraw-Hill Book Company, New York, 1950.

13. Savarit: *Arts et métiers*, "Definition of Distillation, Simple Discontinuous Distillation, Theory and Operation of Distillation Column, and Exhausting and Concentrating Columns for Liquid and Gaseous Mixtures and Graphical Methods for Their Determination," (1922), pp. 65, 142, 178, 241, 266, 307.

14. E. Sorel: *La Rectification de l'Alcool*, Gauthier-Villars et fils, Paris, France, 1893.

TWO

DEVELOPMENT AND APPLICATION OF THE THETA METHOD OF CONVERGENCE TO CONVENTIONAL DISTILLATION COLUMNS

Three general methods for solving distillation problems are presented in this book. The first of these, called the "θ method of convergence" (Refs. 11, 12) is recommended for solving problems involving any type of distillation column; provided that the mixtures do not deviate too widely from ideal solutions. For such columns, the θ method is one of the fastest known methods (Refs. 2, 3). For columns which do not have both a reboiler and an overhead condenser, such as absorbers and reboiled absorbers, the $2N$ Newton-Raphson method is recommended for separations which either form ideal solutions or do not deviate widely from them. If the mixture to be separated forms an ideal solution, the $2N$ Newton-Raphson method constitutes an exact application of the Newton-Raphson method. The convergence characteristics of the Newton-Raphson method are presented in App. A. For solving problems involving any type of column in the service of separating highly nonideal solutions, the third method, called the "Almost Band Algorithm," is recommended. The $2N$ Newton-Raphson method is presented in Chap. 4 and the Almost Band Algorithm is presented in Chap. 5.

In this chapter, the θ method is developed and applied to conventional distillation columns. In Sec. 2-1, the equations required to describe conventional distillation columns are presented. The formulation and application of the θ method of convergence, the K_b method for computing temperatures, and the constant-composition method for solving the enthalpy balances for the total-flow rates are presented in Sec. 2-2.

2-1 EQUATIONS REQUIRED TO DESCRIBE CONVENTIONAL DISTILLATION COLUMNS

A conventional distillation column is defined as one that has one feed and two product streams, the distillate D and the bottoms B. Such a column has the same configuration as the one shown in Fig. 1-5. First consider the case where the following specifications are made for a column at steady-state operation: (1) the number of plates in each section of the column, (2) the quantity, composition, and thermal condition of the feed at the column pressure, (3) the type of over-head condenser (total or partial), (4) the column pressure (or the pressure at a given point in the column where the variation of the pressure throughout the column is not negligible), (5), the reflux of ratio, L_1/D, or V_2 or L_1, and (6) the temperature of the distillate or the total distillate rate. (The first three of these are specifications of the geometry of the column and the feed, and the second three are the specification of operating variables.) Steady-state operation means that no process variable changes with time. For this set of operating conditions, the problem is to find the compositions of the top and bottom products. The set of equations required to represent such a system for all components $(i = 1, 2, \ldots, c)$ are as follows

$$
\text{Equilibrium relationships}
\begin{cases}
y_{ji} = K_{ji} x_{ji} & (j = 1, 2, \ldots, N) \\[2mm]
\sum_{i=1}^{c} y_{ji} = 1 & (j = 1, 2, \ldots, N) \\[2mm]
\sum_{i=1}^{c} x_{ji} = 1 & (j = 1, 2, \ldots, N)
\end{cases}
$$

$$
\text{Material balances}
\begin{cases}
V_{j+1} y_{j+1,\,i} = L_j x_{ji} + D X_{Di} & (j = 1, 2, \ldots, f-2) \\[2mm]
V_f y_{fi} + V_F y_{Fi} = L_{f-1} x_{f-1,\,i} + D X_{Di} \\[2mm]
V_{j+1} y_{j+1,\,i} = L_j x_{ji} - B x_{Bi} & (j = f, f+1, \ldots, N-1) \\[2mm]
F X_i = D X_{Di} + B x_{Bi}
\end{cases} \quad (2\text{-}1)
$$

$$
\text{Enthalpy balances}
\begin{cases}
V_{j+1} H_{j+1} = L_j h_j + D H_D + Q_C & (j = 1, 2, \ldots, f-2) \\[2mm]
V_f H_f + V_F H_F = L_{f-1} h_{f-1} + D H_D + Q_c \\[2mm]
V_{j+1} H_{j+1} = L_j h_j - B h_B + Q_R & (j = f, f+1, \ldots, N-1) \\[2mm]
F H = B h_B + D H_D + Q_C - Q_R
\end{cases}
$$

Inspection of this set of equations shows that they are a logical extension of those stated in Chap. 1, Eq. (1-43), for the binary system. A schematic representation of the component-material balances is shown in Fig. 2-1. The behavior assumed on the feed plate is demonstrated by model 2, which is shown in Fig. 2-2.

The above enthalpy balances may be represented by the same enclosures

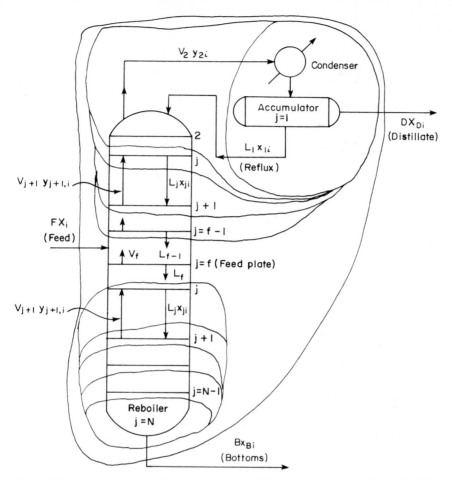

Figure 2-1 Representation of the component-material balances given by Eq. (2-1). (*Taken from Holland: Introduction to the Fundamentals of Distillation, Proc. 4th Ann. Educ. Symp. Instrument Soc. Am., Apr. 5–7, 1972, Wilmington, Delaware.*)

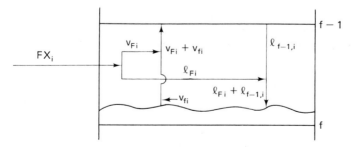

Figure 2-2 Model 2 for the behavior of the feed plate. (*Taken from Holland: Introduction to the Fundamentals of Distillation, Proceedings of the Fourth Annual Education Symposium of the Instrument Society of America, April 5–7, 1972, Wilmington, Delaware.*)

shown in Fig. 2-1. As in the case of the material balances for any one component, the number of independent energy balances is equal to the number of stages $(j = 1, 2, \ldots, N - 1, N)$. In this case the total number of independent equations is equal to $N(2c + 3)$, as might be expected from the fact that an adiabatic flash is represented by $2c + 3$ equations.

For a column whose geometry [the total number of stages, the feed plate locations, and the type of condenser (partial or total)] and feed have been specified, the remaining variables to be specified are as follows:

Variable	Number
Vapor and liquid mole fractions	$2cN$
Total-flow rates	$2N$
Temperatures	N
Reboiler and condenser duties	2
Column pressure	1
	$N(2c + 3) + 3$

Since the number of variables exceeds the number of equations by three, it is necessary to fix three variables in order to obtain a solution to the $N(2c + 3)$ equations. For example, the distillate rate D, reflux rate L_1, and the column pressure may be specified.

When it is supposed that the vapor and liquid streams form ideal solutions, the enthalpy per mole of vapor and the enthalpy per mole of liquid leaving plate j are given by the following expressions (as shown in Chap. 14)

$$H_j = \sum_{i=1}^{c} H_{ji} y_{ji} \quad \text{(vapor)}$$

$$h_j = \sum_{i=1}^{c} h_{ji} x_{ji} \quad \text{(liquid)} \tag{2-2}$$

where the enthalpy of each pure component i in the vapor and liquid streams leaving plate j are represented by H_{ji} and h_{ji}, respectively. These enthalpies are of course evaluated at the temperature and pressure of plate j. The meaning of H_D depends upon the type of condenser employed. For a total condenser, D is withdrawn from the accumulator as a liquid at its bubble-point temperature T_1 at the column pressure, and $y_{2i} = x_{1i} = X_{Di}$. Thus

$$H_D = \sum_{i=1}^{c} h_{1i} X_{Di} = \sum_{i=1}^{c} h_{1i} x_{1i} = h_1 \tag{2-3}$$

For a partial condenser, D is withdrawn from the accumulator as a vapor at its dew-point temperature T_1 at the column pressure, and $y_{1i} = X_{Di}$. Thus

$$H_D = \sum_{i=1}^{c} H_{1i} X_{Di} = \sum_{i=1}^{c} H_{1i} y_{1i} = H_1 \tag{2-4}$$

The enthalpy per mole of bottoms has double representation, h_B and h_N, that is,

$$h_B = \sum_{i=1}^{c} h_{Bi} x_{Bi} = \sum_{i=1}^{c} h_{Ni} x_{Ni} = h_N \qquad (2\text{-}5)$$

The symbols Q_C and Q_R are used to denote the condenser and reboiler duties, respectively. The condenser duty Q_C is equal to the net amount of heat removed per unit time by the condenser, and the reboiler duty Q_R is equal to the net amount of heat introduced to the reboiler per unit time.

A wide variety of numerical methods have been proposed for solving the set of equations represented by Eq. (2-1). Two fundamentally different iterative procedures have been proposed for solving these equations; namely the Lewis and Matheson method[10] and the Thiele and Geddes method.[14] In the Lewis and Matheson method, the terminal compositions $\{X_{Di}\}$ and $\{x_{Bi}\}$ are taken to be the independent variables, and in the Thiele and Geddes method, the temperatures (the temperature of each stage) are taken to be the independent variables. Up until about 1963, the Lewis and Matheson choice of independent variables was used almost exclusively, and since then, the Thiele and Geddes choice of the independent variables has become the most popular.

Merely the statement that the Thiele and Geddes choice of independent variables (or the Thiele and Geddes method) has been employed to solve a problem is not sufficient to describe the calculational procedure. In the solution of a set of nonlinear equations by iterative techniques, the convergence or divergence of a given calculational procedure depends not only upon the initial choice of the independent variables but also upon the precise arrangement and order in which each equation of the set is solved. Over a period of several years, the author has investigated a variety of arrangements and combinations of the expressions given by Eq. (2-1). Of these, the calculational procedure described below was found to converge for almost all problems involving distillation columns. To achieve this result, it was necessary to include the θ method of convergence in the calculational procedure.

2-2 FORMULATION AND APPLICATION OF THE θ METHOD OF CONVERGENCE, THE K_b METHOD, AND THE CONSTANT-COMPOSITION METHOD

The order of presentation of the topics in this section is the same order in which the combined set of methods listed above are applied in the calculational procedure. First, the component-material balances given by Eq. (2-1) are restated in terms of the component-flow rates. The component-flow rates for the liquid phase are eliminated from this set of equations by use of the equilibrium relationships given by Eq. (2-1). Then the θ method is presented. The θ method is used to compute an improved set of compositions on the basis of the most recent set of calculated values of the component-flow rates. The compositions so obtained are used to compute a new set of temperatures by use of the K_b method. The new sets of compositions and temperatures are then used to compute a new set of total flow rates by use of the constant-composition method for solving the enthalpy balances. Numerical examples are used to demonstrate the application of these methods.

Statement of the Component-Material Balances and Equilibrium Relationships as a Tridiagonal Matrix Equation

Although the equations utilized in this procedure differ in form from those presented by Eq. (2-1), they are an equivalent independent set. In the case of the component-material balances, a new set of variables—the component-flow rates in the vapor and liquid phases—are introduced, namely,

$$v_{ji} = V_j y_{ji} \quad \text{and} \quad l_{ji} = L_j x_{ji} \tag{2-6}$$

Also, the flow rates of component i in the distillate and bottoms are represented by

$$d_i = D X_{Di} \quad \text{and} \quad b_i = B x_{Bi} \tag{2-7}$$

and the flow rates of component i in the vapor and liquid parts of the feed by

$$v_{Fi} = V_F y_{Fi} \quad \text{and} \quad l_{Fi} = L_F x_{Fi} \tag{2-8}$$

The equilibrium relationship $y_{ji} = K_{ji} x_{ji}$ may be restated in an equivalent form in terms of the component-flow rates v_{ji} and l_{ji} as follows. First, observe that through the use of Eq. (2-6), the expression $y_{ji} = K_{ji} x_{ji}$ may be restated in the form

$$\frac{v_{ji}}{V_j} = K_{ji} \frac{l_{ji}}{L_j} \tag{2-9}$$

or

$$v_{ji} = S_{ji} l_{ji} \quad \text{and} \quad l_{ji} = A_{ji} v_{ji} \tag{2-10}$$

where the absorption factor A_{ji} and the stripping factor S_{ji} are defined as follows

$$A_{ji} = 1/S_{ji} = L_j/(K_{ji} V_j) \tag{2-11}$$

Instead of enclosing the ends of the column and the respective plates in each section of the column as demonstrated by Eq. (2-1) and Fig. 2-1, an equivalent set of component-material balances is obtained by enclosing each stage ($j = 1, 2, \ldots, N - 1, N$) by a component-material balance as demonstrated in Fig. 2-3. The corresponding set of material balances for each component i are as follows

Material balances
$$\left\{\begin{array}{l} -l_{1i} - d_i + v_{2i} = 0 \\[4pt] l_{1i} - v_{2i} - l_{2i} + v_{3i} = 0 \\[4pt] l_{j-1,\,i} - v_{ji} - l_{ji} + v_{j+1,\,i} = 0, \quad (j = 3, 4, \ldots, f-2) \\[4pt] l_{f-2,\,i} - v_{f-1,\,i} - l_{f-1,\,i} + v_{fi} = -v_{Fi} \\[4pt] l_{f-1,\,i} - v_{fi} - l_{fi} + v_{f+1,\,i} = -l_{Fi} \\[4pt] l_{j-1,\,i} - v_{ji} - l_{ji} + v_{j+1,\,i} = 0, \quad (j = f+1, \\[4pt] \qquad\qquad\qquad\qquad\qquad\qquad\qquad\qquad f+2, \ldots, N-1) \\[4pt] l_{N-1,\,i} - v_{Ni} - b_i = 0 \end{array}\right. \tag{2-12}$$

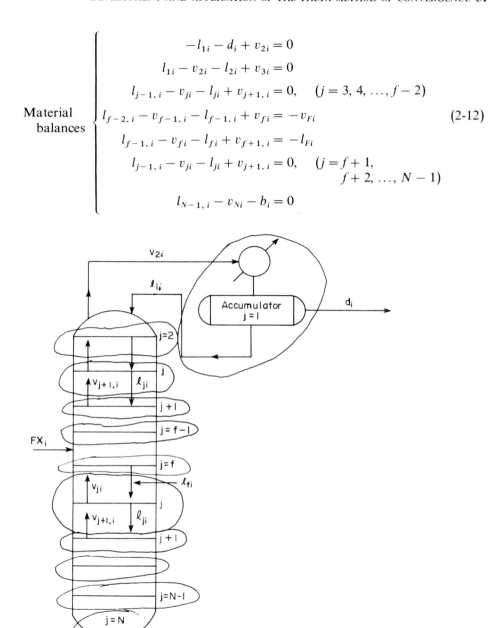

Figure 2-3 Representation of the component-material balances given by Eq. (2-12). (*Taken from Holland: An Introduction to the Fundamentals of Distillation, Proc. 4th Ann. Educ. Symp. Instrument Soc. Am., Apr. 5–7, 1972, Wilmington, Delaware.*)

Except for the first expression of Eq. (2-12), the l_{ji}'s may be eliminated by use of the equilibrium relationship, Eq. (2-11). For the case of a total condenser, l_{1i} and d_i have the same composition, and thus

$$l_{1i} = \left(\frac{L_1}{D}\right)d_i \tag{2-13}$$

For a partial condenser, $y_{1i} = X_{Di}$, and hence $y_{1i} = K_{1i}x_{1i}$ may be restated as follows

$$DX_{Di} = \left(\frac{DK_{1i}}{L_1}\right)L_1 x_{1i} \tag{2-14}$$

or

$$l_{1i} = A_{1i}d_i \tag{2-15}$$

where

$$A_{1i} = L_1/(K_{1i}D)$$

The expression given by Eq. (2-15) may be used to represent both a partial condenser and a total condenser, provided that A_{1i} is set equal to L_1/D for a total condenser.† Also, the form of A_{Ni} differs slightly from that for A_{ji} because of the double representation of the reboiler by the subscripts N and B. Thus, the equilibrium relationship $y_{Ni} = K_{Ni}x_{Ni}$ may be restated in the form

$$V_N y_{Ni} = \left(\frac{K_{Ni}V_N}{B}\right)Bx_{Bi} \tag{2-16}$$

or

$$b_i = A_{Ni}v_{Ni}$$

where

$$A_{Ni} = B/(K_{Ni}V_N)$$

When the l_{ji}'s and b_i are eliminated from Eq. (2-12) by use of Eqs. (2-10), (2-15), and (2-16), the following result is obtained

Material balances and Equilibrium relationships

$$\left\{ \begin{array}{l} -(A_{1i} + 1)d_i + v_{2i} = 0 \\[2mm] A_{1i}d_i - (A_{2i} + 1)v_{2i} + v_{3i} = 0 \\[2mm] A_{j-1,i}v_{j-1,i} - (A_{ji} + 1)v_{ji} + v_{j+1,i} = 0 \qquad (j = 3, \ldots, f-2) \\[2mm] A_{f-2,i}v_{f-2,i} - (A_{f-1,i} + 1)v_{f-1,i} + v_{fi} = -v_{Fi} \\[2mm] A_{f-1,i}v_{f-1,i} - (A_{fi} + 1)v_{fi} + v_{f+1,i} = -l_{Fi} \\[2mm] A_{j-1,i}v_{j-1,i} - (A_{ji} + 1)v_{ji} + v_{j+1,i} = 0 \\[1mm] \qquad\qquad\qquad (j = f+1, f+2, \ldots, N-1) \\[2mm] A_{N-1,i}v_{N-1,i} - (A_{Ni} + 1)v_{Ni} = 0 \qquad (2\text{-}17) \end{array} \right.$$

† This notation of convenience should not be taken to mean that $K_{1i} = 1$ for a total condenser. For the boiling-point temperature, T_1 of the distillate leaving a total condenser is computed by use of the equation

$$\sum_{i=1}^{c} y_{1i} = 1 = \sum_{i=1}^{c} K_{1i}X_{Di}$$

This set of equations may be stated in the matrix form

$$\mathbf{A}_i \mathbf{v}_i = -\mathbf{\textit{f}}_i \tag{2-18}$$

where

$$
\mathbf{A}_i =
\begin{bmatrix}
-\rho_{1i} & 1 & 0 & 0 & \cdots & \cdots & \cdots & 0 \\
A_{1i} & -\rho_{2i} & 1 & 0 & \cdots & \cdots & \cdots & 0 \\
\hdotsfor{8} \\
0 & 0 & A_{f-2,i} & -\rho_{f-1,i} & 1 & 0 & 0 & 0 \\
0 & 0 & 0 & A_{f-1,i} & -\rho_{fi} & 1 & 0 & 0 \\
\hdotsfor{8} \\
0 & \cdots & \cdots & \cdots & 0 & A_{N-2,i} & -\rho_{N-1,i} & 1 \\
0 & \cdots & \cdots & \cdots & 0 & 0 & A_{N-1,i} & -\rho_{Ni}
\end{bmatrix}
$$

$$\mathbf{v}_i = [d_i \; v_{2i} \; v_{3i} \; \cdots \; v_{f-1,i} \; v_{fi} \; \cdots \; v_{N-1,i} \; v_{Ni}]^T$$

$$\mathbf{\textit{f}}_i = [0 \; 0 \; \cdots \; 0 \; v_{Fi} \; l_{Fi} \; 0 \; \cdots \; 0 \; 0]^T$$

$$\rho_{ji} = (1 + A_{ji})$$

The remainder of the development of the calculational procedure is ordered in the same sequence in which the calculations are carried out. The calculational procedure is initiated by the assumption of a set of temperatures $\{T_j\}$ and a set of vapor rates $\{V_j\}$ from which the corresponding set of liquid rates $\{L_j\}$ is found by use of the total material balances presented below. This particular choice of independent variables was first proposed by Thiele and Geddes.[14] On the basis of the assumed temperatures and total-flow rates, the absorption factors $\{A_{ji}\}$ appearing in Eq. (2-18) may be evaluated for component i on each plate j. Since matrix $\mathbf{A}_i$ in Eq. (2-18) is of tridiagonal form, this matrix equation may be solved for the calculated values of the vapor rates for component i [denoted by $(v_{ji})_{ca}$] by use of the Thomas algorithm[4] which follows. Consider the following set of linear equations in the variables $x_1, x_2, \ldots, x_{N-1}, x_N$ whose coefficients form a tridiagonal matrix.

$$b_1 x_1 + c_1 x_2 = d_1$$

$$a_2 x_1 + b_2 x_2 + c_2 x_3 = d_2$$

$$a_3 x_2 + b_3 x_3 + c_3 x_4 = d_3$$

$$\cdots\cdots\cdots\cdots\cdots\cdots\cdots\cdots\cdots\cdots \tag{2-19}$$

$$a_{N-1} x_{N-2} + b_{N-1} x_{N-1} + c_{N-1} x_N = d_{N-1}$$

$$a_N x_{N-1} + b_N x_N = d_N$$

These equations may be solved by use of the following recurrence formulas, which are applied in the order stated

$$f_1 = c_1/b_1 \qquad g_1 = d_1/b_1$$

$$f_k = \frac{c_k}{b_k - a_k f_{k-1}} \qquad (k = 2, 3, \ldots, N - 1)$$

$$g_k = \frac{d_k - a_k g_{k-1}}{b_k - a_k f_{k-1}} \qquad (k = 2, 3, \ldots, N)$$

(2-20)

After the f's and g's have been computed, the values of $x_N, x_{N-1}, \ldots, x_2, x_1$ are computed as follows

$$x_N = g_N$$

$$x_k = g_k - f_k x_{k+1} \qquad (k = N - 1, N - 2, \ldots, 2, 1)$$

(2-21)

The development of the recurrence formulas is outlined in Prob. 2-3. An improved form of these expressions was recently proposed by Boston and Sullivan.[1] For the special case of a conventional distillation column in which model 2 (see Fig. 2-2) for the feed plate is assumed, the procedure proposed by Boston and Sullivan (see Prob. 2-3) may be used to reduce the above formulas to the following form

$$f_1 = -\frac{1}{m_1} \qquad m_1 = 1 + A_{1i}$$

$$f_2 = -\frac{m_1}{m_2} \qquad m_2 = A_{2i}m_1 + 1$$

$$f_k = -\frac{m_{k-1}}{m_k} \qquad m_k = A_{ki}m_{k-1} + 1 \qquad (k = 2, 3, \ldots, N - 1)$$

$$g_1 = g_2 = \cdots = g_{f-2} = 0$$

$$g_{f-1} = v_{Fi}\frac{m_{f-2}}{m_{f-1}}$$

$$g_f = (l_{Fi} + A_{f-1, i}g_{f-1})\frac{m_{f-1}}{m_f}$$

$$g_k = A_{k-1, i}g_{k-1}\frac{m_{k-1}}{m_k} \qquad (k = f + 1, f + 2, \ldots, N)$$

(2-22)

Again, after the f's and g's have been computed, the values of $x_N, x_{N-1}, \ldots, x_2, x_1$ are computed by use of Eq. (2-21).

After the recurrence formulas have been applied for each component i and the set of component vapor rates $\{(v_{ji})_{ca}\}$ have been found, the corresponding set of liquid rates $\{(l_{ji})_{ca}\}$ are then found by use of Eq. (2-10). These sets of calculated flow rates are used in conjunction with the θ method of convergence and the K_b method in the determination of an improved set of temperatures.

Formulation of the θ Method of Convergence

In this application of the θ method of convergence, it is used to weight the mole fractions which are employed in the K_b method for computing a new temperature profile. The corrected product rates are used as weight factors in the calculation of improved sets of mole fractions. The corrected terminal rates are selected such that they are both in overall component-material balance and in agreement with the specified value of D, that is,

$$FX_i = (d_i)_{co} + (b_i)_{co} \tag{2-23}$$

and

$$\sum_{i=1}^{c} (d_i)_{co} = D \tag{2-24}$$

These two conditions may be satisfied simultaneously by suitable choice of the multiplier θ, which is defined by

$$\left(\frac{b_i}{d_i}\right)_{co} = \theta \left(\frac{b_i}{d_i}\right)_{ca} \tag{2-25}$$

(The subscripts *co* and *ca* are used throughout this discussion to distinguish between the corrected and calculated values of a variable, respectively.) Elimination of $(b_i/d_i)_{ca}$ from Eqs. (2-23) and (2-25) yields the formula for $(d_i)_{co}$, namely,

$$(d_i)_{co} = \frac{FX_i}{1 + \theta(b_i/d_i)_{ca}} \tag{2-26}$$

Since the specified values of $(d_i)_{co}$ are to have a sum equal to the specified value of D, the desired value of θ is that $\theta > 0$ that makes $g(\theta) = 0$, where

$$g(\theta) = \sum_{i=1}^{c} (d_i)_{co} - D \tag{2-27}$$

A graph of this function for $\theta > 0$ is shown in Fig. 2-4.

In the determination of θ by Newton's method, the following formula for the first derivative, $g'(\theta)$, is needed

$$g'(\theta) = -\sum_{i=1}^{c} \frac{(b_i/d_i)_{ca} FX_i}{[1 + \theta(b_i/d_i)_{ca}]^2} \tag{2-28}$$

After the desired value of θ has been obtained, $(b_i)_{co}$ may be computed by use of Eq. (2-25). (Note, Newton's method converges to the positive root of $g(\theta)$, provided $\theta = 0$ is taken to be the first trial value; see Prob. 2-11). For the case where the dew-point temperature of the distillate is specified instead of the distillate rate D, the g function has the form shown in Prob. 2-13.

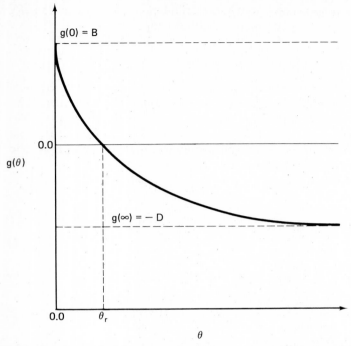

Figure 2-4 Geometrical Representation of the function $g(\theta)$ in the neighborhood of the positive root θ.

The corrected mole fractions for the liquid and vapor phases are computed as follows

$$x_{ji} = \frac{(l_{ji}/d_i)_{ca}(d_i)_{co}}{\sum\limits_{i=1}^{c}(l_{ji}/d_i)_{ca}(d_i)_{co}}$$

$$y_{ji} = \frac{(v_{ji}/d_i)_{ca}(d_i)_{co}}{\sum\limits_{i=1}^{c}(v_{ji}/d_i)_{ca}(d_i)_{co}}$$

(2-29)

The development of these formulas as well as the proof of the fact that they are consistent with the definition of θ is left as an exercise for the student (see Prob. 2-15).

Determination of a Set of Improved Temperatures by Use of the K_b Method

On the basis of the mole fractions given by Eq. (2-29) and the last temperature profile (the one assumed to make the nth trial), the new temperature profile is found by use of the K_b method[13] in the following manner. For any plate j,

Eq. (1-21) or (1-22) may be applied as follows

$$K_{jb}\Bigg|_{T_{j,\,n+1}} = \frac{1}{\displaystyle\sum_{i=1}^{c} \alpha_{ji}\Bigg|_{T_{jn}} x_{ji}} \qquad K_{jb}\Bigg|_{T_{j,\,n+1}} = \sum_{i=1}^{c} \frac{y_{ji}}{\alpha_{ji}}\Bigg|_{T_{jn}} \tag{2-30}$$

where $\alpha_{ji} = K_{ji}/K_{jb}$, the relative volatility of component i at the temperature of plate j. The quantity K_{jb} is the K value of the base component, evaluated at the temperature of plate j.

It can be shown that the x_{ji}'s and y_{ji}'s defined by Eq. (2-29) form a consistent set in that they give the same value of K_{jb} (see Prob. 2-7). Component b represents a hypothetical base component whose K value is given by

$$\ln K_{jb} = \frac{a}{T_j} + b \tag{2-31}$$

where the constants a and b are evaluated on the basis of the values of K at the upper and lower limits of the curve fits of the midboiling component of the mixture or one just lighter. Thus, after K_{jb} has been computed by use of Eq. (2-30), the temperature $T_{j,\,n+1}$ to be assumed for the next trial is calculated directly by use of Eq. (2-31).

The corrected compositions and the new temperatures are used in the enthalpy balances to determine the total flow rates to be used for the next trial through the column.

Determination of a Set of Improved Total-Flow Rates by Use of the Constant-Composition Method

In the constant-composition method, one of the total-flow rates (V_j or L_j) is eliminated from the enthalpy balance given by Eq. (2-1) for each stage by use of the component-material balances for the respective stage. The restatement of the enthalpy balances given by Eq. (2-1) in the form called the constant-composition method may be initiated by first observing that

$$V_j H_j = V_j \sum_{i=1}^{c} H_{ji} y_{ji} = \sum_{i=1}^{c} H_{ji} v_{ji}$$

and

$$L_j h_j = L_j \sum_{i=1}^{c} h_{ji} x_{ji} = \sum_{i=1}^{c} h_{ji} l_{ji}$$

Use of relationships of this type permits the enthalpy balance to be restated in terms of the component-flow rates. For example, the enthalpy balance enclosing plate j

$$V_{j+1} H_{j+1} - L_j h_j - DH_D - Q_c = 0 \qquad (j = 1, 2, \ldots, f-2) \tag{2-32}$$

may be restated in terms of the component-flow rates as follows

$$\sum_{i=1}^{c} [H_{j+1, i} v_{j+1, i} - h_{ji} l_{ji} - H_{Di} d_i] - Q_C = 0 \tag{2-33}$$

where $H_{Di} = h_{1i}$ for a distillation column having a total condenser and, $H_{Di} = H_{1i}$ for a distillation column having a partial condenser. When the component-material balance enclosing plate j [see Eq. (2-12)]

$$v_{j+1, i} = l_{ji} + d_i \qquad (j = 1, 2, ..., f - 2)$$

is used to eliminate $v_{j+1, i}$ from Eq. (2-33), the following result is obtained

$$\sum_{i=1}^{c} [(H_{j+1, i} - h_{ji}) l_{ji} + (H_{j+1, i} - H_{Di}) d_i] - Q_C = 0$$

or

$$L_j \sum_{i=1}^{c} (H_{j+1, i} - h_{ji}) x_{ji} + D \sum_{i=1}^{c} (H_{j+1, i} - H_{Di}) X_{Di} - Q_C = 0$$

The desired expression for calculating L_j is then given by

$$L_j = \frac{Q_C - D \sum_{i=1}^{c} (H_{j+1, i} - H_{Di}) X_{Di}}{\sum_{i=1}^{c} (H_{j+1, i} - h_{ji}) x_{ji}} \qquad (j = 1, 2, ..., f - 2) \tag{2-34}$$

Similarly

$$L_{f-1} = \frac{Q_C - D \sum_{i=1}^{c} (H_{fi} - H_{Di}) X_{Di} + V_F \sum_{i=1}^{c} (H_{fi} - H_{Fi}) y_{Fi}}{\sum_{i=1}^{c} (H_{fi} - h_{f-1, i}) x_{f-1, i}} \tag{2-35}$$

and

$$Q_C = L_1 \sum_{i=1}^{c} (H_{2i} - h_{1i}) x_{1i} + D \sum_{i=1}^{c} (H_{2i} - H_{Di}) X_{Di} \tag{2-36}$$

The flow rates in the stripping section may be determined by use of the enthalpy balances which enclose either the top or the bottom of the column and the given plate. When the reboiler is enclosed, the following formula is obtained

$$V_{j+1} = \frac{Q_R - B \sum_{i=1}^{c} (h_{Ni} - h_{ji}) x_{Bi}}{\sum_{i=1}^{c} (H_{j+1, i} - h_{ji}) y_{j+1, i}} \qquad (j = f, f + 1, ..., N - 1) \tag{2-37}$$

This expression is developed in a manner analogous to that demonstrated above for Eq. (2-34). The above formulas are given the name "constant-composition method" because each of the summations appearing in Eqs. (2-34) through (2-37) may be represented by a thermodynamic process which occurs at constant composition. The reboiler duty Q_R is found by use of the overall enthalpy balance [the last expression given by Eq. (2-1)].

The total-flow rates of the vapor and liquid streams are related by the following total material balances

$$V_{j+1} = L_j + D \qquad (j = 1, 2, \ldots, f - 2)$$

$$V_f + V_F = L_{f-1} + D$$

$$L_j = V_{j+1} + B \qquad (j = f, f + 1, \ldots, N - 1) \tag{2-38}$$

$$F = D + B$$

After the L_j's for the rectifying section and the V_j's for the stripping section have been determined by use of the enthalpy balances, the remaining total-flow rates are found by use of Eq. (2-38). These most recent sets of values of the variables $\{T_{j,\,n+1}\}$, $\{V_{j,\,n+1}\}$, and $\{L_{j,\,n+1}\}$ are used to make the next trial through the column. The procedure described is repeated until values of the desired accuracy have been obtained. A summary of the steps of the proposed calculational procedure follow.

Calculational Procedure for Ideal Solutions

1. Assume a set of temperatures $\{T_j\}$ and a set of vapor rates $\{V_j\}$. [The set of liquid rates corresponding to the set of assumed vapor rates are found by use of the total-material balances; see Eq. (2-38)].
2. On the basis of the temperatures and flow rates assumed in step 1, compute the component-flow rates by use of Eqs. (2-18) through (2-21) [or (2-22)] for each component i.
3. Find the $\theta > 0$ that makes $g(\theta) = 0$; see Eqs. (2-26) through (2-28). (Newton's method[4] always converges to the desired θ, provided that the first assumed value of θ is taken to be equal to zero.)
4. Use Eq. (2-29) to compute the corrected x_{ji}'s or y_{ji}'s for each component i and plate j.
5. Use the results of step 4 to compute the K_{jb} for each stage j by use of either one of the expressions given by Eq. (2-30). Use the K_{jb}'s so obtained to compute a new set of temperatures $\{T_{j,\,n+1}\}$ by use of Eq. (2-31).
6. Use the results of steps 4 and 5 to compute new sets of total-flow rates, $\{V_{j,\,n+1}\}$ and $\{L_{j,\,n+1}\}$, by use of Eqs. (2-34) through (2-38).
7. If θ, the T_j's, and V_j's are within the prescribed tolerances, convergence has been achieved; otherwise, repeat steps 2 through 6 on the basis of the most recent set of T_j's and V_j's.

In the above calculational procedure, it is supposed that the pressure drop from plate to plate is negligible relative to the total pressure. If this assumption is not valid, the calculational procedure is modified as described in Sec. 2-4.

The solution of the component-material balances and equilibrium relationships by use of the above recurrence formulas is demonstrated by the following numerical example.

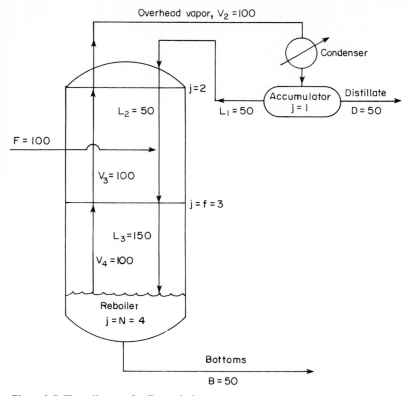

Figure 2-5 Flow diagram for Example 2-1.

Example 2-1 (*a*) On the basis of the initial set of temperatures $(T_1 = T_2 = T_3 = T_4 = 560°R)$ and the total-flow rates displayed in Fig. 2-5, solve Eq. (2-18) for the component-flow rates by use of the above recurrence formulas given by Eqs. (2-20) and (2-21). (*b*) Repeat (*a*) by use of the recurrence formulas given by Eqs. (2-22) and (2-21).

		$K_i = C_i \exp(-E_i/T\dagger)$		
Component	X_i	C_i	E_i	Specifications
1	1/3	$4 \times 10^3/P\ddagger$	4.6447×10^3	Total condenser, $P = 1$ atm, boiling point
2	1/3	$8 \times 10^3/P$	4.6447×10^3	liquid feed $(l_{Fi} = FX_i)$, $N = 4$, $f = 3$,
3	1/3	$12 \times 10^3/P$	4.6447×10^3	$F = 100$ lb mol/h, $D = L_1 = L_2 = 50$ lb mol/h, $L_3 = 150$ lb mol/h, $V_2 = V_3 = V_4 = 100$ lb mol/h.

† T is in °R.
‡ P is in atm.

SOLUTION (a) *Use of Eqs. (2-20) and (2-21)* The correspondence of the symbols in the recurrence formulas and the elements of $\mathbf{A}_i$ and f_i follow

$$b_1 = -(A_{1i} + 1) \qquad c_1 = 1 \qquad d_1 = 0$$

$$a_2 = A_{1i} \qquad b_2 = -(A_{2i} + 1) \qquad c_2 = 1 \qquad d_2 = 0$$

$$a_3 = A_{2i} \qquad b_3 = -(A_{3i} + 1) \qquad c_3 = 1 \qquad d_3 = -FX_i$$

$$a_4 = A_{3i} \qquad b_4 = -(A_{4i} + 1) \qquad d_4 = 0$$

Calculation of the A_{ji}'s follows

Component	$A_{1i} = \dfrac{L_1}{D}$	K_{2i}, K_{3i}, K_{4i} @ 560°R and 1 atm	$A_{2i} = \dfrac{L_2}{K_{2i}V_2}$	$A_{3i} = \dfrac{L_3}{K_{3i}V_3} = \dfrac{3}{2K_{3i}}$
1	1	1	1/2	3/2
2	1	2	1/4	3/4
3	1	3	1/6	1/2

Component	$A_{4i} = \dfrac{B}{K_{4i}V_4} = \dfrac{1}{2K_{4i}}$
1	1/2
2	1/4
3	1/6

Application of the recurrence formulas for tridiagonal matrix equations follows:

Component	b_1	c_1	d_1	a_2	b_2	c_2	d_2	a_3	b_3
1	-2	1	0	1	-1.50000	1	0	0.50000	-2.5
2	-2	1	0	1	-1.25000	1	0	0.25000	-1.75
3	-2	1	0	1	-1.16667	1	0	0.16667	-1.5

Component	c_3	d_3	a_4	b_4	c_4	$f_1 = \dfrac{c_1}{b_1}$	g_1	$a_2 f_1$
1	1	-33.33333	1.5	-1.50000	0	-0.5	0	-0.5
2	1	-33.33333	0.75	-1.25000	0	-0.5	0	-0.5
3	1	-33.33333	0.5	-1.16667	0	-0.5	0	-0.5

Component	$b_2 - a_2 f_1$	$f_2 = \dfrac{c_2}{b_2 - a_2 f_1}$	$d_2 - a_2 g_1$	$g_2 = \dfrac{d_2 - a_2 g_1}{b_2 - a_2 f_1}$	$a_3 f_2$	$b_3 - a_3 f_2$	a_3
1	-1.00000	-1.00000	0	0	-0.50000	-2.00000	0
2	-0.75000	-1.33333	0	0	-0.33333	-1.41667	0
3	-0.66667	-1.50000	0	0	-0.25000	-1.25000	0

Component	$d_3 - a_3 g_2$	$f_3 = \dfrac{c_3}{b_3 - a_3 f_2}$	$g_3 = \dfrac{d_3 - a_3 g_2}{b_3 - a_3 f_2}$	$a_4 f_3$	$b_4 - a_4 f_3$
1	−33.33333	−0.50000	16.6667	−0.75000	−0.75000
2	−33.33333	−0.70588	23.52941	−0.52941	−0.72059
3	−33.33333	−0.80000	26.66667	−0.40000	−0.76667

Component	$a_4 g_3$	d_4	$x_4 = g_4 = \dfrac{d_4 - a_4 g_3}{b_4 - a_4 f_3}$	$f_3 x_4$	$x_3 = g_3 - f_3 x_4$
1	24.99999	0	33.33333	−16.66667	33.33333
2	17.64706	0	24.48974	−17.28687	40.81629
3	13.33333	0	17.39131	−13.91305	40.57914
			(Note: $v_{4i} = x_4$)		$(v_{3i} = x_3)$

Component	$x_2 = g_2 - f_2 x_3$	$x_1 = g_1 - f_1 x_2$	$b_i = A_{4i} v_{4i}$	b_i/d_i
1	33.33333	16.66667	16.66667	1.000000
2	54.42162	27.21087	6.12243	0.225000
3	60.86956	30.43477	2.89855	0.095238
	$(v_{2i} = x_2)$	$(d_i = x_1)$		

(b) Use of Eqs. (2-22) and (2-21)

Component	A_{1i}	$m_1 = A_{1i} + 1$	$f_1 = -\dfrac{1}{m_1}$	A_{2i}	$A_{2i}m_1$	$m_2 = A_{2i}m_1 + 1$	$f_2 = -\dfrac{m_1}{m_2}$
1	1	2	−1/2	1/2	1	2	−1
2	1	2	−1/2	1/4	1/2	3/2	−4/3
3	1	2	−1/2	1/6	1/3	4/3	−3/2

Component	A_{3i}	$A_{3i}m_2$	$m_3 = A_{3i}m_2 + 1$	$f_3 = -\dfrac{m_2}{m_3}$	A_{4i}	$A_{4i}m_3$	$m_4 = A_{4i}m_3 + 1$
1	3/2	3	4	−1/2	1/2	2	3
2	3/4	9/8	17/8	−12/17	1/4	17/32	49/32
3	1/2	2/3	5/3	−4/5	1/6	5/18	23/18

$g_1 = 0$, and since $v_{Fi} = 0$, $g_2 = 0$, and $l_{Fi} = FX_i$.

Component	$g_3 = FX_i \dfrac{m_2}{m_3}$	$A_{3i}g_3$	$\dfrac{m_3}{m_4}$	$v_{4i} = x_4 = g_4 = A_{3i}g_3 \dfrac{m_3}{m_4}$
1	16.66666	24.99999	1.333333	33.33333
2	23.52941	17.647057	1.387755	24.48978
3	26.66666	13.33333	1.304347	17.39130

Component	$f_3 x_4$	$v_{3i} = x_3 = g_3 - f_3 x_4$	$v_{2i} = x_2 = g_2 - f_2 x_3$	$d_i = x_1 = g_1 - f_1 x_2$
1	−16.66666	33.33333	33.33333	16.66666
2	−17.28690	40.81631	54.42175	27.21087
3	−13.91304	40.57970	60.86956	30.43478

The tridiagonal formulation of the component-material balances and equilibrium relationships is generally preferred in computer applications because the method is readily applied to other types of columns such as complex distillation columns as shown in Chap. 3. However, for making calculations for conventional distillation columns by hand, the use of nesting equations as originally suggested by Thiele and Geddes[14] is generally the most convenient method to use.

Solution of the Component-Material Balances and Equilibrium Relationships by Use of Nesting Equations

The nesting equations are obtained by first restating those given by Eq. (2-1) in terms of the component-flow rates as follows

$$
\begin{aligned}
v_{j+1,\,i} &= l_{ji} + d_i && (j = 1, 2, \ldots, f-2) \\
v_{fi} + v_{Fi} &= l_{f-1,\,i} + d_i && \\
v_{j+1,\,i} &= l_{ji} - b_i && (j = f, f+1, \ldots, N-1) \\
FX_i &= d_i + b_i &&
\end{aligned}
\tag{2-39}
$$

Elimination of l_{ji} from the first expression of Eq. (2-39) by means of the equilibrium relationship $l_{ji} = A_{ji} v_{ji}$ [Eq. (2-10)] yields the following expression upon rearrangement

$$
\frac{v_{j+1,\,i}}{d_i} = A_{ji}\left(\frac{v_{ji}}{d_i}\right) + 1
\tag{2-40}
$$

for $j = 2, 3, \ldots, f-1$. For $j = 1$ (the condenser-accumulator section) and for a total condenser, the first expression of Eq. (2-39) becomes

$$
\frac{v_{2i}}{d_i} = \frac{l_{1i}}{d_i} + 1 = \frac{L_1 x_{1i}}{DX_{Di}} + 1 = \frac{L_1}{D} + 1
\tag{2-41}
$$

since $x_{1i} = X_{Di}$.

For a partial condenser, $y_{1i} = X_{Di}$, $y_{1i} = K_{1i} x_{1i}$ or $l_{1i} = A_{1i} d_i$, and the first expression of Eq. (2-39) reduces to

$$
\frac{v_{2i}}{d_i} = A_{1i} + 1
\tag{2-42}
$$

where $A_{1i} = L_1/K_{1i} D$. By use of Eq. (2-41) or (2-42) and Eq. (2-40), the nesting calculations are initiated at the top of the column and continued down toward the feed plate. For the case of boiling-point liquid and subcooled feeds, the nesting calculations are discontinued as soon as v_{fi}/d_i has been obtained. For the case of dew-point vapor and superheated feeds, the nesting calculations are discontinued as soon as $l_{f-1,i}/d_i$ has been obtained. (An expression for the calculation of b_i/d_i for a feed of any thermal condition is developed below.)

The nesting equations for the stripping section are initiated at the reboiler. Since $y_{Ni} = K_{Ni}x_{Ni} = K_{Ni}x_{Bi}$ or $v_{Ni} = S_{Ni}b_i$, the last expression of Eq. (2-39) reduces to

$$\frac{l_{N-1,i}}{b_i} = \left(\frac{S_{Ni}}{b_i}\right)b_i + 1 = S_{Ni} + 1 \tag{2-43}$$

where $S_{Ni} = K_{Ni}V_N/B$). After a number value has been obtained for $l_{N-1,i}/b_i$, it is used to compute $l_{N-2,i}/b_i$ by use of the following equation which is obtained by eliminating $v_{j+1,i}/b_i$ from next the last expression of Eq. (2-39) by use of the equilibrium relationship $v_{ji} = S_{ji}l_{ji}$, that is,

$$\frac{l_{ji}}{b_i} = S_{j+1,i}\left(\frac{l_{j+1,i}}{b_i}\right) + 1 \tag{2-44}$$

which holds for $j = f, f + 1, \ldots, N - 2$. After l_{fi}/b_i has been computed, the nesting calculations are ceased and the quantity v_{fi}/b_i is computed by use of the equilibrium relationship, namely,

$$\frac{v_{fi}}{b_i} = S_{fi}\left(\frac{l_{fi}}{b_i}\right)$$

For the case of a boiling-point liquid or subcooled feed, $v_{Fi} = 0$, $l_{Fi} = FX_i$, and hence the moles of vapor entering plate $f - 1$ is equal to the moles of vapor leaving plate j. Thus, b_i/d_i may be computed from the number values found for v_{fi}/d_i and v_{fi}/b_i as follows

$$\frac{b_i}{d_i} = \frac{v_{fi}/d_i}{v_{fi}/b_i} \tag{2-45}$$

Next, the overall component-material balance of Eq. (2-39) may be solved for d_i in terms of b_i/d_i in the following manner

$$FX_i = d_i(1 + b_i/d_i) \quad \text{and} \quad d_i = \frac{FX_i}{1 + b_i/d_i} \tag{2-46}$$

After d_i has been obtained, the complete set of component-flow rates $\{b_i, v_{ji}, l_{ji}\}$ may be obtained from previously calculated results in an obvious manner.

For the general case of a partially vaporized feed, the expression for computing b_i/d_i is obtained by commencing with the second expression of Eq. (2-39) and rearranging it to give

$$\left(\frac{v_{fi}}{b_i}\right)\left(\frac{b_i}{d_i}\right) + \left(\frac{v_{Fi}}{FX_i}\right)\left(\frac{FX_i}{d_i}\right) = \frac{l_{f-1,i}}{d_i} + 1 \tag{2-47}$$

Since

$$\frac{v_{Fi}}{FX_i} = 1 - \frac{l_{Fi}}{FX_i} \qquad \text{and} \qquad \frac{FX_i}{d_i} = 1 + \frac{b_i}{d_i}$$

Eq. (2-47) may be solved for b_i/d_i to give

$$\frac{b_i}{d_i} = \frac{(l_{f-1,i}/d_i) + (l_{Fi}/FX_i)}{(v_{fi}/b_i) + (v_{Fi}/FX_i)} \tag{2-48}$$

When the appropriate values for l_{Fi} and v_{Fi} are employed, Eq. (2-48) may be used to calculate b_i/d_i for a feed of any thermal condition. For bubble-point liquid and subcooled feeds, $l_{Fi} = FX_i$ and $v_{Fi} = 0$. For feeds that enter the column as dew-point and superheated vapors, $v_{Fi} = FX_i$ and $l_{Fi} = 0$.

Example 2-2 Use the above nesting equations to solve Example 2-1 for the component-flow rates.

SOLUTION 1. Calculations for the Rectifying Section

Component	$\dfrac{L_1}{D}$	$\dfrac{v_{2i}}{d_i} = \dfrac{L_1}{D} + 1$	K_{2i} @ 560°F and 1 atm	$A_{2i} = \dfrac{L_2}{K_{2i}V_2}$	$\dfrac{l_{2i}}{d_i} = A_{2i}\left(\dfrac{v_{2i}}{d_i}\right)$	$\dfrac{v_{3i}}{d_i} = \dfrac{l_{2i}}{d_i} + 1$
1	1.0	2.0	1.0	0.5	1.0	2.0
2	1.0	2.0	2.0	0.25	0.5	1.5
3	1.0	2.0	3.0	0.1666667	0.33333333	1.33333333

2. Calculations for the Stripping Section

Component	K_{4i} @ 560°R and 1 atm	$S_{4i} = \dfrac{K_{4i}V_4}{B}$	$\dfrac{l_{3i}}{b_i} = S_{4i} + 1$	K_{3i} @ 560°R and 1 atm	$S_{3i} = \dfrac{K_{3i}V_3}{L_3}$	$\dfrac{v_{3i}}{b_i} = S_{3i}\dfrac{l_{3i}}{b_i}$
1	1.0	2.0	3.0	1.0	0.6666666	1.999999
2	2.0	4.0	5.0	2.0	1.3333333	6.666666
3	3.0	6.0	7.0	3.0	1.9999999	13.999999

Component	$\dfrac{b_i}{d_i} = \dfrac{v_{3i}/d_i}{v_{3i}/b_i}$	$d_i = \dfrac{FX_i}{1 + b_i/d_i}$
1	1.000000	16.66666
2	0.225000	27.21088
3	0.095238	30.43478
		74.31232

The next illustrative example is presented for the purpose of demonstrating the determination of a corrected set of mole fractions by use of the θ method of convergence as indicated by step 2 of the proposed calculational procedure.

Example 2-3 On the basis of the calculated set of values for b_i/d_i which were found in Example 2-2, find the θ that makes $g(\theta) = 0$ by use of Newton's method [see Eqs. (2-25) through (2-28)].

SOLUTION *Determination of θ by Use of Newton's Method*
   Trial 1. Assume $\theta_1 = 0$

Component	$\left(\dfrac{b_i}{d_i}\right)_{ca}$	$\theta_1\left(\dfrac{b_i}{d_i}\right)_{ca}$	$1 + \theta_1\left(\dfrac{b_i}{d_i}\right)_{ca}$	$(d_i)_{co} = \dfrac{FX_i}{1 + \theta_1\left(\dfrac{b_i}{d_i}\right)_{ca}}$	$\dfrac{FX_i}{\left[1 + \theta_1\left(\dfrac{b_i}{d_i}\right)_{ca}\right]^2}$	$\dfrac{FX_i(b_i/d_i)_{ca}}{\left[1 + \theta_1\left(\dfrac{b_i}{d_i}\right)_{ca}\right]^2}$
1	1	0	1	33.33333	33.33333	33.33333
2	0.225	0	1	33.33333	33.33333	7.49999
3	0.095238	0	1	33.33333	33.33333	3.17459
				100.00000		44.00791

$$g(\theta) = \sum_{i=1}^{c}(d_i)_{co} - D \qquad g'(\theta) = -\sum_{i=1}^{c}\frac{FX_i(b_i/d_i)_{ca}}{[1 + \theta(b_i/d_i)_{ca}]^2}$$

Therefore, $g(\theta) = 100 - 50 = 50$, $g'(\theta) = -44.00791$
Then by Newton's method

$$\theta_2 = \theta_1 - \frac{g(\theta_1)}{g'(\theta_1)} = 0.0 - \frac{50}{(-44.00791)} = 1.13616$$

   Trial 2. Assume $\theta_2 = 1.13616$

Component	$\left(\dfrac{b_i}{d_i}\right)_{ca}$	$\theta_2\left(\dfrac{b_i}{d_i}\right)_{ca}$	$1 + \theta_2\left(\dfrac{b_i}{d_i}\right)_{ca}$	$(d_i)_{co} = \dfrac{FX_i}{1 + \theta_2\left(\dfrac{b_i}{d_i}\right)_{ca}}$	$\dfrac{FX_i}{\left[1 + \theta_2\left(\dfrac{b_i}{d_i}\right)_{ca}\right]^2}$	$\dfrac{FX_i(b_i/d_i)_{ca}}{\left[1 + \theta_2\left(\dfrac{b_i}{d_i}\right)_{ca}\right]^2}$
1	1	1.13616	2.13616	15.60432	7.30485	7.30485
2	0.225	0.25564	1.25564	26.54688	21.14211	4.75697
3	0.095238	0.10821	1.10821	30.07853	27.14154	2.58490
				72.22973		14.64672

$$\sum_{i=1}^{c} (d_i)_{co} = 72.22973 \qquad g'(\theta_2) = -14.64672 \qquad g(\theta_2) = 22.22973$$

Then by Newton's method

$$\theta_3 = \theta_2 - \frac{g(\theta_2)}{g'(\theta_2)} = 1.13616 - \frac{22.22973}{(-14.64672)} = 2.65389$$

Continuation of this process gives a $\theta = 3.687276$ for which $|g(\theta)| < 0.0001$.

The solution value of θ found in Example 2-3 is used to compute the corrected compositions. Then these compositions are used in the K_b method to compute a new set of temperatures as demonstrated below.

Example 2-4 Use the θ found in Example 2-3 to compute the corrected compositions for plates 2 and 3. Then determine the new temperatures T_1, T_2, and T_3.

Component	$\theta\left(\dfrac{b_i}{d_i}\right)_{ca}$	$1 + \theta\left(\dfrac{b_i}{d_i}\right)_{ca}$	$(d_i)_{co} = \dfrac{FX_i}{1 + \theta\left(\dfrac{b_i}{d_i}\right)_{ca}}$	$X_{Di} = \dfrac{(d_i)_{co}}{D}$	α_{1i} @ 560°R	$\alpha_{1i}X_{Di}$
1	3.68728	4.68728	7.11145	0.14223	1	0.14223
2	0.82964	1.82964	18.21854	0.36437	2	0.72874
3	0.35117	1.35117	24.67001	0.49340	3	1.48020
			50.00000			2.35117

$$K_{1b} = \frac{1}{\sum\limits_{i=1}^{c} \alpha_{1i} X_{Di}} = \frac{1}{2.35117} = 0.42532 = C_b \exp\left(-E_b/T_1\right)$$

Thus

$$T_1 = \frac{E_b}{\ln C_b/K_{1b}} = \frac{4.6447 \times 10^3}{\ln (4.0 \times 10^3)/(0.42532)} = 507.68°R = 47.68°F$$

To calculate the y_{2i}'s, the values of the v_{ji}'s and d_i's found in Example 2-1 are used to obtain the (v_{ji}/d_i)'s, and then these ratios are multiplied by the $(d_i)_{co}$'s found above.

Component	$\left(\dfrac{v_{2i}}{d_i}\right)_{ca}(d_i)_{co}$	$y_{2i} = \dfrac{(v_{2i}/d_i)_{ca}(d_i)_{co}}{\sum\limits_{i=1}^{c}(v_{2i}/d_i)_{ca}(d_i)_{co}}$	α_{2i} @ 560°R	$\dfrac{y_{2i}}{\alpha_{2i}}$	$\left(\dfrac{v_{3i}}{d_i}\right)_{ca}(d_i)_{co}$
1	14.22290	0.14223	1	0.14223	14.2229
2	36.43709	0.36437	2	0.18219	27.3278
3	49.34002	0.49340	3	0.16447	32.8933
	100.00001			0.48889	74.4440

$$K_{2b} = \sum_{i=1}^{c} \frac{y_{2i}}{\alpha_{2i}} = 0.48889 = C_b \exp\left(-E_b/T_2\right)$$

$$T_2 = \frac{E_b}{\ln C_b/K_{2b}} = \frac{4.6447 \times 10^3}{\ln (4.0 \times 10^3)/(0.48889)} = 515.52°R = 55.52°F$$

Component	$y_{3i} = \dfrac{(v_{3i}/d_i)_{ca}(d_i)_{co}}{\displaystyle\sum_{i=1}^{c} (v_{3i}/d_i)_{ca}(d_i)_{co}}$	α_{3i} @ 560°R	$\dfrac{y_{3i}}{\alpha_{3i}}$
1	0.19105	1	0.19105
2	0.36709	2	0.18355
3	0.44185	3	0.14728
			0.52188

$$K_{3b} = \sum_{i=1}^{c} \frac{y_{3i}}{\alpha_{3i}} = 0.52188 = C_b \exp\left(-E_b/T_3\right)$$

$$T_3 = \frac{E_b}{\ln C_b/K_{3b}} = \frac{4.6447 \times 10^3}{\ln (4.0 \times 10^3)/(0.52188)} = 519.29°R = 59.29°F$$

Example 2-5 Use the compositions and temperatures found in Examples 2-3 and 2-4 and the enthalpy functions given in Example 1-5 to compute Q_C, L_2, and V_3.

SOLUTION (a) For a total condenser: $h_{1i} = H_{Di}$, $x_{1i} = X_{Di}$, $y_{2i} = X_{Di}$.
    Calculation of Q_C

$$Q_C = L_1 \sum_{i=1}^{c} (H_{2i} - h_{1i})x_{1i} + D \sum_{i=1}^{c} (H_{2i} - H_{Di})X_{Di}$$

$$= (L_1 + D) \sum_{i=1}^{c} (H_{2i} - h_{1i})X_{Di}$$

Compo-nent	H_{2i} @ 55.52°F	h_{1i} @ 47.68°F	$H_{2i} - h_{1i}$	$(H_{2i} - h_{1i})X_{Di}$	H_{3i} @ 59.29°F	h_{2i} @ 55.52°F
1	18,665.60	11,430.40	7235.20	1029.06	18,778.7	11,665.60
2	14,110.40	8,953.60	5156.80	1878.98	14,185.8	9,110.40
3	855.52	547.68	307.84	151.89	859.29	555.52
				3059.93		

$$Q_C = (100)(3059.93) = 305,993 \text{ Btu/h}$$

Calculation of L_2

$$L_2 = \frac{Q_C - D\left[\sum_{i=1}^{c} (H_{3i} - h_{1i})X_{Di}\right]}{\sum_{i=1}^{c} (H_{3i} - h_{2i})x_{2i}}$$

Component	$\left(\dfrac{l_{2i}}{d_i}\right)_{ca}(d_i)_{co}$	$x_{2i} = \dfrac{(l_{2i}/d_i)_{ca}(d_i)_{co}}{\sum_{i=1}^{c}(l_{2i}/d_i)_{ca}(d_i)_{co}}$	$(H_{3i} - h_{1i})$	$(H_{3i} - h_{1i})X_{Di}$	$H_{3i} - h_{2i}$
1	7.11145	0.290927	7348.30	1045.15	7113.10
2	9.10927	0.372657	5232.20	1906.46	5075.40
3	8.22336	0.336415	311.61	153.75	303.77
	24.44408			3105.36	

Component	$(H_{3i} - h_{2i})x_{2i}$
1	2069.39
2	1891.38
3	102.19
	4062.96

Thus

$$L_2 = \frac{(-50)[3105.36] + 305,993}{[4062.96]} = \textbf{37.10 lb mol/h}$$

Then

$$V_3 = L_2 + D = \textbf{87.10 lb mol/h}$$

Example 2-6 Repeat the iterative procedure demonstrated by Examples 2-1 through 2-5 until convergence has been achieved. The column has a total condenser, $P = 1$ atm, boiling-point liquid feed, $N = 4$, $f = 3$, the feed composition $\{X_i\}$ is given in Example 2-1, $F = 100$, $L_1 = D = 50$ lb mol/h. The K data are given in Example 2-1 and the enthalpy data are given in Example 1-5.

SOLUTION The convergence characteristics of the proposed calculational procedure and final results obtained for this example are shown in Table 2-1.

Table 2-1 Solution of Example 2-6 by use of the θ method, K_b method, and the constant-composition for making enthalpy balances

	Calculated values of the temperature, °F Trial					
Stage	1	2	3	4	5	6
1	47.67480	47.57593	47.64648	47.68652	47.70581	47.71484
2	55.52295	55.35938	55.48169	55.55078	55.58423	55.59985
3	59.28833	60.60938	60.54272	60.50366	60.48169	60.46973
4	68.15991	68.31323	68.20361	68.14209	68.11208	68.09814
D (calcu- lated)	74.31046	49.53528	50.02797	50.01648	50.00912	50.00485
θ	3.686828	0.9772162	1.001385	1.000813	1.000451	1.000240

	Calculated values of the temperature, °F Trial					
Stage	7	8	9	10	11	12
1	47.71924	47.72119	47.72217	47.72241	47.72266	47.72266
2	55.60718	55.61060	55.61279	55.61279	55.61328	55.61328
3	60.46313	60.45972	60.45801	60.45728	60.45679	60.45654
4	68.09155	68.08862	68.08716	68.08643	68.08618	68.08618
D (calcu- lated)	50.00258	50.00131	50.00060	50.00290	50.00014	50.00005
θ	1.000128	1.000063	1.000031	1.000014	1.000008	1.000004

Final flow rates

Component	b	d
1	26.15640	7.176938
2	15.14437	18.18893
3	8.699208	24.63411

2-3 CONVERGENCE CHARACTERISTICS OF THE θ METHOD OF CONVERGENCE, THE K_b METHOD, AND THE CONSTANT-COMPOSITION METHOD

The θ method of convergence may be classified as an iterative procedure, and the success of such a method depends upon the proper arrangement of all equations involved in the entire calculational procedure. The complete calculational procedure was developed by trial by finding what appeared to be the best form for all equations. In particular, it was found that the method was most stable when the temperatures were determined by the K_b method and the total-flow rates by the constant-composition method. The use of other procedures for the determination of either the temperatures or the total-flow rates caused some problems to diverge which could be solved by use of the K_b method and the constant-composition method as discussed below.

In summary, it was found that if the combination of the θ method of convergence, the K_b method, and the constant-composition method were used, convergence of almost all problems involving conventional and complex columns as well as systems of distillation columns could be achieved. This combination of calculational procedures is also one of the fastest known methods.[2, 3]

Although a proof of the convergence of the combination of the θ method, K_b method, and constant-composition method cannot be constructed, it can be shown (and is shown below) that the θ method constitutes an exact solution to certain total reflux problems.

To illustrate the characteristics of the combination of these three calculational procedures, Example 2-7 was selected. The statement of this example is presented in Table 2-2. The temperature profiles obtained by use of the K_b method on the basis of the corrected compositions found by use of the θ method [Eqs. (2-27) and (2-28)] are presented in Table 2-3. The constants a and b in the expression for K_b [Eq. (2-31)] were found by use of the K values for $i\text{-}C_4H_{10}$ at 510 and 960°R. The values of θ and the calculated values of D at the end of each trial are also listed in this table. The vapor rates computed by use of the constant-composition method, the temperatures found by the K_b method, and the corrected compositions are presented in Table 2-4. The solution sets $\{d_i\}$ and $\{b_i\}$ are presented in Table 2-5. To satisfy the convergence criterion, twelve trials were required and 2.60 seconds of computer time on an AMDAHL 470 V/6 computer using a WATFIV compiler.

Table 2-2 Statement of Example 2-7

Compo-nent	FX_i	Specifications
CH_4	2.0	$D = 31.6$, $V_2 = 94.8$ (all flow rates are in lb mol/h) boiling-point liquid feed,
C_2H_6	10.0	partial condenser, column pressure $= 300$ lb/in² abs, $N = 12$ and $f = 5$.
C_3H_6	6.0	Equilibrium and enthalpy data for all components are given in Tables B-1
C_3H_8	12.5	and B-2 of the Appendix. The initial temperature profile is to be taken linear
$i\text{-}C_4H_{10}$	3.5	with plate number between $T_1 = 610°R$ and $T_{13} = 910°R$. Take the initial
$n\text{-}C_4H_{10}$	15.0	vapor rate profiles to be $V_j = 94.8$ $(j = 2, 3, \ldots, 13)$, and the corresponding
$n\text{-}C_5H_{12}$	15.2	liquid rate profile is given by material balance. Component $i\text{-}C_4H_{10}$ was
$n\text{-}C_6H_{14}$	11.3	taken as the base component and a and b in Eq. (2-30) were determined on
$n\text{-}C_7H_{16}$	9.0	the basis of the values for the K of $i\text{-}C_4H_{10}$ at 510 and 960°R.
$n\text{-}C_8H_{18}$	8.5	
400†	7.0	

† Commonly referred to as the 400°F—normal boiling fraction.

Table 2-3 Temperature profiles, theta, and calculated values of D

	Temperature profiles (°R) Trial					
Stage	1	2	3	4	5	6
1 (distillate)	573.59	567.84	567.56	567.56	567.56	567.57
2	596.60	590.67	594.54	594.13	594.42	594.33
3	607.01	606.81	612.61	611.63	611.99	611.89
4	612.65	627.85	631.10	629.91	630.26	630.25
5 (feed)	617.02	695.72	658.24	670.53	665.95	668.04
6	647.02	709.68	681.56	691.68	687.53	689.20
7	670.05	717.92	697.73	705.76	702.37	703.69
8	689.59	724.20	709.70	716.11	713.40	714.46
9	707.47	730.07	719.35	724.47	722.32	723.17
10	725.93	737.58	728.94	732.96	731.25	731.92
11	747.73	749.93	742.09	745.17	743.78	744.32
12	777.47	774.19	766.54	768.84	767.70	768.13
13 (bottoms)	833.68	833.41	825.47	827.21	826.36	826.68
D (calculated)	41.5475	30.4606	32.3414	31.37940	31.6940	31.5598
θ	22.4131	0.303676	2.34622	0.835224	1.07461	0.968934

	Temperature profiles (°R) Trial					
Stage	7	8	9	10	11	12
1 (distillate)	567.57	567.57	567.57	567.57	567.57	567.57
2	594.38	594.36	594.37	594.37	594.37	594.37
3	611.95	611.92	611.93	611.93	611.93	611.93
4	630.27	630.26	630.27	630.27	630.27	630.26
5 (feed)	667.14	667.53	667.37	667.44	667.41	667.41
6	688.47	688.79	688.65	688.71	688.68	688.69
7	703.11	703.36	703.25	703.30	703.28	703.28
8	714.00	714.19	714.11	714.14	714.13	714.13
9	722.79	722.95	722.88	722.91	722.90	722.90
10	731.63	731.75	731.70	731.72	731.71	731.71
11	744.08	744.18	744.13	744.15	744.15	744.15
12	767.94	768.02	767.99	768.00	767.99	768.00
13 (bottoms)	826.53	826.59	826.56	826.58	826.57	826.57
D (calculated)	31.6175	31.5925	31.6031	31.5986	31.60057	31.59977
θ	1.01356	0.994182	1.00243	0.998940	1.00042	1.00000

Table 2-4 Vapor rates obtained by the constant-composition method for Example 2-7

	Vapor rates (lb-mol/h) Trial					
Stage	1	2	3	4	5	6
2	94.80	94.80	94.80	94.80	94.80	94.80
3	94.63	93.23	93.06	93.29	93.29	93.29
4	94.13	88.11	89.31	89.41	89.45	89.41
5	93.13	72.10	82.71	79.25	80.51	79.94
6	140.71	108.19	117.18	109.35	111.52	110.55
7	143.80	125.43	129.81	125.40	126.79	126.13
8	146.69	137.24	138.81	136.34	137.20	136.78
9	148.52	145.60	145.39	144.16	144.60	144.38
10	148.03	150.07	149.13	148.83	148.94	148.87
11	144.63	150.08	149.09	149.59	149.43	149.50
12	137.74	143.88	142.75	143.97	143.60	143.77
13	120.86	123.86	122.24	123.93	123.33	123.60

	Vapor rates (lb mol/h) Trial					
Stage	7	8	9	10	11	12
2	94.80	94.80	94.80	94.80	94.80	94.80
3	93.29	93.29	93.29	93.29	93.29	93.29
4	89.41	89.41	89.42	89.42	89.42	89.42
5	80.18	80.08	80.12	80.10	80.11	80.11
6	110.92	110.75	110.82	110.79	110.81	110.80
7	126.36	126.25	126.29	126.24	126.28	126.27
8	136.91	136.84	136.87	136.86	136.86	136.86
9	144.44	144.41	144.42	144.42	144.42	144.41
10	148.89	148.87	148.88	148.88	148.88	148.87
11	149.47	149.48	149.47	149.48	149.47	149.47
12	143.70	143.73	143.72	143.72	143.72	143.72
13	123.49	123.54	123.52	123.53	123.52	123.52

By solving a wide variety of examples it was found that the boiling range, the thermal condition of the feed, the number of plates, the initial temperature, and vapor rate profiles had no appreciable effect on the number of trials required to achieve convergence. However, as the boiling range of the feed is increased, it may become necessary to average the temperature profiles as well as the vapor-rate profiles for successive trials.

The use of a component much lighter than the midboiling component of the mixture can introduce significant overcorrections of the temperature profiles while the use of a component much heavier than the midboiling component can lead to undercorrections of the temperature profiles. The use of the midboiling

Table 2-5 Solution sets of the component flow rates in the distillate and bottoms of Example 2-7

Component	d_i	b_i
CH_4	0.200000×10	0.11163×10^{-8}
C_2H_6	0.999990×10	0.11627×10^{-3}
C_3H_6	0.597230×10	0.27665×10^{-1}
C_3H_8	0.1234600×10^2	0.15358
$i\text{-}C_4H_{10}$	0.74216	0.27578×10
$n\text{-}C_4H_{10}$	0.53699	0.14462×10^2
$n\text{-}C_5H_{12}$	0.20153×10^{-2}	0.15197×10^2
$n\text{-}C_6H_{12}$	0.94035×10^{-5}	0.11299×10^2
$n\text{-}C_7H_{14}$	0.94025×10^{-5}	0.89999×10
$n\text{-}C_8H_{16}$	0.63427×10^{-7}	0.84999×10
400	0.65162×10^{-12}	0.69999×10

$Q_C = 3.9628 \times 10^5$ Btu/h
$Q_R = 1.3278 \times 10^6$ Btu/h
Convergence criterion: $|g(1)| \leq 10^{-5}$

component of the feed mixture or the component just lighter for the calculation of the constants a and b of the K_b method generally give satisfactory results. Recently, however, improvements in the K_b method were proposed by Billingsley[2] wherein a different K_b was used for each plate and composition effects were also included.

To obtain the convergence characteristics exhibited in Table 2-3 through 2-5, it cannot be overemphasized that the complete set of calculational procedures (the θ method, the K_b method, and the constant-composition method) must be used. When the θ method was introduced initially,[11] other procedures were used for the determination of the temperatures and total-flow rates, and the convergence characteristics differed from those presented in Tables 2-3 through 2-5.

Comparison of the θ Method with the Method of Direct Iteration

Some of the convergence characteristics of the θ method are demonstrated by comparing the θ method with the method of direct iteration. The method of direct iteration differs from the θ method only by the procedure used to compute the compositions. Instead of the expressions given by Eq. (2-29), the following expressions are used in the method of direct iteration

$$x_{ji} = \frac{(l_{ji}/d_i)_{ca}(d_i)_{ca}}{\sum\limits_{i=1}^{c} (l_{ji}/d_i)_{ca}(d_i)_{ca}} \tag{2-49}$$

$$y_{ji} = \frac{(v_{ji}/d_i)_{ca}(d_i)_{ca}}{\sum\limits_{i=1}^{c} (v_{ji}/d_i)_{ca}(d_i)_{ca}} \tag{2-50}$$

By comparison of these expressions with those given by Eqs. (2-25), (2-26), and (2-29), it is evident that the method of direct iteration amounts to setting $\theta = 1$ in Eq. (2-26) for all trials. The results obtained for Example 2-2 with the flow rates held fixed at the set stated in this example are presented in Table 2-1. When the method of direct iteration was used, a calculated value of 52.14 was obtained for D at the end of the third trial, and 11 trials were required to obtain temperatures which were correct to eight digits.

At a set of fixed L/V's, the calculated values of D follow the same type of variation with respect to temperature as might be expected for an actual column. If the assumed temperature profile is too high, then the calculated value of D will be greater than the specified value; and if the assumed temperature profile is too low, the calculated value of D will be less than the specified value. More precisely, by use of the expressions presented in Prob. 2-4, it is readily shown that b_i/d_i decreases with an increase in the temperature of any one plate of the column, and conversely. Thus, by Eq. (2-46), d_i increases as any one temperature is increased, and conversely.

Now consider Eqs. (2-26) and (2-27). If the sum of the calculated values of d_i is greater than the specified value of D, a positive value of θ that is greater than unity is required to satisfy Eq. (2-27) (that is, make $g(\theta) = 0$), and if the sum of the calculated values of d_i is less than the specified value of D, a positive value of θ less than unity is required to satisfy Eq. (2-27). Thus, if the temperature profile for the previous trial was too low, a value of θ less than unity will be obtained; and if too high, a value of θ greater than unity will be obtained.

The use of corrected d_i's based on values of $\theta < 1$ and $\theta > 1$ gives lower and higher temperatures, respectively, than those predicted by the method of direct iteration. This is readily shown by consideration of the variation of the corrected d_i's with θ for very light and very heavy components. For a very light component b_i/d_i is very small so that Eq. (2-26) reduces to

$$(d_i)_{co} \cong F X_i \tag{2-51}$$

For a very heavy component, b_i/d_i is very large. Thus

$$(d_i)_{co} \cong \frac{F X_i}{\theta(b_i/d_i)_{ca}} \tag{2-52}$$

Now consider the case where the temperature profile of the previous trial was too high. This condition leads to a value of θ greater than unity. In view of Eqs. (2-51) and (2-52), it is seen that the formulas for the mole fractions for each plate [Eq. (2-29)] give sets of compositions with a relatively smaller proportion of heavies than those obtained by direct iteration. Therefore, the temperatures computed on the basis of the corrected compositions are less than the corresponding temperatures calculated by the method of direct iteration, since the latter are calculated on the basis of the calculated d_i's.

An Exact Solution Given by the θ Method of Convergence

If a set of b_i/d_i's corresponding to a given D are known for a column at total reflux and a system for which the α_i's are constant, then the θ method of convergence may be used to compute the solution set of b_i/d_i's corresponding to any other value of D.

That the θ method of convergence is applicable to a problem of this type is demonstrated in the following manner. Consider the case of a distillation column with N stages and a total condenser which is operated at two different distillate rates, D_1 and D_2 [or at two different specified values of b_b/d_b, denoted by $(b_b/d_b)_1$ and $(b_b/d_b)_2$]. For these two different sets of operating conditions at total reflux, the Fenske equation[4] [see Eq. (1-58)] gives

$$\ln (b_i/d_i)_2 = \ln (b_b/d_b)_2 - (N - 1) \ln \alpha_i$$

$$\ln (b_i/d_i)_1 = \ln (b_b/d_b)_1 - (N - 1) \ln \alpha_i \qquad (2\text{-}53)$$

Elimination of $\ln \alpha_i$ from these two equation gives

$$\ln (b_i/d_i)_2 = \ln \theta + \ln (b_i/d_i)_1$$

or

$$\left(\frac{b_i}{d_i}\right)_2 = \theta \left(\frac{b_i}{d_i}\right)_1 \qquad (2\text{-}54)$$

where

$$\theta = \frac{(b_b/d_b)_2}{(b_b/d_b)_1}$$

If the subscripts 1 and 2 in Eq. (2-54) are replaced by ca and co, respectively, then Eq. (2-25) is obtained. A graphical representation of the expressions given by Eqs. (2-53) and (2-54) is presented in Fig. 2-6. That the θ method is an exact

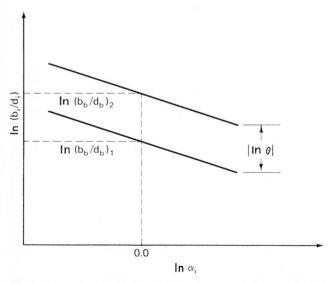

Figure 2-6 A graphical representation of θ is obtained by considering two arbitrarily specified values for a base component b in a column at total reflux.

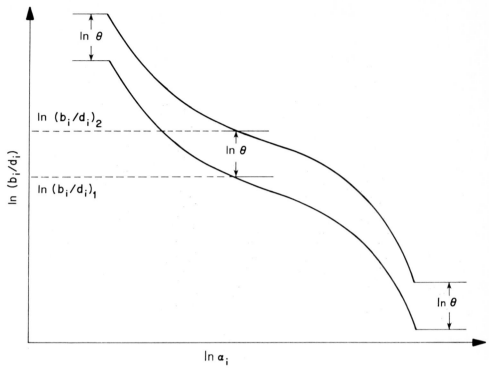

Figure 2-7 Shift of the b_i/d_i profiles by the θ method.

solution for certain problems involving columns at total reflux is demonstrated by Prob. 2-5.

While the θ method constitutes an exact solution to certain total reflux problems, it represents only an approximate solution to problems wherein operating conditions other than total reflux are employed. The θ method for a conventional distillation column at any operating condition other than total may be represented with the aid of Fig. 2-7 as the shifting of the most recently calculated b_i/d_i profile up or down the same distance ($|\ln \theta|$) for each component as required to obtain a new set of b_i/d_i's which are in agreement with the specified value of D.

2-4 OTHER TOPICS: PRESSURE EFFECTS, NONIDEAL SOLUTIONS, AND OTHER SPECIFICATIONS

Pressure Effects

In the calculational procedure demonstrated above, the effect of the variation of the column pressure from plate to plate on the K_{ji}'s, H_{ji}'s, and h_{ji}'s was neglected. The effect of pressure on these variables may be taken into account as described below.

For mixtures which behave as ideal solutions, the K values and the vapor and liquid enthalpies of the pure components $\{K_{ji}, H_{ji}, h_{ji}\}$ depend upon the temperature T_j and pressure P_j of each plate j. Methods for taking the effect of pressure into account in the evaluation of the K_{ji}'s, H_{ji}'s, and h_{ji}'s are to be found in standard thermodynamic texts, handbooks, and the scientific literature as described in Chap. 14. The effort required to account for the effect of pressure on the K values and enthalpies may be reduced by taking advantage of the following observations. First, for columns operating at relatively high pressure, the pressure drop across the column is small relative to the total pressure. Thus, in many instances, it is possible to use a single column pressure in the evaluation of the K_{ji}'s, H_{ji}'s, and h_{ji}'s, and these quantities may be curve fit as a function of temperature alone.

For columns operated at relatively low pressures, such as vacuum distillations, the effect of pressure on the enthalpies can be neglected because the vapor tends to behave as a perfect gas, but the effect of pressure on the K values cannot be neglected. However, the fact that the vapor behavior approaches that of a perfect gas gives rise to a simple method for including the effect of pressure on the K values. In particular, at low pressures,

$$K_i \cong \frac{P_i}{P} \tag{2-55}$$

Now suppose that the K values for component i are evaluated at some pressure P_b at each of several temperatures and that these values of K_i are curve fit as a function of T at the pressure P_b. The value of K_{ji} at P_j and T_j $[K_{ji}(P_j, T_j)]$ may be approximated with good accuracy by use of the value given by the curve fit $[K_{ji}(P_b, T_j)]$ and the following relationship which is suggested by Eq. (2-55)

$$K_{ji}(P_j, T_j) = \frac{P_b}{P_j} K_{ji}(P_b, T_j) \tag{2-56}$$

Nonideal Solutions

Most problems involving the separation of nonideal solutions may be solved by use of the θ method of convergence. When used to solve such problems, the θ method does become slower and it may be necessary to place certain restraints on the calculational procedure. [The "Almost Band Algorithm" which is presented in Chap. 5 may be used to solve any problem for which the θ method fails.]

To apply the θ method to problems involving nonideal solutions, minor modifications of the equations presented above for ideal solutions are necessary. The development of the appropriate expressions for the component-material balances, the K_b method, and the constant-composition method is outlined in Probs. 2-6 through 2-10.

Modification of the Formulas to Account for Separated Components and Inert Components

The formulas for the θ method were developed for distributed components, where a distributed component is defined as one which appears in both the distillate and bottoms; that is, for a distributed component $d_i > 0$ and $b_i > 0$.

A separated component is one which appears in only one product stream. A separated light component is defined as one which appears only in the distillate $(b_i = 0, d_i = FX_i)$, and a separated heavy component is one which appears only in the bottoms $(d_i = 0, b_i = FX_i)$.

In order to avoid numerical difficulties resulting from the divisions by small values of $(d_i)_{ca}$ in Eqs. (2-26) and (2-29), the following expressions for $(d_i)_{co}$ should be used. These expressions are obtained by a simple rearrangement of Eq. (2-26), namely,

$$(d_i)_{co} = (d_i)_{ca} p_i \tag{2-57}$$

where p_i is defined by

$$p_i = \frac{FX_i}{(d_i)_{ca} + \theta(b_i)_{ca}} \tag{2-58}$$

Then for a separated heavy component for which $(d_i)_{ca} = 0$, it follows from Eq. (2-58) that p_i is finite. Since p_i is finite and $(d_i)_{ca} = 0$, Eq. (2-57) gives

$$(d_i)_{co} = 0$$

Consequently, it follows by material balance that for a separated heavy component

$$(b_i)_{ca} = (b_i)_{co} = FX_i$$

Thus, if separated heavies are present, the ratios $(d_i)_{co}/(d_i)_{ca}$ in the formulas given by Eq. (2-29) for x_{ji} and y_{ji} become indeterminate. As suggested by Tetlow,[15] these indeterminate terms may be eliminated by use of Eqs. (2-57) and (2-58). The resulting expressions may be used for all types of components, separated lights, separated heavies, and distributed components

$$x_{ji} = \frac{(l_{ji})_{ca} p_i}{\sum_{i=1}^{c} (l_{ji})_{ca} p_i}$$

$$y_{ji} = \frac{(v_{ji})_{ca} p_i}{\sum_{i=1}^{c} (v_{ji})_{ca} p_i} \tag{2-59}$$

where p_i is to be computed by use of Eq. (2-58).

Similarly, the expression for $g(\theta)$ may be expressed in terms of the p_i's and used to avoid numerical difficulties

$$g(\theta) = \sum_{i=1}^{c} (d_i)_{ca} p_i - D \tag{2-60}$$

Although inert gases and inert liquids may be treated as two separate classes of components, as demonstrated in Chap. 1, it is perhaps simpler from a computational point of view to treat them as volatile components. This treatment may be carried out by assigning the inert gases a K value which is large relative to the light components and the inert liquid a K value which is small relative to the heavy components.

Other Types of Specifications

In all of the formulations considered, it is supposed that the following specifications have been made: the number of plates in each section; the quantity, composition, and thermal condition of the feed; and the column pressure. Two other specifications may be made in addition to these, and the formulation of the functions corresponding to these two additional specifications may be divided into three classes in addition to the class considered above (the specification of the reflux rate L_1 and the distillate rate D or the temperature of the distillate; see Prob. 2-13).

CLASS 1. SPECIFICATION OF THE REFLUX RATE L_1 (OR THE BOILUP RATE V_N) AND ONE ADDITIONAL SPECIFICATION SUCH AS D, A PARTIAL SUM OF d_i's OR X_{Di}'s

When any one of several specifications are made in lieu of the distillate rate D, a corresponding g function may be formulated. Some care must be given, however, to the formulation of this function in order to obtain one which has desirable behavior, such as monotonic behavior with respect to the variable θ. Suitable functions are given in Probs. 2-12 through 2-14 for a number of specifications.

CLASS 2. TWO PURITY SPECIFICATIONS

If instead of L_1 and D, two purity specifications such as $\{X_{Dl}, x_{Bh}\}$, $\{X_{Dh}, x_{Bl}\}$, $\{d_l, b_h\}$, or $\{d_h, d_l\}$ where the subscript l is used to denote the light key and h the heavy key component. To solve problems of this type, the optimization procedure described in Chap. 9 may be applied as described in Chap. 11.

CLASS 3. SPECIFICATION OF THE REFLUX RATIO L_1/D AND THE BOILUP RATIO V_N/B

This particular set of specifications gives rise to a set of g functions which involve the material balances and the energy balances enclosing the column as well as the equilibrium relationships for the terminal streams. This application of the θ method to single columns is deferred until Chap. 7 because of its similarity to calculational procedures developed therein.

NOTATION

A_{ji}	absorption factor; defined by Eq. (2-11)
$\mathbf{A}_i$	a square matrix for each component i; defined below Eq. (2-18)
f_i	a feed vector in the component-material balances; defined below Eq. (2-18)
$g(\theta)$	a function of θ; defined by Eq. (2-27)
H	enthalpy per mole of feed, regardless of state
H_D	enthalpy per mole of distillate, regardless of state. For a total condenser $H_D = h_1$ and $H_{Di} = h_{1i}$. For a partial condenser, $H_D = H_1$ and $H_{Di} = H_{1i}$
h_B	enthalpy per mole of bottoms
$\bar{H}_{ji}$	partial molar enthalpy of component i, evaluated at the temperature and vapor composition of plate j
$\hat{H}_{ji}$	virtual value of the partial molar enthalpy, evaluated at the temperature and vapor composition of plate j
l_{ji}	molar flow rate at which component i in the liquid phase leaves the jth plate
N	total number of plates
p_i	ratio of distillate rates; defined by Eq. (2-58)
P_i	vapor pressure of component i
P	total pressure
S_{ji}	stripping factor for component i; defined by Eq. (2-11)
v_{ji}	molar flow rate at which component i in the vapor leaves plate j
$\mathbf{v}_i$	column vector of components flow rates in the vapor phase; defined by Eq. (2-18)

Subscripts

ca	calculated value
co	corrected value
f	feed plate
F	variables associated with a partially vaporized feed
i	component number, $i = 1, 2, \ldots, c$
j	stage number; for the accumulator $j = 1$; for the top plate $j = 2$, for the feed plate $j = f$, for the bottom plate $j = N - 1$, and for the reboiler $j = N$; that is, $j = 1, 2, 3, \ldots, f, \ldots, N - 1, N$
k	trial number
n	trial number

Greek Symbols

ρ_{ji}	defined beneath Eq. (2-18)
θ	a multiplier defined by Eq. (2-25)

Mathematical Symbols

$[x_1 \ x_2 \ x_3]^T$ transpose of a row vector. The transpose of a row vector is

equal to a column vector; that is, $[x_1 \ x_2 \ x_3]^T = \begin{bmatrix} x_1 \\ x_2 \\ x_3 \end{bmatrix}$

$\displaystyle\sum_k$ summation over all components k, where k denotes any set of

components less than the total number c, say $k = 1, 2, \ldots, c-1$

PROBLEMS

2-1 (a) Complete the trial calculations initiated by Example 2-2 by obtaining the solution value of θ shown.

(b) Complete the trial calculations initiated in Example 2-4 by finding T_4, Q_R, V_4, and L_3.

2-2 If the following set of K values

Component	K_i†
1	$0.01 T/P$
2	$0.0002 T^2/P$
3	$0.03 T/P$

   † T is in °F and P is in atm.

is used instead of the set given in Example 2-1, show that (a) $\theta = 3.687276$, and (b) $T_1 = 42.53$°F, $T_2 = 48.89$°F, $T_3 = 52.19$°F, $T_4 = 60.65$°F. Use the K_b method and take $K_b = K_1$.

2-3 The recurrence formulas given by Eqs. (2-20) and (2-21) for solving equations which are tridiagonal in form may be developed as outlined below by use of the gaussian elimination. The system of linear equations given by Eq. (2-19) is represented by the following matrix equation

$$
\begin{bmatrix}
b_1 & c_1 & 0 & 0 & 0 & \cdots & 0 \\
a_2 & b_2 & c_2 & 0 & 0 & \cdots & 0 \\
0 & a_3 & b_3 & c_3 & 0 & \cdots & 0 \\
& & & \cdots & & & \\
0 & \cdots & 0 & a_{N-1} & b_{N-1} & c_{N-1} & \\
0 & \cdots & 0 & 0 & a_N & b_N &
\end{bmatrix}
\begin{bmatrix}
x_1 \\ x_2 \\ x_3 \\ \vdots \\ x_{N-1} \\ x_N
\end{bmatrix}
=
\begin{bmatrix}
d_1 \\ d_2 \\ d_3 \\ \vdots \\ d_{N-1} \\ d_N
\end{bmatrix}
\tag{A}
$$

(a) By use of the following definitions of f_1, g_1, f_k, and g_k given in the text, show that Eq. (A) may be transformed to the following form

$$
\begin{bmatrix}
1 & f_1 & 0 & 0 & 0 & \cdots & 0 \\
0 & 1 & f_2 & 0 & 0 & \cdots & 0 \\
0 & 0 & 1 & f_3 & 0 & \cdots & 0 \\
& & & \cdots & & & \\
0 & \cdots & & 0 & 1 & f_{N-1} & \\
0 & \cdots & & 0 & 0 & 1 &
\end{bmatrix}
\begin{bmatrix}
x_1 \\ x_2 \\ x_3 \\ \vdots \\ x_{N-1} \\ x_N
\end{bmatrix}
=
\begin{bmatrix}
g_1 \\ g_2 \\ g_3 \\ \vdots \\ g_{N-1} \\ g_N
\end{bmatrix}
\tag{B}
$$

(b) Commencing with the bottom row of Eq. (B), show that the matrix multiplication rule may be applied to give

$$
x_N = g_N \qquad x_k = g_k - f_k x_{k+1} \qquad (k = N-1, N-2, \ldots, 2, 1) \tag{C}
$$

(c) For the special case of a conventional distillation column, $b_j = -(1 + A_{ji})$, $c_j = 1$, $a_j = A_{j-1,i}$, $d_{f-1} = -v_{Fi}$, $d_f = -l_{Fi}$, and $d_j = 0 (j \neq f, f-1)$, show that when these quantities are substituted successively into the formulas for f_j and g_j [Eq. (2-20)], the recurrence formulas given by Eq. (2-22) are obtained.

2-4 Suppose that the solution set of operating conditions are known. For definiteness, suppose that the column has a total condenser and that the feed enters the column as a liquid at its boiling-point temperature at the column pressure.

(a) Suppose that the set $\{b_i/d_i\}$ and the total distillate rate D are computed by using the correct set of L_j/V_j's and the correct T_j's except for one particular plate, say plate k, and for this particular plate, a temperature less than the correct one is used in making the calculations. Show that the calculated value of D so obtained is less than the correct value.

(b) Repeat (a) for the case where the temperature used for plate k in the calculation of the b_i/d_i's is greater than the correct temperature, and show that in this case the calculated value of D is greater than the correct value.

Hint: Show that

$$v_{fi}/d_i = A_{1i}A_{2i} \cdots A_{f-1,i} + A_{2i} \cdots A_{f-1,i} + \cdots + A_{f-2,i}A_{f-1,i} + A_{f-1,i} + 1$$

and that

$$v_{fi}/b_i = S_{fi}S_{f+1,i} \cdots S_{Ni} + S_{fi}S_{f+1,i} \cdots S_{N-1,i} + \cdots + S_{fi}S_{f+1,i} + S_{fi}$$

2-5 The following set of b_i/d_i's were found by use of Fenske's equation in Prob. 1-24(a), and the corresponding distillate rate was $D = 46.4632$. On the basis of this known set of b_i/d_i's, compute the b_i/d_i's and d_i's at a $D = 56.2629$ for the following example by use of the θ method of convergence.

Given:

Component	X_i	α_i	$\dfrac{b_i}{d_i}$	Other specifications
1	1/3	1	16	$N = 3$, partial condenser,
2	1/3	2	1	$F = 100$ mol/h, and
3	1/3	3	0.1975	total reflux operation

This problem is based on material given in Refs. 6, and 9. For the first trial, assume $\theta = 0.5$, and use Eq. (2-54) to explain the results obtained.

2-6 Begin with the equilibrium relationship given by Eq. (1-8) and the definition of the ideal solution K value given by Eq. (1-10) and obtain the following formulas for the K_b method for nonideal solutions

$$K_{jb} = \frac{1}{\sum\limits_{i=1}^{c} [(\gamma_{ji}^L/\gamma_{ji}^V)\alpha_{ji}]x_{ji}} \quad \text{and} \quad K_{jb} = \sum\limits_{i=1}^{c} \frac{y_{ji}}{[(\gamma_{ji}^L/\gamma_{ji}^V)\alpha_{ji}]}$$

2-7 Show that the same value of K_{jb} is obtained regardless of whether the corrected x_{ji}'s or the corrected values of the y_{ji}'s given by Eq. (2-29) are used. A more precise statement of the problem follows.

Given:

(1) $K_{jb|T_{j,n+1}} = \dfrac{1}{\sum\limits_{i=1}^{c} \alpha_{ji|T_{jn}} x_{ji}}$

(2) x_{ji}'s and y_{ji}'s are defined by Eq. (2-29).

(3) $(l_{ji})_{ca} = \left(\dfrac{L_j}{K_{ji} V_j}\right)_n (v_{ji})_{ca}$

(4) $\alpha_{ji|T_{jn}} = \left(\dfrac{K_{ji}}{K_{jb}}\right)_{T_{jn}} = \left(\dfrac{K_{ji}}{K_{jb}}\right)_n$

Show that

$$K_{jb|T_{j,\,n+1}} = \sum_{i=1}^{c} \frac{y_{ji}}{\alpha_{ji|T_{jn}}}$$

where n denotes the values of the variables assumed to make the nth trial.

2-8 Repeat Prob. 2-7 for the case where both the vapor and liquid phases form nonideal solutions. That is, given:

(1) $K_{jb|T_{j,\,n+1}} = \dfrac{1}{\displaystyle\sum_{i=1}^{c} (\gamma_{ji}^{L}\alpha_{ji}/\gamma_{ji}^{V})_n x_{ji}}$

(2) x_{ji}'s and y_{ji}'s are defined by Eq. (2-29)

(3) $(l_{ji})_{ca} = \left(\dfrac{\gamma_{ji}^{V} L_j}{\gamma_{ji}^{L} K_j V_j}\right)_n (v_{ji})_{ca}$

(4) $(\alpha_{ji})_n = \left(\dfrac{K_{ji}}{K_{jb}}\right)_{T_{jn}}$

show that

$$K_{jb|T_{j,\,n+1}} = \sum_{i=1}^{c} \frac{y_{ji}}{(\gamma_{ji}^{L}\alpha_{ji}/\gamma_{ji}^{V})_n}$$

2-9 (a) By use of the definition of the ideal solution K value

$$K_{ji} = f_{ji}^{L}/f_{ji}^{V}$$

(where f_{ji}^{L} and f_{ji}^{V} are the fugacities of pure component i, evaluated at the temperature and pressure of plate j) show that the equilibrium relationship given by Eq. (1-8)

$$\gamma_{ji}^{V} f_{ji}^{V} y_{ji} = \gamma_{ji}^{L} f_{ji}^{L} x_{ji}$$

may be restated in the following forms

$$l_{ji} = A_{ji} v_{ji} \qquad \text{and} \qquad v_{ji} = S_{ji} l_{ji}$$

where

$$A_{ji} = \frac{L_j}{(\gamma_{ji}^{L}/\gamma_{ji}^{V})K_{ji} V_j} = 1/S_{ji}$$

(b) Show that the component-material balances are of the same general form for nonideal solutions as those for ideal solutions, provided that the absorption (or stripping) factor has the definition given in (a).

2-10 For the case of nonideal solutions, show that the expressions for the enthalpy balances are of the same form as those given by Eqs. (2-33) through (2-37) except for the fact that the ideal solution enthalpies of the pure components $\{h_{ji}, H_{ji}\}$ are replaced by their partial molar values $\{\bar{h}_{ji}, \bar{H}_{ji}\}$ or the virtual values of their partial molar enthalpies $\{\hat{h}_{ji}, \hat{H}_{ji}\}$; see Chap. 14.

2-11 (a) Show that the function $g(\theta)$, defined by Eq. (2-27) has c real roots, one positive root, and $c - 1$ negative roots.

(b) Show that if $\theta = 0$ is selected as the first assumed value of θ, then Newton's method always converges to the positive root of $g(\theta)$.

2-12 For the case where the reflux rate and a partial sum of the component distillation rates, $\sum_k (d_i)_{\text{spec}}$, are specified, show that the corresponding g function,

$$g(\theta) = \frac{\sum\limits_k (d_i)_{co}}{\sum\limits_k (d_i)_{\text{spec}}} - 1$$

decreases monotonically with θ. The corrected distillate rates $\{(d_i)_{co}\}$ are defined by Eq. (2-57).

2-13 Show that when reflux rate L_1 and the dew-point temperature T_1 of the distillate are specified, the corresponding form of the g function is obtained

$$g(\theta) = \frac{1}{D_{co}} \left[\sum_{i=1}^{c} (d_i)_{co}/K_{1i} \right] - 1$$

and that it decreases monotonically with θ. The corrected distillate rate $\{(d_i)_{co}\}$ are defined by Eq. (2-57) and

$$D_{co} = \sum_{i=1}^{c} (d_i)_{co}$$

2-14 If the reflux ratio L_1 and a partial sum of mole fractions are specified, show that the following form of the corresponding g function varies monotonically with θ

$$g(\theta) = \frac{\sum\limits_k X_{DI}}{\sum\limits_k (X_{Di})_{\text{spec}}} - 1$$

Sketch the function $g(\theta)$ for the case where the k components are relatively light and for the case when the k components are relatively heavy.

  Note: $X_{Di} = (d_i)_{co}/D_{co}$.

2-15 Let the corrected component flow rates be defined as follows

$$(l_{ji})_{co} = \eta_j \left(\frac{l_{ji}}{d_i} \right)_{ca} (d_i)_{co}$$

$$(v_{ji})_{co} = \sigma_j \left(\frac{v_{ji}}{d_i} \right)_{ca} (d_i)_{co}$$

where η_j and σ_j are undetermined multipliers which are to be picked such that

$$\sum_{i=1}^{c} (l_{ji})_{co} = (L_j)_{co}$$

$$\sum_{i=1}^{c} (v_{ji})_{co} = (V_j)_{co}$$

  (a) Use the definition of the mole fraction and the above definitions of the corrected flow rates to obtain the expression given by Eq. (2-29) for the calculation of the mole fractions by the θ method.

  (b) Show that if the undetermined multiplier η_N for the reboiler is denoted by θ, then the defining equation for θ [Eq. (2-25)] follows from the above definition of $(l_{ji})_{co}$ for $j = N$, the reboiler.

REFERENCES

1. J. F. Boston and S. L. Sullivan, Jr.: "An Improved Algorithm for Solving Mass Balance Equations in Multistage Separation Processes," *Can. J. Chem. Eng.* **50**:663 (1972).
2. D. S. Billingsley: "On the Equations of Holland in the Solution of Problems in Multicomponent Distillation," *IBM J. Res. Dev.*, **14**:33 (1970).

3. ———: "On the Numerical Solution of Problems in Multicomponent Distillation at Steady State II," *AIChE J.* **16**:441 (1970).
4. Brice Carnahan, H. A. Luther, and J. O. Wilkes: *Applied Numerical Methods*. John Wiley & Sons, Inc., New York, 1964.
5. M. R. Fenske: "Fractionation of Straight-Run Pennsylvania Gasoline," *Ind. Eng. Chem.*, **24**:482 (1932).
6. C. D. Holland and G. P. Pendon: "Solve More Distillation Problems: Part I. Improvements Give Exact Answers," *Hydrocarbon Process.*, **53**:148 (1974).
7. ——— and P. T. Eubank: "Solve More Distillation Problems: Part II. Partial Molar Enthalpies Calculated," *Hydrocarbon Process.*, **53**:176 (1974).
8. ———: "Introduction to the Fundamentals of Distillation," *Proceedings of the Fourth Annual Educational Symposium of the ISA*, Apr. 5–7, 1972, Wilmington, Delaware.
9. ———: "Exact Solutions Given by the θ-Method of Convergence for Complex and Conventional Distillation Columns," *Chem. Eng. J.*, Loughborough University of Technology, **7**:10 (1972).
10. W. K. Lewis and G. L. Matheson, "Studies in Distillation—Design of Rectifying Columns for Natural and Refinery Gasoline," *Ind. Eng. Chem.*, **24**:494 (1932).
11. W. N. Lyster, S. L. Sullivan, Jr., D. S. Billingsley, and C. D. Holland: "Figure Distillation This New Way: Part I—New Convergence Method Will Handle Many Cases," *Pet. Refiner*, **38**(6):221 (1959).
12. ———, ———, ———, and ———: "Figure Distillation This New Way: Part II—Product Purity Can Set Conditions for Column," *Pet. Refiner*, **38**(7):151 (1959).
13. T. A. Nartker, J. M. Srygley, and C. D. Holland, "Solution of Problems Involving Systems of Distillation Columns," *Can. J. Chem. Eng.*, **44**:217 (1966).
14. E. W. Thiele and R. L. Geddes, "Computation of Distillation Apparatus for Hydrocarbon Mixtures," *Ind. Eng. Chem.*, **25**:289 (1933).
15. N. J. Tetlow (of Dow Chemical Company) Personal Communication.

THREE

APPLICATION OF THE THETA METHOD OF CONVERGENCE TO COMPLEX COLUMNS AND TO SYSTEMS OF COLUMNS

The calculation procedures [the θ method, K_b method, and constant composition method] developed in Chap. 2 for conventional distillation columns are applied to complex distillation columns in Sec. 3-1. For solving problems involving systems of columns interconnected by recycle streams, a variation of the theta method, called the "capital Θ method" of convergence is presented in Secs. 3-2 and 3-3. For the case where the terminal flow rates are specified, the capital Θ method is used to pick a set of corrected component-flow rates which satisfy the component-material balances enclosing each column and the specified values of the terminal rates simultaneously. For the case where other specifications are made in lieu of the terminal rates, sets of corrected terminal rates which satisfy the material and energy balances enclosing each column as well as the equilibrium relationships of the terminal streams are found by use of the capital Θ method of convergence as described in Chap. 7.

3-1 COMPLEX DISTILLATION COLUMNS

A complex distillation column is defined as one which has either more feeds introduced or streams withdrawn or a combination of these than does a conventional distillation column. To demonstrate the application of the θ method and associated calculational procedures, the complex column shown in Fig. 3-1

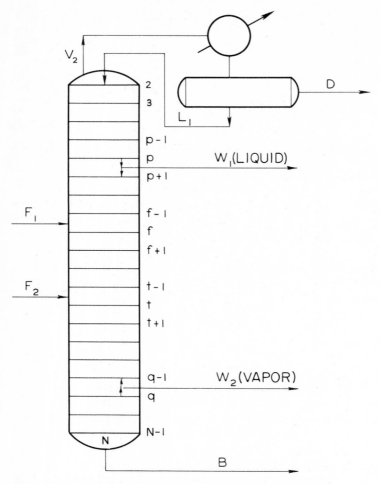

Figure 3-1 Complex column with two feed plates and two sidestreams.

which has two feed plates and two sidestreams (in addition to the top and bottom products) was selected.

In the application of the aforementioned calculational procedures, the specifications commonly made are as follows: the column pressure; the number of plates; the rate, composition, and thermal condition of each feed; as well as the locations of the feed plates and sidestreams. The number of additional specifications that may be made is equal to the total number of streams withdrawn (the distillate, bottoms, and sidestreams). For the column shown in Fig. 3-1, the additional specifications V_2 (or L_1), D, W_1, and W_2 may be made. These in turn fix the dependent variable B.

In order to make the first trial, temperature and L/V profiles for the column are assumed. This allows one to solve the component-material balances for the component-flow rates. After the θ method of convergence for the complex column has been applied, the compositions are computed. The compositions so obtained are used to compute an improved set of temperatures by use of the K_b method. These temperatures and compositions are used to find a new set of total-flow rates by solving the enthalpy balances after they have been stated in the form of the constant-composition method.

Component-Material Balances

The component-material balances are stated in terms of a single set of flow rates, the vapor rates, by use of the equilibrium relationship $l_{ji} = A_{ji} v_{ji}$ [Eq. (2-10)]. Except for the plates from which the sidestreams are withdrawn, the equations for the remaining stages of the complex column shown in Fig. 3-1 are formulated in precisely the same manner which was shown in Chap. 2 for conventional columns.

To demonstrate the notation employed, an enlarged view of plate p (see Fig. 3-2) from which the liquid sidestream W_1 is withdrawn as shown in Fig. 3-1. The component-material balance enclosing plate p is seen to be

$$v_{p+1, i} + l_{p-1, i} - v_{pi} - l_{pi} - w_{1i} = 0 \tag{3-1}$$

Since w_{1i} has the same composition as l_{pi} and since $l_{pi} = A_{pi} v_{pi}$ [Eq. (2-10)], it follows that

$$w_{1i} = \frac{W_1}{L_p} l_{pi} = \frac{W_1}{L_p} A_{pi} v_{pi} \tag{3-2}$$

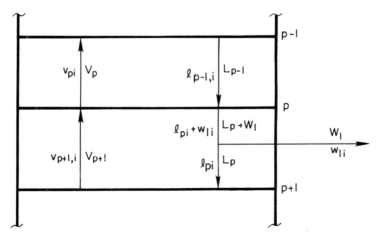

Figure 3-2 Notation used for the streams entering and leaving a plate from which a liquid sidestream is withdrawn.

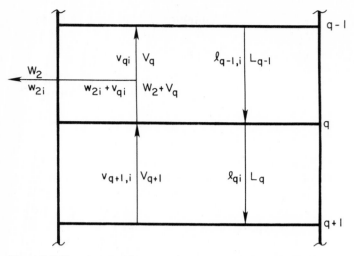

Figure 3-3 Notation used for the streams entering and leaving a plate from which a vapor sidestream is withdrawn.

Then after the equilibrium relationship [Eq. (2-10)] has been used to state liquid flow rates of Eq. (3-1) in terms of the vapor rates, one obtains

$$A_{p-1,i}v_{p-1,i} - \left[1 + A_{pi}\left(1 + \frac{W_1}{L_p}\right)\right]v_{pi} + v_{p+1,i} = 0 \tag{3-3}$$

An enlarged view of plate q from which the vapor sidestream W_2 is withdrawn is shown in Fig. 3-3. The component-material balance for plate q is given by

$$v_{q+1,i} + l_{q-1,i} - v_{qi} - l_{qi} - w_{2i} = 0 \tag{3-4}$$

Since W_2 and V_q have the same composition, it follows that

$$w_{2i} = \frac{W_2}{V_q}v_{qi} \tag{3-5}$$

Then by use of this relationship and the equilibrium relationship [Eq. (2-10)], it is possible to restate Eq. (3-4) in terms of the vapor rates as follows

$$A_{q-1,i}v_{q-1,i} - \left[1 + A_{qi} + \frac{W_2}{V_q}\right]v_{qi} + v_{q+1,i} = 0 \tag{3-6}$$

The complete set of component-material balances may be represented by a matrix equation of the same form as the one used to represent these balances for a conventional column, namely,

$$\mathbf{A}_i \mathbf{v}_i = -\mathbf{f}_i \tag{3-7}$$

The elements of A_i are the same as those displayed beneath Eq. (2-18) except that in this case, the following expressions for ρ_{ji} must be used for plates p and q from which sidestreams W_1 and W_2 are withdrawn,

$$\rho_{pi} = 1 + A_{pi}\left(1 + \frac{W_1}{L_p}\right)$$

$$\rho_{qi} = 1 + A_{qi} + \frac{W_2}{V_q}$$

(3-8)

When model 2, shown in Fig. 2-2, is used for the feed plates, the column vector f_i contains the vapor and liquid rates (v_{Fi} and l_{Fi}) for each feed. The elements of v_i and f_i may be displayed as follows

$$v_i = [d_i \; v_{2i} \; v_{3i} \; \cdots \; v_{f-1, i} \; v_{fi} \; \cdots \; v_{t-1, i} \; v_{ti} \; \cdots \; v_{Ni}]^T$$

$$f_i = [0 \cdots 0 \; v_{F1i} \, l_{F1i} \, 0 \cdots 0 \; v_{F2i} \, l_{F2i} \, 0 \cdots 0]^T$$

(3-9)

The elements $v_{f-1, i}$ and v_{F1i}, v_{fi} and l_{F1i}, $v_{t-1, i}$ and v_{F2i}, v_{ti} and l_{F2i} lie in rows $f - 1, f, t - 1, t$, respectively.

The component-material balances $A_i v_i = -f_i$, may be solved for the vapor rates and d_i by use of the recurrence formulas given by Eqs. (2-20) and (2-21) in the same manner as was demonstrated in Example 2-1.

After the component-material balances have been solved for the component-flow rates d_i, v_{2i}, v_{3i}, ..., v_{Ni}, the corresponding set of flow rates for the liquid may be calculated by use of the equilibrium relationship given by Eq. (2-10). Then the θ method is applied for the purpose of finding a set of terminal-component flow rates which are in component-material balance and in agreement with the specified values of the total-flow rates of the terminal streams.

Formulation of the θ Method of Convergence

The formulation of the θ method of convergence for a complex column follows that originally proposed by Lyster et al.[9] First, a θ multiplier is defined for each stream withdrawn from the column which may be specified independently. For the column shown in Fig. 3-1, any three of the four streams D, W_1, W_2, and B may be specified independently. For definiteness, suppose that D, W_1, W_2 are specified. Then B may be found by an overall material balance. The θ multipliers are defined by the following equations

$$\left(\frac{b_i}{d_i}\right)_{co} = \theta_0 \left(\frac{b_i}{d_i}\right)_{ca}$$

$$\left(\frac{w_{1i}}{d_i}\right)_{co} = \theta_1 \left(\frac{w_{1i}}{d_i}\right)_{ca}$$

$$\left(\frac{w_{2i}}{d_i}\right)_{co} = \theta_2 \left(\frac{w_{2i}}{d_i}\right)_{ca}$$

(3-10)

where $\theta_0 > 0$, $\theta_1 > 0$, and $\theta_2 > 0$. Again the corrected rates are those which satisfy simultaneously, the specifications D, W_1, and W_2 (where B is taken to be the dependent variable) and the overall material balance

$$FX_i = (d_i)_{co} + (w_{1i})_{co} + (w_{2i})_{co} + (b_i)_{co} \tag{3-11}$$

where

$$FX_i = F_1 X_{1i} + F_2 X_{2i}$$

A combination of Eqs. (3-10) and (3-11) yields the following expression for the corrected distillate rate

$$(d_i)_{co} = \frac{FX_i}{1 + \theta_0(b_i/d_i)_{ca} + \theta_1(w_{1i}/d_i)_{ca} + \theta_2(w_{2i}/d_i)_{ca}} \tag{3-12}$$

Again, in a manner analogous to that shown for conventional columns, numerical problems resulting from the divisions by small values for $(d_i)_{ca}$ may be avoided by use of the following expression for $(d_i)_{co}$

$$(d_i)_{co} = (d_i)_{ca} p_i \tag{3-13}$$

where p_i is defined by

$$p_i = \frac{FX_i}{(d_i)_{ca} + \theta_0(b_i)_{ca} + \theta_1(w_{1i})_{ca} + \theta_2(w_{2i})_{ca}} \tag{3-14}$$

When the defining equations for the θ's given by Eq. (3-10) are restated in terms of p_i, the following expressions are obtained for the calculation of the corrected terminal rates, namely,

$$(b_i)_{co} = \theta_0(b_i)_{ca} p_i$$
$$(w_{1i})_{co} = \theta_1(w_{1i})_{ca} p_i \tag{3-15}$$
$$(w_{2i})_{co} = \theta_2(w_{2i})_{ca} p_i$$

The requirement that the corrected rates satisfy the specifications D, W_1, and W_2 leads to the g functions

$$g_0(\theta_0, \theta_1, \theta_2) = \sum_{i=1}^{c} (d_i)_{co} - D$$

$$g_1(\theta_0, \theta_1, \theta_2) = \sum_{i=1}^{c} (w_{1i})_{co} - W_1 \tag{3-16}$$

$$g_2(\theta_0, \theta_1, \theta_2) = \sum_{i=1}^{c} (w_{2i})_{co} - W_2$$

The desired solution is the set of positive values of θ_0, θ_1, and θ_2 that makes $g_0 = g_1 = g_2 = 0$, simultaneously. When the solution set of θ's has been found, the set of corrected d_i's (or p_i's) will have been computed. These values of $(d_i)_{co}$ and p_i may be used to compute the set of improved compositons. In order to reduce numerical difficulties, in the application of the θ method, the g functions should be stated in normalized form. For example

$$g_0 = \frac{1}{D} \sum_{i=1}^{c} (d_i)_{co} - 1$$

Corrected Mole Fractions

Expressions for the corrected mole fractions for the column shown in Fig. 3-1 may be developed from first principles as outlined in Prob. 2-15. The expressions so obtained are the same as those given in Chap. 2, namely

$$x_{ji} = \frac{(l_{ji}/di)_{ca}(d_i)_{co}}{\sum\limits_{i=1}^{c} (l_{ji}/d_i)_{ca}(d_i)_{co}} \quad \text{or} \quad x_{ji} = \frac{(l_{ji})_{ca}p_i}{\sum\limits_{i=1}^{c} (l_{ji})_{ca}p_i} \tag{3-17}$$

where $(d_i)_{co}$ is given by Eq. (3-13) and p_i by Eq. (3-14). The formulas for y_{ji} are obtained by replacing l_{ji} in the above expressions by v_{ji}.

Calculation of a Corrected Set of Temperatures by Use of the K_b Method

After either a corrected set of x_{ji}'s or y_{ji}'s has been computed as described above, a new set of temperatures may be computed by use of the K_b method in the same manner that was described in Chap. 2 for conventional distillation columns.

Calculation of a Corrected Set of Total-Flow Rates by Use of the Constant-Composition Method and the Total-Material Balances

The development of the expressions for the constant-composition form of the enthalpy balances is carried out in the same manner demonstrated in Chap. 2 for conventional distillation columns. For example, the energy balance enclosing the top of the column and some plate j located between plates p and $f - 1$ of the column shown in Fig. 3-1 can be expressed as follows

$$\sum_{i=1}^{c} [H_{j+1, i} v_{j+1, i} - h_{ji} l_{ji} - h_{pi} w_{1i} - H_{Di} d_i] - Q_C = 0 \tag{3-18}$$

Elimination of $v_{j+1, i}$ by use of the component-material balance

$$v_{j+1, i} = l_{ji} + w_{1i} + d_i \tag{3-19}$$

yields

$$\sum_{i=1}^{c} [(H_{j+1, i} - h_{ji}) l_{ji} + (H_{j+1, i} - h_{pi}) w_{1i} + (H_{j+1, i} - H_{Di}) d_i] - Q_C = 0$$

Since $l_{ji} = L_j x_{ji}$, $w_{1i} = W_1 x_{pi}$, $d_i = DX_{Di}$, the above form of the energy balance may be solved for L_j to give

$$L_j = \frac{-W_1 \sum\limits_{i=1}^{c} (H_{j+1,\,i} - h_{pi})x_{pi} - D \sum\limits_{i=1}^{c} (H_{j+1,\,i} - H_{Di})X_{Di} + Q_C}{\sum\limits_{i=1}^{c} (H_{j+1,\,i} - h_{ji})x_{ji}} \tag{3-20}$$

The complete set of enthalpy balances for the column which are obtained in a manner analogous to that demonstrated above. After the L_j's have been computed by use of these enthalpy-balance expressions, the corresponding vapor rates may be computed by use of the total-material balances.

Geometrical Representation of the g Functions

To simplify the geometrical representation of the g functions, consider the case where only one sidestream W_1 is withdrawn in addition to the distillate D and the bottoms B as shown in Fig. 3-4. For this column, only two θ multipliers exist, θ_0 and θ_1. These two multipliers are defined by the first two expressions of Eq. (3-10). If D and W_1 are specified, then the corresponding g functions are

$$g_0(\theta_0, \theta_1) = \sum_{i=1}^{c} (d_i)_{co} - D \tag{3-21}$$

$$g_1(\theta_0, \theta_1) = \sum_{i=1}^{c} (w_{1i})_{co} - W_1 \tag{3-22}$$

where

$$(d_i)_{co} = \frac{FX_i}{1 + \theta_0(b_i/d_i)_{ca} + \theta_1(w_{1i}/d_i)_{ca}}$$

Again, $(d_i)_{co}$ may be stated in terms of p_i as demonstrated above by Eqs. (3-13) and (3-14).

Traces of the functions g_0 in the $\overline{g_0\theta_0}$, $\overline{g_0\theta_1}$, and the $\overline{\theta_0\theta_1}$, planes are shown in Fig. 3-5, and traces of g_1 in the $\overline{g_1\theta_0}$, $\overline{g_1\theta_1}$, and $\overline{\theta_1\theta_0}$ planes are shown in Fig. 3-6. The desired solution is the intersection of the traces of g_0 and g_1 in the $\overline{\theta_0\theta_1}$ plane where $g_0 = g_1 = 0$ as shown in Fig. 3-7. Examination of these graphs reveals that g_0 and g_1 are continuous, monotonic functions for all positive values of θ_0 and θ_1. Such behavior is desirable in the numerical solution of problems.

The set of θ's that makes $g_0 = g_1 = 0$ may be found by use of the Newton-Raphson method[3] as described in App. A. The Newton-Raphson method consists of the successive solution of the equations corresponding to the linear terms

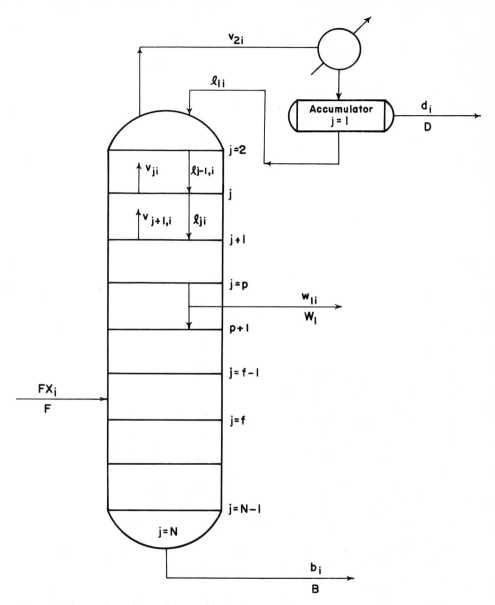

Figure 3-4 A complex column with one sidestream.

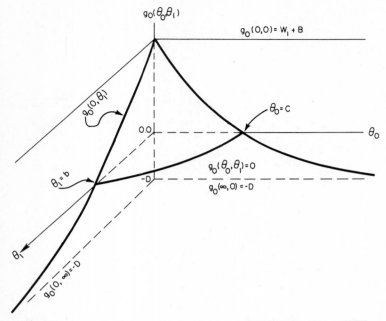

Figure 3-5 Traces of the function $g_0(\theta_0, \theta_1)$ in the $\overline{g_0\theta_0}$, $\overline{g_0\theta_1}$, and the $\overline{\theta_0\theta_1}$ planes.

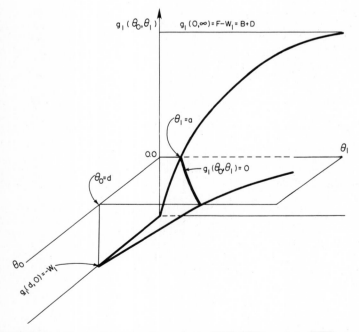

Figure 3-6 Traces of the function g_1 in the $\overline{g_1\theta_1}$, $\overline{g_1\theta_0}$, and the $\overline{\theta_1\theta_0}$ planes.

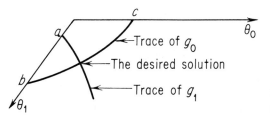

Figure 3-7 Traces of the functions g_0 and g_1 in the $\theta_0 \theta_1$ plane.

of the Taylor series expansions of the functions g_0 and g_1 about the assumed values of the variables θ_0 and θ_1, namely

$$0 = g_0 + \frac{\partial g_0}{\partial \theta_0} \Delta\theta_0 + \frac{\partial g_0}{\partial \theta_1} \Delta\theta_1$$

$$0 = g_1 + \frac{\partial g_1}{\partial \theta_0} \Delta\theta_0 + \frac{\partial g_1}{\partial \theta_1} \Delta\theta_1$$

(3-23)

Expressions for the partial derivatives are given in Prob. 3-1. The Newton-Raphson method is illustrated by the following example.

Example 3-1 Make two trials by use of the Newton-Raphson method for the set of values x and y which make $f_1 = f_2 = 0$ simultaneously

$$f_1(x, y) = x^2 - y^2 + 1$$
$$f_2(x, y) = x^2 + y^2 - 2$$

Take the initial values of x and y to be $x_0 = y_0 = 1$.

SOLUTION

    Trial 1. For $x_0 = y_0 = 1$

$$f_1(1, 1) = 1 - 1 + 1 = 1$$
$$f_2(1, 1) = 1 + 1 - 2 = 0$$

and

$$\frac{\partial f_1}{\partial x} = 2x \qquad \frac{\partial f_1}{\partial y} = -2y \qquad \frac{\partial f_2}{\partial x} = 2x \qquad \frac{\partial f_2}{\partial y} = 2y$$

Thus, for $x_0 = y_0 = 1$, the Newton-Raphson equations become

$$0 = 1 + 2\,\Delta x - 2\,\Delta y$$
$$0 = 0 + 2\,\Delta x + 2\,\Delta y$$

Addition of these equations yields

$$0 = 1 + 4\,\Delta x \qquad \Delta x = -1/4$$

and

$$x_1 = x_0 + \Delta x = 1 - 1/4 = 3/4$$

Substitution of this value of Δx into either of the Newton-Raphson equations yields

$$\Delta y = 1/4$$

and

$$y_1 = y_0 + \Delta y = 1 + 1/4 = 5/4$$

Trial 2. For $x_1 = 3/4$ and $y_1 = 5/4$

$$f_1(3/4, 5/4) = (3/4)^2 - (5/4)^2 + 1 = 0$$
$$f_2(3/4, 5/4) = (3/4)^2 + (5/4)^2 - 2 = 1/8$$

Thus

$$\frac{\partial f_1}{\partial x} = (2)(3/4) = 3/2 \qquad \frac{\partial f_2}{\partial x} = (2)(3/4) = 3/2$$

$$\frac{\partial f_1}{\partial y} = (-2)(5/4) = -5/2 \qquad \frac{\partial f_2}{\partial y} = (2)(5/4) = 5/2$$

and

$$0 = 0 + (3/2)\,\Delta x - (5/2)\,\Delta y$$
$$0 = 1/8 + (3/2)\,\Delta x + (5/2)\,\Delta y$$

Solution of these equations for Δx and Δy yields

$$\Delta x = -1/24 \qquad x_2 = 3/4 - 1/24 = 17/24 = 0.7083$$
$$\Delta y = -1/40 \qquad y_2 = 5/4 - 1/40 = 49/40 = 1.225$$

These values compare well with the solution set of values which are

$$x = \frac{1}{\sqrt{2}} = 0.7071 \qquad \text{and} \qquad y = \sqrt{3/2} = 1.2247$$

Convergence Characteristics of the θ Method of Convergence for Complex Columns

To demonstrate some of the numerical characteristics of the θ method for complex columns, Example 3-2 was selected. The statement of this example is given in Table 3-1 and the geometry of the column is depicted in Fig. 3-4. In addition to the distillate and bottoms, the column has one sidestream withdrawn. Thus, the θ method has two θ multipliers, θ_0 and θ_1, which are defined by the first two expressions given by Eq. (3-10). Since D and W_1 are specified (see Table 3-1) the two g functions to be used for computing these θ's are given by Eqs. (3-21) and (3-22).

The convergence characteristics of this example are presented in Table 3-2, and the solution sets of T_j's, V_j's, d_i's, w_{1i}'s and b_i's are presented in Table 3-3. To

Table 3-1 Statement of Example 3-2, a complex distillation problem

Component	FX_i (lb mol/h)
CH_4	2.0
C_2H_6	10.0
C_3H_6	6.0
C_3H_8	12.5
$i\text{-}C_4H_{10}$	3.5
$n\text{-}C_4H_{10}$	15.0
$n\text{-}C_5H_{12}$	15.2
$n\text{-}C_6H_{14}$	11.3
$n\text{-}C_7H_{16}$	9.0
$n\text{-}C_8H_{17}$	8.5
400	7.0

Other specifications

The column pressure is 300 lb/in² abs. The K values and enthalpies are given in Tables B-1 and B-2. The feed enters the column on plate $f = 6$ at its bubble-point temperature at the column pressure. The column has 11 plates, a reboiler, and a partial condenser. The distillate is withdrawn at the rate $D = 32.298$ lb mol/h. The sidestream is withdrawn from plate $p = 10$ as a liquid at the rate $W_1 = 25.0$ lb mol/h. A reflux ratio $L_1/D = 2.25$ is to be used. Take the initial temperatures to be linear between $T_2 = 200°F$ and $T_{12} = 300°F$, and take $T_1 = 200°F$ and $T_{13} = 300°F$. Take the initial values of $L_j/V_j = 1$ for $j = 2, 3, \ldots, 12$, and take $V_{13}/B = 1.0$.

Table 3-2 Convergence characteristics exhibited by the θ method in the solution of Example 3-2

Trial	θ_0	θ_1	Temp. of D, °F	Temp. of B, °F	Calculated value of D	Calculated value of W_1
1	0.7365	0.8632	125.79	410.82	31.9304	24.4257
2	1.16464	1.16219	115.13	443.44	32.6193	24.7670
3	0.984057	0.962613	114.25	441.59	32.2084	24.7236
4	1.03172	1.02741	114.13	446.36	32.3555	25.0021
5	0.984476	0.989619	114.13	445.68	32.22745	24.95864
6	1.00754	1.006039	114.13	446.53	32.3110	25.00688
7	0.996285	0.997367	114.13	446.28	32.29217	24.99218
8	1.001739	1.001348	114.13	446.45	32.30094	25.00218
9	0.9991422	0.999384	114.13	446.39	32.29664	24.99829
10	1.000344	1.000269	114.13	446.42	32.29861	25.00039
11	0.9998121	0.9998789	114.13	446.41	32.29772	24.99945
12	1.0000	1.0000	114.13	446.41	32.29811	24.99993

Convergence criterion: (1) For each trial, values of θ_0 and θ_1 were found such that both $|g_0|$ and $|g_1|$ were equal to or less than 10^{-5}. (2) Convergence for the problem was said to have been achieved when $|g_0(1, 1)|$ and $|g_1(1, 1)|$ were each equal to or less than 10^{-5}.

The convergence criteria were satisfied at the end of the 12th trial and 2.81 seconds on an AMDAHL 470 V/6 computer with the WATFIV compiler.

Table 3-3 Solution of Example 3-2

1. Final temperature, liquid, and vapor profiles

Stage	T_j (°F)	L_j (lb mol/h)	V_j (lb mol/h)
1	114.13	72.671	32.298
2	145.45	61.302	104.97
3	165.40	68.080	103.60
4	183.47	58.993	100.38
5	217.13	194.95	91.291
6	239.30	212.35	127.25
7	253.55	222.45	144.65
8	264.68	227.36	154.75
9	276.18	227.38	159.66
10	292.27	196.97	159.67
11	319.29	186.10	154.27
12	363.18	164.71	143.39
13	456.41	42.702	122.01

2. Final product flow rates

Component	d_i (lb mol/h)	w_{1i} (lb mol/h)	b_i (lb mol/h)
CH_4	0.20000×10^1	0.32125×10^{-6}	0.30273×10^{-9}
C_2H_6	0.99985×10^1	0.15077×10^{-2}	0.19926×10^{-4}
C_3H_6	0.59356×10^1	0.61044×10^{-1}	0.33848×10^{-2}
C_3H_8	0.12224×10^2	0.24742	0.18230×10^{-1}
$i\text{-}C_4H_{10}$	0.11119×10^1	0.19577×10^1	0.43043
$n\text{-}C_4H_{10}$	0.10242×10^1	0.10428×10^2	0.35483×10^1
$n\text{-}C_5H_{12}$	0.36429×10^{-2}	0.64835×10^1	0.87129×10^1
$n\text{-}C_6H_{14}$	0.18929×10^{-4}	0.25252×10^1	0.87747×10^1
$n\text{-}C_7H_{16}$	0.15019×10^{-6}	0.13819×10^1	0.76181×10^1
$n\text{-}C_8H_{17}$	0.12374×10^{-8}	0.10969×10^1	0.74031×10^1
400	0.14162×10^{-11}	0.80727	0.61927×10^1

$Q_C = 0.47243 \times 10^6$ Btu/h
$Q_R = 0.15519 \times 10^7$ Btu/h

satisfy the convergence criterion stated in Table 3-2, a total of 12 complete trials and 2.81 seconds of computer time (AMDAHL 470 V/6 WATFIV) were required.

Exact Solutions Given by the θ Method of Convergence for Complex Columns

For a distillation column such as the one shown in Fig. 3-4, the θ method constitutes an exact solution at total reflux; that is, the sets of b_i/d_i's and w_{1i}/d_i's at two different sets of values for D and W_1 are related by the single multipliers

θ_0 and θ_1 as indicated by the first two expressions of Eq. (3-15). The proof of this statement is established by solving Probs. 3-3 and 3-4. That the θ method is an exact solution to certain problems at total reflux is further illustrated by solving Probs. 3-5 and 3-6.

3-2 SYSTEMS OF INTERCONNECTED DISTILLATION COLUMNS

The particular difficulty associated with the solution of problems involving systems of distillation columns becomes apparent upon close examination of the system shown in Fig. 3-8. First observe that the composition of the recycled bottoms from the second column to the first is not known until a solution to the

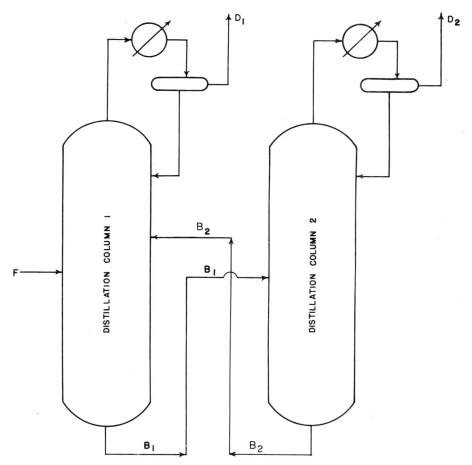

Figure 3-8 A system of two interconnected distillation columns.

second column has been obtained. On the other hand, the composition of the feed to the second column is not known until a solution has been found for the first column, since the feed to the second column is the bottoms of the first column.

For solving problems of this type, Tomme and Holland[10] and Tomme[11] proposed an extension of the θ method of convergence, called the "Capital Θ Method for Systems." The "capital Θ" is used to distinguish this method for systems from the θ method for single columns. The capital Θ method for systems is similar to the θ method for complex columns in that for each external stream which may be specified independently, there exists a Θ multiplier. These Θ multipliers are to be picked such that the overall component-material balances and the specifications are satisfied simultaneously. For each component i, there exists one overall component-material balance per column which is independent. The strong convergence characteristics of the Θ method for systems of interconnected distillation columns can be attributed to the fact that there exists many problems for which the Θ method for systems constitutes an exact solution for a system of columns at total reflux.

The capital Θ method for systems is introduced by the formulation of the equations for this method for the simple system of two conventional distillation columns shown in Fig. 3-8. In a subsequent section, the formulation is generalized for the case of any number of columns with any number of sidestreams withdrawn.

Formulation of the Equations for the Θ Method for a System of Two Distillation Columns

As shown in Fig. 3-8, the number of the column from which a stream is withdrawn is denoted by a subscript on D and B. To identify the variables of each column, the column number is carried as the last subscript. For example, $d_{i,\,1}$ and $d_{i,\,2}$ denote the flow rates of component i in the distillate streams withdrawn from columns 1 and 2, respectively.

For the system of columns shown in Fig. 3-8, there exists one independent component-material balance per column, namely

$$FX_i + b_{i,\,2} - d_{i,\,1} - b_{i,\,1} = 0$$
$$b_{i,\,1} - d_{i,\,2} - b_{i,\,2} = 0$$

(3-24)

In the interest of simplicity, the subscript co used to identify the corrected flow rates for conventional and complex distillation columns has been omitted in the above expressions and from those which follow.

For each total-flow rate of the set D_1, B_1, D_2, and B_2 which may be fixed independently, there exists a capital Θ multiplier. If D_1 is fixed, then D_2 is seen to be dependent since it is uniquely determined by a total material balance which encloses the entire system. From the total material balances enclosing each column, it is seen that if B_1 is fixed, then either D_1, D_2, or B_2 may be fixed independently. Thus, two Θ multipliers may be defined as follows

$$
\left(\frac{b_{i,1}}{d_{i,1}}\right)_{co} = \Theta_1 \left(\frac{b_{i,1}}{d_{i,1}}\right)_{ca}
$$

$$
\left(\frac{b_{i,2}}{d_{i,2}}\right)_{co} = \Theta_2 \left(\frac{b_{i,2}}{d_{i,2}}\right)_{ca}
$$

(3-25)

where the calculated values are those found by use of the calculational procedure described below. The subscript co is dropped in the interest of simplicity in the remainder of the development. The g functions may be stated in terms of any two of the flow rates D_1, D_2, B_1, and B_2 which may be specified independently.

Thus, if B_1 and D_2 are elected as the specified flow rates, then

$$
g_1(\Theta_1, \Theta_2) = \sum_{i=1}^{c} b_{i,1} - B_1
$$

$$
g_2(\Theta_1, \Theta_2) = \sum_{i=1}^{c} d_{i,2} - D_2
$$

(3-26)

The desired set of Θ's is that set of positive numbers that makes $g_1 = g_2 = 0$, simultaneously. These Θ's may be found by use of the Newton-Raphson method in a manner analogous to that described in Sec. 3-1 for complex distillation columns.

Formulas for $\{d_{i,1}\}$ and $\{d_{i,2}\}$ are obtained as follows. Let the expressions given by Eq. (3-25) be restated in the form

$$
b_{i,1} = r_{i,1} d_{i,1}
$$

$$
b_{i,2} = r_{i,2} d_{i,2}
$$

(3-27)

where

$$
r_{i,1} = \Theta_1 (b_{i,1}/d_{i,1})_{ca}
$$

$$
r_{i,2} = \Theta_2 (b_{i,2}/d_{i,2})_{ca}
$$

After $b_{i,1}$ and $b_{i,2}$ have been eliminated from the component-material balances by use of the relationships given by Eq. (3-27), the equations so obtained may be rearranged to give

$$
-(1 + r_{i,1}) d_{i,1} + r_{i,2} d_{i,2} = -FX_i
$$

$$
r_{i,1} d_{i,1} - (1 + r_{i,2}) d_{i,2} = 0
$$

(3-28)

This set of equations may be solved for the corrected distillate rates to give

$$
d_{i,1} = \frac{FX_i(1 + r_{i,2})}{1 + r_{i,1} + r_{i,2}}
$$

(3-29)

$$
d_{i,2} = \frac{FX_i r_{i,1}}{1 + r_{i,1} + r_{i,2}}
$$

(3-30)

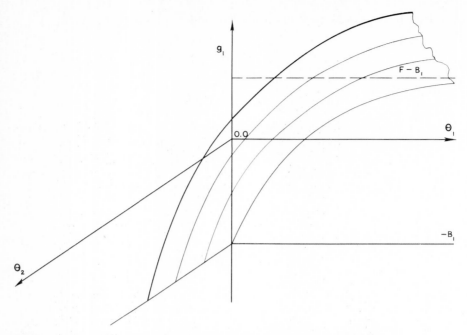

Figure 3-9 Behavior of the function g_1 in the neighborhood of the positive roots.

Graphs of the functions g_1 and g_2 in the neighborhood of the solution set of Θ's are shown in Figs. 3-9, 3-10, and 3-11. An outline of the proof of the existence of a set of positive Θ's that satisfy the g functions simultaneously is presented in Prob. 3-13. A proof for a more general case has been given by Billingsley.[1, 2]

Calculational Procedure

The calculational procedure recommended by Nartker et al.[12] consists of applying the θ method of convergence (for conventional and complex distillation columns) one time to each column of the system in succession. The terminal rates so obtained for each column of the system are called the "calculated values for the system," and denoted by the subscript *ca*. To initiate the calculational procedure for the system, the compositions for the minimum number of streams are selected as the independent variables for the system. For example, to initiate the calculational procedure for column 1 of the system shown in Fig. 3-9 the composition of the stream B_2 is assumed. One complete column trial is made on column 1. A complete column trial consists of the application of the θ method for single columns, the calculation of a new set of temperatures by use of the K_b method, and the calculation of a new set of total-flow rates for use of constant-composition method. These temperatures and total-flow rates are stored for use in the next trial calculation for the system. The set of $b_{i,1}$'s obtained by the

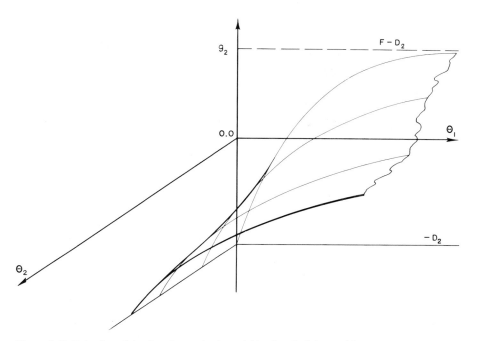

Figure 3-10 Behavior of the function g_2 in the neighborhood of the positive roots.

column trial on column 1 become the set of assumed flow rates for making the column trial on column 2. After the column trial on column 2 has been completed, the capital Θ method for systems is applied, and the set of $b_{i,\,2}$'s so obtained is used to initiate the next system trial by making one column trial on column 1.

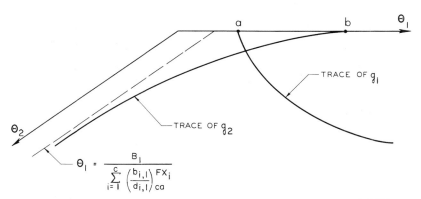

Figure 3-11 Intersections of the surfaces of the functions g_1 and g_2 with the $\overline{\Theta_1 \Theta_2}$ plane for the case where $F - B_1 > 0$.

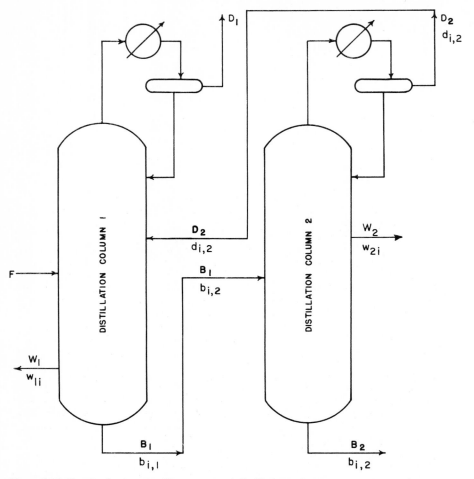

Figure 3-12 Sketch of a system of interconnected distillation columns.

Convergence Characteristics of the Capital Θ Method

To demonstrate the convergence characteristics of the capital Θ method, Example 3-3 was selected. This example involves the system of two complex columns shown in Fig. 3-12. A statement of this example is given in Table 3-4 and the solution is given in Table 3-5. The variation of the θ's and Θ's with trial number is presented in Table 3-6.

Exact Solutions Given by the Θ Method of Convergence for Systems of Distillation Columns

For systems of interconnected distillation columns, the Θ method constitutes an exact solution for certain problems; provided that the α_i's are constant in each

Table 3-4 Statement of Example 3-3

Component	No.	FX_i (kg mol/h)
Toluene	1	0.80000
Ethylbenzene	2	0.5100×10^2
Styrene	3	0.4777×10^2
Isopropylbenzene	4	0.5000×10^{-1}
1-methyl-3-ethylbenzene	5	0.1000×10^{-1}
α-methylstyrene	6	0.1300
cis-1-propylbenzene	7	0.2000

Other specifications

Column 1. The column has 50 plates, a total condenser (stage $j = 1$) and a reboiler (stage $j = N = 52$). The feed F enters on plate $j = 10$ as a subcooled liquid at 317.75 K at 270 mmHg. The distillate is withdrawn as a liquid at its boiling point at the rate $D_1 = 52$ g mol/h. The sidestream is withdrawn as a liquid from plate $j = 21$ at the rate $W_1 = 11$ g mol/h. The top product stream of column 2 is recycled to plate $j = 45$ in column 1. The pressure in the condenser is 40 mmHg, and it may be assumed to vary linearly with plate number between stage $j = 2$ at a pressure of 50 mmHg and the reboiler at 270 mmHg. A reflux ratio $L_1/D = 2.5$ is to be used. Use the equilibrium and enthalpy data given in Tables B-3 and B-4.

Initial temperature profile. Linear between 325.15 K (the boiling point of pure ethylbenzene at 40 mmHg) and 381.143 K (the boiling point of pure styrene at 270 mmHg).

Initial vapor rates. $V_j = 182.0$ g mol/h ($j = 2, 3, \ldots, 52$).

Initial liquid rates. $L_j = 130.0$ g mol/h ($j = 1, 2, \ldots, 9$), $L_j = 230.0$ g mol/h ($j = 10, 11, \ldots, 21$), $L_j = 219.0$ g mol/h ($j = 22, 23, \ldots, 44$), $L_j = 235.0$ g mol/h ($j = 45, 46, \ldots, 51$).

Base component. Base the values of a and b in the expression for K_b given by Eq. (2-37) on the K values for styrene at 298.15 K and 40 mmHg and at 700 K and 270 mmHg.

Column 2. The "other specifications" for column 2 are the same as those stated for column 1 except for the following items. $D_2 = 16$ g mol/h, $W_2 = 7$ g mol/h. The sidestream W_2 is withdrawn as a liquid from plate $j = 21$. The bottoms B_1 of column 1 is fed to column 2 on plate $j = 31$. The pressure in the condenser is 40 mmHg, and it may be assumed to vary linearly with plate number between stage 2 at a pressure of 50 mm and the reboiler at 250 mmHg. A reflux ratio of L_1/D of 2.5 is to be used.

Initial temperature, vapor, and liquid rate profiles. Same as stated for column 1.

Initial composition profiles for the recycle stream D_2. To initiate the first trial for column 1, assume that essentially all of the ethylbenzene in the feed leaves in the stream D_1, and take

$$d_{2,2} = D_1 - FX_2 = 52 - 51 = 1$$

Assume the recycle stream D_2 is composed mostly of styrene, and

$$d_{3,2} = D_2 - d_{2,2} = 16 - 1 = 15$$

Take $d_{1,2} = d_{4,2} = d_{5,2} = d_{6,2} = d_{7,2} = 0$.

Table 3-5 Solution of Example 3-3

I. Final temperature and vapor rate profiles for column 1†

Stage	T_j (K)	V_j (kg mol/h)	Stage	T_j (K)	V_j (kg mol/h)
1	326.69	52.0	27	363.32	204.10
2	332.41	182.0	28	364.08	204.42
3	334.80	183.56	29	364.84	204.73
4	336.93	183.38	30	365.57	205.03
5	338.86	183.24	31	366.30	205.32
6	340.64	183.12	32	367.01	205.60
7	342.30	183.04	33	367.74	205.87
8	343.83	182.98	34	368.41	206.13
9	345.28	182.92	35	369.10	206.38
10	346.66	182.87	36	369.78	206.62
11	347.96	196.73	37	370.46	206.86
12	349.19	197.35	38	371.14	207.08
13	350.35	197.95	39	371.81	207.28
14	351.48	198.52	40	372.49	207.48
15	352.56	199.06	41	373.17	207.66
16	353.60	199.59	42	373.85	207.83
17	354.61	200.09	43	374.54	207.98
18	355.59	200.58	44	375.24	208.12
19	356.54	201.05	45	375.94	208.25
20	357.47	201.50	46	376.65	211.97
21	358.36	201.95	47	377.38	212.10
22	359.24	202.37	48	378.13	212.22
23	360.10	202.74	49	378.91	212.30
24	360.93	203.10	50	379.71	212.36
25	361.75	203.44	51	380.55	212.39
26	362.54	203.78	52	381.44	212.37

II. Solution sets of product rates for column 1

Component	$d_{i,1}$ (kg mol/h)	w_{1i} (kg mol/h)	$b_{i,1}$ (kg mol/h)
1	0.8000	0.54160×10^{-5}	0.14787×10^{-14}
2	0.40954×10^{2}	0.57815×10^{1}	0.10105×10^{2}
3	0.10244×10^{2}	0.51932×10^{1}	0.42530×10^{2}
4	0.20858×10^{-2}	0.45960×10^{-2}	0.45430×10^{-1}
5	0.20724×10^{-4}	0.72101×10^{-3}	0.92581×10^{-2}
6	0.51283×10^{-4}	0.87917×10^{-2}	0.12115
7	0.11536×10^{-5}	0.11275×10^{-1}	0.18871

† The convergence criteria for the g functions g_0 and g_1 for each column were the same as stated in Table 3-3 for Example 3-1, except that in this case the criterion of 10^{-4} instead of 10^{-5} was used.

$$Q_{C,1} = 0.18264 \times 10^{10} \text{ cal/h}$$
$$Q_{R,1} = 0.20182 \times 10^{10} \text{ cal/h}$$

Table 3-5 (*continued*)

III. Final temperature and vapor rate profiles for column 2

Stage	T_j (K)	V_j (kg mol/h)	Stage	T_j (K)	V_j (kg mol/h)
1	330.55	16.000	27	364.39	56.304
2	336.13	56.000	28	365.11	56.270
3	338.27	56.515	29	365.83	56.236
4	240.16	56.486	30	566.54	56.200
5	341.85	56.475	31	367.24	56.161
6	343.41	56.472	32	367.91	52.425
7	344.86	56.474	33	368.57	52.564
8	346.22	56.478	34	369.21	52.700
9	347.51	56.482	35	369.84	52.834
10	348.74	56.485	36	370.47	52.966
11	349.92	56.488	37	371.08	53.095
12	351.05	56.490	38	371.68	53.221
13	352.14	56.492	39	372.28	53.345
14	353.19	56.492	40	362.86	53.467
15	354.20	56.492	41	373.44	53.587
16	355.19	56.490	42	374.02	53.703
17	356.14	56.489	43	374.59	53.818
18	357.07	56.486	44	375.15	53.930
19	357.96	56.483	45	375.72	54.039
20	358.84	56.479	46	376.29	54.145
21	359.64	56.475	47	376.86	54.247
22	360.52	56.470	48	377.44	54.345
23	361.33	56.437	49	378.03	54.439
24	362.12	56.404	50	378.64	54.527
25	362.89	56.370	51	379.29	54.606
26	363.65	56.337	52	380.01	54.665

IV. Solution sets of product rates for column 2

Component	$d_{i,2}$ (g mol/h)	$w_{2,i}$ (g mol/h)	$b_{i,2}$ (g mol/h)
1	0.13988×10^{-14}	0.79899×10^{-16}	0.18892×10^{-20}
2	0.57999×10^{1}	0.12848×10	0.30208×10
3	0.10198×10^{2}	0.57079×10	0.26624×10^{2}
4	0.21138×10^{-2}	0.58192×10^{-2}	0.37496×10^{-1}
5	0.38839×10^{-6}	0.21680×10^{-3}	0.90409×10^{-2}
6	0.13979×10^{-6}	0.12231×10^{-2}	0.11993
7	0.30294×10^{-12}	0.23966×10^{-4}	0.18869

$Q_{C,2} = 0.57000 \times 10^{9}$ cal/h
$Q_{R,2} = 0.52253 \times 10^{9}$ cal/h

Table 3-6 Convergence characteristics of the Θ method for Example 3-3

1. Column θ's

Trial	Column 1		Column 2	
	θ_0	θ_1	θ_0	θ_1
1	435.43	29.111	94.863	3.0438
2	0.091353	0.11463	0.39932	0.48084
3	2.9271	2.4237	1.4243	1.2706
4	0.56381	0.62113	0.89824	0.91891
5	1.3090	1.3016	1.0657	1.0492
6	0.88564	0.89622	0.96894	0.97597
7	1.0402	1.0449	1.0165	1.0126
8	0.98519	0.98711	0.99377	0.99512
9	1.0066	1.0045	1.0022	1.0017
10	0.99671	0.99741	0.99913	0.99935
11	1.0012	1.0014	1.0004	1.0004
12	0.99953	0.99935	0.99973	0.99980
13	0.99995	1.0001	1.0001	1.0001
14	1.0	1.0	1.0	1.0

2. System Θ's

Trial	Θ_1	Θ_2	Θ_3	Θ_4
1	1.1727	1.2217	1.0173	1.0521
2	1.0446	1.0755	1.0235	1.0276
3	0.99716	0.99252	0.99827	0.99714
4	1.0161	1.0308	1.0099	1.0140
5	1.0006	1.0013	1.0004	1.0006
6	0.99918	0.99817	0.9942	0.99916
7	1.0008	1.0018	1.0006	1.0009
8	1.0	1.0	1.0	1.0
9	0.99984	0.99965	0.99989	0.99985
10	1.0	1.0	1.0	1.0
11	0.99953	0.9935	0.99973	0.99980
12	1.0	1.0	1.0	1.0
13	1.0	1.0	1.0	1.0

3. The convergence criteria were satisfied at the end of the 14th trial. 16.84 seconds of computer time were required on an AMDAHL 470 V/6 computer with a WATFIV compiler.

The convergence criterion for the system was that the euclidean norm of the g functions for the system $\leq 10^{-4}$, that is,

$$\left[\frac{(g_1)^2 + (g_2)^2 + (g_3)^2 + (g_4)^2}{4} \right]^{1/2} \leq 10^{-4}$$

when the g functions are evaluated at $\Theta_1 = \Theta_2 = \Theta_3 = \Theta_4 = 1$.

column. A proof of this statement is established by solving Probs. 3-7 and 3-8, and a demonstration of the truth of the statement is afforded by solving Probs. 3-9 and 3-10.

Systems Containing Mixers and Proportional Dividers

The Θ method of convergence for systems is readily extended to include other types of units such as mixers and proportional dividers which are commonly found in systems of distillation columns. To demonstrate the application of the Θ method to systems containing units such as these, consider the system shown in Fig. 3-13. Suppose the specifications for the system are taken to be D_2 and B_3. In addition to these, specifications such as the total-flow rate, the thermal condition and composition of the feed F, the reflux rate, the column pressure, the type of condenser, and the plate configuration for unit 2 are made. The remaining flow rates, D_1, D_3, and B_2 are computed from the set of three overall material balance equations. For any component i, the component-material balances are as follows

$$\begin{bmatrix} -1 & 0 & r_{i,\,3} \\ 1 & -(1 + r_{i,\,2}) & 0 \\ 0 & r_{i,\,2} & -(1 + r_{i,\,3}) \end{bmatrix} \begin{bmatrix} d_{i,\,1} \\ d_{i,\,2} \\ d_{i,\,3} \end{bmatrix} = - \begin{bmatrix} FX_i \\ 0 \\ 0 \end{bmatrix} \tag{3-31}$$

When only one stream is withdrawn from a unit such as a mixer, there exists no question about the product distribution for the given unit. For any choice of inputs to the mixer, the corresponding output is uniquely determined by the component-material balance for the mixer $(FX_i = d_{i,\,1} - r_{i,\,3}d_{i,\,3})$. Thus, for a mixer with a single output, a Θ does not exist. Furthermore, Θ_3 for the proportional divider is unity. Thus, $r_{i,\,3}$ may be set equal to $(b_{i,\,3}/d_{i,\,3})_{ca}$, and the unknown Θ_2 found by use of the following g function

$$g = \sum_{i=1}^{c} (d_{i,\,2})_{co} - D_2 \tag{3-32}$$

where the value of $(d_{i,\,2})_{co}$ is found by solving Eq. (3-31).

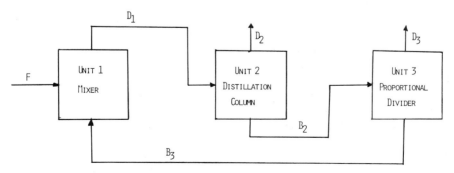

Figure 3-13 A system containing a mixer, a distillation column, and a proportional divider.

3-3 A GENERALIZED FORMULATION FOR A SYSTEM OF COLUMNS IN WHICH THE TOTAL-FLOW RATES OF THE EXTERNAL STREAMS ARE SPECIFIED

To illustrate the formulation of the capital Θ method for the general case of a system having any number of columns interconnected by any number of recycle streams, consider the system of four columns shown in Fig. 3-14. The formulation is simplified by the introduction of a new system of notation.

For each column added to the system, there exists an additional capital Θ multiplier, and each sidestream withdrawn from a column leads to an additional Θ multiplier. Thus, the number of Θ multipliers is equal to the number of columns plus the number of sidestreams. The number of g functions is equal to the number of Θ multipliers, and the number of overall component-material balances to be satisfied by the solution set of the Θ's is equal to the number of columns.

The treatment presented below is restricted to systems of distillation columns. Systems containing both absorbers, strippers, or reboiled absorbers and distillation columns may also be solved by use of the capital Θ method of convergence as demonstrated in Chap. 4.

As shown in Fig. 3-14, the column from which any stream is withdrawn is carried as a superscript, and the column to which an independent feed is introduced is also carried as a superscript. This notation follows closely that developed by Nartker et al.[12] For the system of four columns, there are four independent overall material balances for each component, namely

$$F^1 X_i^1 + d_i^2 - d_i^1 - w_{1i}^1 - w_{2i}^1 = 0$$

$$w_{2i}^3 - d_i^2 - w_{1i}^2 - w_{2i}^2 = 0$$

$$F^3 X_i^3 + d_i^4 - d_i^3 - w_{1i}^3 - w_{2i}^3 = 0 \qquad (3\text{-}33)$$

$$w_{2i}^1 - d_i^4 - w_{1i}^4 - w_{2i}^4 = 0$$

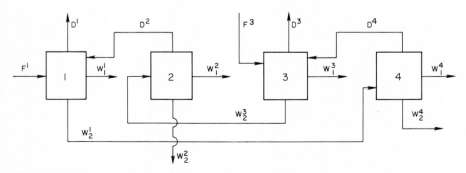

Figure 3-14 A generalized system of four interconnected separation units.

For the system of columns shown in Fig. 3-14, two independent multipliers per column exist, and they are defined as follows

$$\frac{w_{1i}^k}{d_i^k} = \Theta_1^k \left(\frac{w_{1i}^k}{d_i^k}\right)_{ca} \qquad (k = 1, 2, 3, 4)$$

$$\frac{w_{2i}^k}{d_i^k} = \Theta_2^k \left(\frac{w_{2i}^k}{d_i^k}\right)_{ca} \qquad (k = 1, 2, 3, 4)$$

(3-34)

or, more compactly,

$$\frac{w_{si}^k}{d_i^k} = \Theta_s^k \left(\frac{w_{si}^k}{d_i^k}\right)_{ca} \qquad (k = 1, 2, 3, 4; \ s = 1, 2)$$

(3-35)

where k denotes the column number and s denotes the sidestream number.

To avoid the numerical problems resulting from the division of the w_i's by small to zero values for one or more of the d_i's, the following definitions of the r_i's are used instead of those given by Eq. (3-27)

$$r_{si}^k = \Theta_s^k (w_{si}^k)_{ca} \qquad (k = 1, 2, 3, 4; \ s = 1, 2)$$

(3-36)

The corrected values of the w_{si}'s for each column may be restated as follows by combining Eqs. (3-35) and (3-36)

$$w_{si}^k = r_{si}^k p_i^k \qquad (k = 1, 2, 3, 4; \ s = 1, 2)$$

(3-37)

where

$$p_i^k = d_i^k / (d_i^k)_{ca} \qquad (k = 1, 2, 3, 4)$$

By use of the relationships given by Eq. (3-37), the component-material balances [Eq. (3-33)] may be restated in terms of the p_i's as follows

$$F^1 X_i^1 + (d_i^2)_{ca} p_i^2 - [(d_i^1)_{ca} + r_{1i}^1 + r_{2i}^1] p_i^1 = 0$$

$$r_{2i}^3 p_i^3 - [(d_i^2)_{ca} + r_{1i}^2 + r_{2i}^2] p_i^2 = 0$$

$$F^3 X_i^3 + (d_i^4)_{ca} p_i^4 - [(d_i^3)_{ca} + r_{1i}^3 + r_{2i}^3] p_i^3 = 0$$

$$r_{2i}^1 p_i^1 - [(d_i^4)_{ca} + r_{1i}^4 + r_{2i}^4] p_i^4 = 0$$

(3-38)

These equations may be represented by the matrix equation

$$\mathbf{r}_i \mathbf{p}_i = -\boldsymbol{f}_i$$

(3-39)

where

$$\mathbf{r}_i = \begin{bmatrix} -R_i^1 & (d_i^2)_{ca} & 0 & 0 \\ 0 & -R_i^2 & r_{2i}^3 & 0 \\ 0 & 0 & -R_i^3 & (d_i^4)_{ca} \\ r_{2i}^1 & 0 & 0 & -R_i^4 \end{bmatrix} \qquad \mathbf{p}_i = \begin{bmatrix} p_i^1 \\ p_i^2 \\ p_i^3 \\ p_i^4 \end{bmatrix} \qquad \boldsymbol{f}_i = \begin{bmatrix} F^1 X_i^1 \\ 0 \\ F^3 X_i^3 \\ 0 \end{bmatrix}$$

$$R_i^k = (d_i^k)_{ca} + r_{1i}^k + r_{2i}^k \qquad (k = 1, 2, 3, 4)$$

Again the number of external streams which may be specified independently is equal to the number of columns plus the number of sidestreams withdrawn (in addition to the distillate and bottoms). Thus, for the system shown in Fig. 3-14, only eight of the twelve external streams may be specified independently. Furthermore, care must be exercised in the selection of the streams to be specified in order to obtain an independent set. For example, an examination of this system shows that all of the W's may be specified independently, and for this set of specifications, the normalized g functions become

$$g_s^k = \sum_{i=1}^{c} (w_{si}^k/W_s^k) - 1 \qquad (k = 1, 2, 3, 4; \; s = 1, 2) \qquad (3\text{-}40)$$

and in view of Eq. (3-37), the expression represented by Eq. (3-40) may be restated in terms of the r_i's and p_i's to give

$$g_s^k = \sum_{i=1}^{c} [(r_{si}^k p_i^k)/W_s^k] - 1 \qquad (k = 1, 2, 3, 4; \; s = 1, 2) \qquad (3\text{-}41)$$

After the last column trial of a given system trial has been performed, the capital Θ method for systems is applied to find a new set of product-component-flow rates which satisfy all of the system component-material balances simultaneously while being in agreement with the specified values of the terminal flow rates W_1^k and W_2^k ($k = 1, 2, 3, 4$). The solution set of p_i's is determined by finding the set of positive Θ's that make $g_1^k = g_2^k = 0$, ($k = 1, 2, 3, 4$), simultaneously. This set of Θ's may be found by use of the Newton-Raphson method; the equations for this method are represented by

$$\mathbf{J} \, \Delta\Theta = -\mathbf{g} \qquad (3\text{-}42)$$

where the jacobian matrix $\mathbf{J}$ and the column vectors $\mathbf{g}$ and $\Delta\Theta$ have the following matrix representations

$$\mathbf{J} = \begin{bmatrix} \dfrac{\partial g_1^1}{\partial \Theta_1^1} & \cdots & \dfrac{\partial g_1^1}{\partial \Theta_2^4} \\ \vdots & & \vdots \\ \dfrac{\partial g_2^4}{\partial \Theta_1^1} & \cdots & \dfrac{\partial g_2^4}{\partial \Theta_2^4} \end{bmatrix}$$

$$\mathbf{g} = [g_1^1 \; \cdots \; g_1^4 g_2^1 \; \cdots \; g_2^4]^T$$

$$\Delta\Theta = [\Delta\Theta_1^1 \; \cdots \; \Delta\Theta_1^4 \, \Delta\Theta_2^1 \; \cdots \; \Delta\Theta_2^4]^T$$

After the functions and their partial derivatives with respect to the Θ's have been evaluated, Eq. (3-42) may be solved for $\Delta\Theta$ by use of any one of many procedures such as gaussian elimination.[3, 5]

The functions and their partial derivatives which appear in $\mathbf{g}$ and $\mathbf{J}$, respectively, may be evaluated by any one of the techniques described in Chap. 4. Also, the matrix equation [Eq. (3-42)] may be solved by use of any one of the matrix techniques described in Chap. 4.

NOTATION

(See also Chapters 1 and 2)

$d_i^k, d_{i,k}$	molar flow rate of component i in the distillate from unit number k
D^k, D_k	total molar flow rate of the distillate from unit k
F^k	total molar flow rate of the independent feeds to unit k
g_0, g_1	g functions of the θ method for a single column
g_s^k	g function for terminal stream number s of unit k
$\mathbf{g}$	column vector of g functions; see Eq. (3-42)
$\mathbf{J}$	jacobian matrix; defined by Eq. (3-42)
p_i	$(d_i)_{co}/(d_i)_{ca}$; see Eq. (3-13)
p_i^k	$d_i^k/(d_i^k)_{ca}$
$\mathbf{p}_i$	column vector of the p_i's
r_{si}^k	$\Theta_s^k(w_{si}^k)_{ca}$, value of r for stream s of unit k
$\mathbf{r}_i$	a square matrix; defined by Eq. (3-39)
W_s^k	total molar flow rate of sidestream s from unit k
w_{si}^k	molar flow rate of component i in sidestream s leaving unit k
W_s	total molar flow rate of sidestream s leaving a complex column

Subscripts

f	number of the first feed plate; see Fig. 3-1
p	number of the plate from which the first sidestream is withdrawn; see Fig. 3-1
q	number of the plate from which the second sidestream is withdrawn; see Fig. 3-1
t	number of the second feed plate relative to the top of the column; see Fig. 3-1

Superscripts

k	column or unit number

Greek Letters

θ	multiplier for the θ method for single columns
Θ_s^k	multiplier for sidestream s of unit k of a system

Mathematical Symbols

$[x_1 \ x_2 \ x_3]^T$	transpose of a row vector; the transpose of a row vector is equal to a column vector, $[x_1 \ x_2 \ x_3]^T = \begin{bmatrix} x_1 \\ x_2 \\ x_3 \end{bmatrix}$

PROBLEMS

3-1 For the case of a complex distillation column which has one sidestream withdrawn in addition to the distillate and bottoms, show that the partial derivatives are given by the following expressions.

(a) $\dfrac{\partial g_0}{\partial \theta_0} = -\sum_{i=1}^{c} \dfrac{(b_i)_{ca}(d_i)_{ca}FX_i}{[(d_i)_{ca} + \theta_0(b_i)_{ca} + \theta_1(w_{1i})_{ca}]^2}$

(b) $\dfrac{\partial g_0}{\partial \theta_1} = -\sum_{i=1}^{c} \dfrac{(w_{1i})_{ca}(d_i)_{ca}FX_i}{[(d_i)_{ca} + \theta_0(b_i)_{ca} + \theta_1(w_{1i})_{ca}]^2}$

(c) $\dfrac{\partial g_1}{\partial \theta_0} = -\sum_{i=1}^{c} \dfrac{\theta_1(w_{1i})_{ca}(b_i)_{ca}FX_i}{[(d_i)_{ca} + \theta_0(b_i)_{ca} + \theta_1(w_{1i})_{ca}]^2}$

(d) $\dfrac{\partial g_1}{\partial \theta_1} = \sum_{i=1}^{c} \dfrac{(w_{1i})_{ca}[(d_i)_{ca} + \theta_0(b_i)_{ca}]FX_i}{[(d_i)_{ca} + \theta_0(b_i)_{ca} + \theta_1(w_{1i})_{ca}]^2}$

3-2 (a) Formulate the constant-composition form of the enthalpy balances for the complex column shown in Fig. 3-1.

(b) Develop the total material balances for the complex column shown in Fig. 3-1.

3-3 Suppose that the complex column shown in Fig. 3-4 is to be operated at total reflux at finite and nonzero values of D, B, and W_1. Further suppose that the column has a partial condenser and that the relative volatilities remain constant throughout the column. By use of the same approach used to derive Eq. (1-58), show that the Fenske-type relationships are given by

$$\frac{w_{1i}}{d_i} = \left(\frac{w_{1b}}{d_b}\right)\alpha_i^{-p}$$

and

$$\frac{b_i}{d_i} = \left(\frac{b_b}{d_b}\right)\alpha_i^{-N}$$

where b refers to the base component, the component used to compute the α_i's ($\alpha_i = K_i/K_b$).

3-4 Use the Fenske equations of Prob. 3-3 to show that the values of $\{(b_i/d_i)_1\}$ and $\{(w_{1i}/d_i)_1\}$ for one set of total flow rates D_1 and W_1 are related to the values $\{(b_i/d_i)_2\}$ and $\{(w_{1i}/d_i)_2\}$ for a second set of total flow rates $(D_2$ and $W_2)$ by the multipliers θ_0 and θ_1, that is

$$\left(\frac{b_i}{d_i}\right)_2 = \theta_0\left(\frac{b_i}{d_i}\right)_1$$

$$\left(\frac{w_{1i}}{d_i}\right)_2 = \theta_1\left(\frac{w_{1i}}{d_i}\right)_1$$

where

$$\theta_0 = \frac{(b_b/d_b)_2}{(b_b/d_b)_1}$$

$$\theta_1 = \frac{(w_{1b}/d_b)_2}{(w_{1b}/d_b)_1}$$

3-5 This problem is based on material given in Refs. 7 and 8. A complex column having one sidestream W_1 is operating at total reflux. The feed composition, relative volatilities, and other specifications for this column follow

Component	X_i	α_i	Other specifications
1	1/3	1	$N = 5$, $p = 3$, total condenser,
2	1/3	2	$F = 100$ mol/h, and
3	1/3	3	total reflux operation

(a) Compute the values of b_i/d_i, w_{1i}/d_i, D, and W_1 when it is specified that

$$b_b/d_b = 16 \quad \text{and} \quad w_{1b}/d_b = 10$$

(b) Repeat (a) for the case when it is specified $b_b/d_b = 8$ and $w_{1b}/d_b = 5$.

Answer:

(a)	Component	b_i/d_i	w_{1i}/d_i
	1	16	10
	2	1	2.5
	3	0.1975	1.11111

$D = 23.0808$ $W_1 = 46.9075$
$D = 34.6522$ $W_1 = 38.2498$

3-6 This problem is also based on material presented in Refs. 6 and 7. On the basis of the following sets of b_i/d_i's and w_{1i}/d_i's at $D = 23.0808$ and $W_1 = 46.9078$, compute the b_i/d_i's and w_{1i}/d_i's at $D = 34.6522$ and $W_1 = 38.2494$ by use of the θ method of convergence for complex columns.

Given:

Component	X_i	α_i	$(b_i/d_i)_1$	$(w_{1i}/d_i)_1$	Other specifications
1	1/3	1	16	10	$N = 5$, $p = 3$, total
2	1/3	2	1	2.5000	condenser, $F = 100$, and
3	1/3	3	0.1975	1.1111	total reflux operation

Hint: For the first trial assume that $\theta_0 = 0.5$ and $\theta_1 = 0.5$.

Answer:

Component	b_i/d_i	w_{1i}/d_i
1	8.000	5.0000
2	0.500	1.2500
3	0.09875	0.5555

3-7 If each column in the system of columns shown in Fig. 3-8 is at total reflux of the type in which all of the total-flow rates of the terminal streams (F, D_1, D_2, B_1, B_2) are nonzero, finite, and positive, show that the Fenske-type equations for columns 1 and 2 are as follows

$$\frac{b_{i,1}}{d_{i,1}} = \left(\frac{b_{b,1}}{d_{b,1}}\right)\alpha_i^{-N_1}$$

$$\frac{b_{i,2}}{d_{i,2}} = \left(\frac{b_{b,2}}{d_{b,2}}\right)\alpha_i^{-N_2}$$

Thus, when both columns are operated at the condition of total reflux, the above Fenske equations for columns 1 and 2 give the values of $r_{i,1}$ and $r_{i,2}$ of Eq. (3-27) exactly, that is, $r_{i,1} = b_{i,1}/d_{i,1}$ and $r_{i,2} = b_{i,2}/d_{i,2}$.

3-8 Use the Fenske equations of Prob. 3-7 to show that the values of $\{(b_{i,1}/d_{i,1})_1\}$ and $\{(b_{i,2}/d_{i,2})_1\}$ for one set of total-flow rates D_1 and B_2 are related to the values $\{(b_{i,1}/d_{i,1})_2\}$ and $\{(b_{i,2}/d_{i,2})_2\}$ for a second set of total-flow rates $(D_1$ and $B_2)$ by the multipliers Θ_1 and Θ_2, that is

$$\left(\frac{b_{i,1}}{d_{i,1}}\right)_2 = \Theta_1\left(\frac{b_{i,1}}{d_{i,1}}\right)_1$$

$$\left(\frac{b_{i,2}}{d_{i,2}}\right)_2 = \Theta_2\left(\frac{b_{i,2}}{d_{i,2}}\right)_1$$

where

$$\Theta_1 = \frac{(b_{b,1}/d_{b,1})_2}{(b_{b,1}/d_{b,1})_1}$$

$$\Theta_2 = \frac{(b_{b,2}/d_{b,2})_2}{(b_{b,2}/d_{b,2})_1}$$

3-9 The following example was also formulated on the basis of information given in Refs. 7 and 8.

(a) For the system of conventional distillation columns shown in Fig. 3-8, compute D_1, D_2, B_1, and B_2 for the case where component 1 is selected as the base component b, and it is specified that the columns are at total reflux and that

$$b_{b,1}/d_{b,1} = 16 \qquad b_{b,2}/d_{b,2} = 8$$

Given:

Component	Column 1, X_i	α_i columns 1 and 2	Other specifications
1	1/3	1	Columns 1 and 2: $N = 5$, total condenser,
2	1/3	2	and total reflux operation. External feed
3	1/3	3	F to column 1 is 100 mol/h

(b) Repeat part (a) for the case where

$$b_{b,1}/d_{b,1} = 8 \qquad b_{b,2}/d_{b,2} = 4$$

Answer:
(a) $D_1 = 60.2543$, $D_2 = 39.7448$, $B_1 = 217.5755$, and $B_2 = 177.8305$ mol/h.
(b) $D_1 = 67.0963$, $D_2 = 32.9036$, $B_1 = 117.4772$, and $B_2 = 84.5736$ mol/h.

3-10 On the basis of the product distributions found in Prob. 3-9(a), compute the product distributions at the total-flow rates specified in Prob. 3-9(b) by use of the capital Θ method of convergence.

Hint: Assume $\Theta_1 = 0.5$, $\Theta_2 = 0.5$ for the first trial.

Answer:

Component	$b_{i,1}/d_{i,1}$	$b_{i,2}/d_{i,2}$
1	8	102.5640
2	0.5	11.9047
3	0.09875	3.0085

3-11 Tomme[11] has shown that it is always possible to find a set of positive θ's that make all of the functions go to zero for a column having one sidestream. By use of the following outline, construct the proof of the existence of a set of positive θ's such that $g_0 = g_1 = 0$, simultaneously, for a column with one sidestream. It is also to be understood that D and W_1 are specified positive numbers such that $D + W_1 < F$; that is, $F = D + W_1 + B$, where $B > 0$.

Step 1. Construct the graphs shown in Figs. 3-5 and 3-6 for g_0 and g_1 in the neighborhood where both θ_0 and θ_1 are positive.

Step 2. Show that g_0 always has a trace in the $\overline{\theta_0 \theta_1}$ plane.

Step 3. Show that g_1 always has a trace in the $\overline{\theta_0 \theta_1}$ plane.

Step 4. In order to prove that a point of intersection exists for every pair of functions, it is sufficient to show that

$$a < b$$

In order to prove this, verify the following relationships and then employ them as required.

$$|g_0(0, 0)| = B + W_1$$

$$|g_1(0, 0)| = W_1$$

$$\left| \frac{\partial g_0(0, \theta_1)}{\partial \theta_1} \right| = \left| \frac{\partial g_1(0, \theta_1)}{\partial \theta_1} \right|$$

3-12 Obtain expressions for the partial derivatives of the g functions given Eq. (3-26) for the system shown in Fig. 3-8.

3-13 Show that it is always possible to find a set of positive Θ's for the g functions of the capital Θ method for the system shown in Fig. 3-8.

Hints:

1. Show that the trace of the function g_1 in the $\overline{\Theta_1 \Theta_2}$ plane exists for all sets of positive and finite values for Θ_1 and Θ_2.
2. Show that the trace of g_2 exists for all finite and positive values of Θ_1 and for all positive values of Θ_2 greater than the asymptotic value.
3. Show that point a in Fig. 3-11 is always to the left of point b, that is $a < b$.

3-14 For the proportional divider [unit 3 of Fig. 3-13], show that $\Theta_3 = 1$ by use of the following information

1. $(b_{i,3}/d_{i,3})_{co} = \Theta_3 (b_{i,3}/d_{i,3})_{ca}$.
2. The composition of stream B_3 is always equal to the composition of stream D_3.
3. The calculated values of $b_{i,3}$ and $d_{i,3}$ are based on the specified (or corrected) values of B_3 and D_3, that is, $(B_3/D_3)_{co} = (B_3/D_3)_{ca}$.

REFERENCES

1. D. S. Billingsley: "On the Equations of Holland in the Solution of Problems in Multicomponent Distillation," *IBM J. Res. Develop.*, **14**: 33 (1970).
2. ———: "Existence and Uniqueness of Solutions to Holland's Equations for a Case of Multicolumn Distillation Systems," *IBM J. Res. Develop.* **16**(5): 482 (1972).
3. Brice Carnahan, H. A. Luther, and J. O. Wilkes: *Applied Numerical Methods*, John Wiley and Sons, Inc., New York, 1964.
4. M. R. Fenske: "Fractionation of Straight-Run Pennsylvania Gasoline," *Ind. Eng. Chem.*, **24**: 482 (1932).
5. C. D. Holland: *Unsteady State Processes with Applications in Multicomponent Distillation*, Prentice-Hall, Inc., Englewood Cliffs, N.J., 1966.
6. ———: *Fundamentals and Modeling of Separation Processes—Absorption, Distillation, Evaporation, and Extraction*, Prentice-Hall, Inc., Englewood Cliffs, N.J., 1975.
7. ———: "Exact Solutions Given by the θ Method of Convergence for Complex and Conventional Distillation Columns," *Chem. Eng. J.*, Loughborough University of Technology, **7**: 10 (1972).
8. ——— and G. P. Pendon: "Solve More Distillation Problems: Part 1. Improvements Give Exact Answers," *Hydrocarbon Process.*, **53**(7): 148 (1974).
9. W. N. Lyster, S. L. Sullivan, Jr., D. S. Billingsley, and C. D. Holland: "Figure Distillation This New Way: Part 3—Consider Multi-Feed Columns with Sidestreams," *Pet. Refiner*, **38**(10): 139 (1959).
10. W. J. Tomme and C. D. Holland: "Figure Distillation This New Way: Part II—When Columns are Operated as a Unit," *Pet. Refiner*, **41**(6): 139 (1962).
11. ———: "A Convergence Method for Distillation Systems," Ph.D. Dissertation, Texas A&M University, College Station, Texas, 1963.
12. T. A. Nartker, J. M. Srygley, and C. D. Holland: "Solution of Problems Involving Systems of Distillation Columns," *Can. J. Chem. Eng.*, **44**: (1966).

FOUR

THE 2N NEWTON-RAPHSON METHOD

When the Newton-Raphson method[7] is formulated in terms of two independent variables per stage, the temperature T_j and the flow ratio L_j/V_j, the resulting procedure is called the 2N Newton-Raphson method. (Originally, this procedure was called the multi-θ method.[16, 17]) When both the vapor and liquid phases form ideal solutions on each stage, this procedure is an exact application of the Newton-Raphson method.

The 2N Newton-Raphson method may be applied to any type of distillation column or to any system of interconnected columns. Absorbers, strippers, reboiled absorbers, and distillation columns are treated in Sec. 4-1. Selected numerical methods for solving the 2N Newton-Raphson equations are presented in Sec. 4-2. In Sec. 4-3, two methods for solving problems involving systems of columns interconnected by recycle streams are presented.

Recommended procedures for solving problems involving single columns as well as systems of columns interconnected by recycle streams are summarized in Table 4-1 for the case where mixtures to be separated form ideal or near solutions throughout the column. As shown there, the 2N Newton-Raphson method is recommended for solving problems involving absorber-type columns (any column which does not possess both an overhead condenser and a reboiler such as absorbers, strippers, and reboiled absorbers).

For solving problems involving distillation-type columns (any column possessing both an overhead condenser and a reboiler), the θ method presented in Chaps. 2 and 3 is recommended. For systems of columns containing both absorber-type and distillation-type columns, it is recommended that the 2N Newton-Raphson method be used for the absorber-type columns and that the θ

Table 4-1 Summary of methods and their applications for separating mixtures which form ideal solutions

Type of column	Recommended method	Description
Conventional and complex distillation columns	θ method	Chaps. 2 and 3
Absorbers, strippers, reboiled absorbers, and any column which does not have both a condenser and a reboiler	$2N$ Newton-Raphson	This chapter
Systems of distillation-type columns with recycle streams	θ method for the individual columns and the capital Θ method for the system	Chaps. 2 and 3
Systems of absorber-type and distillation-type columns with recycle streams	θ method for the distillation-type columns, the $2N$ Newton-Raphson method for absorber-type columns, and the capital Θ method for the system	This chapter and Chaps. 2 and 3

method be used for the distillation-type columns. After one trial has been made on each column of the system, the system is placed in component-material balance and in agreement with the specified values of the total-flow rates by use of the capital Θ method of convergence.

4-1 FORMULATION OF THE 2N NEWTON-RAPHSON METHOD FOR SINGLE COLUMNS

The development of this application of the Newton-Raphson method is presented first for an absorber (or stripper)—see Fig. 4-1. Then the method is applied to conventional and complex distillation columns.

Absorber and strippers may be classified as complex columns because they possess two feeds and because they possess neither an overhead condenser nor a reboiler. The sketch of the absorber in Fig. 4-1 depicts an historic application of absorbers in the natural gas industry. From a light gas stream such as natural gas that contains primarily methane plus small quantities of, say, ethane through n-pentane, the desired quantities of the components heavier than methane may be removed by contacting the natural gas stream with a heavy oil stream (say n-octane or heavier) in a countercurrent, multiple-stage column such as the one shown in Fig. 4-1. Since absorption is a heat-liberating process, the lean oil is customarily introduced at a temperature below the average temperature at which the column is expected to operate. The flow rate of the lean oil is denoted by L_0, and the lean oil enters at the top of the column as implied by Fig. 4-1. The rich gas (which is sometimes called the wet gas) enters at the bottom of the

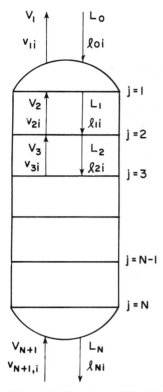

Figure 4-1 Absorber and identifying symbols.

column at a temperature equal to or above its dew-point temperature at the column pressure but generally below the average operating temperature of the column. The total-flow rate of the *rich gas* is denoted by V_{N+1}. The absorber oil plus the material that it has absorbed leave at the bottom of the column; this stream is called the *rich oil*. The treated gas leaving the top of the column is called the *lean gas* (or the stripped gas).

Strippers are used to remove relatively light gases from a heavy oil stream by contacting it with a relatively light gas stream such as steam. Figure 4-1 is also used to depict a typical stripper.

Formulation of the $2N$ Newton-Raphson Method for an Absorber (or Stripper) with Any Number of Equilibrium Stages

In the following formulation of the Newton-Raphson equations, the independent variables are taken to be the N-stage temperatures $\{T_j\}$ and the N-ratios of the total-flow rates $\{L_j/V_j\}$. When the gas and liquid phases form ideal solutions, the procedure described is an exact application of the Newton-Raphson method.

The absorber (or stripper) shown in Fig. 4-1 is described by the $N(2c + 3)$ independent equations. When these equations are stated in terms of component and total-flow rates instead of mole fractions, one obtains

Equilibrium
relationships
$$
\begin{cases}
\dfrac{v_{ji}}{V_j} = K_{ji}\dfrac{l_{ji}}{L_j} & \begin{aligned} &(j = 1, 2, \ldots, N)\\ &(i = 1, 2, \ldots, c) \end{aligned}\\[2ex]
\displaystyle\sum_{i=1}^{c} l_{ji} = L_j & (j = 1, 2, \ldots, N)\\[2ex]
\displaystyle\sum_{i=1}^{c} v_{ji} = V_j & (j = 1, 2, \ldots, N)
\end{cases}
$$

Component-
material
balances
$$
v_{j+1,\,i} + l_{j-1,\,i} - v_{ji} - l_{ji} = 0 \qquad \begin{aligned} &(j = 1, 2, \ldots, N)\\ &(i = 1, 2, \ldots, c) \end{aligned}
$$

Energy
balances
$$
\sum_{i=1}^{c} [v_{j+1,\,i} H_{j+1,\,i} + l_{j-1,\,i} h_{j-1,\,i} - v_{ji} H_{ji} - l_{ji} h_{ji}] = 0
$$

$$(j = 1, 2, \ldots, N)$$

(4-1)

When the column pressure P, the rich gas rates $\{v_{N+1,\,i}\}$ and the temperature T_{N+1}, the lean gas rates $\{l_{0i}\}$ and the temperature T_0, and the number of plates N are fixed (or specified), the set of equations represented by Eq. (4-1) contains $N(2c + 3)$ unknowns: $\{v_{ji}\}$, $\{l_{ji}\}$, $\{V_j\}$, $\{L_j\}$, and $\{T_j\}$.

The N equilibrium functions are formulated by first restating the second and third expressions of Eq. (4-1) as follows

$$
0 = \frac{\displaystyle\sum_{i=1}^{c} l_{ji}}{L_j} - \frac{\displaystyle\sum_{i=1}^{c} v_{ji}}{V_j}
\tag{4-2}
$$

Elimination of the l_{ji}'s by use of the first expression of Eq. (4-1) and restatement of the result so obtained in functional form yields

$$
F_j = \frac{1}{V_j} \sum_{i=1}^{c} \left(\frac{1}{K_{ji}} - 1\right) v_{ji} \qquad (j = 1, 2, \ldots, N)
\tag{4-3}
$$

The N enthalpy functions are obtained by dividing the sum of the input terms of the last expression of Eq. (4-1) by the sum of the output terms. When the result so obtained is restated in functional form, one obtains

$$
G_j = \frac{\displaystyle\sum_{i=1}^{c} [v_{ji} H_{ji} + l_{ji} h_{ji}]}{\displaystyle\sum_{i=1}^{c} [v_{j+1,\,i} H_{j+1,\,i} + l_{j-1,\,i} h_{j-1,\,i}]} - 1
\tag{4-4}
$$

The functions F_j and G_j contain the dependent variables $\{v_{ji}\}$, $\{l_{ji}\}$, and $\{V_j\}$. For any choice of values of the independent variables $\{T_j\}$ and $\{L_j/V_j\}$, expressions are needed for computing the corresponding values of the dependent variables.

First, an equation for computing the $\{V_j\text{'s}\}$ for any set of assumed $L_j/V_j\text{'s}$ is developed. Summation of the component-material balances over all components followed by the elimination of the summations through the use of the second and third expressions of Eq. (4-1) yields

$$V_2 - V_1 - L_1 = -L_0$$
$$V_{j+1} + L_{j-1} - V_j - L_j = 0 \qquad (j = 2, 3, \ldots, N - 1) \qquad (4\text{-}5)$$
$$L_{N-1} - V_N - L_N = -V_{N+1}$$

For any given set of $L_j/V_j\text{'s}$ it is desired to solve the total-material balances for the corresponding set of vapor rates $\{V_j\}$. In the restatement of the total-material balances, it is convenient to define the new variable θ_j as follows

$$\frac{L_j}{V_j} = \theta_j \left(\frac{L_j}{V_j}\right)_a \qquad (j = 1, 2, \ldots, N) \qquad (4\text{-}6)$$

where $(L_j/V_j)_a$ is any arbitrary value of L_j/V_j. Taking this assumed ratio equal to the most recently assumed value of L_j/V_j serves to normalize the $\theta_j\text{'s}$ so that at convergence θ_j approaches unity for all j. Let Eq. (4-6) be restated as follows

$$L_j = R_j V_j \qquad (4\text{-}7)$$

where R_j is defined by

$$R_j = \theta_j \left(\frac{L_j}{V_j}\right)_a \qquad (4\text{-}8)$$

Equation (4-7) may be used to restate the total material balances in terms of either the vapor or the liquid rates. For any interior plate j ($j = 2, 3, \ldots, N - 1$), the total material balance may be restated in terms of the vapor rates as follows

$$R_{j-1} V_{j-1} - (1 + R_j)V_j + V_{j+1} = 0 \qquad (4\text{-}9)$$

The complete set of total-material balances may be represented by the matrix equation

$$\mathbf{RV} = -\mathscr{F} \qquad (4\text{-}10)$$

where

$$\mathbf{R} = \begin{bmatrix} -(1 + R_1) & 1 & 0 & 0 & 0 \\ R_1 & -(1 + R_2) & 1 & 0 & 0 \\ 0 & 0 & R_{N-2} & -(1 + R_{N-1}) & 1 \\ 0 & 0 & 0 & R_{N-1} & -(1 + R_N) \end{bmatrix}$$

$$\mathbf{V} = [V_1 \ V_2 \ \cdots \ V_N]^T$$
$$\mathscr{F} = [L_0 \ 0 \ \cdots \ 0 \ V_{N+1}]^T$$

For a given set of values of the independent variables $\{\theta_j\}$ and $\{T_j\}$, the corresponding sets of values of the component-flow rates $\{v_{ji}\}$ and $\{l_{ji}\}$ are needed in order to evaluate the functions $\{F_j\}$ and $\{G_j\}$. These rates may be computed through the use of the first and fourth expressions of Eq. (4-1) which may be rearranged to the form

$$l_{ji} = A_{ji} v_{ji} \tag{4-11}$$

where

$$A_{ji} = \frac{L_j}{K_{ji} V_j} = \frac{\theta_j}{K_{ji}} \left(\frac{L_j}{V_j} \right)_a$$

After the l_{ji}'s ($j = 1, 2, \ldots, N$) have been eliminated from the component-material balances by use of Eq. (4-11), the resulting set of equations may be restated in the matrix form of Eq. (2-18), $A_i v_i = -\mathcal{f}_i$, where each absorption factor appearing in A_i is given by Eq. (4-11) and v_i and $\mathcal{f}_i$ have the following elements

$$\mathbf{v}_i = [v_{1i} \; v_{2i} \; \cdots \; v_{Ni}]^T$$

$$\mathcal{f}_i = [l_{0i} \; 0 \; \cdots \; 0 \; v_{N+1, i}]^T \tag{4-12}$$

Now observe that for any given set of θ_j's and T_j's (and some arbitrary set of $(L_j/V_j)_a$'s), sets of numerical values may be found for the V_j's and the v_{ji}'s by solving Eqs. (4-10) and (2-18) respectively. After the V_j's have been found, the L_j's may be computed by use of Eq. (4-7). Similarly, after the v_{ji}'s have been computed, the corresponding l_{ji}'s may be computed by use of the equilibrium relationship, Eq. (4-11). In summary, it is desired to find the set of $2N$ independent variables

$$\mathbf{X} = [\theta_1 \; \theta_2 \; \cdots \; \theta_N \; T_1 \; T_2 \; \cdots \; T_N]^T$$

which satisfy the $2N$ independent functions

$$\mathbf{f} = [F_1 \; F_2 \; \cdots \; F_N \; G_1 \; G_2 \; \cdots \; G_N]^T$$

simultaneously.

The Newton-Raphson equations for solving the $2N$ functions $\{F_j, G_j\}$ for the $2N$ independent variables $\{\theta_j, T_j\}$ may be represented by the matrix equation

$$\mathbf{J} \, \Delta \mathbf{X} = -\mathbf{f} \tag{4-13}$$

where the jacobian $\mathbf{J}$ has the representation

$$\mathbf{J} = \begin{bmatrix} \dfrac{\partial F_1}{\partial \theta_1} & \cdots & \dfrac{\partial F_1}{\partial \theta_N} & \dfrac{\partial F_1}{\partial T_1} & \cdots & \dfrac{\partial F_1}{\partial T_N} \\ \vdots & & \vdots & \vdots & & \vdots \\ \dfrac{\partial F_N}{\partial \theta_1} & \cdots & \dfrac{\partial F_N}{\partial \theta_N} & \dfrac{\partial F_N}{\partial T_1} & \cdots & \dfrac{\partial F_N}{\partial T_N} \\ \dfrac{\partial G_1}{\partial \theta_1} & \cdots & \dfrac{\partial G_1}{\partial \theta_N} & \dfrac{\partial G_1}{\partial T_1} & \cdots & \dfrac{\partial G_1}{\partial T_N} \\ \vdots & & \vdots & \vdots & & \vdots \\ \dfrac{\partial G_N}{\partial \theta_1} & \cdots & \dfrac{\partial G_N}{\partial \theta_N} & \dfrac{\partial G_N}{\partial T_1} & \cdots & \dfrac{\partial G_N}{\partial T_N} \end{bmatrix}$$

and

$$\mathbf{\Delta X} = [\Delta\theta_1, \Delta\theta_2, \ldots, \Delta\theta_N, \Delta T_1, \Delta T_2, \ldots, \Delta T_N]^T$$

$$\mathbf{f} = [F_1, F_2, \ldots, F_N, G_1, G_2, \ldots, G_N]$$

Next, the 2N Newton-Raphson method is applied to reboiled absorbers, conventional distillation columns, and complex distillation columns, and then a procedure which makes use of the calculus of matrices for solving these equations is presented.

In the formulation of the Newton-Raphson equations, each of the functions to be employed may be obtained by any combination of the independent equations which produces an independent function. One of the most important steps in the application of the Newton-Raphson method is the formulation of the functions because the precise form of the functions determines the region of convergence. To illustrate this concept, consider the formulation of the isothermal flash function. Although a variety of flash functions may be developed by different combinations of the $2c + 2$ equilibrium and component-material balances, many of these functions could prove unsatisfactory for solving the adiabatic flash problem by use of a formulation of the Newton-Raphson method which involves two independent variables and two independent functions. In general it is desirable to construct functions which are monotonic in the independent variables throughout the region of convergence. For example, the flash function given by Eq. (1-30) is not monotonic in the independent variable Ψ throughout the solution domain $0 \leq \Psi \leq 1$. Although the function $P(\Psi)$ may be used satisfactorily to find the solution to the isothermal flash problem by starting at $\Psi = 1$, its use could lead to difficulties in the solution of the adiabatic flash problem by a Newton-Raphson formulation in terms of Ψ and T. Examination of $P(\Psi)$ shown in Fig. 1-8 shows that when any value of $\Psi > 0$ to the left of the minimum is used, Newton's method will predict a negative Ψ. The customary procedure used in n-dimensional space consists of the successive reduction of the corrections [$\Delta\Psi$ for the function $P(\Psi)$] by factors of $1/2$ until positive values of the variables are obtained. For the function $P(\Psi)$, this procedure fails because all values of Ψ to the left of the minimum are outside of the region of convergence for the positive nonzero root. In this case, trials would be made at successively smaller values of Ψ and the trivial solution $\Psi = 0$ would be approached as this procedure is applied indefinitely. To obtain the desired solution (the $\Psi > 0$ which makes $P(\Psi) = 0$), a new starting value which is to the right of the minimum value $P(\Psi)$ must be selected. Selection of an initial set of the variables which are in the region of convergence can prove difficult for n-dimensional problems unless the functions are very carefully formulated.

Another type of serious difficulty arises when $f(x)$ exhibits the following type of behavior. Suppose that after having passed through the x axis at some $x > 0$, it then passes through a minimum (or a maximum) and then approaches zero asymptotically. Thus, for larger and larger values of x to the right of the minimum (or maximum), the function $f(x)$ becomes smaller and smaller. The function in Prob. 4-5(b) behaves in this manner. Although it is difficult to deduce the behavior of a function in n-dimensional space, the traces of the function can be examined in two-dimensional space and an attempt should be made to formulate functions whose traces are monotonic.

Reboiled Absorbers

The sketch for a typical reboiled absorber is shown in Fig. 4-2. To demonstrate the formulation of the $2N$ Newton-Raphson method for reboiled absorbers, two different sets of specifications are considered.

Specification set 1 P, F, $\{X_i\}$, **thermal condition of** F, L_0, $\{x_{0i}\}$, T_0, f, N **and** Q_R
In order to solve a problem of this type by the $2N$ Newton-Raphson method,

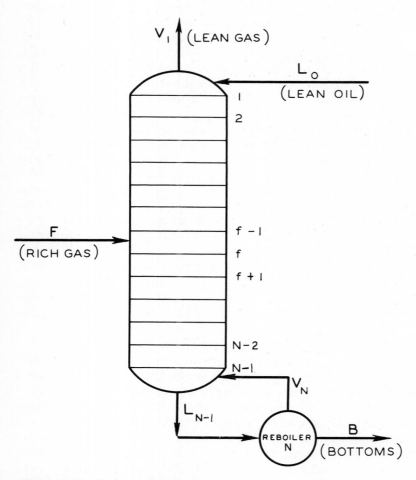

Figure 4-2 Sketch of a reboiled absorber.

the following sets of $2N$ independent functions are selected

$$\mathbf{X} = [\theta_1 \ \theta_2 \ \cdots \ \theta_N \ T_1 \ T_2 \ \cdots \ T_N]^T$$

$$\mathbf{f} = [F_1 \ F_2 \ \cdots \ F_N \ G_1 \ G_2 \ \cdots \ G_N]^T$$

The matrix $\mathbf{A}_i$ of the component-material balances are of the same general form as the $\mathbf{A}_i$ given by Eq. (2-18) for distillation columns. When plate 1 is assumed to behave according to model 1 (see Fig. 1-13) and plate f is assumed to have the behavior characterized by model 2 (see Fig. 2-2), the component-material balances may be represented by Eq. (2-18); provided that the elements of f_i are taken to be the following set

$$f_i = [l_{0i} \ 0 \ 0 \ \cdots \ 0 \ v_{Fi} \ l_{Fi} \ 0 \ \cdots \ 0]^T \tag{4-14}$$

where v_{Fi} lies in row $f - 1$ and l_{Fi} lies in row f.

The second set of specifications differs from the first in that the boilup ratio V_N/B is specified instead of the reboiler duty.

Specification set 2 P, F, $\{X_i\}$, **thermal condition of** F, L_0, $\{x_{0i}\}$, T_0, f, N, V_N/B
For this set of specifications, the $2N$ independent variables are given by

$$\mathbf{X} = [\theta_1 \ \theta_2 \ \cdots \ \theta_{N-1} \ T_1 \ T_2 \ \cdots \ T_{N-1} \ T_N \ Q_R]^T \tag{4-15}$$

and the $2N$ independent functions $\mathbf{f}$ are the same set listed for Specification Set 1. The N vapor-liquid equilibrium functions are given by Eq. (4-3), and the enthalpy balance functions are given by Eq. (4-4) for all stages except $j = f - 1, f,$ N. For stage $f - 1$ and f, the functions G_{f-1} and G_f contain the additional term

$$\sum_{i=1}^{c} v_{Fi} H_{Fi} \qquad \text{and} \qquad \sum_{i=1}^{c} l_{Fi} h_{Fi}$$

in their denominators, respectively. For stage N, the normalized form of G_N is given by

$$G_N = \frac{\displaystyle\sum_{i=1}^{c} [v_{Ni} H_{Ni} + b_i h_{Ni}]}{\displaystyle\sum_{i=1}^{c} [l_{N-1, i} h_{N-1, i}] + Q_R} - 1 \tag{4-16}$$

Conventional Distillation Columns

From the sketches of a conventional distillation column (Fig. 2-1) and a reboiled absorber (Fig. 4-2), it is seen that the geometrical configuration of a conventional distillation column is obtained by replacing plate 1 of the reboiled absorber by a condenser-accumulator section (stage 1) and by eliminating the feed L_0. The condenser-accumulator section is assigned the stage number 1, and when the condenser duty Q_C is specified the independent variables corresponding to this stage are θ_1 (where $\theta_1 = L_1/D$) and T_1.

The matrix equation representing the component-material balance is again given by Eq. (2-18), and the elements of the matrices A_i, v_i, and f_i have the meanings stated below Eq. (2-18).

For a column having a partial condenser, the dew-point functions are given by Eq. (4-3) for $j = 2, 3, \ldots, N$. For $j = 1$, D and d_i play the same role as V_j and v_{ji} in Eq. (4-3) and the dew-point function F_1 is given by

$$F_1 = \frac{1}{D} \sum_{i=1}^{c} \left(\frac{1}{K_{1i}} - 1 \right) d_i \tag{4-17}$$

For a column having a total condenser, the bubble-point function for the distillate is used for F_1, namely,

$$F_1 = \frac{1}{D} \sum_{i=1}^{c} (K_{1i} - 1) \, d_i \tag{4-18}$$

Except for stage 1, the enthalpy balance functions are given by Eqs. (4-4) and (4-16). For stage $j = 1$, the normalized form of the enthalpy balance function is given by

$$G_1 = \frac{\sum_{i=1}^{c} [d_i H_{Di} + l_{1i} h_{1i}] + Q_C}{\sum_{i=1}^{c} v_{2i} H_{2i}} - 1 \tag{4-19}$$

where $H_{Di} = H_{1i}$ for a partial condenser and $H_{Di} = h_{1i}$ for a total condenser. For stages $j = 2, 3, \ldots, f - 2, f + 1, f + 2, \ldots,$ and $N - 1$, the enthalpy balance functions G_j are given by Eq. (4-4). The functions G_{f-1} and G_f are formulated as described below Eq. (4-15) and G_N is given by Eq. (4-16).

The independent variables for different sets of specifications are listed in Table 4-2.

Complex Columns

Complex columns were defined in Chap. 3 and illustrated by Figs. 3-1 and 3-4. To illustrate the application of the $2N$ Newton-Raphson method to the solution of problems involving complex columns, consider the simple case where the sidestream W_1 is withdrawn in the liquid phase from some interior plate p. The withdrawal of the sidestream W_1 gives rise to one specification in addition to those stated for conventional columns, in items 1 through 4 of Table 4-2. When this additional specification is taken to be either the total-flow rate W_1 or the ratio W_1/L_p, the sets of specifications, independent variables, and functions for this complex column are the same as those stated in Table 4-2 except that either W_1 or W_1/L_p should be added to each set of specifications.

When W_1/L_p is specified, the total-material balance for plate p, $V_{p+1} + L_{p-1} - V_p - L_p - W_1 = 0$ may be restated as follows

$$R_{p-1} V_{p-1} - \left(1 + R_p + \frac{W_1}{L_p} R_p \right) V_p + V_{p+1} = 0 \tag{4-20}$$

Table 4-2 Specifications, independent variables, and functions for conventional distillation columns

1. *Specifications:* P, F, $\{X_i\}$, thermal condition of F, f, N, Q_C, Q_R, a partial condenser, and the model of the feed plate.
   Independent variables: θ_1, θ_2, ..., θ_{N-1}, θ_N, T_1, T_2, ..., T_{N-1}, T_N.
   Functions: F_1, F_2, ..., F_{N-1}, F_N, G_1, G_2, ..., G_{N-1}, G_N. F_1 is given by Eq. (4-17).

2. *Specifications:* P, F, $\{X_i\}$, thermal condition of the feed F, f, N, L_1/D, V_N/B, and a partial condenser.
   Independent variables: Q_C, θ_2, θ_3, ..., θ_{N-1}, Q_R, T_1, T_2, ..., T_{N-1}, T_N.
   Functions: Same as item 1.

3. *Specifications:* Same as item 1 above except that a total condenser instead of a partial condenser is to be used.
   Independent variables: Same as item 1.
   Functions: Same as item 1 except F_1 is given by Eq. (4-18).

4. *Specifications:* Same as item 2 above except that a total condenser instead of a partial condenser is to be used.
   Independent variables: Same as item 2.
   Functions: Same as item 3.

Thus, the matrix equation $\mathbf{RV} = -\mathscr{F}$ [Eq. (4-10)] applies, provided that the element lying on the central diagonal of row p of $\mathbf{R}$ is changed from $(1 + R_p)$ to $[1 + R_p + (W_1/L_p)R_p]$.

When W_1 is specified, the total-material balance for plate p takes the form

$$R_{p-1}V_{p-1} - (1 + R_p)V_p + V_{p+1} = W_1 \tag{4-21}$$

In this case, the constant W_1 appears in row p of the column vector $\mathscr{F}$. For example, if W_1 is withdrawn from plate p above the feed plate f, then $\mathscr{F}$ is of the form

$$\mathscr{F} = [0 \;\cdots\; 0 \;(-W_1)\; 0 \;\cdots\; 0 \; V_F \; L_F \; 0 \;\cdots\; 0]^T \tag{4-22}$$

The component-material balances are formulated in precisely the same manner as demonstrated in Chap. 3 for complex columns.

Three procedures are presented in this chapter for solving the Newton-Raphson equations. Procedure 1 is presented below and procedures 2 and 3 are presented in Sec. 4-3.

Procedure 1. Solution of the 2N Newton-Raphson Equations by Use of the Calculus of Matrices and LU Factorization

The analytical expressions for the partial derivatives of the F_j's and G_j's with respect to the θ_j's and T_j's are readily obtained by termwise differentiation of these functions. For example

$$\frac{\partial F_j}{\partial \theta_k} = \frac{1}{V_j}\left[\sum_{i=1}^{c}\left(\frac{1}{K_{ji}} - 1\right)\frac{\partial v_{ji}}{\partial \theta_k} - F_j\frac{\partial V_j}{\partial \theta_k}\right] \qquad \begin{array}{l}(k = 1, 2, ..., N)\\(j = 1, 2, ..., N)\end{array} \tag{4-23}$$

The partial derivatives of the flow rates which appear in these expressions may be evaluated by solving the matrix equations given below. The partial derivatives of F_j's with respect to the θ_j's may be computed by use of the matrix equation obtained by partial differentiation of the members of Eq. (4-10) with respect to any θ, say θ_k

$$A_i \frac{\partial \mathbf{v}_i}{\partial \theta_k} = -\frac{\partial \mathbf{A}_i}{\partial \theta_k} \mathbf{v}_i \qquad (k = 1, 2, \ldots, N) \tag{4-24}$$

where

$$\frac{\partial \mathbf{v}_i}{\partial \theta_k} = \left[\frac{\partial v_{1i}}{\partial \theta_k} \frac{\partial v_{2i}}{\partial \theta_k} \cdots \frac{\partial v_{Ni}}{\partial \theta_k} \right]^T$$

Similarly, partial differentiation of the component-material balances [Eq. (2-18)], $\mathbf{A}_i \mathbf{v}_i = -\mathcal{f}_i$, with respect to T_k gives

$$\mathbf{A}_i \frac{\partial \mathbf{v}_i}{\partial T_k} = -\frac{\partial \mathbf{A}_i}{\partial T_k} \mathbf{v}_i \tag{4-25}$$

where

$$\frac{\partial \mathbf{v}_i}{\partial T_k} = \left[\frac{\partial v_{1i}}{\partial T_1} \frac{\partial v_{2i}}{\partial T_2} \cdots \frac{\partial v_{Ni}}{\partial T_k} \right]^T$$

Since the matrix $\mathbf{A}_i$ has at most only two elements which depend upon a particular θ or T, the right-hand sides of Eqs. (4-24) and (4-25) reduce to relatively simple column vectors as shown in Table 4-3 for absorbers. The elements in these column vectors are found by carrying out the matrix operations indicated on the right-hand sides of Eqs. (4-24) and (4-25).

The partial derivatives of the total-flow rates with respect to the θ_j's may be found by use of the following expression which is obtained by termwise partial differentiation of Eq. (4-10) to give

$$\mathbf{R} \frac{\partial \mathbf{V}}{\partial \theta_k} = -\frac{\partial \mathbf{R}}{\partial \theta_k} \mathbf{V} \tag{4-26}$$

After the matrix operations implied by the right-hand side of this equation have been performed, the elements of the column matrix shown in Table 4-3 are obtained.

The partial derivatives of the V_j's with respect to the temperatures are all equal to zero because every element of the matrix $\mathbf{R}$ is independent of temperature. The truth of this statement is demonstrated by first differentiation of each member of Eq. (4-10) with respect to any T, say T_k, to obtain

$$\mathbf{R} \frac{\partial \mathbf{V}}{\partial T_k} = -\frac{\partial \mathbf{R}}{\partial T_k} \mathbf{V} = \mathbf{0}\mathbf{V} = \mathbf{0}$$

and thus

$$\frac{\partial \mathbf{V}}{\partial T_k} = \mathbf{R}^{-1}\mathbf{0} = \mathbf{0}$$

Table 4-3 Elements of the column vectors on the right-hand sides of Eqs. (4-24), (4-25), and (4-26) for absorbers

---

1. Elements of $\mathbf{C}_{ki} = -(\partial \mathbf{A}_i / \partial \theta_k)\mathbf{v}_i$, $(k = 1, 2, \ldots, N)$

$\mathbf{C}_{1i} = [C_{1i} \ (-C_{1i}) \ 0 \ \cdots \ 0)]^T$

$\mathbf{C}_{2i} = [0 \ C_{2i} \ (-C_{2i}) \ 0 \ \cdots \ 0]^T$

$\vdots$

$\mathbf{C}_{N-1, i} = [0 \ \cdots \ 0 \ C_{N-1, i} \ (-C_{N-1, i})]^T$

$\mathbf{C}_{Ni} = [0 \ \cdots \ 0 \ C_{Ni}]^T$

$$C_{ki} = \frac{\partial A_{ki}}{\partial \theta_k} v_{ki} = \frac{v_{ki}}{K_{ki}} \left(\frac{L_k}{V_k} \right)_a$$

2. Elements of $\mathbf{D}_{ki} = -(\partial \mathbf{A}_i / \partial T_k)\mathbf{v}_i$, $(k = 1, 2, \ldots, N)$

$\mathbf{D}_{1i} = [D_{1i} \ (-D_{1i}) \ 0 \ \cdots \ 0]^T$

$\mathbf{D}_{2i} = [0 \ D_{2i}(-D_{2i}) \ 0 \ \cdots \ 0]^T$

$\vdots$

$\mathbf{D}_{N-1, i} = [0 \ \cdots \ 0 \ D_{N-1, i} \ (-D_{N-1, i})]^T$

$\mathbf{D}_{Ni} = [0 \ \cdots \ 0 \ D_{Ni}]^T$

$$D_{ki} = \frac{\partial A_{ki}}{\partial T_k} v_{ki} = -\frac{v_{ki}}{K_{ki}} A_{ki} \frac{\partial K_{ki}}{\partial T_k}$$

3. Elements of $\mathbf{E}_k = -(\partial \mathbf{R}/\partial \theta_k)\mathbf{V}$

$\mathbf{E}_1 = [E_1 \ (-E_1) \ 0 \ \cdots \ 0]^T$

$\mathbf{E}_2 = [0 \ E_2 \ (-E_2) \ 0 \ \cdots \ 0]^T$

$\vdots$

$\mathbf{E}_{N-1} = [0 \ \cdots \ 0 \ E_{N-1} \ (-E_{N-1})]^T$

$\mathbf{E}_N = [0 \ \cdots \ 0 \ E_N]^T$

$$E_k = \frac{\partial R_k}{\partial \theta_k} V_k = \left(\frac{L_k}{V_k} \right)_a V_k$$

Consequently

$$\frac{\partial V_j}{\partial T_k} = 0 \qquad \begin{array}{l} (j = 1, 2, \ldots, N) \\ (k = 1, 2, \ldots, N) \end{array} \qquad (4\text{-}27)$$

Comparison of Eqs. (3-7), (4-24), and (4-25) shows that the same tridiagonal matrix $\mathbf{A}_i$ appears in the expressions for computing the component-flow rates as appears in the expressions for calculating the partial derivatives of the v_{ji}'s with respect to the temperatures and the θ's. Similarly, the tridiagonal matrix $\mathbf{R}$ which appears in the total-material balances [Eq. (4-10)] also appears in the expression for calculating the partial derivatives of the V_j's with respect to the θ's [Eq. (4-26)]. This characteristic gives rise to several possible procedures for solving for the component-flow rates, the total-flow rates, and their partial derivatives. Three of the possible methods for solving these equations are (1) the use of the recurrence formulas for tridiagonal matrices [see Eqs. (2-20) through (2-22)], (2) the use of inverse matrices, and (3) the use of the procedure called "**LU** factorization." Only the fastest of these, **LU** factorization, is described. The method of **LU** factorization is best described by use of the following numerical example.

Example 4-1 (Hess et al.,[13] by courtesy *Hydrocarbon Process.*) Lean oil at the rate $L_0 = 1$ g mol/s enters the top plate of a three-plate absorber. Rich gas at the rate $V_4 = 2$ g mol/s enters the absorber at the bottom of the column on plate 3 (see Fig. 4-3). (*a*) For the first trial, take $(L_j/V_j)_a = 1/2$ for all j and take the assumed values for the θ_j's to be $\theta_1 = \theta_2 = \theta_3 = 1$. Find the corresponding values of the total flow rates V_1, V_2, and V_3. Solve by gaussian elimination, and store the negative values of each multiplier in the same location as the zero produced by the given multiplier. (*b*) Show that the resulting matrix produced by the operations described in part (*a*) may be restated in terms of a lower triangular matrix **L** and an upper triangular matrix **U** which are related to the original matrix **R** as follows

$$\mathbf{LU} = \mathbf{R}$$

(*c*) Show that the multipliers saved in part (*a*) may be used to operate on the elements of the original matrix $(-\mathscr{F})$ to produce the form possessed by this matrix at the end of the gaussian elimination process. (*d*) Show that the multipliers saved in part (*a*) and the final form of the **R** matrix at the end of the gaussian elimination process may be used to compute the partial derivatives for the V_j's with respect to θ_1.

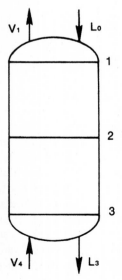

Figure 4-3 The three-plate absorber used in Example 4-1.

SOLUTION (a) The total-material balances may be stated as follows in terms of the vapor rates and the R_j's

$$-(1 + R_1)V_1 + V_2 = -L_0$$
$$R_1 V_1 - (1 + R_2)V_2 + V_3 = 0$$
$$R_2 V_2 - (1 + R_3)V_3 = -V_4$$

These equations may be stated as a matrix equation of the form

$$\mathbf{RV} = \mathbf{C}$$

where $-\mathscr{F}$ of Eq. (4-10) is replaced by $\mathbf{C}$ in this example to give the following

$$\begin{bmatrix} -(1 + R_1) & 1 & 0 \\ R_1 & -(1 + R_2) & 1 \\ 0 & R_2 & -(1 + R_3) \end{bmatrix} \begin{bmatrix} V_1 \\ V_2 \\ V_3 \end{bmatrix} = \begin{bmatrix} -L_0 \\ 0 \\ -V_4 \end{bmatrix}$$

On the basis of the values given above, $R_1 = R_2 = R_3 = 1/2$, and the fact that $L_0 = 1$ and $V_4 = 2$, matrices $\mathbf{R}$ and $\mathbf{C}$ become

$$\begin{bmatrix} -3/2 & 1 & 0 \\ 1/2 & -3/2 & 1 \\ 0 & 1/2 & -3/2 \end{bmatrix} \begin{bmatrix} -1 \\ 0 \\ -2 \end{bmatrix} \tag{A}$$

Step 1. To eliminate $1/2$ (the first element in the second row) multiply row 1 by $1/3$ and add this result to row 2. Then replace row 2 by the result obtained by this addition to give

$$\begin{bmatrix} -3/2 & 1 & 0 \\ (-1/3) & -7/6 & 1 \\ 0 & 1/2 & -3/2 \end{bmatrix} \begin{bmatrix} -1 \\ -1/3 \\ -2 \end{bmatrix}$$

The negative value of the multiplier $1/3$ has been stored in the location where a zero was obtained by use of this multiplying factor. To emphasize that a zero and not $-1/3$ is to be used in subsequent row operations, the multiplier is enclosed by parentheses.

Step 2. To eliminate $1/2$ (the second element of row 3) multiply row 2 by $3/7$ and add this result to row 3. Then replace row 3 by the result obtained by this addition and store the multiplier $3/7$ as before to give the final matrix equation

$$\begin{bmatrix} -3/2 & 1 & 0 \\ (-1/3) & -7/6 & 1 \\ 0 & (-3/7) & -15/14 \end{bmatrix} \begin{bmatrix} V_1 \\ V_2 \\ V_3 \end{bmatrix} = \begin{bmatrix} -1 \\ -1/3 \\ -15/7 \end{bmatrix} \tag{B}$$

which may be represented symbolically as

$$\mathbf{R'V} = \mathbf{C'} \tag{C}$$

where $\mathbf{R}'$ is the upper triangular form of the matrix $\mathbf{R}$, and $\mathbf{C}'$ is the resulting form of $\mathbf{C}$ obtained by transforming $\mathbf{R}$ to $\mathbf{R}'$.

Application of the multiplication rule[7] to row 3 [with the understanding that the actual element where $(-3/7)$ is stored is zero] gives

$$-(15/14)V_3 = -15/7$$

Thus

$$V_3 = 2$$

Application of the multiplication rule to row 2 gives

$$-(7/6)V_2 + V_3 = -1/3$$

After $V_3 = 2$ has been substituted into this equation, one finds

$$V_2 = 2$$

Similarly, application of the multiplication rule to the top row gives

$$-(3/2)V_1 + V_2 = -1$$

and thus

$$V_1 = 2$$

(b) Let $\mathbf{U}$ denote the upper triangular form of $\mathbf{R}$ obtained in part (a).

$$\mathbf{U} = \begin{bmatrix} -3/2 & 1 & 0 \\ 0 & -7/6 & 1 \\ 0 & 0 & -15/14 \end{bmatrix}$$

Let $\mathbf{L}$ be the lower triangular matrix formed by use of the multipliers of part (a) and by the use of elements of unity along the central diagonal, that is,

$$\mathbf{L} = \begin{bmatrix} 1 & 0 & 0 \\ -1/3 & 1 & 0 \\ 0 & -3/7 & 1 \end{bmatrix}$$

Now carry out the multiplication of $\mathbf{L}$ times $\mathbf{U}$

$$\mathbf{LU} = \begin{bmatrix} 1 & 0 & 0 \\ -1/3 & 1 & 0 \\ 0 & -3/7 & 1 \end{bmatrix} \begin{bmatrix} -3/2 & 1 & 0 \\ 0 & -7/6 & 1 \\ 0 & 0 & -15/14 \end{bmatrix}$$

$$= \begin{bmatrix} -3/2 & 1 & 0 \\ 1/2 & -3/2 & 1 \\ 0 & 1/2 & -3/2 \end{bmatrix}$$

The matrix obtained by multiplication is seen to be the original matrix $\mathbf{R}$. Thus the calculational procedure described in part (a) and the formulation of the matrices $\mathbf{L}$ and $\mathbf{U}$ from these results does produce a factorization of $\mathbf{R}$, and for this reason the process is called $\mathbf{LU}$ factorization.

(c) It will now be demonstrated that most of the operations which would be required to solve any other set of equations that differ from the original set by only the elements of the **C** matrix have already been performed and need not be repeated. The results of these operations are contained in the triangularized form of **R** in which the multipliers have been saved as demonstrated by Eq. (B).

First observe that the operations required to transform **R** into an upper triangular matrix are the same regardless of the particular set of elements appearing in the **C** matrix. Next, examine steps 1 and 2 of part (a) and observe that the elements of the **C** matrix are operated on by only the multipliers used to transform **R** into the upper triangular matrix **R′**. Thus, the final set of elements shown in the derived matrix **C′** in Eqs. (B) and (C) can be obtained by commencing with the original set of elements in **C** and the multipliers stored in **R′** and performing the operations shown below. Let the elements of **C** and **C′** be denoted by C_1, C_2, C_3 and C_1', C_2', C_3', respectively. Examination of Eqs. (A) and (B) shows that

$$C_1 = C_1' = -1$$

By examination of the operations in step 1, it is seen that C_2' may be computed as follows

$$C_2' = C_2 - (-1/3)C_1' = 0 - (-1/3)(-1) = -1/3$$

Similarly, an examination of the operations in step 2 shows that

$$C_3' = C_3 - (-3/7)C_2' = -2 - (-3/7)(-1/3) = -15/7$$

Thus, for any set of initial values of C_1, C_2, and C_3, the final set of values C_1', C_2', and C_3' may be found by performing the above calculations. The method described is readily extended for the general case where the matrix **R** contains any number of elements.

(d) By use of the value obtained for V_1 in part (a) and the formulas given in Table 4-3 for the matrix **E₁** of the matrix equation **R** $\partial V/\partial \theta = E_1$, the elements of **E₁** are evaluated as follows

$$
\mathbf{E}_1 =
\begin{bmatrix}
\dfrac{\partial R_1}{\partial \theta_1} V_1 \\[2ex]
-\dfrac{\partial R_1}{\partial \theta_1} V_1 \\[2ex]
0
\end{bmatrix}
=
\begin{bmatrix}
\left(\dfrac{L_1}{V_1}\right)_a V_1 \\[2ex]
-\left(\dfrac{L_1}{V_1}\right)_a V_1 \\[2ex]
0
\end{bmatrix}
=
\begin{bmatrix}
1 \\[2ex]
-1 \\[2ex]
0
\end{bmatrix}
$$

To determine the values which the elements of **E₁** would take on by the gaussian elimination process required to transform **R** into the upper triangular matrix **R′**, one may perform on **E₁** the same set of operations performed on **C** in part (a). Let the elements of **E₁** be denoted by $E_{1,1}$, $E_{1,2}$, $E_{1,3}$ and the elements of **E₁′** by $E_{1,1}'$, $E_{1,2}'$, $E_{1,3}'$. By use of the elements of **E₁** and the multipliers stored in **R′** [see Eq. (B)] the elements of $E_{1,1}'$, $E_{1,2}'$, and $E_{1,3}'$

may be computed as follows

$$E'_{1,1} = E_{1,1} = 1$$

$$E'_{1,2} = E_{1,2} - (-1/3)E'_{1,1} = -1 - (-1/3)(1) = -2/3$$

$$E'_{1,3} = E_{1,3} - (-3/7)E'_{1,2} = 0 - (-3/7)(-2/3) = -2/7$$

Observe that the form of $\mathbf{R}'$ obtained by performing gaussian elimination on $\mathbf{RV} = \mathbf{C}$ to give $\mathbf{R'V} = \mathbf{C}'$ is the same as that obtained by performing gaussian elimination on $\mathbf{R} \, \partial\mathbf{V}/\partial\theta_1 = \mathbf{E}_1$ to give $R' \, \partial\mathbf{V}/\partial\theta_1 = \mathbf{E}'_1$. Thus, the final result may be obtained by use of $\mathbf{R}'$ as given by Eq. (B) and the elements of $\mathbf{E}'_1$ found above

$$\begin{bmatrix} -3/2 & 1 & 0 \\ (-1/3) & -7/6 & 1 \\ 0 & (-3/7) & -15/14 \end{bmatrix} \begin{bmatrix} \dfrac{\partial V_1}{\partial \theta_1} \\ \dfrac{\partial V_2}{\partial \theta_1} \\ \dfrac{\partial V_3}{\partial \theta_1} \end{bmatrix} = \begin{bmatrix} 1 \\ -2/3 \\ -2/7 \end{bmatrix}$$

Application of the multiplication rule to the bottom row gives

$$\frac{\partial V_3}{\partial \theta_1} = (-2/7)(-14/15) = 4/15$$

Similarly, for the second row

$$-7/6 \frac{\partial V_2}{\partial \theta_1} + \frac{\partial V_3}{\partial \theta_1} = -2/3$$

and

$$\frac{\partial V_2}{\partial \theta_1} = (-2/3 - 4/15)(-6/7) = 4/5$$

and for the first row

$$\frac{\partial V_1}{\partial \theta_1} = (1 - 4/5)(-2/3) = -2/15$$

Use of the **LU** factorization technique (demonstrated above) for the calculation of the partial derivatives of the v_{ji}'s with respect to the θ_j's and T_j's materially reduces the time required to compute the partial derivatives of the F_j's and G_j's which appear in the jacobian matrix. Additional speed is also achieved by performing numerical operations on only those elements lying on the principal and two adjacent diagonals of the tridiagonal matrices. The remaining elements are zero at the outset of the gaussian elimination process and are not altered by this process. A summary of the steps of the proposed calculational procedure follows.

Calculational Procedure

1. Assume a set of temperatures $\{T_j\}$ and flow ratios $\{(L_j/V_j)_a\}$. For each trial after the first one, take the assumed values of the T_j's and the (L_j/V_j)'s to be equal to the most recent values found in step 6. Assume $\theta_1 = \theta_2 = \cdots = \theta_N = 1$.

2. Solve the total-material balances Eq. (4-10) for the V_j's. Save the upper triangular matrix $\mathbf{R}'$ and the multipliers used to obtain it.

3. Use $\mathbf{R}'$ and the vectors $\mathbf{E}_k$ ($k = 1, 2, \ldots, N$) given in Table 4-3 to compute the partial derivatives $\partial V/\partial \theta_k$ ($k = 1, 2, \ldots, N$) by use of Eq. (4-26) and $\mathbf{LU}$ factorization.

4. Solve the component-material balances [Eq. (2-18)] for the v_{ji}'s. Then compute the l_{ji}'s by use of the equilibrium relationship, $l_{ji} = A_{ji} v_{ji}$, where A_{ji} is defined by Eq. (4-11). Save the upper triangular matrix $\mathbf{A}'_i$ for each component and the multipliers used to obtain it.

5. Use the upper triangular matrix $\mathbf{A}'_i$ and the expressions given for the column vectors $\mathbf{C}_{ki}$ ($k = 1, 2, \ldots, N$) in Table 4-3 to compute the partial derivatives $\partial \mathbf{v}_i/\partial \theta_k$ ($k = 1, 2, \ldots, N$) by $\mathbf{LU}$ factorization. The partial derivatives $\partial \mathbf{v}_i/\partial T_k$ ($k = 1, 2, \ldots, N$) may be computed by $\mathbf{LU}$ factorization by use of the upper triangular matrix $\mathbf{A}'_i$ and the expressions given for the column vectors $\mathbf{D}_{ki}$ ($k = 1, 2, \ldots, N$) in Table 4-3.

6. Use the results obtained by steps 1 through 5 to evaluate the functions $\{F_j\}$, $\{G_j\}$, and their partial derivatives. Then solve Eq. (4-13) for the set of corrections $\Delta \mathbf{X}$. If the convergence criteria are satisfied by the values of the functions or the values of the variables so obtained, convergence is said to have been achieved. If the convergence criteria are not satisfied, then correct the assumed set of variables $\mathbf{X}_n$ to give the next assumed set $\mathbf{X}_{n+1}$ for step 1 as follows

$$\mathbf{X}_{n+1} = \mathbf{X}_n + \beta \Delta \mathbf{X}$$

First a value of $\beta = 1$ is used, and if any one of the corresponding values of the independent variables is not positive or if any one of the temperatures is not within the range of the curve fits, a value of $\beta = 1/2$ is used. The reduction of β by a factor of $1/2$ is continued until the θ's are all positive and all of the temperatures are within the range of the curve fits.

A variety of methods were tested for adjusting the calculated values of the variables, but none of them appeared to be significantly better than the relatively simple halving method described above in step 6.

In all of the applications of the $2N$ Newton-Raphson method which follow, both the functions and the variables were normalized for the purpose of reducing roundoff error. The functions $\{F_j\}$ and $\{G_j\}$ are stated in a normalized form. The definition of the θ_j's contains a normalizing factor, namely, $(L_j/V_j)_a$. Temperatures were normalized by dividing each temperature by some base temperature. Although other more precise methods of normalization may be used such as the one presented in Chap. 15, the above relatively simple procedure was satisfactory for the examples presented in this chapter.

When both the F_j's and G_j's are normalized, the sum of their squares is meaningful as a measure of how far a given set of trial values of the variables are from the solution set. In the examples presented in this chapter, the convergence criterion used was that ϕ, the square root of the average of the euclidean norm, must be less than some small preassigned number ε, say $\varepsilon = 10^{-4}$.

$$\phi = \frac{1}{2N} \sum_{j=1}^{N} [(F_j^2 + G_j^2)]^{1/2} \tag{4-28}$$

where N is equal to the total number of stages.

Numerical examples The statement of a typical absorber problem is presented as Example 4-2 in Table 4-4. To demonstrate the characteristics exhibited by procedure 1, the calculated values of the temperatures and flow rates are presented in Table 4-5. A typical stripper problem is presented as Example 4-3 in Table 4-6, and the intermediate trial results as well as the solution sets of temperatures, flow rates, and product flow rates are presented in Table 4-7.

To demonstrate the application of the $2N$ Newton-Raphson method to reboiled absorbers, Example 4-4 was solved. The statement and solution of this example appears in Table 4-8. The application of the $2N$ Newton-Raphson method to conventional and complex distillation columns is illustrated by Examples 4-5 and 4-6. Example 4-5 which involves a conventional distillation column, is a restatement of Example 2-7 (see Table 2-2). The

Table 4-4 Statement of an absorber problem, Example 4-2

Component	Rich gas $v_{N+1, i}$(lb mol/h)	Lean oil l_{0i}(lb mol/h)	Other specifications
CO_2	0.4703	0.0	$T_0 = 2.9°F$, $T_{N+1} = 0°F$, $N = 8$, and
N_2	0.1822	0.0	$P = 800$ lb/in^2 abs. Initial
CH_4	88.7000	0.0	temperature profile to be constant
C_2H_6	6.6747	0.0	at $T_j = 25°F$ for all j ($j = 1, 2,$
C_3H_8	2.7786	0.0015	$..., N$). The initial vapor rate
$i\text{-}C_4H_{10}$	0.6375	0.0006	profile is to be constant at
$n\text{-}C_4H_{10}$	0.3655	0.0013	$V_j = 90.88$ ($j = 1, 2, ..., 8$), and
$i\text{-}C_5H_{12}$	0.1158	0.0067	the liquid rates are $L_j = 6.3092$
$n\text{-}C_5H_{12}$	0.0505	0.0061	($j = 1, 2, ..., 7$), and $L_8 = 15.42$
C_6H_{14}	0.0146	0.1495	Use the K values and enthalpies
C_7H_{16}	0.0081	0.5736	given in Tables B-5 and B-6
C_8H_{18}	0.0020	1.8214	
C_9H_{20}	0.0	1.6866	
$C_{10}H_{22}$	0.0	2.0619	
	100.00	6.3092	

Table 4-5 Convergence characteristics exhibited by procedure 1 in the solution of the absorber problem, Example 4-2

I. Temperatures, °R

Plate	Initial profile	Trial			
		2	3	4	5
1	485.0	484.54	487.52	487.92	487.94
2	....	485.72	490.41	491.03	491.06
3	....	484.97	490.06	490.87	490.92
4	....	483.51	488.62	489.56	489.61
5	....	481.85	486.61	487.62	487.68
6	....	480.08	484.08	485.04	485.11
7	....	478.13	480.67	481.42	481.48
8	485.0	476.54	475.33	475.32	475.33

II. θ_j's Computed at the end of the trial indicated

Plate	Trial			
	1	2	3	4
1	1.4709	1.2128	1.0397	1.0013
2	1.4669	1.1944	1.0348	1.0010
3	1.4829	1.2031	1.0374	1.0011
4	1.4974	1.2142	1.0413	1.0013
5	1.5124	1.2291	1.0472	1.0018
6	1.5296	1.2516	1.0575	1.0028
7	1.5587	1.2888	1.0779	1.0056
8	1.0353	1.0255	1.0006	0.99988

III. Solution sets of product rates

Component	v_{1i}	l_{Ni}
CO_2	0.34840	0.12190
N_2	0.18037	0.19104×10^{-2}
CH_4	0.82841×10^2	0.58600×10^1
C_2H_6	0.46875×10^1	0.19872×10^1
C_3H_8	0.67786	0.21022×10^1
$i\text{-}C_4H_{10}$	0.90314×10^{-2}	0.62915
$n\text{-}C_4H_{10}$	0.11957×10^{-2}	0.36556
$i\text{-}C_5H_{12}$	0.17761×10^{-2}	0.12069
$n\text{-}C_5H_{12}$	0.11812×10^{-2}	0.55469×10^{-1}
C_6H_{14}	0.11098×10^{-1}	0.15298
C_7H_{16}	0.17115×10^{-1}	0.56458
C_8H_{18}	0.23692×10^{-1}	0.12997×10^1
C_9H_{20}	0.10240×10^{-1}	0.16763×10^1
$C_{10}H_{22}$	0.60846×10^{-2}	0.20558×10^1
Total	88.816	17.494

Table 4-6 Statement of a stripper problem, Example 4-3

Component	$v_{N+1, i}$ (lb mol/h)	l_{0i} (lb mol/h)
Steam	13.47	0
CH_4	0	0.01
C_2H_6	0	1.17
C_3H_8	0	1.30
$n\text{-}C_4H_{10}$	0	2.38
$n\text{-}C_5H_{12}$	0	1.75
C_6H_{14}	0	2.35
C_7H_{16}	0	2.55
500	0	82.24
	13.47	93.20

Other specifications

The stripper has eight plates and it is to be operated at a column pressure of 50 lb/in^2 abs. The K values and the enthalpies of the hydrocarbons were taken from Tables B-7 and B-2. The K values for steam given by $(K/T)^{1/3} = 0.191302 - 0.692641 \times 10^{-4}T + 0.8664775 \times 10^{-6}T^2 - 0.6587865 \times 10^{-9}T^3$ (where T is in °R) were used. This curve fit is based on the K values given by Hadden.[12] The following curve fits were used for the vapor and liquid enthalpies of steam:

$$h^{1/2} = 27.89976 + 0.9787701 \times 10^{-2}T - 0.1599299 \times 10^{-5}T^2$$

$$H^{1/2} = 118.3686 + 0.4152569 \times 10^{-1}T - 0.6785253 \times 10^{-5}T^2$$

(where T is in °R and h and H are in Btu per lb mol). These curve fits were based on data presented by Smith and Van Ness.[21]

The rich oil enters at 370°F and the steam at 500°F. Take the initial temperature profile to be linear between $T_1 = 340$°F and $T_8 = 375$°F. Take the vapor profile to be constant at $V_j = 80$ ($j = 1, 2, \ldots, 8$).

corresponding solution sets of temperatures, vapor rates, and product rates presented in Tables 2-3 through 2-5 were obtained. Similarly, Example 4-6 (see Table 4-10) involves a complex distillation column and consists of a restatement of Example 3-2 (see Table 3-1) in a form which is convenient for the application of the $2N$ Newton-Raphson method. The solution sets of temperatures, total-flow rates, and component-flow rates are the same as those given in Table 3-3.

Also, to demonstrate the effect of the number of plates on the computer time required to obtain a solution, the first column of the system of columns of Example 3-3 was solved as a complex column and referred to as Example 4-7. The recycle stream B_2 (see Fig. 3-12) was taken to be an independent feed and its composition was taken to be the solution set of values given in Table 3-5. The solution sets of temperatures, total-flow rates, and composition were the same as those presented for Example 3-3 in Table 3-5.

A comparison of the computer times and number of trials required to

Table 4-7 Convergence characteristics exhibited by the Newton-Raphson method (procedure 1) in the solution of the stripper problem, Example 4-3

	I. Temperature profiles (°R)			
		Trial		
Plate	Initial profile	2	7	8
1	800.0	827.60	827.26	826.28
2	805.0	827.87	827.46	827.52
3	810.0	827.68	827.17	827.48
4	815.0	827.55	827.26	827.42
5	820.0	827.62	827.41	827.37
6	825.0	828.79	827.33	827.32
7	830.0	832.48	827.29	827.26
8	835.0	822.20	827.36	827.34

	II. θ_j's computed at the end of trial indicated			
		Trial		
Plate	1	2	7	8
1	1.8715	1.6876	0.00877	0.99998
2	1.5446	2.3695	0.99669	0.99996
3	1.2804	1.8663	0.99158	0.99990
4	1.2064	1.4493	0.95722	0.99946
5	1.1858	1.2503	0.92400	0.00067
6	1.1806	1.1696	1.0065	1.0001
7	1.1135	1.1266	1.0047	1.0000
8	1.3040	1.1583	1.0016	1.0000

	III. Solution sets of product rates	
Component	v_{1i}	l_{Ni}
Steam	11.336	2.1339
CH_4	0.010	0.0
C_2H_6	0.17	0.0
C_3H_8	1.30	0.0
$n\text{-}C_4H_{10}$	2.8149	0.01506
$n\text{-}C_5H_{12}$	1.5479	0.20204
C_6H_{14}	2.3499	0.0
C_7H_{16}	2.5499	0.0
500	1.0992	81.140

Table 4-8 Statement and solution of the reboiled absorber problem, Example 4-4

I. Statement

Component	FX_i (lb mol/h)	l_{oi} (lb mol/h)
CH_4	65	0
C_2H_6	13	0
C_3H_8	1	0
$i\text{-}C_4H_{10}$	1	0
$n\text{-}C_5H_{12}$	20	0
$n\text{-}C_8H_{18}$	0	100

Other specifications

$P = 300$ lb/in^2 abs. $f = 5$, $N = 11$, $T_0 = 100°F$. The feed F enters on plate $f = 5$ as a vapor at its dew-point temperature at the column pressure. The column has 10 plates plus a reboiler. The reboiler duty is specified at $Q_R = 3 \times 10^6$ Btu/h. The equilibrium data are given in Table B-1 and the enthalpy data are given in Table B-2. For the first trial, take the vapor rates to be: $V_1 = 75$, $V_j = 150$ ($j = 2, 3, 4$), $V_j = 50$ ($j = 5, 6, \ldots, 10$), $V_{11} = 77.925$ and take the corresponding liquid rates to be those obtained by making total material balances. Take the temperatures to be linear between $T_1 = 200°F$ and $T_{11} = 400°F$.

II. Solution

Stage	T_j (°R)	V_j (lb mol/h)	v_{1i}	b_i
1	562.25	73.537	64.995	0.44694×10^{-2}
2	568.57	86.627	8.4004	4.59966
3	578.54	88.924	0.72798×10^{-1}	0.92720
4	610.38	92.336	0.46200×10^{-2}	0.99538
5	612.82	114.002	0.39219×10^{-3}	19.999
6	618.34	153.312	0.15292	99.847
7	624.67	24.786		
8	631.12	36.248		
9	655.64	43.403		
10	758.21	51.182		
11	938.74	78.840		

solve these examples by procedure 1 and by procedures 2 and 3 (presented below) is given as a summary at the end of this chapter.

The results shown for these examples are in good agreement with the fact that the Newton-Raphson method is said to exhibit *quadratic convergence*. For a single variable problem, *quadratic convergence* means that the error for the nth trial is proportional to the square of the error for the previous trial. The error for the nth trial is defined as the correct value of the variable minus the value predicted by the nth trial. For a multivariable problem, quadratic convergence means that the norm of the errors given by

Table 4-9 Statement of Example 4-5: a conventional distillation column

1. *Feed.* The components, composition, and flow rates are the same as stated in Table 2-2 for Example 2-7.

2. *Other specifications.* Same as stated in Table 2-2 for Example 2-7 except for the following changes. Instead of specifying D as was done in Example 2-6, the ratio B/V_N is specified at $B/V_N = 0.55365$, which is the solution value obtained for Example 2-2. Again, as in Example 2-7 the reflux ratio is specified at $L_1/D = 2$. For the first trial, the temperatures are taken to be linear between $T_2 = 135°F$ and the reboiler temperature $T_{13} = 300°F$. The values of $(L_j/V_j)_a$ for the first trial are to be taken equal to unity for stages $j = 2, 3, \ldots, 12$.

the nth trial is proportional to the square of the norm for the $(n-1)$st trial. In summary, the Newton-Raphson method, under the mild restrictions stated in App. A, converges quadratically provided that the initial choice of variables lies within the region of convergence. The latter condition is commonly replaced by the less precise statement "provided that the initial choice of variables is close enough to the solution set."

Since the $2N$ Newton-Raphson method is an exact application of the Newton-Raphson method for all columns in the service of separating ideal solutions, convergence for all such problems can be assured; provided that suitable starting values for the T_j's and L_j/V_j's are selected (see App. A). The fact that no difficulty was encountered in the selection of the starting values of the variables of the $2N$ Newton-Raphson method suggests that the functions and variables selected possess a wide region of convergence. The use of other functions in the $2N$ Newton-Raphson method could reduce its region of convergence, and it could become difficult to find a suitable set of starting values. For example, the alternative form of the dew-point function given in Prob. 4-5(b) did result in a smaller region of convergence than that resulting from the use of the form of the dew-point function given by Eq. (4-4).

Applications of approximate formulations of the Newton-Raphson method such as the ones proposed by Sujata[22] and Holland[17] may fail to converge for some examples. For example, the $2N$ Newton-Raphson method converged for the test problem called Example 3 by Boyum[4] while the approximate methods of Sujata and Holland failed.

Table 4-10. Statement of Example 4-6: a complex column

1. *Feed.* The components, composition, and flow rates are the same as stated in Table 3-1 for Example 3-2.

2. *Other specifications.* Same as stated in Table 3-1 for Example 3-2 except for the following changes. The sidestream W_1 is withdrawn from plate $p = 11$. The reflux ratio $L_1/D = 2.25$. For B/V_N and W_1/L_p, the solution values obtained for Example 3-2 are specified; namely, $B/V_N = 0.31$, and $W_1/L_p = 0.12692$.

   For the first trial, assume that the temperatures are linear between $T_2 = 200°F$ and $T_{12} = 300°F$, and take $T_1 = 200°F$ and $T_{13} = 300°F$. Take the initial values of $(L_j/V_j)_a = 1$ for $j = 2, 3, \ldots, 12$.

Other Formulations of the Newton-Raphson Method for Ideal Solutions

A number of formulations of the Newton-Raphson method for the solution of distillation problems has been proposed. A brief description of each of several of these formulations wherein the number of independent variables ranges from N to $2N$ is given below. (Other formulations of the Newton-Raphson method in which compositions or component-flow rates are among the independent variables are described in Chap. 5.)

Greendstadt et al.[11] were among the first to apply the Newton-Raphson method to the solution of distillation problems. In 1961, Sujata[22] proposed an approximate application of the Newton-Raphson method for the solution of absorber and stripper problems. Sujata regarded the temperatures as the independent variables. For each set of assumed temperatures, the component-material balances and equilibrium relationships were solved by direct iteration (or successive iteration) for the solution sets of compositions and flow rates. These compositions and flow rates were then used in the enthalpy balances in the calculation of an improved set of temperatures by use of the Newton-Raphson method. In Sujata's application of the Newton-Raphson method to the enthalpy balances, the dependency of the flow rates and compositions on the temperature was neglected in the calculation of the partial derivatives. A version of the θ method, called the "single θ method" has been proposed.[17]

Another formulation of the Newton-Raphson method was proposed by Newman[20] in 1963 in which the total-flow rates of the liquid $\{L_j\}$ were taken as the independent variables and the corresponding sets of temperatures needed to satisfy the component-material balances and equilibrium relationships was found by successive application of the Newton-Raphson equations. The compositions and temperatures so obtained were used to solve the enthalpy balances explicitly for a new set of liquid rates. The procedure was then repeated by commencing with this most recent set of liquid rates $\{L_j\}$.

Following Newman, Boynton[3] also took the liquid rates $\{L_j\}$ to be the independent variables, and for each choice of these variables the temperatures required to satisfy the component-material balances, and equilibrium relationships were found by successive application of the Newton-Raphson method. The results so obtained were then used in the enthalpy balances to compute a new set of liquid rates by use of one application of the Newton-Raphson method. For the case where both the vapor and liquid phases form ideal solutions, Boynton's method constitutes an exact application of the Newton-Raphson method. All of the matrix equations solved by Boynton were of order N.

Another exact formulation of the Newton-Raphson method, called the multi-θ method, was proposed by Holland.[17] Two procedures for applying this method were presented. In both of the procedures, the matrices were all of order N. All derivatives were evaluated by use of the calculus of matrices.

In a series of papers, Tierney et al.[23, 24] proposed an application of the Newton-Raphson method in which the vapor rates $\{V_j\}$ and the temperatures $\{T_j\}$ were taken to be independent variables. The formulation was exact and the partial derivatives were evaluated by use of the calculus of matrices.

4-2 NUMERICAL METHODS FOR SOLVING THE $2N$ NEWTON-RAPHSON EQUATIONS

Instead of using analytical expressions for the evaluation of the partial derivatives of the F_j's and G_j's, they may be evaluated numerically. Procedures 2 and 3, presented below, make use of the numerical approximations of the partial derivatives. Also presented is a calculational procedure for systems of absorber-type and distillation-type columns which are interconnected by recycle streams.

Procedure 2. Solution of the Newton-Raphson Equations by Use of Broyden's Method[6]

Tomich[25] was the first to apply Broyden's method (developed in chap. 15) to the solution of distillation problems. Broyden's method is based on the use of numerical approximations of the partial derivatives appearing in the jacobian matrix. The approach proposed by Broyden permits the inverse of the jacobian matrix to be updated each trial after the first through the use of Householder's formula.[6] Thus, it is necessary to invert the jacobian matrix only once. Since approximate values for the partial derivatives are used, procedure 2 generally requires more trials than does procedure 1. However, since the evaluation of the partial derivatives and the inversion of the jacobian matrix are not generally required after the first trial of procedure 2, it requires less computer time per trial than does procedure 1.

For the general case of n independent equations in n unknowns, the Newton-Raphson method may be formulated in terms of n functions in n unknowns. For the kth trial, the resulting set of Newton-Raphson equations may be represented as follows

$$\mathbf{J}_k \, \Delta \mathbf{X}_k = -\mathbf{f}_k \tag{4-29}$$

where $\mathbf{J}_k$ is the square jacobian matrix of order n and $\Delta \mathbf{X}_k$ and $\mathbf{f}_k$ are conformable column vectors

$$\mathbf{J}_k = \begin{bmatrix} \dfrac{\partial f_1}{\partial x_1} & \cdots & \dfrac{\partial f_1}{\partial x_n} \\ \vdots & & \vdots \\ \dfrac{\partial f_n}{\partial x_1} & \cdots & \dfrac{\partial f_n}{\partial x_n} \end{bmatrix}$$

$$\Delta \mathbf{X}_k = \mathbf{X}_{k+1} - \mathbf{X}_k \qquad \mathbf{X}_k = [x_{1k}, x_{2k}, \ldots, x_{nk}]^T$$
$$\mathbf{f}_k = [f_{1k}, f_{2k}, \ldots, f_{nk}]^T$$

In the class of methods proposed by Broyden, the partial derivatives $\partial f_i / \partial x_j$ in the jacobian matrix $\mathbf{J}_k$ of Eq. (4-29) are generally evaluated only once. In each successive trial, the elements of the inverse of the jacobian matrix are corrected by use of the computed values of the functions. An algebraic example will be given after the calculational procedure proposed by Broyden has been presented.

Step 1. Assume an initial set of values for the variables, $\mathbf{X}_0$, and compute

$$\mathbf{f}_0 = \mathbf{f}(\mathbf{X}_0)$$

Step 2. Approximate the elements of $\mathbf{H}_0$, where $\mathbf{H}_0$ is defined as follows

$$\mathbf{H}_0 = -\mathbf{J}_0^{-1}$$

Broyden obtained a first approximation of the elements of $\mathbf{J}_0$ by use of the formula

$$\frac{\partial f_i}{\partial x_j} \cong \frac{f_i(x_j + h_j) - f_i(x_j)}{h_j}$$

where h_j was taken to be equal to $0.001\ x_j$.

Step 3. On the basis of the most recent values of $\mathbf{H}$ and $\mathbf{f}$, say $\mathbf{H}_k$ and $\mathbf{f}_k$, compute

$$\Delta\mathbf{X}_k = \mathbf{H}_k\mathbf{f}_k$$

Step 4. Find the s_k such that the euclidean norm of $\mathbf{f}(\mathbf{X}_k + s_k\ \Delta\mathbf{X}_k)$ is less than that of $\mathbf{f}(\mathbf{X}_k)$. First try $s_{k,\,1} = 1$ and if the following inequality is satisfied

$$\left[\sum_{i=1}^{N} f_i^2(\mathbf{X}_k + s_k\ \Delta\mathbf{X}_k)\right]^{1/2} < \left[\sum_{i=1}^{N} f_i^2(\mathbf{X}_k)\right]^{1/2}$$

proceed to step 5. Otherwise, compute $s_{k,\,2}$ by use of the following formula which was developed by Broyden

$$s_{k,\,2} = [(1 + 6\eta)^{1/2} - 1]/3\eta$$

where

$$\eta = \frac{\displaystyle\sum_{i=1}^{N} f_i^2(\mathbf{X}_k + s_k\ \Delta\mathbf{X}_k)}{\displaystyle\sum_{i=1}^{N} f_i^2(\mathbf{X}_k)}$$

If the norm is not reduced by use of $s_{k,\,2}$ after a specified number of trials through the complete procedure, return to step 2 and reevaluate the partial derivatives of $\mathbf{J}_k$ on the basis of $\mathbf{X}_k$. As pointed out by Broyden, other methods for picking s_k may be used. For example, s_k may be picked such that the euclidean norm of $\mathbf{f}$ is minimized.

Step 5. In the course of making the calculations in step 4, the following vectors will have been evaluated

$$\mathbf{X}_{k+1} = \mathbf{X}_k + s_k\ \Delta\mathbf{X}_k$$

$$\mathbf{f}_{k+1} = f(\mathbf{X}_{k+1})$$

Test $\mathbf{f}_{k+1}$ for convergence. If convergence has not been achieved, compute

$$\mathbf{Y}_k = \mathbf{f}_{k+1} - \mathbf{f}_k$$

Step 6. Compute

$$\mathbf{H}_{k+1} = \mathbf{H}_k - \frac{(\mathbf{H}_k \mathbf{Y}_k + s_k \, \Delta\mathbf{X}_k) \, \Delta\mathbf{X}_k^T \mathbf{H}_k}{\Delta\mathbf{X}_k^T \mathbf{H}_k \mathbf{Y}_k}$$

and return to step 3.

Example 4-8 consists of a simple algebraic example which illustrates the application of this method.

Example 4-8 (Hess et al.,[13] by courtesy *Hydrocarbon Process.*) It is desired to find the pair of positive roots that make $f_1(x, y) = 0$ and $f_2(x, y) = 0$, simultaneously

$$f_1(x, y) = x^2 - xy^2 - 2$$
$$f_2(x, y) = 2x^2 - 3xy^2 + 3$$

Take $x_0 = 1$ and $y_0 = 1$, and make one complete trial calculation as prescribed by steps 1 through 6.

SOLUTION:

Step 1. Since $x_0 = 1$, $y_0 = 1$

$$\mathbf{X}_0 = [1, \, 1]^T$$

and

$$f_{1,0} = f_1(\mathbf{X}_0) = f_1(1, \, 1) = 1 - 1 - 2 = -2$$
$$f_{2,0} = f_2(\mathbf{X}_0) = f_2(1, \, 1) = 2 - 3 + 3 = 2$$

Step 2. Take the increment h for computing the derivatives with respect to x to be

$$h = (0.001)x_0 = 0.001$$

Then

$$\frac{\partial f_1}{\partial x} = \frac{f_1(1.001, \, 1) - f_1(1, \, 1)}{0.001} = 1.001$$

For computing the derivatives with respect to y, take

$$h = (0.001)y_0 = 0.001$$

Then

$$\frac{\partial f_1}{\partial y} = \frac{f_1(1, \, 1.001) - f_1(1, \, 1)}{0.001} = -2.001$$

and

$$\frac{\partial f_2}{\partial x} = \frac{f_2(1.001, \, 1) - f_2(1, \, 1)}{0.001} = 1.002$$

$$\frac{\partial f_2}{\partial y} = \frac{f_2(1, \, 1.001) - f_2(1, \, 1)}{0.001} = -6.003$$

Then

$$\mathbf{J}_0 = \begin{bmatrix} 1.001 & -2.001 \\ 1.002 & -6.003 \end{bmatrix}$$

The inverse of $\mathbf{J}_0$ is found by gaussian elimination as follows. Begin with

$$\begin{bmatrix} 1.001 & -2.001 \\ 1.002 & -6.003 \end{bmatrix} \begin{bmatrix} 1 & 0 \\ 0 & 1 \end{bmatrix}$$

and carry out the necessary row operations to obtain

$$\begin{bmatrix} 1 & 0 \\ 0 & 1 \end{bmatrix} \begin{bmatrix} 1.4992 & -0.49975 \\ 0.25023 & -0.25000 \end{bmatrix}$$

Then

$$\mathbf{J}_0^{-1} = \begin{bmatrix} 1.4992 & -0.49975 \\ 0.25023 & -0.25000 \end{bmatrix}$$

and

$$\mathbf{H}_0 = -\mathbf{J}_0^{-1} = \begin{bmatrix} -1.4992 & 0.49975 \\ -0.25023 & 0.25000 \end{bmatrix}$$

Step 3. On the basis of the most recent values $\mathbf{H}$ and $\mathbf{f}$, the correction $\Delta \mathbf{X}$ is computed as follows

$$\Delta \mathbf{X}_0 = \mathbf{H}_0 \mathbf{f}_0 = \begin{bmatrix} -1.4992 & 0.49975 \\ -0.25023 & 0.25000 \end{bmatrix} \begin{bmatrix} -2 \\ 2 \end{bmatrix} = \begin{bmatrix} 3.9979 \\ 1.0005 \end{bmatrix}$$

Step 4. Take $s_{0,1} = 1$. Then

$$\mathbf{X}_0 + \Delta \mathbf{X}_0 = \begin{bmatrix} 1 \\ 1 \end{bmatrix} + \begin{bmatrix} 3.9979 \\ 1.0005 \end{bmatrix} = \begin{bmatrix} 4.9979 \\ 2.0005 \end{bmatrix}$$

and

$$f_1(\mathbf{X}_0 + \Delta \mathbf{X}_0) = f_1(4.9979, 2.0005) = 2.9774$$
$$f_2(\mathbf{X}_0 + \Delta \mathbf{X}_0) = f_2(4.9979, 2.0005) = -7.0468$$

Since

$$(2.9774)^2 + (-7.0468)^2 > (-2)^2 + (2)^2$$

compute

$$\eta = \frac{f_1^2(\mathbf{X}_0 + \Delta \mathbf{X}_0) + f_2^2(\mathbf{X}_0 + \Delta \mathbf{X}_0)}{f_1^2(\mathbf{X}_0) + f_2^2(\mathbf{X}_0)} = \frac{(2.9774)^2 + (-7.0468)^2}{(-2)^2 + (2)^2} = 7.31529$$

and

$$s_{0,2} = \frac{(1 + 6\eta)^{1/2} - 1}{3\eta} = 0.25974$$

Then

$$X_0 + 0.25974\ \Delta X_0 = \begin{bmatrix} 1 \\ 1 \end{bmatrix} + \begin{bmatrix} 1.03839 \\ 0.259865 \end{bmatrix} = \begin{bmatrix} 2.03839 \\ 1.259865 \end{bmatrix}$$

and

$$f_1(X_0 + s_{0,2}\ \Delta X_0) = f_1(2.03839, 1.259865) = -1.08040$$
$$f_2(X_0 + s_{0,2}\ \Delta X_0) = f_2(2.03839, 1.259865) = 1.60372$$

Thus, the criterion on f_k, namely,

$$(-1.08040)^2 + (1.6037)^2 < (-2)^2 + (2)^2$$

has been satisfied.

Step 5. If the convergence criterion is taken to be that the sum of the squares f_1 and f_2 is to be reduced to some small preassigned number ε, say $\varepsilon = 10^{-10}$, then this criterion has not been satisfied by $x = 2.0384$ and $y = 1.25986$. Then compute

$$Y_0 = f_1 - f_0 = \begin{bmatrix} -1.08040 \\ 1.60372 \end{bmatrix} - \begin{bmatrix} -2 \\ 2 \end{bmatrix} = \begin{bmatrix} 0.91960 \\ -0.39628 \end{bmatrix}$$

Step 6. Compute the following products which are needed to find H_1.

$$H_0 Y_0 = \begin{bmatrix} -1.4992 & 0.49975 \\ -0.25023 & 0.25000 \end{bmatrix} \begin{bmatrix} 0.91960 \\ -0.39628 \end{bmatrix} = \begin{bmatrix} -1.5767 \\ -0.32918 \end{bmatrix}$$

$$\Delta X_0^T H_0 = [3.9979, 1.0005] \begin{bmatrix} -1.4992 & 0.49975 \\ -0.25023 & 0.25000 \end{bmatrix} = [-6.2440, 2.2481]$$

$$H_0 Y_0 + s_{0,2}\ \Delta X_0 = \begin{bmatrix} -1.5767 \\ -0.32918 \end{bmatrix} + \begin{bmatrix} 1.03839 \\ 0.259865 \end{bmatrix} = \begin{bmatrix} -0.53831 \\ -0.06931 \end{bmatrix}$$

$$\Delta X_0^T H_0 Y_0 = [-6.2440, 2.2481] \begin{bmatrix} 0.91960 \\ -0.39628 \end{bmatrix} = -6.6328$$

Since

$$H_1 = H_0 - \frac{(H_0 Y_0 + s_0\ \Delta X_0)\ \Delta X_0^T H_0}{\Delta X_0^T H_0 Y_0}$$

it follows that

$$H_1 = \begin{bmatrix} -1.4992 & 0.49975 \\ -0.25023 & 0.25000 \end{bmatrix} - \begin{bmatrix} -0.50674 & 0.18245 \\ -0.06525 & 0.023491 \end{bmatrix}$$

$$= \begin{bmatrix} -0.99246 & 0.31730 \\ -0.18498 & 0.22651 \end{bmatrix}$$

and the next trial is commenced by returning to step 3 with H_1.

Examples 4-3 through 4-7 as well as examples presented below were solved by this procedure. These results are discussed in Sec. 4-4.

Procedure 3. Broyden-Bennett Algorithm

In the method proposed by Broyden, Householder's formula was used to obtain the formula for the inverted matrix shown in step 6. The second term on the right-hand side of the expression in step 6 contains the correction proposed by Broyden.

Instead of applying Householder's formula, the calculation of an inverse of the jacobian may be avoided altogether by use of the algorithm proposed by Bennett for updating the **LU** factors of the jacobian matrix. Example 4-9 will show that fewer numerical operations are required to compute the **LU** factors than are required to compute the inverse of a matrix. Bennett's algorithm is applied to the Broyden equations as follows.

Broyden's algorithm consists of successively updating of the jacobian matrix of the Newton-Raphson equations by use of the correction matrix $\mathbf{x}C\mathbf{y}^T$, that is,

$$\mathbf{J}_{k+1} = \mathbf{J}_k + \mathbf{x}_k C\mathbf{y}_k^T$$

where

$$C = \frac{1}{s_k \, \Delta\mathbf{X}_k^T \, \Delta\mathbf{X}_k} \qquad \text{(a scalar)}$$

$$\mathbf{x}_k = \mathbf{f}_{k+1} - (1 - s_k)\mathbf{f}_k$$

$$\mathbf{y}_k = \Delta\mathbf{X}_k$$

Since the jacobian matrix **J** may be stated in terms of its factors **L** and **U** as demonstrated in procedure 1, the above expression for $\mathbf{J}_{k+1}$ may be restated in the following form

$$\mathbf{L}_{k+1}\mathbf{U}_{k+1} = \mathbf{L}_k\mathbf{U}_k + \mathbf{x}_k C\mathbf{y}_k^T$$

Bennett proposed the algorithm presented in Fig. 4-4 for updating the matrices $\mathbf{L}_k$ and $\mathbf{U}_k$ to obtain the updated matrices $\mathbf{L}_{k+1}$ and $\mathbf{U}_{k+1}$. When Bennett's algorithm is used to make the Broyden correction, the following calculational procedure is used.

Step 1. Same as step 1 of procedure 2.

Step 2. The partial derivatives of $\mathbf{J}_0$ are found in the same manner as shown in step 2 of procedure 2. Then find the factors $\mathbf{L}_0$ and $\mathbf{U}_0$ of $\mathbf{J}_0$ such that

$$\mathbf{J}_0 = \mathbf{L}_0\mathbf{U}_0$$

as described by Hess et al.;[14] see also Conte and de Boor.[9]

Step 3. On the basis of $\mathbf{L}_k$, $\mathbf{U}_k$, and $\mathbf{f}_k$ (the most recent values of **L**, **U**, and **f**) compute $\Delta\mathbf{X}_k$ as follows

$$\mathbf{L}_k\mathbf{U}_k \, \Delta\mathbf{X}_k = -\mathbf{f}_k$$

Step 4. Same as step 4 of procedure 2.

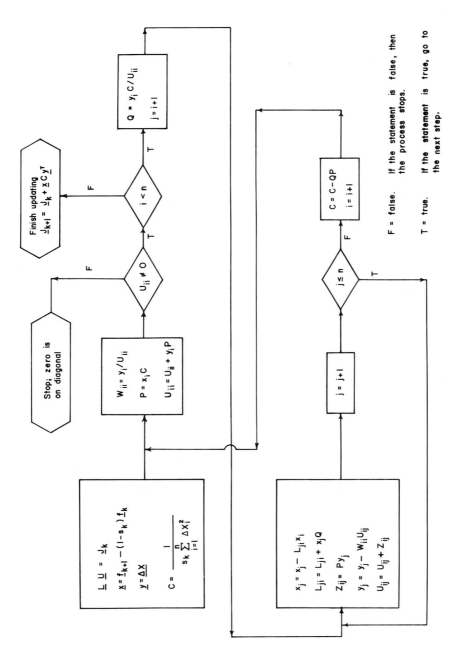

Figure 4-4 The Bennet algorithm for updating an LU factorization.[1]

Step 5. Test f_{k+1} for convergence. If convergence has not been achieved, compute

$$C = \frac{1}{s_k \, \Delta X_k^T \, \Delta X_k}$$

$$x_k = f_{k+1} - (1 - s_k) f_k$$

$$y_k^T = \Delta X_k^T$$

Step 6. Use the algorithm proposed by Bennett (see Fig. 4-4) to obtain the updated matrices L_{k+1} and U_{k+1} of L_k and U_k, where

$$L_{k+1} U_{k+1} = L_k U_k + x_k C y_k^T$$

Then return to step 3.

The use of Bennett's algorithm is illustrated by the following numerical example:

Example 4-9 (Hess et al.,[13] by courtesy *Hydrocarbon Process.*) Given the matrix A_k and its factors L_k and U_k

$$A_k = L_k U_k = \begin{bmatrix} 2 & 4 & -6 \\ -4 & -7 & 10 \\ 2 & 7 & -9 \end{bmatrix}$$

where

$$L_k = \begin{bmatrix} L_{1,1} & 0 & 0 \\ L_{2,1} & L_{2,2} & 0 \\ L_{3,1} & L_{3,2} & L_{3,3} \end{bmatrix} = \begin{bmatrix} 1 & 0 & 0 \\ -2 & 1 & 0 \\ 1 & 3 & 1 \end{bmatrix}$$

$$U_k = \begin{bmatrix} U_{1,1} & U_{1,2} & U_{1,3} \\ 0 & U_{2,2} & U_{2,3} \\ 0 & 0 & U_{3,3} \end{bmatrix} = \begin{bmatrix} 2 & 4 & -6 \\ 0 & 1 & -2 \\ 0 & 0 & 3 \end{bmatrix}$$

The upper and lower triangular matrices U_k and L_k may be found by use of the technique demonstrated in Example 4-1. It is desired to find L_{k+1} and U_{k+1} when A_k is corrected by adding the matrix $x_k \, C y_k^T$ to it; that is, find L_{k+1} and U_{k+1}, where

$$L_{k+1} U_{k+1} = L_k U_k + x_k C y_k^T$$

and

$$x_k = \begin{bmatrix} x_1 \\ x_2 \\ x_3 \end{bmatrix} = \begin{bmatrix} 1 \\ 2 \\ 3 \end{bmatrix} \qquad y_k = \begin{bmatrix} y_1 \\ y_2 \\ y_3 \end{bmatrix} = \begin{bmatrix} 3 \\ 2 \\ 1 \end{bmatrix} \qquad C = 1$$

SOLUTION The calculations are carried out according to the algorithm shown in Fig. 4-4. For this example, $n = 3$. Begin by making the calculations shown in the second square for $i = 1$.

When $i = 1$

$$W_{1,1} = y_1/U_{1,1} = 3/2$$
$$P = x_1 C = (1)(1) = 1$$
$$U_{1,1} = U_{1,1} + Py_1 = 2 + (1)(3) = 5$$
$$Q = y_1 C/U_{1,1} = (3)(1)/5 = 3/5$$

Since $i < n(1 < 3)$, take $j = i + 1$.
Then for $j = 2$, $i = 1$

$$x_2 = x_2 - L_{2,1}x_1 = 2 - (-2)(1) = 4$$
$$L_{2,1} = L_{2,1} + x_2 Q = -2 + (4)(3/5) = 2/5$$
$$Z_{1,2} = Py_2 = (1)(2) = 2$$
$$y_2 = y_2 - W_{1,1}U_{1,2} = 2 - (3/2)(4) = -4$$
$$U_{1,2} = U_{1,2} + Z_{1,2} = 4 + 2 = 6$$

Take $j = j + 1 = 3$.
Then for $j = 3$, $i = 1$

$$x_3 = x_3 - L_{3,1}x_1 = 3 - (1)(1) = 2$$
$$L_{3,1} = L_{3,1} + x_3 Q = 1 + (2)(3/5) = 11/5$$
$$Z_{1,3} = Py_3 = (1)(1) = 1$$
$$y_3 = y_3 - W_{1,1}U_{1,3} = 1 - (3/2)(-6) = 10$$
$$U_{1,3} = U_{1,3} + Z_{1,3} = -6 + 1 = -5$$

Since $j > n$, compute

$$C = C - QP = 1 - (3/5)(1) = 2/5$$

and take $i = i + 1 = 2$.
Then for $i = 2$

$$W_{2,2} = y_2/U_{2,2} = -4/1 = -4$$
$$P = x_2 C = (4)(2/5) = 8/5$$
$$U_{2,2} = U_{2,2} + Py_2 = 1 + (8/5)(-4) = -27/5$$
$$Q = y_2 C/U_{2,2} = (-4)(2/5)/(-27/5) = 8/27$$

Since $i < n$, take $j = i + 1 = 3$.

Then for $j = 3$, $i = 2$

$$x_3 = x_3 - L_{3,2}x_2 = 2 - (3)(4) = -10$$

$$L_{3,2} = L_{3,2} + x_3 Q = 3 + (-10)(8/27) = 1/27$$

$$Z_{2,3} = Py_3 = (8/5)(10) = 16$$

$$y_3 = y_3 - W_{2,2}U_{2,3} = 10 - (-4)(-2) = 2$$

$$U_{2,3} = U_{2,3} + Z_{2,3} = -2 + 16 = 14$$

Since $j > n$, compute

$$C = C - QP = 2/5 - (8/27)(8/5) = -2/27$$

and take $i = i + 1 = 3$.
Then for $i = 3$

$$W_{3,3} = y_3/U_{3,3} = 2/3$$

$$P = x_3 C = (-10)(-2/27) = 20/27$$

$$U_{3,3} = U_{3,3} + y_3 P = 3 + (2)(20/27) = 121/27$$

Since $i = 3$, the process has been completed and the final result is given by

$$A + \mathbf{x}C\mathbf{y}^T = \mathbf{LU} = \begin{bmatrix} 1 & 0 & 0 \\ 2/5 & 1 & 0 \\ 11/5 & 1/27 & 1 \end{bmatrix} \begin{bmatrix} 5 & 6 & -5 \\ 0 & -27/5 & 14 \\ 0 & 0 & 121/27 \end{bmatrix}$$

Note that, since the elements of the lower triangular factor $\mathbf{L}$ along the diagonal are always equal to unity, it was not necessary to make any correction to them in the algorithm.

Less time is consumed by procedure 3 than by procedure 1. Calculation of the $\mathbf{LU}$ factors of the matrix $\mathbf{J}$ in step 2 of procedure 3 requires approximately $n^3/3$ operations, whereas the calculation of the inverse of $\mathbf{J}$ in step 2 of procedure 2 requires approximately n^3 operations, where the matrix $\mathbf{J}$ is a square matrix of order n. To update the $\mathbf{LU}$ factors in step 6 of procedure 3 by use of Bennett's algorithm requires approximately $2n^2$ operation, whereas approximately $3n^2$ operations are required to update the inverse of $\mathbf{J}$ by use of Householder's formula as proposed by Broyden in step 6 of procedure 2.

4-3 SYSTEMS OF COLUMNS

Two approaches exist for solving problems involving systems of columns, the "column modular method" and the "system modular method" proposed by Hess.[15] In the column modular approach, the equations for each column of a system are solved in succession, and in the system modular approach, the complete set of equations for the system are solved simultaneously. While the system modular approach may be the ultimate method for solving problems involving systems of interconnected columns, the column modular approach appears to be the most realistic approach at the present time.

Solution of Systems of Columns by Use of the Column Modular Method

In the column modular approach, the equations for each column are solved by use of the most efficient procedure for each column. After one trial has been made on each column, the terminal flow rates are placed in component-material balance and in agreement with the specified values of the terminal flow rates by use of the " capital Θ method " for systems. The entire calculational process is repeated until convergence has been achieved.

To illustrate the application of the column modular approach, consider the particular system of columns shown in Fig. 4-5, which consists of a reboiled absorber (column 1) and a distillation column (column 2). For such a system, a combination of the θ method and the $2N$ Newton-Raphson method is recommended. The $2N$ Newton-Raphson method is used for solving the reboiled absorber and the θ method is recommended for the distillation column. At the end of one complete trial for each of the two columns, the capital Θ method is applied to place the system in component-material balance and in agreement with the specified values of the terminal flow rates.

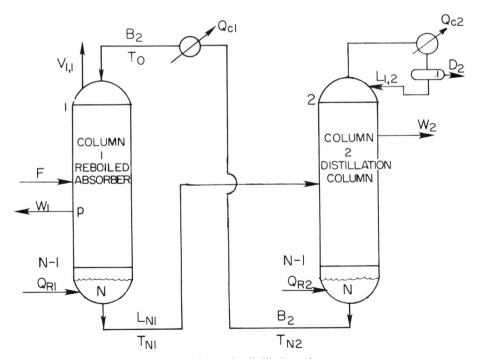

Figure 4-5 A system of a reboiled absorber and a distillation column.

Suppose that the total flow rates L_{N1}, D_2, W_1, and W_2 are specified. These specifications in turn fix V_1 and B_2. The specifications for the reboiled absorber are as follows: F, $\{X_i\}$, the thermal condition of F, f, N, and the column pressure. The specifications for the distillation column are as follows: N, type of condenser, f, L_1, D, and the common pressure.

In the application of the $2N$ Newton-Raphson method, V_1 is regarded as specified and the temperature T_0 of B_2 entering column 1 or the intercooler duty Q_C, is taken to be the new independent variable. In this case the $2N + 1$ variables X and the $2N + 1$ functions f are as follows

$$X = [\theta_1, \theta_2, \ldots, \theta_N, T_0, T_1, T_2, \ldots, T_N]^T$$
$$f = [f_0, F_1, F_2, \ldots, F_N, G_1, G_2, \ldots, G_N]^T \tag{4-30}$$

Corresponding to the new independent variable T_0 for the absorber, the new function f_0 is introduced to express the condition that the specified value of V_1, denoted by $(V_1)_{sp}$, must be equal to the calculated value at convergence, that is

$$f_0 = \frac{V_1}{(V_1)_{sp}} - 1 \tag{4-31}$$

Thus, the $2N + 1$ functions denoted by f, consist of f_0, the N dew-point functions, and the N energy balance functions given by Eqs. (4-4) and (4-5).

The proposed calculational procedure consists of first making one trial calculation on the reboiled absorber by use of the $2N$ Newton-Raphson method and then one trial on the distillation column by use of the θ method. Then the capital Θ method is applied one time to the system in order to place it in overall component-material balance and in agreement with the specified values of the terminal flow rates. To initiate the calculational procedure the composition of any recycle stream which is needed is assumed. After the first trial through the system, the composition of such recycle streams found by the Θ method are used. The steps of the proposed calculational procedure follow.

Step 1. On the basis of the most recent set of values $\{(b_{i,\,2})_{co}\}$ found in step 3, make one trial on the reboiled absorber by use of the $2N$ Newton-Raphson method. [In order to initiate the first trial, a set of $b_{i,\,2}$'s is assumed.] Save the resulting values of T_j and L_j/V_j for making the next trial on this column. After one trial, go to step 2.

Step 2. Use the $l_{Ni,\,1}$'s found in step 1 as the feed to the distillation column. Make one trial by use of the θ method. Save the resulting values of the T_j's and V_j's.

Step 3. Apply the capital Θ method to the system. Find the set of Θ multipliers required to place the system in component-material balance and in agreement with the specified terminal rates L_{N1}, D_2, W_1, and W_2. Return to step 1.

To illustrate the use of the combination of the $2N$ Newton-Raphson method, the θ method, and the capital Θ method, Example 4-10 which involves the system shown in Fig. 4-5 was solved. The statement and solution of this example are presented in Tables 4-11 and 4-12, respectively. After one trial had been made on the reboiled absorber by use of the $2N$ Newton-Raphson method (procedures 1, 2, or 3), one trial was made on the distillation column by use of the θ method for single columns (see Chap. 2). Then the system was placed in component-material balance and in agreement with the specified values of the total-flow rates by use of the capital Θ method of convergence (see Chap. 3). The formulation of the four g functions and the material balances of the capital Θ method for Example 4-10 are left as an exercise; see Prob. 4-2.

Solution of Systems of Columns by Use of the System Modular Method

In this formulation of the Newton-Raphson method, the equations for each column of the system are solved simultaneously.[19] To demonstrate this approach, the pipestill example is used; see Fig. 4-6. To solve problems involving

Table 4-11 Statement of Example 4-10

Component	FX_i (lb mol/h)	Initial values $b_{i,2}$ (lb mol/h)
CH_4	35.0	
C_2H_6	2.0	
C_3H_6	3.0	
C_3H_8	5.0	
$i\text{-}C_4H_8$	1.0	
$i\text{-}C_4H_{10}$	5.0	3.5
$n\text{-}C_4H_{10}$	15.0	3.5
$i\text{-}C_5H_{12}$	10.0	14.0
$n\text{-}C_5H_{12}$	20.0	14.0
$n\text{-}C_6H_{14}$	2.0	14.0
$n\text{-}C_7H_{16}$	1.0	10.5
$n\text{-}C_8H_{18}$	1.0	10.5
	100.0	70.0

Other specifications

Column 1—Reboiled Absorber $N = 12$, $f = 7$, column pressure $= 300$ lb/in^2 abs, $V_1 = 60$, $W_1 = 10$ lb mol/h, W_1 is to be withdrawn as a liquid from plate $j = 10$, and $B_1/V_{N1} = 1.435$. The feed F enters as a vapor at 724.77°R at 300 lb/in^2 abs. For the initial values of the temperatures $\{T_j\}$ and vapor rates $\{V_j\}$ use the following. $T_1 = 666°R$, $T_2 = 624.45°R$, $T_j = T_{j-1} + 18.46$ ($j = 3, 4, \ldots, N - 1$), $T_{12} = 809°R$. $V_1 = 60$, $V_j = 120$ ($j = 2, 3, \ldots, 6$), $V_N = 55.68$ lb mol/h.

Column 2—Distillation Column $N = 11$, $f = 5$, the column has a total condenser and the column pressure $= 300$ lb/in^2 abs, $D_2 = 20$, $W_2 = 10$ lb mol/h, $L_{1,2}/D_2 = 6.5$, and W_2 is to be withdrawn as a liquid from plate $j = 2$. For this initial values of $\{T_j\}$ and $\{V_j\}$, use the following. $T_1 = 725°R$, $T_j = T_{j-1} + 3.4$ ($j = 2, 3, \ldots, N - 1$), $T_N = 858°R$. $V_1 = 20$, $V_j = 150$ ($j = 2, 3, \ldots, N - 1$). $V_{11} = 36.41$ lb mol/h. Use the equilibrium and enthalpy data given in Tables B-1 and B-2.

System specifications. $W_1 = 10$, $B_1 = 100$, $D_2 = 20$, and $W_2 = 10$ lb mol/h.

Table 4-12 Solution sets of temperatures, vapor rates, and product distributions for Example 4-10

1. Final temperature and vapor rate profiles

	Column 1		Column 2	
Stage	T_j (K)	V_j (lb mol/h)	T_j (K)	V_j (lb mol/h)
1	669.38	60.000	723.08	20.00
2	698.52	92.519	745.73	150.00
3	712.68	108.84	762.89	149.13
4	719.95	117.73	776.93	147.36
5	723.99	122.38	795.85	139.57
6	726.51	124.85	800.90	141.47
7	747.96	25.341	805.36	145.01
8	759.51	46.852	809.79	147.70
9	765.88	59.501	815.63	148.91
10	772.32	65.105	826.90	146.62
11	782.41	68.706	855.12	135.80
12	805.33	69.687		

2. Product distributions for column 1

	Column 1		
	Flow rates (lb mol/h)		
Component	v_{1i}	$w_{i,1}$	$b_{i,1}$
CH_4	0.34999×10^2	0.47620×10^{-3}	0.10515×10^{-3}
C_2H_6	0.19963×10	0.19172×10^{-2}	0.17459×10^{-2}
C_3H_6	0.26196×10	0.99014×10	0.28156
C_3H_8	0.40833×10	0.21911	0.69825
$i\text{-}C_4H_8$	0.26137	0.18099	0.65269
$i\text{-}C_4H_{10}$	0.16655×10	0.51893	0.28765×10
$n\text{-}C_4H_{10}$	0.29343×10	0.16889×10	0.11110×10^2
$i\text{-}C_5H_{12}$	0.27728×10	0.15129×10	0.14529×10^2
$n\text{-}C_5H_{12}$	0.71043×10	0.36612×10	0.37907×10^2
$n\text{-}C_6H_{14}$	0.97548	0.80332	0.10623×10^2
$n\text{-}C_7H_{16}$	0.36977	0.60741	0.90232×10
$n\text{-}C_8H_{18}$	0.21819	0.77794	0.12298×10^2
	$\overline{V_1 = 60.0}$	$\overline{W_1 = 10.000}$	$\overline{B_1 = 100.00}$

$Q_{C1} = 0.10389 \times 10^7$ Btu/h
$Q_{R1} = 0.62979 \times 10^6$ Btu/h

Table 4-12 (*continued*)

3. Product distribution for column 2

	Column 2		
	Flow rates (lb mol/h)		
Component	$d_{i,2}$	$w_{i,2}$	$b_{i,2}$
CH_4	0.10182×10^{-3}	0.33282×10^{-5}	0.23697×10^{-11}
C_2H_6	0.16214×10^{-2}	0.12448×10^{-2}	0.75617×10^{-8}
C_3H_6	0.24057	0.40839×10^{-1}	0.15194×10^{-3}
C_3H_8	0.58828	0.10930	0.66903×10^{-3}
i-C_4H_8	0.47254	0.15710	0.23051×10^{-1}
i-C_4H_{10}	0.21553×10	0.66034	0.60867×10^{-1}
n-C_4H_{10}	0.75806×10	0.27965×10	0.73261
i-C_5H_{12}	0.35072×10	0.22071×10	0.88146×10
n-C_5H_{12}	0.53558×10	0.38787×10	0.28672×10^2
n-C_6H_{14}	0.91142×10^{-1}	0.13006	0.10402×10^2
n-C_7H_{16}	0.61908×10^{-2}	0.16632×10^{-1}	0.90003×10
n-C_8H_{18}	0.61185×10^{-3}	0.32479×10^{-2}	0.12294×10^2
	$D_2 = 20.0$	$W_2 = 10.0$	$B_2 = 70.0$

$Q_{C2} = 0.10509 \times 10^7$ Btu/h
$Q_{R2} = 0.13928 \times 10^7$ Btu/h

systems of columns, the column modular method is generally recommended over the system modular method except for the possible exception of pipestills. Since, however, each sidestripper is small relative to the main column, it is perhaps easier to treat this particular system by the column modular method rather than the system modular method. To illustrate the system modular method, an abbreviated development of the pipestill problem, Example 4-11, is given.

The pipestill problem solved herein was originally solved by Cecchetti et al.[8] by use of the original θ method of convergence which is described in Chap. 3. This problem is based on data from field tests which were made on the pipestill shown in Fig. 4-6. The θ method for distillation columns may fail to converge for some absorber-type problems, such as the pipestill. The pipestill is classified as an absorber-type problem because the main column has a condenser but no reboiler; the first sidestripper has a reboiler but no condenser; and all of the remaining strippers are of the conventional type.

For the pipestill shown in Fig. 4-6, Cecchetti et al.[8] found a theoretical analogue column by trial. The theoretical analogue column is defined as that column having perfect plates which gives calculated results that are in good agreement with field test results for the pipestill. The theoretical analogue column shown in Fig. 4-7 was proposed by Hess et al.[13] This is essentially the same as the analogue proposed by Cecchitti et al.

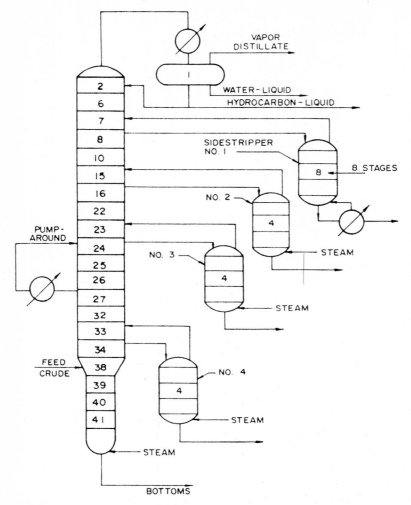

Figure 4-6 Actual stage numbers in the main column and the side strippers of the pipestill. [*Hess et al., Hydrocarbon Process.*, **56**(5): 241 (1977), by courtesy Hydrocarbon Processing.]

The minor differences between the theoretical analogue shown in Fig. 4-7 and the one used by Cecchetti et al. are a reflection of the different manner in which water was treated. In the present analysis, which follows that of Hess et al.,[13] water was regarded as being distributed between the vapor and liquid phases on stages 2 through 37, whereas Cecchetti et al. regarded water to be in the vapor phase alone on these stages. On stage 1 (the accumulator), two immiscible liquid phases (a water and a hydrocarbon phase) are assumed to be in equilibrium with the vapor phase. In Fig. 4-7, the withdrawal rate of the liquid water phase is W_0, and the withdrawal rate of the liquid hydrocarbon phase is W_1.

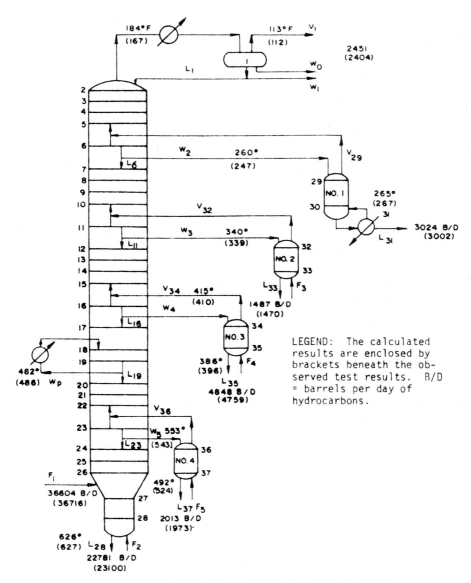

Figure 4-7 Theoretical analogue of the pipestill. [*Hess et al., Hydrocarbon Process., **56**(5): 241 (1977), Courtesy Hydrocarbon Processing.*]

The physical properties (normal boiling points, densities, and molecular weights) of the 34 pseudo components selected to represent the true boiling-point curves of the feed, distillate, and sidestreams are presented by Cecchetti et al.[8] On the basis of these data, a feed composed of 34 pseudo components and having the molar compositions and total flow rate shown in Table 4-13 was selected by Cecchetti et al. The specifications for the theoretical analogue column are also given in this table.

The calculated sets of temperatures, flow rates, and compositions obtained by the calculational procedure developed below are presented in Tables 4-14 through 4-16. These results were obtained by use of the data of Cecchetti et al. for the K values and enthalpies of the pseudo components. Curve fits of these data are presented in Tables B-14 and B-16. The K values for water $i = 35$ was taken from nomographs given by Hadden and Grayson,[12] and the enthalpies from the steam tables given by Smith and Van Ness.[21] In the condenser, the K_i was taken equal to its vapor pressure (as given in the steam tables) divided by the total pressure in the condenser.

A SYSTEM MODULAR FORMULATION OF THE $(2N + 1)$ NEWTON-RAPHSON METHOD FOR A PIPESTILL

The pipestill shown in Fig. 4-7 is used for the purpose of separating the hydrocarbon feed F_1 into seven fractions (V_1, W_1, W_2, W_3, W_4, W_5, L_{28}). The remaining feeds, F_2 through F_5, consist of steam which is used as the stripping medium. The stages are numbered down from the top of the main column. The condenser-accumulator section is assigned the number 1, the top plate the number 2, and the bottom plate of the main column the number 28. The stages of the sidestrippers are numbered consecutively 29 through 37, as shown in Fig. 4-7.

For the theoretical analog column, the following variables are regarded as fixed.

1. The number of theoretical plates in the main column and side-strippers as well as the locations of all stream withdrawals and return positions
2. Quantity, composition, and thermal conditions of all feeds
3. Column pressure and the pressure drop per stage
4. The reflux ratio L_1/V_1 (or alternatively, the condenser duty Q_{C1})
5. Distillate and sidestream withdrawal rates, W_1, W_2, W_3, W_4, W_5
6. The boilup ratio V_{31}/L_{31} for the sidestripper No. 1 which has a reboiler (or alternatively, the reboiler duty Q_{R31} for this stripper could be specified)
7. The intercooler duty Q_p for the pumparound stream W_p
8. The pumparound rate W_p and its withdrawal and return positions

The numerical values for the specifications are listed in Table 4-13.

Table 4-13 Composition of the feed stream F_1 and other specifications for the pipestill, Example 4-11 (*Taken from Ref. 13 by courtesy Hydrocarbon Process.*)

Component	$F_1 X_1$	Other specifications for the theoretical analogue column
1	0.73000×10^1	The hydrocarbon feed F_1 enters the
2	0.25700×10^2	column at 637°F with 69.038% of
3	0.38000×10^2	the feed vaporized and a total
4	0.43800×10^2	enthalpy of 219.890×10^6 Btu/h.
5	0.95700×10^2	The steam enters the main column
6	0.71400×10^2	and the sidestrippers as super-
7	0.63300×10^2	heated steam at 572°F and with
8	0.63300×10^2	the rates $F_2 = 66$, $F_3 = 6.94$,
9	0.76250×10^2	$F_4 = 26.8$, and $F_5 = 15.8$ lb mol/h.
10	0.72250×10^2	The sidestreams are to be with-
11	0.43950×10^2	drawn at the rates $W_1 = 135$,
12	0.43950×10^2	$W_2 = 313.44$, $W_4 = 136.08$, and
13	0.86500×10^2	$W_5 = 368.40$ lb mol/h. The pump-
14	0.29400×10^2	around stream is to be withdrawn
15	0.29400×10^2	at the rate of $W_p = 823$ lb mol/h
16	0.51000×10^2	and the intercooler duty $Q_p =$
17	0.34000×10^2	18.0×10^6 Btu/h. The reflux ratio
18	0.34000×10^2	$L_1/V_1 = 10.95$ and the boilup ratio
19	0.30640×10^2	$V_{31}/L_{31} = 0.13245$. The pressure in
20	0.30650×10^2	the accumulator is 23.1 lb/in^2 abs,
21	0.67600×10^2	and the pressure on plate 28 is
22	0.65600×10^2	29.24 lb/in^2 abs. An equal
23	0.42400×10^2	pressure drop per plate may be
24	0.71200×10^2	assumed for the main column and
25	0.67500×10^2	the sidestrippers. The pressure
26	0.12780×10^3	on the top plate of each side-
27	0.11360×10^3	stripper may be taken equal to
28	0.97100×10^2	the pressure of the plate in the
29	0.81200×10^2	main column where the sidestream
30	0.67800×10^2	feed to the stripper originated.
31	0.47700×10^2	
32	0.57300×10^2	
33	0.29600×10^2	
34	0.28300×10^2	
35	0.26400×10^3	

$F_1 = 0.22032 \times 10^4$ lb mol/h

CHOICE OF INDEPENDENT VARIABLES

Although the temperatures and the L/V's are commonly selected as the independent variables in the $2N$ Newton-Raphson method, a heating or cooling duty may be selected as an independent variable, and the L/V for that stage specified. For example, by specification of the reflux ratio L_1/V_1, the condenser duty Q_{C1} and the temperature T_1 become the independent variables for the first stage of the main column. The variable L_j/V_j is replaced by the new variable θ_j, which is defined by Eq. (4-8).

For the set of specifications stated above and in Table 4-13, the corresponding set of $2N + 1$ independent variables (where N is equal to the total number of stages) is: θ_0, Q_{C1}, θ_2, θ_3, ..., θ_{30}, Q_{R31}, θ_{32}, ..., θ_{37}, T_1, T_2, ..., T_{37}. Except for θ_0, the θ_j's are defined by Eq. (4-6). The new variable θ_0 is introduced to account for the two liquid phases in the accumulator, and it is defined as follows

$$W_0/V_1 = \theta_0(W_0/V_1)_a \qquad (4\text{-}32)$$

TOTAL MATERIAL BALANCES

The total material balances are formulated in a manner analogous to that demonstrated for complex columns. The returning vapor streams from the side-strippers to plates 5, 10, 15, and 22 give rise to elements which lie above the three diagonals of the matrix **R** of Eq. (4-10).

The presence of two perfectly immiscible liquid phases on stage 1 and the typical two-phase (vapor and liquid) behavior on all subsequent stages calls for a special treatment of water on stages 1 and 2. The development of the component-material balances for all components except water are developed for the first and second stages, and then the component-material balances for water on these two stages are developed.

COMPONENT-MATERIAL BALANCES

For stage 1, the component-material balance for any component except water (for water or steam $i = c = 35$) is given by

$$v_{2i} - v_{1i} - l_{1i} - w_{1i} = 0 \qquad (i \neq c)$$

Since W_1 and L_1 have the same composition

$$w_{1i} = (W_1/L_1)l_{1i} \qquad (i \neq c)$$

Use of this relationship and the equilibrium relationship given by Eq. (4-11) permits the component-material balance to be restated in the following form

$$-[1 + A_{1i} + (W_1/L_1)A_{1i}]v_{1i} + v_{2i} = 0 \qquad (i \neq c)$$

For the second stage, the component-material balance is given by

$$v_{3i} + l_{1i} - v_{2i} - l_{2i} = 0 \qquad (i \neq c)$$

and by use of the equilibrium relationship, the component-material balance is restated in the following form

$$A_{1i}v_{1i} - (1 + A_{2i})v_{2i} + v_{3i} = 0 \qquad (i \neq c) \tag{4-33}$$

The component-material balances for the remaining stages are developed in an analogous manner, and the complete set of equations so obtained may be stated in matrix form as follows

$$\mathbf{A}_i \mathbf{v}_i = -\mathbf{f}_i \qquad (i \neq c) \tag{4-34}$$

where

$$\mathbf{v}_i = [v_{1i} \ v_{2i} \ \cdots \ v_{37i}]^T$$

$$\mathbf{f}_i = [0 \ \cdots \ 0 \ v_{F1i} \ l_{F1i} \ 0 \ \cdots \ 0]^T$$

The square matrix $\mathbf{A}_i$ differs from a tridiagonal matrix by the appearance of only four nonzero elements to the right and four nonzero elements to the left of the tridiagonal band of elements. The four nonzero elements to the right of the tridiagonal band result from the return of the vapor streams from the sidestrippers, and the four elements to the left of the tridiagonal band of elements result from the introduction of the sidestreams W_2, W_3, W_4, and W_5 to the sidestrippers. The hydrocarbon feed F_1 enters on plate 27, and v_{F1i} is in row 26 while l_{F1i} is in row 27.

COMPONENT-MATERIAL BALANCES FOR WATER

Since the two liquid phases (the water phase and the hydrocarbon phase) in the accumulator are taken to be immiscible, it follows that

$$w_{0c} = W_0$$

$$l_{1c} = 0$$

Thus the component-material balance for water for stage 1 (the condenser-accumulator section) is given by

$$v_{2c} - v_{1c} - w_{0c} = 0$$

Since the partial pressure of water vapor above the two liquid phases in the accumulator is equal to its vapor pressure, it is evident that

$$P_{1c} = p_{1c} = (v_{1c}/V_1)P \tag{4-35}$$

where P_{1c} is the vapor pressure of water at the temperature T_1 of the accumulator and p_{1c} is the partial pressure of water in the accumulator. Symmetry of the equations is preserved by restating the flow rate of water w_{0c} in terms of the vapor rate v_{1c}. Commencing with

$$w_{0c} = W_0 = (W_0/v_{1c})v_{1c}$$

and making use of Eq. (4-35) gives

$$w_{0c} = \frac{W_0}{V_1(P_{1c}/P)} v_{1c} = A_{0c}v_{1c}$$

Use of this relationship permits the component-material balance for water for stage 1 to be restated as follows

$$-(1 + A_{0c})v_{1c} + v_{2c} = 0 \tag{4-36}$$

For stage 2, the component-material balance is

$$v_{3c} - v_{2c} - l_{2c} = 0$$

since $l_{1c} = 0$. For stages 2 through 37, water is taken to be a two-phase component and the equilibrium relationship given by Eq. (4-11) may be used to restate the component-material balance for the second stage as follows

$$-(1 + A_{2c})v_{2c} + v_{3c} = 0$$

FUNCTIONS FOR THE NEWTON-RAPHSON METHOD

For the general case of any stage j in which a single-liquid phase is in equilibrium with the vapor phase, there exist two independent variables per stage, θ_j (or Q_{Cj} or Q_{Rj}) and T_j for a total of $2N$ independent variables. The total of $2N + 1$ independent variables for the pipestill results from the existence of two liquid phases in the accumulator, which give rise to three independent variables $[\theta_0, \theta_1$ (or Q_{C1}), $T_1]$ for stage 1.

In order to solve for the $2N + 1$ independent variables by use of the Newton-Raphson method, $2N + 1$ independent equations must be selected and expressed in functional form. The equations so selected are the $N + 1$ equilibrium relationships and the N enthalpy balances.

The $N - 1$ equilibrium functional expressions for stages 2 through N are formulated by commencing with the condition that a set of the independent variables is to be found such that $F_j = 0$ for all j ($j = 2, 3, \ldots, N$), where F_j is given by Eq. (4-4).

The existence of two liquid phases on stage 1 leads to two independent equilibrium functional expressions. The functional expression for the hydrocarbon phase is developed by commencing with Eq. (4-2) and the fact that $l_{1c} = 0$ to give

$$F_1 = \frac{1}{V_1} \left\{ \sum_{i=1}^{c-1} \left[\frac{1}{K_{1i}} - 1 \right] v_{1i} - v_{1c} \right\} \tag{4-37}$$

The functional form of the equilibrium expression for the liquid-water phase on plate 1 is obtained by commencing with the equilibrium relationships given by Eq. (4-35). Rearrangement of this equation followed by the statement of the result so obtained in functional form gives

$$F_0 = \left[\frac{v_{1c}}{V_1} \left(\frac{P}{P_{1c}} \right) \right] - 1 \tag{4-38}$$

The enthalpy balance functions are formulated for any stage j in a manner analogous to that for absorbers and distillation columns. The solution values of the variables (the temperatures, vapor rates, and product flow rates) are presented in Tables 4-14 and 4-15.

Table 4-14 Initial and final temperature and flow rate profiles for Example 4-11 (*Taken from Ref. 13*)

Plate	Temp., °F		Liquid, lb mol/h		Vapor, lb mol/h	
	Initial	Final	Initial	Final	Initial	Final
1	100.00	111.64	2178.0	2129.8	198.90	194.50
2	122.22	116.79	2178.0	2108.9	2862.0	2816.1
3	144.44	192.56	2178.0	2053.4	2862.0	2795.2
4	166.77	210.74	2178.0	1985.4	2862.0	2738.7
5	188.89	227.83	2178.0	1894.9	2862.0	2671.6
6	211.11	247.41	1864.6	1463.3	2825.3	2561.5
7	233.33	272.47	1864.6	1359.9	2825.3	2443.3
8	255.56	296.95	1864.6	1303.0	2825.3	2340.0
9	277.78	316.27	1864.6	1281.4	2825.3	2283.1
10	300.00	329.46	1864.6	1265.5	2825.3	2261.4
11	322.22	339.07	1728.5	1102.5	2818.4	2225.0
12	355.55	348.00	1728.5	1061.2	2818.4	2198.2
13	366.67	357.68	1728.5	1003.4	2818.4	2156.8
14	388.89	369.84	1728.5	922.26	2818.4	2099.1
15	411.11	386.60	1728.5	812.16	2818.4	2017.9
16	433.33	410.45	1360.1	329.69	2791.6	1842.2
17	455.55	439.86	1360.0	285.59	2791.6	1728.1
18	477.78	453.90	2183.1	1607.8	2791.6	1684.0
19	500.00	486.45	1360.1	879.87	2791.6	2183.3
20	522.22	508.06	1360.1	859.71	2791.6	2278.3
21	544.44	520.82	1360.1	816.99	2791.6	2258.1
22	566.67	531.37	1360.1	759.57	2791.6	2251.4
23	588.89	542.52	1234.7	554.13	2775.8	2124.4
24	611.11	556.74	1234.7	462.72	2775.8	2044.5
25	633.33	572.98	1234.7	354.27	2775.8	1953.1
26	655.55	593.34	1234.7	154.74	2775.8	1844.6
27	677.78	629.46	1916.8	811.63	1254.7	123.98
28	700.00	626.51	728.12	778.86	1254.7	98.768
29	211.11	256.46	313.4	326.68	36.63	19.775
30	233.33	261.04	313.4	332.53	36.63	33.020
31†	255.56	266.80	276.8	293.66	36.63	38.866
32	322.22	331.89	136.1	130.74	6.94	20.442
33†	344.44	321.95	136.1	122.58	6.94	15.105
34	433.33	404.36	368.4	352.74	26.80	65.626
35†	455.55	395.54	368.4	329.59	26.80	49.969
36	588.89	535.40	125.4	117.97	15.80	33.464
37†	611.11	524.46	125.4	107.70	15.80	26.072

Other variables	Initial	Final
W_0/V_1	1.75	1.838
Q_{C1} (Btu/h)	38.33×10^6	39.179×10^6
Q_{R31} (Btu/h)	0.65×10^6	0.66462×10^6

† Plates 31, 33, 35, and 37, are the bottom plates of sidestrippers 1, 2, 3, and 4, respectively.

Table 4-15 Product distributions for the main column of Example 4-11 (*Taken from Ref. 13*)

Compo-nent	Liquid w_{0i}	Liquid w_{1i}	Vapor v_{1i}	Liquid $l_{28,i}$
1	....	0.36758×10^{-1}	0.72632×10^{1}	0.32644×10^{4}
2	....	0.62566	0.25071×10^{2}	0.26807×10^{2}
3	....	0.29272×10^{1}	0.35048×10^{2}	0.83867×10^{2}
4	....	0.89763×10^{1}	0.34632×10^{2}	0.15499×10^{1}
5	....	0.40897×10^{2}	0.52850×10^{2}	0.85879×10^{1}
6	....	0.44157×10^{2}	0.21337×10^{2}	0.11619
7	....	0.26133×10^{2}	0.47275×10^{1}	0.17989
8	....	0.99476×10^{1}	0.14782×10^{1}	0.22583
9	....	0.10933×10^{1}	0.88609×10^{-1}	0.38925
10	....	0.19251	0.10284×10^{-1}	0.50368
11	....	0.16419×10^{-1}	0.56516×10^{-3}	0.37881
12	....	0.16518×10^{-3}	0.22649×10^{-5}	0.67363
13	....	0.16641×10^{-4}	0.14177×10^{-6}	0.17456×10^{-2}
14	....	0.15145×10^{-6}	0.78507×10^{-9}	0.79277
15	....	0.22765×10^{-8}	0.70243×10^{-11}	0.10205×10^{1}
16	....	0.55656×10^{-10}	0.10392×10^{-12}	0.23381×10^{1}
17	....	0.14330×10^{-12}	0.13288×10^{-15}	0.20558×10^{1}
18	....	0.76684×10^{-15}	0.40886×10^{-18}	0.26460×10^{1}
19	....	0.68740×10^{-16}	0.28692×10^{-19}	0.26962×10^{1}
20	....	0.22190×10^{-18}	0.51428×10^{-22}	0.34258×10^{1}
21	....	0.17505×10^{-20}	0.22792×10^{-24}	0.96953×10^{1}
22	....	0.20760×10^{-24}	0.10708×10^{-28}	0.13251×10^{2}
23	....	0.61138×10^{-29}	0.12121×10^{-33}	0.12496×10^{2}
24	....	....	....	0.30110×10^{2}
25	....	....	....	0.44651×10^{2}
26	....	....	....	0.11928×10^{3}
27	....	....	....	0.11338×10^{3}
28	....	....	....	0.97097×10^{2}
29	....	....	....	0.81199×10^{2}
30	....	....	....	0.67799×10^{2}
31	....	....	....	0.47699×10^{2}
32	....	....	....	0.57299×10^{2}
33	....	....	....	0.29599×10^{2}
34	....	....	....	0.28299×10^{2}
35	0.35684×10^{3}	....	....	0.77105×10^{1}
Total	356.84	135.00	194.50	778.86

Table 4-16 Product distribution for the sidestrippers of Example 4-11 (*Taken from Ref. 13*)

Component	$l_{31,i}$	$l_{33,i}$	$l_{35,i}$	$l_{37,i}$
1	0.69343×10^{-6}	0.25754×10^{-5}	0.61494×10^{-5}	0.94406×10^{-6}
2	0.19430×10^{-3}	0.18269×10^{-3}	0.41215×10^{-3}	0.75324×10^{-4}
3	0.91788×10^{-2}	0.24606×10^{-2}	0.37864×10^{-2}	0.34557×10^{-3}
4	0.14254	0.15945×10^{-1}	0.17375×10^{-1}	0.85512×10^{-3}
5	0.15792×10^{1}	0.14054	0.14165	0.57496×10^{-2}
6	0.51627×10^{1}	0.31236	0.30460	0.10062×10^{-1}
7	0.29863×10^{-2}	0.68628	0.69136	0.20813×10^{-1}
8	0.49639×10^{-2}	0.10066×10^{1}	0.97587	0.27918×10^{-1}
9	0.70439×10^{2}	0.22821×10^{1}	0.19077×10^{1}	0.50768×10^{-1}
10	0.69003×10^{2}	0.38008×10^{1}	0.26698×10^{1}	0.70253×10^{-1}
11	0.36973×10^{2}	0.43218×10^{1}	0.22027×10^{1}	0.57564×10^{-1}
12	0.17744×10^{2}	0.20337×10^{2}	0.50738×10^{1}	0.12158
13	0.12172×10^{2}	0.54715×10^{2}	0.17523×10^{2}	0.34493
14	0.18682	0.16350×10^{2}	0.11270×10^{2}	0.17098
15	0.10401	0.96660×10^{1}	0.18373×10^{2}	0.23646
16	0.16173×10^{-1}	0.65683×10^{1}	0.41485×10^{2}	0.59569
17	0.49269×10^{-3}	0.11589×10^{1}	0.30219×10^{2}	0.56579
18	0.23143×10^{-4}	0.28438	0.30271×10^{2}	0.79865
19	0.50537×10^{-5}	0.12979	0.26966×10^{2}	0.85782
20	0.18022×10^{-6}	0.27553×10^{-1}	0.25961×10^{2}	0.12358×10^{1}
21	0.13224×10^{-7}	0.11837×10^{-1}	0.53601×10^{2}	0.42919×10^{1}
22	0.63624×10^{-10}	0.88605×10^{-3}	0.42437×10^{2}	0.99107×10^{1}
23	0.82631×10^{-13}	0.20398×10^{-4}	0.12174×10^{2}	0.17731×10^{2}
24	0.80735×10^{-16}	0.39169×10^{-6}	0.29254×10^{1}	0.38165×10^{2}
25	0.67213×10^{-21}	0.22248×10^{-9}	0.60178×10^{-1}	0.22788×10^{2}
26	....	0.30389×10^{-14}	0.14531×10^{-3}	0.85209×10^{1}
27	....	0.21479×10^{-22}	0.23489×10^{-8}	0.21711
28	....	....	0.12939×10^{-13}	0.31133×10^{-2}
29	....	....	0.64525×10^{-20}	0.20250×10^{-4}
30	....	....	....	0.15680×10^{-7}
31	....	....	....	0.52034×10^{-9}
32	....	....	....	0.53617×10^{-14}
33	....	....	....	0.26740×10^{-19}
34	....	....	....	....
35	0.26242×10^{-2}	0.75718	0.23097×10^{1}	0.89814
Total	293.66	122.58	329.57	107.70

4-4 COMPARISON OF PROCEDURES 1, 2, 3, AND THE COLUMN MODULAR AND SYSTEM MODULAR METHODS

A summary of the computer times and the number of trials required to solve a variety of examples by the methods enumerated above is presented in Table 4-17. These results show that for absorber-type columns (Examples 4-2, 4-3, and 4-4), procedure 1 requires fewer trials than do procedures 2 and 3. For columns which have a relatively small number of plates, all three procedures require about the same amount of computer time. As the number of plates and components are increased (see Example 4-12), the speed advantage of procedures 2 and 3 over procedure 1 becomes more pronounced.

For distillation-type columns (Examples 4-5 through 4-7) the θ method, which is presented in Chaps. 2 and 3, is seen to be from 5 to 20 times faster than the $2N$ Newton-Raphson method.

Since the number of operations required to invert or to find the **LU** factorization of the jacobian matrix requires n^3 or n^2 operations, respectively, it is fortunate that most absorber-type problems are characterized by mixtures which contain a relatively large number of components and by columns which contain a relatively small number of plates. For problems of this type, the $2N$ Newton-Raphson method is best suited and is recommended. For separations carried out in distillation type columns involving large numbers of components and plates, the θ method is the most efficient and is recommended.

In order to obtain a comparison between the computer times required for the column modular and the system modular methods, Example 3-4 was used by Hess[15] to test the two methods. In the global modular application of the $2N$ Newton-Raphson method to the system shown in Fig. 3-12, stages are numbered consecutively beginning with the accumulator of the first column and terminating with the reboiler of the last column. Example 4-12 consisted of solving the complete set of equations for the system simultaneously by use of the $2N$ Newton-Raphson method in a manner similar to that demonstrated above for pipestills. From the results presented in Table 4-17, it is evident that the column modular method consisting of the θ method for each of the columns and capital Θ method for the system is significantly faster than the system modular approach of the $2N$ Newton-Raphson method. In conclusion, the results presented in Table 4-17 support the recommended combination of procedures presented in Table 4-1.

Table 4-17 Computer times required for various methods

1. Single columns

| | | | | | | Execution times and number of trials on an AMDAHL 470V/6 computer | | | |
| | Example | | | | | | $2N$ Newton-Raphson | | |
Example	Type of column	No. of stages	No. of components	Remarks	θ method	Procedure 1	Procedure 2	Procedure 3	Compiler
4-5	Conventional distillation column	13	11	See Table 4-9	2.6 s 12 trials†	10.74 s 5 trials†	12.45 10 trials† 2 = no. of evaluations of jacobian matrix‡	12.98 s 12 trials 2	WATFIV
4-7	Complex distillation column	52	7	Column 1 of Fig. 3-12; described in text	2.14 19	39.18 11	27.0 11 3	23.42 11 3	FORTRAN H with OPT 2
4-6	Complex distillation column	13	11	See Table 4-10	2.81 13	13.10 6	12.57 9 2	11.93 9 2	WATFIV
4-2	Absorber	8	14	See Tables 4-4 and 4-5	....	6.22 5	6.25 18 1	8.45 15 2	WATFIV
4-4	Reboiler absorber	11	6	See Table 4-8	....	9.91 9	7.46 11 2	7.13 12 2	WATFIV
4-3	Stripper	......	....	See Tables 4-6 and 4-7	....	7.09 9	7.29 14	7.16 14	WATFIV

† For the θ method, the following norm ϕ was used as the convergence criterion, $\phi = 1/n \sum_{i=1}^{n} |g_i(1, 1, ..., 1)|$, where n is equal to the number of g functions. For the $2N$ Newton-Raphson method, $\phi = 1/N[\sum_{j=1}^{N} f_j^2]^{1/2}$, where N is equal to the number of functions f. For all examples except Examples 4-11 and 4-12, ϕ was taken to equal 10^{-5}, and for Examples 4-11 and 4-12, ϕ was taken to equal 10^{-4}.

(Continued on page 174.)

Table 4-17(*continued*)

2. Systems of columns

Example	Type of column	No. of components	No. of stages	Remarks	Method	Compiler
4-12	System of distillation column	7	104	See Example 3-3 and Fig. 3-12. See Ref. 15	1. The θ and Θ methods: 3.65 s, 17 system trials 2. Procedure 3 of the $2N$ Newton-Raphson method: 90 s per iteration—did not run to convergence.	FORTRAN H with OPT. 2
4-11	Pipestill	35	37	See Refs. 8, 13	Procedures 1, 2, and 3 of $2N$ Newton-Raphson method. Procedure 1: 214.59 s, 15 trials, Procedure 2: 145.23 s, 13 trials. Procedure 3: 140.23 s, 12 trials, jacobian matrix evaluated 3 times for procedures 2 and 3.	FORTRAN H with OPT. 2
4-10	Reboiled absorber and distillation column	12	24	See Table 4-11	$2N$ Newton-Raphson method. Procedure 3 for reboiled absorber, θ method for distillation column and Θ method for the system. Procedure 1: 17.8 s, 15 trials, Procedure 2: 14.62 s, 18 trials, Procedure 3: 14.46 s, 18 trials.	FORTRAN H

‡ The number of evaluations of the jacobian matrix includes the initial evaluation called for in step 2 plus any additional evaluations required in step 4 of the calculational procedure described for Broyden's method.

PROBLEMS

4-1 An adiabatic flash process may be regarded as an absorber with one equilibrium stage in which the two entering feeds are combined to give a feed rate F and composition $\{X_i\}$, that is,

$$v_{N+1, i} + l_{0i} = FX_i$$

and

$$X_i = \frac{v_{N+1, i} + l_{0i}}{V_{N+1} + L_0}$$

(a) Restate Eqs. (4-1) through (4-13) for the special case of an absorber with one equilibrium stage.

(b) In the interest of simplicity, let $(L_1/V_1)_a$ be set equal to unity. Show that Eq. (4-10) reduces to a scalar-total material balance which may be solved for V_1 to give

$$V_1 = \frac{F}{1 + \theta}$$

and that $\mathbf{A}_i \mathbf{v}_i = -\mathbf{f}_i$ reduces to a scalar-component-material balance which may be solved for v_{1i} to give

$$v_{1i} = \frac{FX_i}{1 + \theta/K_{1i}}$$

4-2 (a) Obtain expressions for the partial derivatives of v_{1i} and V_1 (see Prob. 4-1) with respect to T_1 and θ_1.

(b) Obtain expressions for the partial derivatives of the functions F_1 and G_1 (see Prob. 4-1) with respect to T_1 and θ_1.

4-3 Show that the function $F_1(\theta_1, T_1)$ (see Prob. 4-1) decreases monotonically with both θ_1 and T_1 for all $\theta_1 > 0$ and all T_1 lying between the bubble-point and the dew-point temperatures of the combined feed.

4-4 Repeat Prob. 4-3 for the function $G_1(\theta_1, T_1)$ (see Prob. 4-1).

4-5 Produce a sketch of the following forms of the phase equilibrium function in the positive domain of θ at a fixed value of T_1 lying between the bubble-point and dew-point temperatures of the feed.

$$(a) \quad F_1 = \frac{1}{V_1} \sum_{i=1}^{c} \frac{v_{1i}}{K_{1i}} - 1$$

$$(b) \quad F_1 = \sum_{i=1}^{c} \left(\frac{1}{K_{1i}} - 1 \right) v_{1i}$$

$$(c) \quad F_1 = \left(\frac{1}{\sum\limits_{i=1}^{c} v_{1i}} \right) \sum_{i=1}^{c} \left(\frac{1}{K_{1i}} - 1 \right) v_{1i}$$

where the formulas for v_{1i} and V_1 are given in Prob. 4-1.

4-6 A single-stage absorber is to be operated adiabatically at 1 atm pressure. The lean oil stream L_0 enters the single-stage absorber as a liquid at $139.6°F$, and the rich gas enters the absorber at its dew-point temperature of $200°F$ at 1 atm. Use the equilibrium and enthalpy data stated in Examples 1-4 and 1-5. The component-flow rates of the rich gas V_{N+1} and lean oil L_0 are as follows.

Component	$v_{N+1, i}$	l_{0i}
1	0	50
2	50	0
3	50	0

It is desired to determine V_1, L_1, and their compositions by use of the Newton-Raphson method. Evaluate the functions F_1 and G_1 and the partial derivatives $\partial F_1 / \partial \theta_1$ and $\partial G_1 / \partial \theta_1$. These are two of the four partial derivatives which are needed to make the first trial and calculation by use of the Newton-Raphson method. For the first trial take $\theta_1 = 1$ and $T_1 = 105°F$.

In practice, both the variables and the functions should be normalized. Let a new normalized temperature be defined as follows

$$T \text{ (normalized)} = \frac{T(°F)}{100}$$

4-7 Begin with first principles and formulate the enthalpy balance functions G_j for a conventional distillation column.

4-8 For the functions $\{F_j\}$ and $\{G_j\}$, given by Eqs. (4-3) and (4-4), respectively, develop the expressions for the following sets of partial derivatives for all j ($j = 1, 2, ..., N$) and all k ($k = 1, 2, ..., N$)

$$\left| \frac{\partial F_j}{\partial \theta_k} \right| \quad \left| \frac{\partial F_j}{\partial T_k} \right| \quad \left| \frac{\partial G_j}{\partial \theta} \right| \quad \left| \frac{\partial G_j}{\partial T_k} \right|$$

4-9 Perform the matrix differentiation implied by the right-hand side of Eq. (4-24) and verify the formulas given in Table 4-3 for $\mathbf{C}_{ki}$.

4-10 Supply the missing steps required to find the inverse of the jacobian matrix $\mathbf{J}_0$ in Example 4-8.

4-11 (a) If

$$\mathbf{A}_k = \begin{bmatrix} 2 & 4 & -6 \\ -4 & -7 & 10 \\ 2 & 7 & -9 \end{bmatrix}$$

$$\mathbf{x}_k = \begin{bmatrix} 1 \\ 2 \\ 3 \end{bmatrix}, \mathbf{y}_k = \begin{bmatrix} 3 \\ 2 \\ 1 \end{bmatrix}, C = 1$$

Compute

$$\mathbf{A}_{k+1} = \mathbf{A}_k + \mathbf{x}_k C \mathbf{y}_k^T$$

(b) Obtain the **LU** factorization of $\mathbf{A}_{k+1}$ and compare the result so obtained with that found in Example 4-9.

4-12 Formulate the four g functions and the matrix equation for the overall component-material balances for Example 4-10.

4-13 Show that the Broyden correction may be stated in the form given in procedure 3, namely,

$$\mathbf{J}_{k+1} = \mathbf{J}_k + \mathbf{x}_k C \mathbf{y}_k^T$$

Hint: Make use of the Broyden relationships given in Chap. 15.

REFERENCES

1. J. M. Bennett: "Triangular Factors of Modified Matrices," *Numerische Mathematik*, **7**:217 (1965).
2. D. S. Billingsley: "Numerical Solution of Problems in Multicomponent at Steady State," *AIChE J.*, **16**:441 (1970).
3. G. W. Boynton: "Iteration Solves Distillation," *Hydrocarbon Process.*, **49**:153 (January 1970).
4. A. A. Boyum: Ph.D. Dissertation, Polytechnic Institute of Brooklyn, Brooklyn, N.Y. (1966).
5. G. G. Brown and M. Souders, Jr.: "Fundamental Design of Absorbing and Stripping Columns for Complex Vapors," *Ind. Eng. Chem.*, **24**:519 (1932).

6. C. G. Broyden: "A Class of Methods for Solving Nonlinear Simultaneous Equations," *Math. Comp.*, **19**:577 (1965).

7. B. Carnahan, H. A. Luther, and J. O. Wilkes: *Applied Numerical Methods*, John Wiley & Sons, Inc., New York, 1969.

8. R. Cecchetti, R. H. Johnston, J. L. Niedswiecki, and C. D. Holland: "Pipestill Products Verify These Computer Estimates," *Hydrocarbon Process.* and *Pet. Refiner*, **42**(9):159 (1963).

9. S. D. Conte and Carl deBoor: *Elementary Numerical Analysis*, 2d ed., McGraw-Hill Book Company, New York, 1972.

10. S. E. Gallun: "An Extension of the Multi-θ Method to Distillation Columns and an Almost Band Solution of Equilibrium Stage Problem," M.S. Thesis, Texas A&M University, 1975.

11. John Greenstadt, Yonathan Bard, and Burt Morse: "Multicomponent Distillation on the IBM 704," *Ind. Eng. Chem.*, **50**:1644 (1958).

12. S. T. Hadden and H. G. Grayson: "New Charts for Hydrocarbon Vapor-Liquid Equilibria," *Hydrocarbon Process.* and *Pet. Refiner*, **40**(9):207 (1961).

13. F. E. Hess, C. D. Holland, Ron McDaniel, and N. J. Tetlow: "Solve More Distillation Problems, Part 7—Absorber-Type Pipestills," *Hydrocarbon Process.*, **56**(5):(1977).

14. —— S. E. Gallun, G. W. Bentzen, C. D. Holland, Ron McDaniel, and N. J. Tetlow: "Solve More Distillation Problems, Part 8—Which Method to Use," *Hydrocarbon Process.*, **56**(6):181 (1977).

15. —— Ph.D. Dissertation, Texas A&M University, 1977.

16. C. D. Holland, G. P. Pendon, and S. E. Gallun: "Solve More Distillation Problems, Part 3—Application to Absorbers," *Hydrocarbon Processing*, **54**(1):101 (1975).

17. —— *Fundamentals and Modeling of Separation Processes—Absorption, Distillation, Evaporation, and Extraction*, Prentice-Hall, Inc., Englewood Cliffs, N.J. 1975.

18. A. Kremser: "Theoretical Analysis of Adsorption Process," *Nat. Pet. News.* **22**:43 (May 21, 1930).

19. M. Kubiček, V. Hlaváček, and F. Procháska: "Global Modular Newton-Raphson Technique for Simulation of an Interconnected Plant Applied to Complex Rectifying Columns," *Chem. Eng. Sci.*, **31**:277 (1976).

20. J. S. Newman: "Temperature Computed for Distillation," *Hydrocarbon Process.*, **43**(4):141 (1963).

21. J. M. Smith and H. C. Van Ness: "Introduction to Chemical Engineering Thermodynamics," McGraw-Hill Book Company, New York, 2d ed., 1959.

22. A. D. Sujata: "Absorber Stripper Calculations Made Easier," *Hydrocarbon Process.*, **40**:137 (1961).

23. J. W. Tierney and J. L. Yanosik: "Simultaneous Flow and Temperature Correction in the Equilibrium Stage Problem," *AIChE J.*, **15**:897 (1969).

24. J. W. Tierney and J. A. Bruno: "Equilibrium Stage Calculations," *AIChE J.*, **13**:556 (1967).

25. J. F. Tomich: "A New Simulation Method for Equilibrium Stage Processes," *AIChE J.*, **16**:229 (1970).

FIVE

ALMOST BAND ALGORITHMS OF THE NEWTON-RAPHSON METHOD

The Newton-Raphson formulations, called the *Almost Band Algorithms* are recommended for solving problems involving columns in the service of separating highly nonideal mixtures. In the Almost Band Algorithms, the independent variables in the Newton-Raphson method are taken to be either one or both sets of the component-flow rates $\{l_{ji}\}$ and $\{v_{ji}\}$, the temperatures $\{T_j\}$, and in some of the formulations one or more of the total flow rates $\{L_j\}$. In Sec. 5-1 the independent variables are taken to be the component-flow rates, $\{l_{ji}\}$, $\{v_{ji}\}$, and the temperatures $\{T_j\}$. The formulation is presented in Sec. 5-1 for absorbers and strippers, and in Sec. 5-2 the formulation for conventional and complex columns is presented. Two modifications of Broyden's method are presented in Sec. 5-2. The modifications of Broyden's method preserve the sparsity of the initial jacobian matrix, whereas the original method as proposed by Broyden does not. The treatment of systems of columns in the service of separating highly nonideal solutions is presented in the next chapter.

Although the Almost Band Algorithms use a large number of independent variables, far less computer time is required to obtain a solution to a given distillation problem than might be expected. The computational speed results from the use of selected techniques of sparse matrices and the characteristics of homogeneous functions.

Sparsity of the jacobian matrix is achieved by a suitable ordering of the variables and functions. The particular choice of variables and functions and their ordering (discussed below) leads to the unique form of the jacobian matrix shown in Fig. 5-1. The well-known method of gaussian elimination may be applied in a stepwise fashion in the transformation of the matrix shown in Fig. 5-1 to the one shown in Fig. 5-2. At any one time, only four of the $(c + 2)$ square matrices along the diagonal and the two corresponding column matrices are considered in the gaussian elimination process instead of the complete $N(c + 2)$ square matrix. No arithmetic is ever performed on any of the zero elements lying outside of the squares in Figs. 5-1 and 5-2.

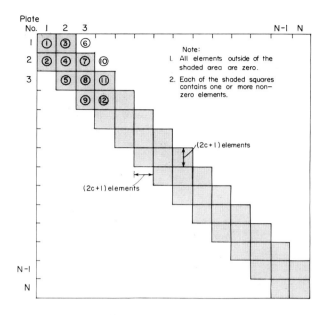

Figure 5-1 Structure of the jacobian matrix for an absorber.

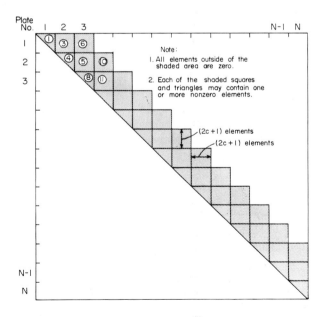

Figure 5-2 Upper triangular matrix for an absorber.

Highly nonideal solutions are characterized by the fact that the activity coefficients and the partial molar enthalpies are strongly dependent upon composition. In order to compute the partial derivatives of these quantities which are needed in the application of the Newton-Raphson method, it is convenient to choose compositions or component-flow rates as members of the set of independent variables. Numerous choices of the independent variables have been made.[6, 7, 8, 13, 15, 17, 19, 20] To demonstrate the formulation of the Newton-Raphson method, the choice of independent variables proposed by Naphtali and Sandholm[17] is used. The Almost Band Algorithm may be formulated for other choices of independent variables as shown by Gallun and Holland.[7, 8, 9]

5-1 ALMOST BAND ALGORITHMS FOR ABSORBERS AND STRIPPERS, INDEPENDENT VARIABLES: $\{l_{ji}\}, \{v_{ji}\},$ AND $\{T_j\}$

As shown in Fig. 4-1, the plates of the absorber are numbered down from the top of the column, the top plate is assigned the number 1 and the bottom plate the number N. The variables regarded as fixed (or specified) in the developments which follow are:

1. $\{l_{0i}\}$, liquid at T_0 and at the inlet pressure P_0
2. $\{v_{N+1, i}\}$, vapor at T_{N+1} and at the inlet pressure P_{N+1}
3. the column pressure or the pressure on each stage

The $N(2c + 3)$ equations required to describe the column may be stated in the following form:

$$\text{Equilibrium relationship} \begin{cases} \dfrac{\gamma_{ji}^V v_{ji}}{V_j} = \dfrac{\gamma_{ji}^L K_{ji} l_{ji}}{L_j} & \begin{aligned} (j &= 1, 2, \ldots, N) \\ (i &= 1, 2, \ldots, c) \end{aligned} \\[2ex] \displaystyle\sum_{i=1}^{c} l_{ji} = L_j & (j = 1, 2, \ldots, N) \\[2ex] \displaystyle\sum_{i=1}^{c} v_{ji} = V_j & (j = 1, 2, \ldots, N) \end{cases}$$

$$\text{(5-1)}$$

$$\text{Component-material balances} \begin{cases} v_{ji, i} + l_{j-1, i} - v_{ji} - l_{ji} = 0 & \begin{aligned} (j &= 1, 2, \ldots, N) \\ (i &= 1, 2, \ldots, c) \end{aligned} \end{cases}$$

$$\text{Energy balance} \begin{cases} \displaystyle\sum_{i=1}^{c} [v_{j+1, i}\hat{H}_{j+1, i} + l_{j-1, i}\hat{h}_{j-1, i} - v_{ji}\hat{H}_{ji} - l_{ji}\hat{h}_{ji}] = 0 \\[1ex] \hspace{4cm} (j = 1, 2, \ldots, N) \end{cases}$$

where $\hat{H}_{ji}$ and $\hat{h}_{ji}$ denote the virtual values of the partial molar enthalpies; see Chap. 14. In the above statement of the equations, the component-material balances and the energy balances enclose each stage j.

Use of the second and third expressions of Eq. (5-1) to eliminate the total-flow rates from the equilibrium relationships yields a total of $N(2c + 1)$ equations for the description of an absorber. When the independent variables are chosen as shown above, it is convenient to state each component-material balance and each energy balance for the enclosure of a single stage. Thus, after the total-flow rates have been eliminated from Eq. (5-1) as described above, the resulting set of $N(2c + 1)$ independent equations required to describe the column may be stated in functional form to give

$$f_{ji} = \frac{\gamma^L_{ji} K_{ji} l_{ji}}{\displaystyle\sum_{i=1}^{c} l_{ji}} - \frac{\gamma^V_{ji} v_{ji}}{\displaystyle\sum_{i=1}^{c} v_{ji}} \qquad \begin{matrix} (j = 1, 2, \ldots, N) \\ (i = 1, 2, \ldots, c) \end{matrix} \qquad (5\text{-}2)$$

$$m_{ji} = v_{j+1, i} + l_{j-1, i} - v_{ji} - l_{ji} \qquad \begin{matrix} (j = 1, 2, \ldots, N) \\ (i = 1, 2, \ldots, c) \end{matrix} \qquad (5\text{-}3)$$

$$G_j = \frac{\displaystyle\sum_{i-1}^{c} [v_{ji} \hat{H}_{ji} + l_{ji} \hat{h}_{ji}]}{\displaystyle\sum_{i=1}^{c} [v_{j+1, i} \hat{H}_{j+1, i} + l_{j-1, i} \hat{h}_{j-1, i}]} - 1 \qquad (j = 1, 2, \ldots, N) \qquad (5\text{-}4)$$

In the expressions for the activity coefficients $\{\gamma^V_{ji}, \gamma^L_{ji}\}$ and the virtual values of the partial molar enthalpies $\{\hat{H}_{ji}, \hat{h}_{ji}\}$, the mole fractions must have the sum of unity. This condition is satisfied by use of the following expressions for these mole fractions

$$x_{ji} = \frac{l_{ji}}{\displaystyle\sum_{i=1}^{c} l_{ji}} \qquad y_{ji} = \frac{v_{ji}}{\displaystyle\sum_{i=1}^{c} v_{ji}} \qquad (5\text{-}5)$$

In order to obtain a jacobian matrix having the form shown in Fig. 5-1 for an absorber, both the functions and the variables must be appropriately ordered. The functions must be ordered as follows

$$\mathbf{f} = [(f_{j, 1} \ f_{j, 2} \ \cdots \ f_{j, c} \ m_{j, 1} \ m_{j, 2} \ \cdots \ m_{j, c} \ G_j)_{j=1, N}]^T \qquad (5\text{-}6)$$

where the subscript $j = 1, N$ means that the argument is to be repeated for $j = 1, 2, \ldots, N$. The variables must be ordered as follows

$$\mathbf{x} = [(l_{j, 1} \ l_{j, 2} \ \cdots \ l_{j, c} \ v_{j, 1} \ v_{j, 2} \ \cdots \ v_{j, c} \ T_j)_{j=1, N}]^T \qquad (5\text{-}7)$$

By the ordering of the variables is meant the order in which the differentiation of each function is carried out in the Newton-Raphson method which is the same as the order in which the variables appear in the vector given by Eq. (5-7).

For example, the Newton-Raphson equation for any one function, say f_{jk} (the equilibrium function for plate j and component k, where k denotes a particular one of the c components), is

$$(\partial f_{jk}/\partial l_{1,1}) \, \Delta l_{1,1} + (\partial f_{jk}/\partial l_{1,2}) \, \Delta l_{1,2} + \cdots + (\partial f_{jk}/\partial l_{1,c}) \, \Delta l_{1,c}$$
$$+ (\partial f_{jk}/\partial v_{1,1}) \, \Delta v_{1,1} + \cdots + (\partial f_{jk}/\partial v_{1,c}) \, \Delta v_{1,c} + (\partial f_{jk}/\partial T_1) \, \Delta T_1$$
$$+ \cdots + (\partial f_{jk}/\partial v_{N,c}) \, \Delta v_{N,c} + (\partial f_{jk}/\partial T_N) \, \Delta T_N$$

The complete set of Newton-Raphson equations may be stated in the following matrix form

$$\mathbf{J} \, \Delta \mathbf{x} = -\mathbf{f} \qquad (5\text{-}8)$$

$$\mathbf{J} = \begin{bmatrix} \dfrac{\partial f_{1,1}}{\partial l_{1,1}} & \dfrac{\partial f_{1,1}}{\partial l_{1,2}} & \cdots & \dfrac{\partial f_{1,1}}{\partial T_N} \\[2ex] \vdots & \vdots & & \vdots \\[1ex] \dfrac{\partial m_{1,c}}{\partial l_{1,1}} & \dfrac{\partial m_{1,c}}{\partial l_{1,2}} & \cdots & \dfrac{\partial m_{1,c}}{\partial T_N} \\[2ex] \dfrac{\partial G_1}{\partial l_{1,1}} & \dfrac{\partial G_1}{\partial l_{1,2}} & \cdots & \dfrac{\partial G_1}{\partial T_N} \\[2ex] \vdots & \vdots & & \vdots \\[1ex] \dfrac{\partial G_N}{\partial l_{1,1}} & \dfrac{\partial G_N}{\partial l_{1,2}} & \cdots & \dfrac{\partial G_N}{\partial T_N} \end{bmatrix}$$

$$\Delta \mathbf{x} = [(\Delta l_{j,1} \; \Delta l_{j,2} \; \cdots \; \Delta l_{j,c} \; \Delta v_{j,1} \; \Delta v_{j,2} \; \cdots \; \Delta v_{j,c} \; \Delta T_j)_{j=1,N}]^T$$

and $\mathbf{f}$ is defined by Eq. (5-6).

The complete set of Newton-Raphson equations may be solved by transforming the matrix shown in Fig. 5-1 into the upper triangular matrix shown in Fig. 5-2. The triangularization procedure, to be described next, is based on gaussian elimination.

The matrix shown in Fig. 5-1 has several desirable features which arise from ordering. First, the elements lying outside of the shaded areas in Figs. 5-1 and 5-2 are always zero. Second, most of the elements below the principal diagonal in Fig. 5-1 are zero, and those which are nonzero are clustered along the diagonal. This characteristic makes it possible to consider only a relatively small number of squares of elements (or submatrices) at any one time.

For example, in the first step of the triangularization of the jacobian matrix in Fig. 5-1, the submatrices 1, 2, 3, 4, 6, and 7 are considered. To initiate the elimination process, the largest element of column 1 of submatrices 1 and 2 of Fig. 5-1 is selected as the pivot element. If this element lies in submatrix 2, then submatrix 6 may be filled in the process of eliminating all elements above the pivot element.

After the procedure has been applied to the last column of submatrix 2, the entire process is repeated for the next set of six submatrices; namely, submatrices 4, 5, 7, 8, 10, and 11. If one or more of the pivot elements lie in submatrix 5, then submatrix 10 may be filled in the elimination process.

Refinements of the gaussian elimination process which were used have been described by others[23] under headings such as *symbolic partial pivoting*. One of the refinements used consisted of keeping an account of the precise number of rows filled in matrices such as 3 and 10 by the triangularization process. No subsequent arithmetic was performed on the empty rows. A second refinement consisted of the use of minimum element sizes whereby each element was examined and if found to be less than some small preassigned number, it was set equal to zero. No subsequent arithmetic was performed on such elements. Also, all of the variables were scaled so that their values generally fell within the range of 0 to 2, and the columns and rows of the Newton-Raphson equations were scaled as recommended by Tewarson[23] (see also Chap. 15) before the gaussian elimination process was initiated.

If the approximations presented in this section are made, the number of nonzero elements in the shaded submatrices of Fig. 5-1 is smaller, and the effort required to evaluate many of the partial derivatives is significantly reduced. However, the general algorithm just presented for solving the Newton-Raphson equations may be used regardless of whether or not any or all of the approximations presented in this section are used.

The first class of approximations are referred to as "mathematical approximations" because they are based upon purely mathematical considerations. The second class of approximations are called "physical approximations" because they are based upon the physical characteristics of a particular system.

MATHEMATICAL APPROXIMATION

The proposed approximation amounts to neglecting the partial derivatives of the enthalpy departure functions, the Ω's, with respect to the component-flow rates. As shown in Chap. 14, Ω appears in the definition of the virtual value of the partial molar enthalpy. For example, for any component i in the liquid phase on plate j, the virtual value of the partial molar enthalpy is given by

$$\hat{h}_{ji} = H_{ji}^{\circ} + \Omega_j^L \tag{5-9}$$

when H_{ji}° is the enthalpy per mole of pure component i in the perfect gas state at the temperature T_j and pressure of one atmosphere, $\hat{h}_{ji}$ is the virtual value of the partial molar enthalpy per mole of component i, $\hat{h}_{ji} = \hat{h}_{ji}(P_j, T_j, \{l_{ji}\})$, and Ω_j^L is called the departure or deviation function per mole of the mixture, $\Omega_j^L = \Omega_j^L(P_j, T_j, \{l_{ji}\})$.

The mathematical approximation which follows is based on the fact that the Ω's are homogeneous functions of degree zero in the component-flow rates.

Thus, when the $\{\partial\Omega_j^L/\partial l_{ji}\}$ are evaluated at the l_{ji}'s assumed to make trial n, Euler's theorem[22] gives the following relationship (see App. A).

$$\sum_{i=1}^{c}(\partial\Omega_j^L/\partial l_{ji})_n(l_{ji})_n = 0 \tag{5-10}$$

Advantage of this relationship may be taken by consideration of the following terms of the Newton-Raphson equation for the function G_j

$$\cdots + \left[\frac{\partial\sum_{i=1}^{c}\hat{h}_{ji}l_{ji}}{\partial l_{j1}}\right]\Delta l_{j1} + \left[\frac{\partial\sum_{i=1}^{c}\hat{h}_{ji}l_{ji}}{\partial l_{j2}}\right]\Delta l_{j2} + \cdots + \left[\frac{\partial\sum_{i=1}^{c}\hat{h}_{ji}l_{ji}}{\partial l_{jc}}\right]\Delta l_{jc} + \cdots$$

Since $\hat{h}_{ji} = H_{ji}^{\circ} + \Omega_j^L$, this series of terms may be rearranged to the following form

$$\cdots + \sum_{i=1}^{c}\hat{h}_{ji}\,\Delta l_{ji} + \left(\sum_{i=1}^{c}l_{ji}\right)\left[\sum_{i=1}^{c}\frac{\partial\Omega_j^L}{\partial l_{ji}}\Delta l_{ji}\right] + \cdots$$

Since $\Delta l_{ji} = (l_{ji})_{n+1} - (l_{ji})_n$, the last term of the above expression may be restated as follows

$$\sum_{i=1}^{c}\left(\frac{\partial\Omega_j^L}{\partial l_{ji}}\right)_n\Delta l_{ji} = \sum_{i=1}^{c}\left(\frac{\partial\Omega_j^L}{\partial l_{ji}}\right)_n(l_{ji})_{n+1} - \sum_{i=1}^{c}\left(\frac{\partial\Omega_j^L}{\partial l_{ji}}\right)_n(l_{ji})_n$$

By Eq. (5-10), the second sum in this series is identically equal to zero. Then, if the approximation that

$$\sum_{i=1}^{c}(\partial\Omega_j^L/\partial l_{ji})_n(l_{ji})_{n+1} \cong 0 \tag{5-11}$$

is made, derivatives of the Ω_j^L's with respect to the component-flow rates may be neglected. Thus, on the basis of the assumption given by Eq. (5-11), the selected terms of the Newton-Raphson equation reduce to the same result which would have been obtained had it been assumed that mixture formed an ideal solution, that is,

$$\sum_{k=1}^{c}\frac{\partial\left[\sum_{i=1}^{c}\hat{h}_{ji}l_{ji}\right]}{\partial l_{jk}}\Delta l_{jk} = \sum_{k=1}^{c}\hat{h}_{jk}\,\Delta l_{jk}$$

A similar argument may be made for neglecting the partial derivatives of Ω_j^V's with respect to the component-flow rates of the vapor.

In support of the approximation given by Eq. (5-11), it should be noted that the relationship becomes exact if the predicted flow rates in this expression are approximated by a linear combination of the assumed set of flow rates, namely, $(l_{ji})_{n+1} = \alpha(l_{ji})_n$. This relationship is also exact if $(\partial\Omega_j^L/\partial l_{ji})_n = (\partial\Omega_j^L/\partial l_{ji})_{n+1}$.

PHYSICAL APPROXIMATIONS

Most approximations of this class involve the relative magnitudes of the partial derivatives of the activity coefficients, fugacities, and the departure function Ω with respect to temperature. If, for example, the Ω is independent of temperature or its variation with temperature is small, then the approximation $\partial\Omega/\partial T = 0$ may be made.

5-2 ALMOST BAND ALGORITHMS FOR CONVENTIONAL AND COMPLEX DISTILLATION COLUMNS

Only minor modifications of the algorithm formulated for absorbers are needed in order to solve problems involving all types of distillation columns.

The $[N(2c + 1) + 2]$ Formulation of the Newton-Raphson Equations for a Conventional Distillation Column

Distillation columns may be described by $N(2c + 3)$ equations similar to those given by Eq. (5-1) for absorbers except that in the case of distillation columns, the energy balances for stages $j = 1$ (the condenser-accumulator section) and $j = N$ (the reboiler) contain two new variables, the condenser duty Q_C and the reboiler duty Q_R. To demonstrate the formulation of the equations for distillation columns, two cases are considered: (1) a conventional distillation column with a partial condenser and the reflux rate L_1 and bottoms rate L_N specified, and (2) a conventional distillation column with a total condenser and L_1 and L_N specified.

In addition to the specification of the type of condenser, it is also supposed in all of the cases considered that the following variables have been fixed: column pressure; number of stages N; complete definition of the feed (the feed rate F, composition X_i and thermal condition); and the feed plate location f.

CONVENTIONAL DISTILLATION COLUMNS WITH PARTIAL CONDENSERS

Consider first the case where a partial condenser is used. Introduction of a partial condenser and a reboiler introduces two new variables (relative to those which were used to describe absorbers), the condenser duty Q_C and the reboiler duty Q_R. If Q_C and Q_R are regarded as independent variables, two others must be fixed. When these are taken to be the reflux rate L_1 and the bottoms rate L_N, the corresponding specification functions take the form

$$S_1 = \frac{\sum_{i=1}^{c} l_{1i}}{L_1} - 1 \tag{5-12}$$

$$S_N = \frac{\sum_{i=1}^{c} l_{Ni}}{L_N} - 1 \tag{5-13}$$

When the $[N(2c + 1) + 2]$ functions are ordered as follows

$$\mathbf{f} = [S_1(f_{j, 1} \; f_{j, 2} \; \cdots \; f_{j, c} \; m_{j, 1} \; m_{j, 2} \; \cdots \; m_{j, c} \; G_j)_{j=1, N} \; S_N]^T \tag{5-14}$$

and the differentiation is carried out with respect to the following $[N(2c + 1) + 2]$ variables in the order listed

$$\mathbf{x} = [Q_c(l_{j, 1} \; l_{j, 2} \; \cdots \; l_{j, c} \; v_{j, 1} \; v_{j, 2} \; \cdots \; v_{j, c} \; T_j)_{j=1, N} \; Q_R]^T \tag{5-15}$$

a jacobian matrix similar to the one shown in Fig. 5-1 is obtained. In this case, however, the first and last squares on the principal diagonal are $(2c + 2)$ by $(2c + 2)$ rather than $(2c + 1)$ by $(2c + 1)$ as shown in Fig. 5-1. In the above listing of the variables, v_{1i} has been used to denote d_i. The corresponding Newton-Raphson equations may be solved by gaussian elimination in a manner analogous to that described for absorbers.

The equilibrium functions are of the same form as those presented for absorbers; provided that the distillates rates $\{d_i\}$ are denoted by $\{v_{1i}\}$ and the bottom rates $\{b_i\}$ by $\{l_{Ni}\}$. For all stages except for $j = 1, f - 1, f$, and N, the material balance functions m_{ji} are of the same form as those shown for absorbers. For the exceptions, the functions take the form

$$m_{1i} = v_{2i} - v_{1i} - l_{1i}$$

$$m_{f-1, i} = v_{fi} + v_{Fi} + l_{f-2, i} - v_{f-1, i} - l_{f-1, i}$$

$$m_{fi} = v_{f+1, i} + l_{f-1, i} + l_{Fi} - v_{fi} - l_{fi} \tag{5-16}$$

$$m_{Ni} = l_{N-1, i} - v_{Ni} - l_{Ni}$$

Likewise, the enthalpy balances G_j are of the same form as those given by Eq. (5-4) for absorbers except for stages $j = 1, f - 1, f$, and N. For stages $f - 1$ and f, the denominators of G_{f-1} and G_f contains the additional terms

$$\sum_{i=1}^{c} v_{Fi} \hat{H}_{Fi} \qquad \text{and} \qquad \sum_{i=1}^{c} l_{Fi} \hat{h}_{Fi}$$

respectively. The functions G_1 (for a column with a partial condenser) and G_N are given by

$$G_1 = \frac{\sum\limits_{i=c}^{c} (v_{1i} \hat{H}_{1i} + l_{1i} \hat{h}_{1i}) + Q_c}{\sum\limits_{i=1}^{c} v_{2i} \hat{H}_{2i}} - 1$$

$$\tag{5-17}$$

$$G_N = \frac{\sum\limits_{i=1}^{c} [v_{Ni} \hat{H}_{Ni} + l_{Ni} \hat{h}_{Ni}]}{\sum\limits_{i=1}^{c} l_{N-1, i} \hat{h}_{N-1, i} + Q_R} - 1$$

For the case where Q_C and Q_R are specified instead of L_1 and L_N, the functions $\mathbf{f}$ are of the same form as shown above, but the set of independent variables x now contain L_1 and L_N instead of Q_C and Q_R

$$x = [L_1 \; (l_{j,1} \; l_{j,2} \; \cdots \; l_{j,c} \; v_{j,1} \; v_{j,2} \; \cdots \; v_{j,c} \; T_j)_{j=1,N} \; L_N]^T \tag{5-18}$$

CONVENTIONAL DISTILLATION COLUMNS WITH TOTAL CONDENSERS

For the case where a total condenser is employed and L_1 and L_N are specified, the problem is formulated as follows. Specification of L_N fixes D, since by total material balance, $D = F - L_N$. Since d_i and l_{1i} have the same composition, they are related by

$$d_i = \left(\frac{D}{L_1}\right) l_{1i}$$

and d_i should be replaced wherever it appears by its equivalent $(D/L_1)l_{1i}$. In the case of a total condenser, the variable v_{1i} does not exist, and it is replaced in $f_{1,i}$ and x by y_{1i}, the mole fraction of component i in the vapor above the liquid in the accumulator, that is,

$$f_{1,i} = \frac{\gamma_{1i}^L K_{1i} l_{1i}}{\sum\limits_{i=1}^{c} l_{1i}} - \frac{\gamma_{1i}^V y_{1i}}{\sum\limits_{i=1}^{c} y_{1i}} \tag{5-19}$$

and

$$x = [Q_C \; l_{1,1} \; \cdots \; l_{1,c} \; y_{1,1} \; \cdots \; y_{1,c} \; T_1 \; (l_{j,1} \; \cdots \; l_{jc} \; v_{j1} \; \cdots \; v_{jc} \; T_j)_{j=2,N} \; Q_R]^T \tag{5-20}$$

Also, wherever y_{1i} appears in the expression for the activity coefficients $\{\gamma_{1i}^V\}$ and the enthalpies $\{\hat{H}_{1i}\}$, it is replaced by its normalized value $y_{1i}/\sum_{i=1}^{c} y_{1i}$.

The set of $N[(2c + 1) + 2]$ functions is again given by Eq. (5-14). However, a different form of the function S_1 must be used because the expression given by Eq. (5-12) is no longer independent. For it may be obtained by a linear combination of the other functions, namely,

$$S_1 = - \left[\sum_{i=1}^{c} \sum_{j=1}^{N} m_{ji} + L_N S_N\right] \Big/ D \tag{5-21}$$

Thus, the following expression should be used for S_1 for columns with total condensers

$$S_1 = \sum_{i=1}^{c} y_{1i} - 1 \tag{5-22}$$

When the condenser duty Q_C and the reboiler duty Q_R are specified rather than L_1 and L_N, then Q_C and Q_R are replaced in the vector x of independent variables by L_1 and L_N, respectively. The specification functions are given by Eqs. (5-12) and (5-22).

The $[N(2c + 1) + 3]$ Formulation of the Almost Band Algorithm for a Complex Column with One Sidestream

Consider the case where a liquid sidestream is withdrawn at the specified rate of W_p moles per hour from plate p as shown in Fig. 3-4. This problem may be solved by introducing a new independent function and a new independent variable ϕ, which is equal to the multiplier of the flow rate L_p required to give the specified flow rate W_p. The new independent function may be stated as follows

$$F_\phi = \frac{\phi \sum\limits_{i=1}^{c} l_{pi}}{W_p} - 1 \tag{5-23}$$

The functions $\mathbf{f}$ and variables $\mathbf{x}$ are ordered as indicated by Eqs. (5-14) and (5-15) for conventional columns except for stage p, and for this stage, the functions and their ordering follow

$$\cdots f_{p1} \; f_{p2} \; \cdots \; f_{pc} \; m_{p1} \; m_{p2} \; \cdots \; m_{pc} \; G_p \; F_\phi \; \cdots$$

The variables for stage p and their ordering are

$$\cdots l_{p,\,1} \; l_{p,\,2} \; \cdots \; l_{p,\,c} \; v_{p,\,1} \; v_{p,\,2} \; \cdots \; v_{p,\,c} \; T_p \; \phi \; \cdots$$

where each w_{pi} is replaced by ϕl_{pi} in the material balances. The $2c + 2$ functions and $2c + 2$ variables for stage p are reflected by the $2c + 2$ by $2c + 2$ squares shown in Fig. 5-3.

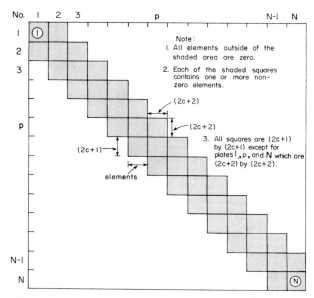

Figure 5-3 Jacobian matrix for a complex column with a liquid sidestream withdrawn from plate p.

In the application of the Almost Band Algorithm to problems involving sharp separations, it was required that all of the corrected flow rates be positive and that the temperatures lie within the range of the curve fits. In this procedure, the vector correction $\Delta\mathbf{x}$ was reduced by an appropriate scalar α until all of the corrected flow rates were positive and the corrected temperatures were within the range of the curve fits, that is,

$$\mathbf{x}_{k+1} = \mathbf{x}_k + \alpha\,\Delta\mathbf{x} \tag{5-24}$$

where $\alpha = 1, 1/2, 1/4, 1/8, \ldots$. First a value of $\alpha = 1$ is tried, and if the conditions enumerated are not satisfied, the value of α is reduced successively by a factor of $1/2$ until all flow rates are positive and the temperatures lie within range of the curve fits.

When only the l_{ji}'s or v_{ji}'s are selected as the independent variables, the picking of α becomes more difficult, particularly in the case of complex columns, because the α must be selected such the dependent component-flow rates given by the constraining equations (the component-material balances) are positive.

The disadvantage of the choice of the $\{l_{ji}\}$, $\{v_{ji}\}$, and $\{T_j\}$ as the independent variables over the choice of the $\{l_{ji}\}$ and $\{T_j\}$ as the independent variables is the additional storage requirement for the $N \times c$ vapor rates $\{v_{ji}\}$. Other formulations involving different sets of flow rates are considered in Prob. 5-5.

Formulations involving the use of the mole fractions as independent variables have been proposed by Bruno et al.[4] and Ishii and Otto.[15] Bruno et al. formulated the problem in terms of $N(c + 1)$ independent variables: the $\{V_j\}$, $\{T_j\}$, and $\{x_{j2}\ x_{j3}\ \cdots\ x_{jc}\}$. To solve an extractive distillation problem which involved nine plates and three components, Bruno et al. reported an execution time of 1.5 minutes on an IBM 360-50. Gallun[6] solved the same problem (except for a minor difference in specifications) in nine seconds of IBM 360-50 execution time with six iterations. The difference in execution time of the two methods was attributed to the efficient matrix solving techniques used by Gallun.

Ishii and Otto[15] presented a very fast algorithm based on the Newton-Raphson method. The problem was formulated in terms of $N(c + 2)$ independent variables, $\{V_j\}$, $\{T_j\}$, $\{x_{ji}\}$. In contrast to the algorithm described herein, their algorithm was based upon making several approximations in the evaluation of the partial derivatives. These approximations could lead to failure in the solution of problems involving highly nonideal solutions. However, if the approximations they proposed are not made, their algorithm for solving the jacobian matrix is no longer applicable. When their approximations were made in the algorithm presented herein, Example 5-1 appeared to be converging but was far from convergence at the end of 20 trials.

NUMERICAL EXAMPLES

Statements of Examples 5-1 and 5-2 are presented in Table 5-1, and the solutions are presented in Tables 5-2 through 5-4. The curve fits of all data used in the

Table 5-1 Statement of Examples 5-1 and 5-2

	Example 5-1	Example 5-2	
Components	FX_i	F_1X_{1i}	F_2X_{2i}
Methanol	15	....	65
Acetone	40	....	25
Ethanol	....	....	5
Water	....	50	5
Methyl acetate	5	....	
Benzene	20	....	
Chloroform	20	....	
Type of column	Conventional distillation column	Complex distillation column with two feeds	
Column pressure, atm	1	1	
Feed plate location	$f = 6$	F_1 enters on plate 6 F_2 enters on plate 21	
N	17	42	
Thermal condition of feed	Liquid at 137.1°F	F_1 is liquid at 120°F F_2 is liquid at 100°F	
Type of condenser	Total	Total	
L_1/D	9.5	3	
B, mol/h	61.91443	124	

Table 5-2 Solutions of Examples 5-1 and 5-2

Initial assumptions	Example 5-1	Example 5-2
Temperature profile, °F	$T_1 = 100$, $T_2 = 110$ $T_{17} = 175$ $T_j = 110 + (175 - 110)/15$, for $j = 3, 4, \ldots, 16$	$T_1 = 100$, $T_2 = 110$ $T_{42} = 175$ $T_j = 110 + (175 - 110)/40$, for $j = 3, 4, \ldots, 41$
L_j, mol/h	$L_j = 570, j = 1, 2, \ldots, 5$ $L_j = 610, j = 6, 7, \ldots, 16$	$L_j = 78, j = 1, 2, \ldots, 5$ $L_j = 128, j = 6, 7, \ldots, 20$ $L_j = 228, j = 21, 22, \ldots, 41$
l_{ji}, mol/h	$l_{ji} = L_jX_i$ for all i and j	$l_{ji} = L_j\left[\dfrac{F_1X_{1i} + F_2X_{2i}}{F_1 + F_2}\right]$ for all i and j
$\{y_{1i}\}$	$y_{1i} = 0.2$ for all i	$y_{1i} = 0.25$ for all i
Q_C, Btu/h	4×10^6	1×10^6
Q_R, Btu/h	4×10^6	1×10^6

Table 5-3 Solution values of selected variables for Example 5-1

Plate	T_j, °F	L_j, mol/h	Plate	T_j, °F	L_j, mol/h
1	132.60	361.81	10	134.66	462.03
2	133.02	360.85	11	134.72	462.43
3	133.40	360.39	12	134.83	463.30
4	133.74	360.65	13	135.95	465.15
5	134.09	361.98	14	135.60	498.89
6	134.56	461.76	15	137.01	475.10
7	134.58	461.65	16	140.39	483.64
8	134.60	461.78	17	145.99	61.91443
9	134.63	461.84			

Component	d_i (mol/h)	y_{1i}
Methanol	13.2015	0.33186
Acetone	17.8976	0.49304
Methyl acetate	2.9490	0.08770
Benzene	2.8971	0.06688
Chloroform	1.1402	0.02052

$Q_C = 5.633271 \times 10^6$ Btu/h
$Q_R = 5.714335 \times 10^6$ Btu/h

solution of these examples are presented in Tables B-11 through B-18. These curve fits were taken from Gallun.[6, 7] The enthalpy of the liquid phase was approximated by the assumption of ideal solution behavior. The enthalpy function Ω for the vapor phase was evaluated by use of the first two terms of the virial equation of state as described in Table B-18. The second virial coefficients were approximated as described by Prausnitz et al.[18] The critical properties and parameters needed are presented in Table B-17. These were taken from Table B-1 of App. B-1, page 213, of Ref. 18. Vapor pressures were expressed by Antoine equations, and the constants for these equations are given in Table B-13. Activity coefficients for each component in the liquid phase were approximated by use of the Wilson equation as described in Chap. 14. The energy terms appearing in this equation are given in Table B-15 for Example 5-1 and in Table B-16 for Example 5-2. The molar volumes appearing in the Wilson equation were curve fit on the basis of the data given in Table B-14. The fugacity coefficients $\gamma_i^V f_i^V/P$ for the vapor phase were approximated by use of Chap. 3 (Eqs. 3-10 through 3-12) and pages 143–144 of App. A of Ref. 18 as described in Table B-18.

The assumption of ideal solution behavior for the calculation of the enthalpy of the liquid phase was made for both examples. For the vapor phase, the

Table 5-4 Solution values of selected variables for Example 5-2

Plate	T_j (°F)	L_j (mol/h)	Plate	T_j (°F)	L_j (mol/h)
1	132.08	78.00	23	145.58	223.00
2	132.38	77.42	24	145.58	223.00
3	132.80	76.58	25	145.59	222.99
4	133.51	75.21	26	145.60	222.98
5	135.09	72.20	27 •	145.62	222.96
6	140.88	125.40	28	145.65	222.94
7	141.02	125.08	29	145.70	222.90
8	141.23	124.68	30	145.77	222.84
9	141.52	124.18	31	145.89	222.74
10	141.97	123.54	32	146.09	222.59
11	142.65	122.71	33	146.39	222.36
12	143.71	121.69	34	146.86	222.03
13	145.27	120.56	35	147.55	221.56
14	147.30	119.51	36	148.55	220.97
15	149.47	118.72	37	149.89	220.29
16	151.35	118.23	38	151.54	219.58
17	152.64	117.99	39	153.38	218.96
18	153.29	117.99	40	155.29	218.35
19	153.14	118.38	41	157.50	217.38
20	151.38	119.87	42	161.91	124.00
21	145.57	223.00			
22	145.57	223.00			

Component	d_i (mol/h)	y_{1i}
Methanol	0.8396	0.03631
Acetone	23.8181	0.29418
Ethanol	0.0008	0.00002
Water	1.3414	0.03849

$Q_C = 1.3887 \times 10^6$ Btu/h
$Q_R = 1.5230 \times 10^6$ Btu/h

Table 5-5 Convergence characteristics for Examples 5-1 and 5-2

Example	Number of iterations	Execution time (IBM 360–65)	Final value of ϕ†
5–1	8	25.8 s	1.0×10^{-11}
5–2	9	45 s	4.3×10^{-12}

† Convergence was said to have been achieved when the value of ϕ computed by

$$\phi = \frac{1}{(N+1)(c+2)} \left\{ \sum_{j=0}^{N} \left[F_j^2 + G_j^2 + \sum_{j=1}^{c} f_{ji}^2 \right] \right\}$$

was equal to or less than 1×10^{-10}. The functions appearing in ϕ were all normalized.

approximation which is analogous to the one represented by Eq. (5-11) was made. It was further assumed that $\partial\Omega_j^L/\partial T \cong 0$.

The initial set of assumed values of the variables shown in Table 5-1 were selected in a relatively arbitrary fashion so that they could not be regarded as good first guesses in the sense that they were close to the solution set. On the basis of these first guesses, for the variables, Examples 5-1 and 5-2 were each solved in less than one minute of IBM 360-65 computer time (see Table 5-5) by use of the calculational procedure presented herein. The computer times listed for all examples presented in this chapter were obtained by use of Almost Band Algorithms involving $[N(c+1)+2]$ independent variables. The computer times required to solve problems by use of these algorithms were essentially the same as those by the algorithm presented in the text.

Comparison of the $2N$ Newton-Raphson Method with the Almost Band Algorithm for Mixtures which Form Ideal Solutions

To compare the characteristics of the $2N$ Newton-Raphson method and the $[N(2c+1)+2]$ Almost Band Algorithm for solving problems involving ideal solutions, the sequence of examples shown in Table 5-6 were solved by both methods. For small numbers of components, procedure 2 of the $2N$ Newton-Raphson method is faster than the Almost Band Algorithm and conversely for large numbers of plates and a small number of components, the Almost Band Algorithm is faster than the $2N$ Newton-Raphson method as indicated by the results shown in Table 5-7. Since absorber-type problems generally involve large numbers of components relative to the number of plates, and since the mixtures encountered in most absorber applications do not deviate significantly from ideal solutions, the $2N$ Newton-Raphson method is recommended for solving such problems.

Table 5-6 Statement of examples used in the comparison of the $2N$ Newton-Raphson and the Almost Band matrix methods

The distillation column had a total condenser and the feed plate was located in the middle of the column, $N/2$, where N is equal to the total number of stages. An equimolar feed was used for each example and the total flow rate of the feed was fixed at 100 moles per hour. Examples were solved with 4, 6, 8, 10, and 12 components. The identity of the particular set of components used for each example is given in tabular form below. The temperature of the feed for each example was 100°F, and a column pressure of 300 lb/in^2 abs was used for all examples. The reflux ratio was held fixed at 2, and the product rates were set at 50% of the feed rate for all examples. The ideal solution K values and enthalpies were taken from Tables B-1 and B-2.

	Number of components				
Component	4	6	8	10	12
C_2H_6				×	×
C_3H_6		×	×	×	×
C_3H_8	×	×	×	×	×
$i\text{-}C_4H_8$					×
$i\text{-}C_4H_{10}$			×	×	×
$n\text{-}C_4H_{10}$	×	×	×	×	×
$i\text{-}C_5H_{12}$	×	×	×	×	×
$n\text{-}C_5H_{12}$	×	×	×	×	×
$n\text{-}C_6H_{14}$			×	×	×
$n\text{-}C_7H_{16}$				×	×
$n\text{-}C_8H_{18}$			×	×	×
400					×

Table 5-7 Comparison of the $2N$ Newton-Raphson and the Almost Band matrix methods

Example		$2N$ Newton-Raphson method†		Almost Band matrix method	
No. of stages	No. of components	No. of trials	Time‡ (s)	No. of trials	Time‡ (s)
12	4	9	1.03	6	0.57
12	6	5	1.32	9	0.92
12	8	5	1.49	9	2.35
12	10	8	1.73	9	3.68
12	12	7	1.87	10	6.11
25	4	6	2.59	10	1.05
25	6	8	4.52	10	1.62
25	8	12	4.32	10	2.47
25	10	16	5.37	12	4.36
25	12	9	7.34	12	6.15
50	4	16	13.13	20	3.08

† These results were obtained by use of procedure 2, Broyden's method as modified by Bennett.
‡ AMDAHL, FORTRAN H OPT 2.

5-3 MODIFICATIONS OF BROYDEN'S METHOD

As originally proposed, the sparsity of the jacobian matrix is destroyed by Broyden's method. Two procedures (or modifications) which preserves the sparsity of the jacobian matrices are presented. The procedures are demonstrated by use of simple algebraic examples and applied to the solution of distillation problems whose jacobian matrices are sparse.

Since all derivatives may be evaluated numerically in Broyden's method,[5] the necessity for programming the expressions needed for the derivatives appearing in the Newton-Raphson equations is avoided by use of these methods. The wide variety of thermodynamic packages which are available make these approaches very attractive.

After the Broyden correction for the independent variables has been computed, Broyden proposed that the inverse of the jacobian matrix of the Newton-Raphson equations be updated by use of Householder's formula. Herein lies the difficulty with Broyden's method. For Newton-Raphson formulations such as the Almost Band Algorithm for problems involving highly nonideal solutions, the corresponding jacobian matrices are exceedingly sparse, and the inverse of a sparse matrix is not necessarily sparse. The sparse characteristic of these jacobian matrices makes the application of Broyden's method (wherein the inverse of the jacobian matrix is updated by use of Householder's formula) impractical.

Two methods have been proposed for retaining the desirable characteristics of Broyden's method and eliminating the undesirable characteristic of the loss of sparsity of the jacobian matrix through the use of inverses. In both of these modifications of Broyden's method, the necessity for the development of analytical expressions for the partial derivations is eliminated. To initiate the calculational procedure in each of these modified versions of Broyden's method, the partial derivatives appearing in the jacobian matrix are evaluated numerically, and the jacobian matrix is updated in subsequent trials through the use of functional evaluations. The first modified form of Broyden's method is the one proposed by Gallun and Holland,[9] and the second modification is the one proposed by Schubert.[21]

Method 1. The Broyden-Householder Algorithm

As shown in Chap. 4, Broyden proposed the following formula for updating the jacobian matrix $\mathbf{J}_k$ to obtain $\mathbf{J}_{k+1}$

$$\mathbf{J}_{k+1} = \mathbf{J}_k + \frac{(\mathbf{f}_{k+1} - (1 - s_k)\mathbf{f}_k)\,\Delta\mathbf{x}_k^T}{s_k\,\Delta\mathbf{x}_k^T\,\Delta\mathbf{x}_k} \tag{5-25}$$

Let the scalar c_{k+1}, and the vectors $\mathbf{u}_{k+1}$ and $\mathbf{p}_{k+1}$ be defined as follows

$$c_{k+1} = \frac{1}{s_k\,\Delta\mathbf{x}_k^T\,\Delta\mathbf{x}_k} \tag{5-26}$$

$$\mathbf{u}_{k+1} = \mathbf{f}_{k+1} - (1 - s_k)\mathbf{f}_k \tag{5-27}$$

$$\mathbf{p}_{k+1}^T = \Delta\mathbf{x}_k^T \tag{5-28}$$

Use of these definitions permits Eq. (5-25) to be restated in the following form

$$J_{k+1} = J_k + u_{k+1} c_{k+1} p_{k+1}^T \tag{5-29}$$

In this algorithm, Broyden's method is applied by updating the jacobian matrices by use of Householder's formula.[13] Let J_0 be the initial approximation of the jacobian matrix with which the iterative procedure is started. Then

$$J_0 \, \Delta x_0 = -f_0 \tag{5-30}$$

and

$$\Delta x_0 = -J_0^{-1} f_0 \tag{5-31}$$

Although the inverse of J_0 appears in Eq. (5-31), it should be noted that the explicit expression of J_0^{-1} need never be developed; only the **LU** factorization is required. If J_0 is sparse, its inverse J_0^{-1} is not necessarily sparse, but its factorization $L_0 U_0$ is sparse. Thus, throughout the remainder of the development, inverses are shown but the actual numerical solutions are to be found by use of the **LU** factorizations rather than the inverses of the jacobian matrices.

After Δx_0 has been used to find x_1 as described above, the updated jacobian matrix J_1 is found as follows

$$J_1 = J_0 + u_1 c_1 p_1^T \tag{5-32}$$

where

$$u_1 = f_1 - (1 - s_0) f_0$$

$$p_1^T = \Delta x_0^T$$

$$c_1 = \frac{1}{s_0 \, \Delta x_0^T \, \Delta x_0}$$

After Δx_1 has been used to find x_2, the updated jacobian matrix J_2 is found as follows

$$J_2 = J_1 + u_2 c_2 p_2^T \tag{5-33}$$

After J_1 in this equation has been replaced by its equivalent as given by Eq. (5-32), one obtains

$$J_2 = J_0 + u_1 c_1 p_1^T + u_2 c_2 p_2^T \tag{5-34}$$

By continuation of this procedure, the matrix J_{k+1} is found as follows

$$J_{k+1} = J_0 + \sum_{i=1}^{k+1} u_i c_i p_i^T \tag{5-35}$$

Thus, it is possible to state the jacobian matrix J_{k+1} in terms of the initial jacobian matrix J_0 and the Broyden corrections for each of the successive iterations.

An algorithm is given below for solving the Newton-Raphson equations by use of only the **LU** factorization of J_0 and the Broyden update terms given by Eqs. (5-29), (5-30), and (5-31). As shown in App. 5-1, this algorithm is based on the successive application of Householder's formula to Eq. (5-29).

THE BROYDEN-HOUSEHOLDER ALGORITHM FOR SPARSE MATRICES

(1) $j = 0$

SOLVE $\mathbf{J}_0 \, \Delta\mathbf{x}_0 = -\mathbf{f}_0$

COMPUTE $\mathbf{u}_1, \mathbf{p}_1^T$

(2) SOLVE $\mathbf{J}_0 \mathbf{w} = -\mathbf{f}_{k+1}$

$\mathbf{J}_0 \mathbf{z} = \mathbf{u}_{k+1}$

(3) IF $k = 0$, GO TO (5)

(4) DO $j = 1, 2, \ldots, k$

$\beta = \alpha_j \mathbf{p}_j^T \mathbf{w}$

$\gamma = \alpha_j \mathbf{p}_j^T \mathbf{z}$

$\mathbf{w} \leftarrow \mathbf{w} + \beta \mathbf{v}_j$

$\mathbf{z} \leftarrow \mathbf{z} + \gamma \mathbf{v}_j$

(5) $\quad \mathbf{v}_{k+1} = \mathbf{z}$

$$\alpha_{k+1} = \frac{1}{(1/c_{k+1}) + \mathbf{p}_{k+1}^T \mathbf{v}_{k+1}}$$

$\beta = \alpha_{k+1} \mathbf{p}_{k+1}^T \mathbf{w}$

$\Delta\mathbf{x}_{k+1} = \mathbf{w} + \beta \mathbf{v}_{k+1}$

(6) COMPUTE $\mathbf{x}_{k+1}, \mathbf{f}_{k+1}, \mathbf{u}_{k+1}$, AND RETURN TO (2)

To demonstrate the application of this algorithm, the following algebraic example is used. In order to reduce the arithmetic required to demonstrate the application of the algorithm, a very simple example was selected whose solution is seen by inspection to be $x_1 = 1$, $x_2 = \sqrt{2}$, and $x_3 = \sqrt{3}$.

Example 5-3 (Gallun and Holland,[7] by courtesy *Comput. Chem. Eng.*) It is desired to find the set of positive values of x_1, x_2, and x_3 which make $f_1(\mathbf{x}) = f_2(\mathbf{x}) = f_3(\mathbf{x}) = 0$, where

$$f_1(\mathbf{x}) = x_1^2 - 1$$
$$f_2(\mathbf{x}) = x_2^2 - 2$$
$$f_3(\mathbf{x}) = x_3^2 - 3$$

On the basis of the initial set of assumed values

$$\mathbf{x}_0 = \begin{bmatrix} 1 & 1 & 1 \end{bmatrix}^T$$

compute $\mathbf{x}_1$ and $\mathbf{x}_2$ by use of the Broyden-Householder algorithm for sparse matrices.

SOLUTION

(1) $j = 0$, SOLVE $\mathbf{J}_0 \, \Delta\mathbf{x}_0 = -\mathbf{f}_0$

Since the analytical expression for the partial derivatives are so easily obtained (namely, $\partial f_n/\partial x_n = 2x_n$ and $\partial f_n/\partial x_m = 0$, $m \neq n$), they are used to evaluate the elements of $\mathbf{J}_0$. It is easily verified that $f_1(\mathbf{x}_0) = 0$, $f_2(\mathbf{x}_0) = -1$, and $f_3(\mathbf{x}_0) = -2$. Thus

$$\begin{bmatrix} 2 & 0 & 0 \\ 0 & 2 & 0 \\ 0 & 0 & 2 \end{bmatrix} \begin{bmatrix} \Delta x_1 \\ \Delta x_2 \\ \Delta x_3 \end{bmatrix} = - \begin{bmatrix} 0 \\ -1 \\ -2 \end{bmatrix}$$

Since $\mathbf{J}_0$ is a diagonal matrix, the solution is seen to be

$$\Delta \mathbf{x}_0 = \begin{bmatrix} \Delta x_1 \\ \Delta x_2 \\ \Delta x_3 \end{bmatrix} = \begin{bmatrix} 0 \\ 1/2 \\ 1 \end{bmatrix}$$

First, try $s_0 = 1$ (see procedure 2, Broyden's method in Chap. 4). Thus

$$\mathbf{x}_1 = \mathbf{x}_0 + \Delta \mathbf{x}_0 = \begin{bmatrix} 1 \\ 1 \\ 1 \end{bmatrix} + \begin{bmatrix} 0 \\ 1/2 \\ 1 \end{bmatrix} = \begin{bmatrix} 1 \\ 3/2 \\ 2 \end{bmatrix}$$

and

$$f_1(\mathbf{x}_1) = 0 \qquad f_2(\mathbf{x}_1) = 1/4 \qquad f_3(\mathbf{x}_1) = 1$$

Thus, the inequality (see procedure 2, Chap. 4) is satisfied, since

$$[(1/4)^2 + (1)^2]^{1/2} < [(-1)^2 + (-2)^2]^{1/2}$$

and the full step size ($s_0 = 1$) may be used to compute $\mathbf{x}_1$. By Eq. (5-27).

$$\mathbf{u}_1 = \mathbf{f}_1 = \begin{bmatrix} 0 & 1/4 & 1 \end{bmatrix}^T$$

and

$$\mathbf{p}_1^T = \Delta \mathbf{x}_0^T = \begin{bmatrix} 0 & 1/2 & 1 \end{bmatrix}^T$$

(2) SOLVE $\mathbf{J}_0 \mathbf{w} = -\mathbf{f}_1$

$$\begin{bmatrix} 2 & 0 & 0 \\ 0 & 2 & 0 \\ 0 & 0 & 2 \end{bmatrix} \begin{bmatrix} w_1 \\ w_2 \\ w_3 \end{bmatrix} = - \begin{bmatrix} 0 \\ 1/4 \\ 1 \end{bmatrix}$$

Thus

$$\mathbf{w} = \begin{bmatrix} w_1 \\ w_2 \\ w_3 \end{bmatrix} = \begin{bmatrix} 0 \\ -1/8 \\ -1/2 \end{bmatrix}$$

Next solve $\mathbf{J}_0 \mathbf{z} = \mathbf{u}_1$

$$\begin{bmatrix} 2 & 0 & 0 \\ 0 & 2 & 0 \\ 0 & 0 & 2 \end{bmatrix} \begin{bmatrix} z_1 \\ z_2 \\ z_3 \end{bmatrix} = \begin{bmatrix} 0 \\ 1/4 \\ 1 \end{bmatrix}$$

$$\mathbf{z} = \begin{bmatrix} z_1 \\ z_2 \\ z_3 \end{bmatrix} = \begin{bmatrix} 0 \\ 1/8 \\ 1/2 \end{bmatrix}$$

(3) Since $k = 0$, go to (5)

(5) $\mathbf{v}_1 = \mathbf{z} = [0 \quad 1/8 \quad 1/2]^T$

$$\alpha_1 = \frac{-1}{(1/c_1) + \mathbf{p}_1^T \mathbf{v}_1}$$

By Eq. (5-46)

$$c_1 = \frac{1}{s_0\,\Delta\mathbf{x}_0^T\,\Delta\mathbf{x}_0} = \frac{1}{(1)[0 \quad 1/2 \quad 1]\begin{bmatrix} 0 \\ 1/2 \\ 1 \end{bmatrix}} = \frac{1}{(1/4)+1} = \frac{4}{5}$$

$$\alpha_1 = \frac{-1}{(5/4) + [0 \quad 1/2 \quad 1]\begin{bmatrix} 0 \\ 1/8 \\ 1/2 \end{bmatrix}} = \frac{-1}{(5/4)+(1/16)+(1/2)} = -\frac{16}{29}$$

$$\beta = \alpha_1 \mathbf{p}_1^T \mathbf{w} = (-16/29)[0 \quad 1/2 \quad 1]\begin{bmatrix} 0 \\ -1/8 \\ -1/2 \end{bmatrix}$$

$$= (-16/29)(-(1/16)-(1/2) = 9/29$$

$$\Delta\mathbf{x}_1 = \mathbf{w} + \beta\mathbf{v}_1 = \begin{bmatrix} 0 \\ -1/8 \\ -1/2 \end{bmatrix} + (9/29)\begin{bmatrix} 0 \\ 1/8 \\ 1/2 \end{bmatrix}$$

$$= (-20/29)\begin{bmatrix} 0 \\ 1/8 \\ 1/2 \end{bmatrix} = \begin{bmatrix} 0 \\ -5/58 \\ -10/29 \end{bmatrix}$$

(6) For $s_1 = 1$

$$\mathbf{x}_2 = \mathbf{x}_1 + \Delta\mathbf{x}_1 = \begin{bmatrix} 1 \\ 3/2 \\ 2 \end{bmatrix} + \begin{bmatrix} 0 \\ -0.0862 \\ -0.345 \end{bmatrix} = \begin{bmatrix} 1 \\ 1.414 \\ 1.655 \end{bmatrix}$$

and

$$f_1(\mathbf{x}_2) = 0 \qquad f_2(\mathbf{x}_2) = 0 \qquad f_3(\mathbf{x}_3) = -0.261$$

Thus, the inequality given of Broyden's method (see procedure 2, Chap. 4) is satisfied, that is,

$$[(0)^2 + (-0.261)^2]^{1/2} < [(1/4)^2 + (1)^2]^{1/2}$$

Hence,

$$\mathbf{u}_2 = \mathbf{f}_2 = \begin{bmatrix} 0 \\ 0 \\ -0.261 \end{bmatrix}$$

Now return to (2)

(2) $k = 1$

SOLVE $\mathbf{J}_0 \mathbf{w} = -\mathbf{f}_2$

$$\begin{bmatrix} 2 & 0 & 0 \\ 0 & 2 & 0 \\ 0 & 0 & 2 \end{bmatrix} \begin{bmatrix} w_1 \\ w_2 \\ w_3 \end{bmatrix} = - \begin{bmatrix} 0 \\ 0 \\ -0.261 \end{bmatrix}$$

Thus

$$\mathbf{w} = \begin{bmatrix} w_1 \\ w_2 \\ w_3 \end{bmatrix} = \begin{bmatrix} 0 \\ 0 \\ 0.130 \end{bmatrix}$$

SOLVE $\mathbf{J}_0 \mathbf{z} = \mathbf{u}_2$

$$\begin{bmatrix} 2 & 0 & 0 \\ 0 & 2 & 0 \\ 0 & 0 & 2 \end{bmatrix} \begin{bmatrix} z_1 \\ z_2 \\ z_3 \end{bmatrix} = \begin{bmatrix} 0 \\ 0 \\ -0.261 \end{bmatrix}$$

Thus

$$\mathbf{z} = \begin{bmatrix} z_1 \\ z_2 \\ z_3 \end{bmatrix} = \begin{bmatrix} 0 \\ 0 \\ -0.130 \end{bmatrix}$$

(3)   Since $k = 1$, go to (4)

(4) $j = 1$, and $k = 1$

$$\beta = \alpha_1 \mathbf{p}_1^T \mathbf{w} = (-16/29)[0 \quad 1/2 \quad 1] \begin{bmatrix} 0 \\ 0 \\ 0.130 \end{bmatrix} = -0.0717$$

$$\gamma = \alpha_1 \mathbf{p}_1^T \mathbf{z} = (-16/29)[0 \quad 1/2 \quad 1] \begin{bmatrix} 0 \\ 0 \\ -0.130 \end{bmatrix} = 0.0717$$

$$\mathbf{w} \leftarrow \mathbf{w} + \beta \mathbf{v}_1 = \begin{bmatrix} 0 \\ 0 \\ 0.130 \end{bmatrix} + (-0.0717) \begin{bmatrix} 0 \\ 1/8 \\ 1/2 \end{bmatrix} = \begin{bmatrix} 0 \\ -0.009 \\ 0.094 \end{bmatrix}$$

$$\mathbf{z} \leftarrow \mathbf{z} + \gamma \mathbf{v}_1 = \begin{bmatrix} 0 \\ 0 \\ -0.130 \end{bmatrix} + (0.0717) \begin{bmatrix} 0 \\ 1/8 \\ 1/2 \end{bmatrix} = \begin{bmatrix} 0 \\ 0.009 \\ -0.094 \end{bmatrix}$$

(5) $\mathbf{v}_2 = \mathbf{z} = \begin{bmatrix} 0 \\ 0.009 \\ -0.094 \end{bmatrix}$

$$\alpha_2 = \frac{-1}{(1/c_2) + \mathbf{p}_2^T \mathbf{v}_2}$$

$$c_2 = \frac{1}{s_1 \, \Delta \mathbf{x}_1^T \, \Delta \mathbf{x}_1} = \frac{1}{(1)[0, \; -0.0862, \; -0.345] \begin{bmatrix} 0 \\ -0.0862 \\ -0.345 \end{bmatrix}} = 7.908$$

Since $\mathbf{p}_2 = \Delta\mathbf{x}_1$

$$\mathbf{p}_2^T\mathbf{v}_2 = [0, \ -0.0862, \ -0.345]\begin{bmatrix} 0 \\ 0.009 \\ -0.094 \end{bmatrix} = 0.0316$$

$$\alpha_2 = \frac{-1}{(1/7.908) + 0.0316} = \frac{-1}{0.158} = -6.33$$

$$\beta = \alpha_2\,\mathbf{p}_2^T\mathbf{w} = (-6.33)[0, \ -0.0862, \ -0.345]\begin{bmatrix} 0 \\ -0.009 \\ 0.094 \end{bmatrix} = 0.200$$

$$\Delta\mathbf{x}_2 = \mathbf{w} + \beta\mathbf{v}_2$$

$$\Delta\mathbf{x}_2 = \begin{bmatrix} 0 \\ -0.009 \\ 0.094 \end{bmatrix} + (0.200)\begin{bmatrix} 0 \\ 0.009 \\ -0.094 \end{bmatrix} = \begin{bmatrix} 0 \\ -0.007 \\ 0.075 \end{bmatrix}$$

(6) Try $s_2 = 1$

$$\mathbf{x}_3 = \mathbf{x}_2 + \Delta\mathbf{x}_2 = \begin{bmatrix} 1 \\ 1.414 \\ 1.655 \end{bmatrix} + \begin{bmatrix} 0 \\ -0.007 \\ 0.075 \end{bmatrix} = \begin{bmatrix} 1 \\ 1.407 \\ 1.73 \end{bmatrix}$$

and

$$f_1(\mathbf{x}_3) = 0 \qquad f_2(\mathbf{x}_3) = -0.02 \qquad f_3(\mathbf{x}_3) = -0.007$$

Thus, the inequality of Broyden's method is satisfied

$$[(-0.02)^2 + (-0.007)^2]^{1/2} < [(0)^2 + (-0.261)^2]^{1/2}$$

Hence

$$\mathbf{u}_3 = \mathbf{f}_3 = \begin{bmatrix} 0 \\ -0.020 \\ -0.007 \end{bmatrix}$$

Return to (2).

Schubert's Modification of Broyden's Method

In the formulation of Schubert's method,[21] it is convenient to denote the kth approximation of the jacobian by $\mathbf{G}^{(k)}$, where the iteration number is carried as a superscript enclosed by parentheses. Then Broyden's formula for computing the next approximation of the jacobian is given by

$$\mathbf{G}^{(k+1)} = \mathbf{G}^{(k)} + \frac{[\mathbf{f}^{(k+1)} - (1 - s^{(k)})\mathbf{f}^{(k)}]\,\Delta\mathbf{x}^{(k)T}}{s^{(k)}\,\Delta\mathbf{x}^{(k)T}\,\Delta\mathbf{x}^{(k)}} \qquad (5\text{-}36)$$

Schubert proposed a modification of Broyden's method which takes advantage of the fact that in the case of sparse jacobian matrices, most of the elements are either equal to zero or fixed constants. These known elements would be modified from trial to trial by Broyden's method. Schubert imposed the condition that these known elements should remain unchanged in the jacobian revision. By use of this and certain other conditions,[21] Schubert obtained the following row-by-row analog of Eq. (5-36)

$$\mathbf{g}_i^{(k+1)} = \mathbf{g}_i^{(k)} + \frac{[f_i^{(k+1)} - (1 - s^{(k)})f_i^{(k)}] \Delta \mathbf{x}_i^{(k)T}}{s^{(k)} \Delta \mathbf{x}_i^{(k)T} \Delta \mathbf{x}_i^{(k)}} \qquad (i = 1, 2, \ldots, n) \qquad (5\text{-}37)$$

where n is the order of the jacobian matrix, and

$\mathbf{g}_i^{(k)}$ = a row vector which contains the elements of the ith row of the jacobian $\mathbf{G}^{(k)}$

$\Delta \mathbf{x}_i^{(k)}$ = a column vector derived from $\Delta \mathbf{x}^{(k)}$ by setting to zero each element of $\Delta \mathbf{x}^{(k)}$ that corresponds to an element of $\mathbf{g}_i^{(k)}$ which is a known constant

$f_i^{(k)}$ = the ith element of $\mathbf{f}^{(k)}$

The application of Schubert's method is demonstrated by the following numerical example.

Example 5-4 [Gallun and Holland,[9] by courtesy *Comput. Chem. Eng.*) Make one trial on the problem stated in Example 5-3 by use of Schubert's method.

SOLUTION The calculation of $\mathbf{x}^{(1)}$ by this method is precisely the same as shown for $\mathbf{x}_1$ in Example 5-3. On the basis of the set of assumed values

$$\mathbf{x}^{(0)} = [1 \quad 1 \quad 1]^T$$

the Newton-Raphson equations

$$\mathbf{G}^{(0)} \Delta \mathbf{x}^{(0)} = -\mathbf{f}^{(0)}$$

$$\begin{bmatrix} 2 & 0 & 0 \\ 0 & 2 & 0 \\ 0 & 0 & 2 \end{bmatrix} \begin{bmatrix} \Delta x_1 \\ \Delta x_2 \\ \Delta x_3 \end{bmatrix} = - \begin{bmatrix} 0 \\ -1 \\ -2 \end{bmatrix}$$

are solved for $\Delta \mathbf{x}^{(0)}$ to give

$$\Delta \mathbf{x}^{(0)} = \begin{bmatrix} 0 \\ 1/2 \\ 1 \end{bmatrix}$$

Again as shown in Example 5-3, the inequality of Broyden's method (see procedure 2, Chap. 4) is satisfied by taking $s^{(0)} = 1$. Thus, as in Example 5-3

$$\mathbf{x}^{(1)} = \mathbf{x}^{(0)} + \Delta \mathbf{x}^{(0)} = \begin{bmatrix} 1 \\ 3/2 \\ 2 \end{bmatrix}$$

and as in Example 5-3

$$\mathbf{f}_1 = \mathbf{f}^{(1)} = \begin{bmatrix} f_1^{(1)} \\ f_2^{(1)} \\ f_3^{(1)} \end{bmatrix} = \begin{bmatrix} 0 \\ 1/4 \\ 1 \end{bmatrix}$$

(1) $i = 1$,

$$\mathbf{g}_1^{(0)} = [2 \quad 0 \quad 0]$$

The known constants of $\mathbf{g}_1^{(0)}$ are the two zeros. (These elements will have the value zero for any choice of $\mathbf{x}$.) Then the element $1/2$ and 1 of $\boldsymbol{\Delta}\mathbf{x}^{(0)}$ are replaced by zeros to give

$$\boldsymbol{\Delta}\mathbf{x}_1^{(0)} = \begin{bmatrix} 0 \\ 0 \\ 0 \end{bmatrix}$$

Since $\boldsymbol{\Delta}\mathbf{x}_1^{(0)T} \boldsymbol{\Delta}\mathbf{x}_1^{(0)} = 0$, the denominator of the corrective term is zero. To avoid this division by zero, set the correction term of Eq. (5-37) equal to zero to give

$$\mathbf{g}_1^{(1)} = [2 \quad 0 \quad 0]$$

Since the first and third elements of $\mathbf{g}_2^{(0)}$ are known to be zero for all choices of $\mathbf{x}$, the corresponding elements of $\boldsymbol{\Delta}\mathbf{x}^{(0)}$ are set equal to zero to give

$$\boldsymbol{\Delta}\mathbf{x}_2^{(0)} = \begin{bmatrix} 0 \\ 1/2 \\ 0 \end{bmatrix}$$

Since $s^{(0)} = 1$, and $f_2^{(1)} = 1/4$

$$\mathbf{g}_2^{(1)} = \mathbf{g}_2^{(0)} + \frac{f_2^{(1)} \boldsymbol{\Delta}\mathbf{x}_2^{(0)T}}{(1) \boldsymbol{\Delta}\mathbf{x}_2^{(0)T} \boldsymbol{\Delta}\mathbf{x}_2^{(0)}} = [0 \quad 2 \quad 0] + \frac{(1/4)[0 \quad 1/2 \quad 0]}{[0 \quad 1/2 \quad 0]\begin{bmatrix} 0 \\ 1/2 \\ 0 \end{bmatrix}}$$

$$= [0 \quad 2 \quad 0] + [0 \quad 1/2 \quad 0]$$

Thus

$$\mathbf{g}_2^{(1)} = [0 \quad 5/2 \quad 0]$$

(3) $i = 3$

$$g_3^{(0)} = [0 \quad 0 \quad 2]$$

$$\boldsymbol{\Delta}\mathbf{x}_3^{(0)} = \begin{bmatrix} 0 \\ 0 \\ 1 \end{bmatrix}$$

Since $s^{(0)} = 1$, and $f_3^{(1)} = 1$

$$\mathbf{g}_3^{(1)} = \mathbf{g}_3^{(0)} + \frac{f_3^{(1)} \boldsymbol{\Delta}\mathbf{x}_3^{(0)T}}{\boldsymbol{\Delta}\mathbf{x}_3^{(0)T} \boldsymbol{\Delta}\mathbf{x}_3^{(0)}} = [0 \quad 0 \quad 2] + \frac{(1)[0 \quad 0 \quad 1]}{[0 \quad 0 \quad 1]\begin{bmatrix} 0 \\ 0 \\ 1 \end{bmatrix}}$$

$$= [0 \quad 0 \quad 2] + [0 \quad 0 \quad 1]$$

Thus

$$\mathbf{g}_3^{(1)} = \begin{bmatrix} 0 & 0 & 3 \end{bmatrix}$$

Since the rows of $\mathbf{G}^{(1)}$ are given by $g_1^{(1)}$, $g_2^{(1)}$, and $g_3^{(1)}$, the required numerical values are available for solving

$$\mathbf{G}^{(1)} \, \Delta\mathbf{x}^{(1)} = -\mathbf{f}^{(1)}$$

Namely

$$\begin{bmatrix} 2 & 0 & 0 \\ 0 & 5/2 & 0 \\ 0 & 0 & 3 \end{bmatrix} \begin{bmatrix} \Delta x_1 \\ \Delta x_2 \\ \Delta x_3 \end{bmatrix} = -\begin{bmatrix} 0 \\ 1/4 \\ 1 \end{bmatrix}$$

Thus

$$\Delta\mathbf{x}^{(1)} = \begin{bmatrix} 0 \\ -1/10 \\ -1/3 \end{bmatrix}$$

Try $s^{(1)} = 1$

$$\mathbf{x}^{(2)} = \mathbf{x}^{(1)} + \Delta\mathbf{x}^{(1)} = \begin{bmatrix} 1 \\ 3/2 \\ 2 \end{bmatrix} + \begin{bmatrix} 0 \\ -1/10 \\ -1/3 \end{bmatrix} = \begin{bmatrix} 1 \\ 1.4 \\ 1.67 \end{bmatrix}$$

Then

$$f_1(\mathbf{x}^{(2)}) = 0 \qquad f_2(\mathbf{x}^{(2)}) = -0.04 \qquad f_3(\mathbf{x}^{(2)}) = -0.21$$

and since the inequality of Broyden's method

$$[(-0.04)^2 + (-0.21)^2]^{1/2} < [(1/4)^2 + (1)^2]^{1/2}$$

is satisfied, the $\mathbf{x}^{(2)}$ obtained by use of $s^{(1)} = 1$ is taken to be the assumed value of $\mathbf{x}^{(2)}$ for the next trial.

Example 5-5 was used by Gallun and Holland[9] to compare Broyden's method implemented with the new algorithm to the Newton-Raphson method and to Schubert's[21] modification of Broyden's method. The statement of this example is given in Table 5-8 and the solution is presented in Table 5-9.

Solution of this example as originally implemented by Broyden[5] would have required an excessive amount of computer time. The example is described by 452 Newton-Raphson equations whose jacobian matrix contains only 3532 nonzero elements out of a possible 204,304. Of these 3532 nonzero elements, 798 are known to be constants, generally 1 or -1 due to the linearity of the equations.

Table 5-8 Statement of Example 5-5

Component	$F_1 X_{1i}$ (g mol/s)	$F_2 X_{2i}$ (g mol/s)	$F_3 X_{3i}$ (g mol/s)
Methyl alcohol	0	0.25	65.0
Acetone	0	0.50	25.0
Ethanol	0	5.0	5.0
Water	5.0	197.5	5.0

Other specifications

The column has a total condenser, 50 stages, F_1 enters on plate 4, F_2 on plate 6, and F_3 on plate 21. The column pressure is 760 mm. Feeds F_1, F_2, and F_3 enter with enthalpies 0.33993877×10^4, 0.2853018×10^6, and 0.08139422 cal/g mol, respectively. A reflux ratio (L_1/D) of 2.5 is to be used, and the bottoms is to be withdrawn at the rate of 285 lb mol per hour. The equilibrium and thermodynamic data to be used are the same as stated for Example 5-3.

Table 5-9 Solution of Example 5-5

1. Final profiles, temperature, and vapor and liquid rates

Plate	T_j (°F)	V_j (lb mol/h)	L_j (lb mol/h)	Plate	T_j (°F)	V_j (lb mol/h)	L_j (lb mol/h)
1	134.26	....	58.13	26	154.61	78.21	36.32
2	136.11	81.38	56.32	27	154.61	78.21	36.32
3	139.12	79.57	53.77	28	154.61	78.21	36.32
4	144.65	77.02	56.31	29	154.61	78.21	36.32
5	152.70	74.56	52.30	30	154.61	78.21	36.32
6	169.06	70.55	25.66	31	154.61	78.21	36.32
7	169.33	71.58	25.66	32	154.61	78.21	36.32
8	169.50	71.60	25.66	33	154.61	78.21	36.32
9	169.62	71.61	25.66	34	154.61	78.21	36.32
10	169.72	71.61	25.66	35	153.62	78.21	36.32
11	169.81	71.61	25.66	36	154.62	78.21	36.32
12	169.92	71.61	25.66	37	154.62	78.21	36.32
13	170.05	71.61	25.66	38	154.62	78.21	36.32
14	170.23	71.60	25.66	39	154.62	78.21	36.32
15	170.45	71.60	25.66	40	154.63	78.21	36.32
16	170.70	71.60	25.66	41	154.65	78.20	36.32
17	170.91	71.61	25.67	42	154.69	78.19	36.32
18	170.84	71.65	25.68	43	154.77	78.17	36.31
19	169.96	71.78	25.68	44	154.91	78.14	36.31
20	166.64	72.15	25.72	45	155.20	78.07	36.29
21	154.61	73.55	25.86	46	155.75	77.94	36.27
22	154.61	78.21	36.32	47	156.77	77.71	36.23
23	154.61	78.21	36.32	48	158.63	77.32	36.17
24	154.61	78.21	36.32	49	161.93	76.73	36.06
25	154.61	78.21	36.32	50	169.48	75.64	28.50

(Continued on page 206.)

Table 5-9 (*continued*)

	2. Product distribution	
Component	b_i (lb mol/h)	d_i (lb mol/h)
Methyl alcohol	0.8832228×10^{-1}	0.6516167×10^2
Acetone	0.1936700×10^2	0.6133089×10
Ethanol	0.1895224×10	0.8104776×10
Water	0.1899549×10	0.2056004×10

$Q_C = 111{,}504.0$ Btu/h
$Q_R = 126{,}206.9$ Btu/h

		3. Comparison of the Broyden-Householder and the Schubert Algorithms				
		Jacobian				
Method	Method of calculation of derivatives	Evaluations	Factorization	Iterations	Final squared norm	Execution time(s)†
Newton-Raphson	Analytical	14	13	13	7.2×10^{-13}	9.97
Newton-Raphson	Numerical	13	12	12	4.0×10^{-11}	59.94
Broyden-Householder	Analytical	4	4	51	1.97×10^{-10}	12.77
Broyden-Householder	Numerical	5	5	56	1.16×10^{-9}	35.90
Schubert	Analytical	1	37	37	8.91×10^{-10}	22.77
Schubert	Numerical	1	34	34	2.40×10^{-10}	26.89

† AMDAHL, FORTRAN H, OPT 2

5-4 THE BOSTON-SULLIVAN ALGORITHM[3]

This algorithm is based on the use of newly defined energy and volatility parameters as the primary successive approximation variables, and Broyden's method is used to iterate on these parameters. A brief review of the Boston-Sullivan Algorithm follows.

The component-material balances are formulated in a manner analogous to that shown in Chap. 2 except for the fact that Boston and Sullivan[3] stated these balances in terms of the liquid rates $\{l_{ji}\}$ rather than the vapor rates $\{v_{ji}\}$. Temperatures were computed by use of a variation of the method wherein a different base component is used for each plate as suggested by Billingsley.[2] Partial molar enthalpies were also expressed in terms of the deviation function Ω.

A formulation of the model equations follows. The S_{ji}'s appearing in the component-material balance for each stage j

$$l_{j-1, i} - (1 + S_{ji})l_{ji} + S_{j+1, i}l_{j+1, i} = 0 \tag{5-38}$$

may be stated in terms of the "S parameters" as follows

$$S_{ji} = \frac{K_{ji}V_j}{L_j} = S_b S_{jR}\alpha_{ji} \tag{5-39}$$

where $K_{ji} = K_{ji}(P_j, T_j, \{x_{ji}\}, \{y_{ji}\})$ (note the vapor- and liquid-phase activity coefficients are contained in this definition of the vapor-liquid equilibrium ratio)

$S_j = K_{jb} V_j / L_j$

$S_{Rj} = S_j / S_b$, the relative S parameter

$$S_b = \left(\prod_{j=1}^{N} S_j\right)^{1/N}, \text{ the base } S \text{ parameter, where } \prod_{j=1}^{N} S_j = S_1, S_2, \ldots, S_N.$$

The relative volatility as used in this development is defined as follows

$$K_{ji}(P_j, T_j, \{x_{ji}\}, \{y_{ji}\}) = K_{jb}(T_j) \cdot \alpha_{ji}(P_j, T_j, \{x_{ji}\}, \{y_{ji}\}) \tag{5-40}$$

Thus

$$y_{ji} = K_{ji}x_{ji} = \alpha_{ji}K_{jb}x_{ji} \tag{5-41}$$

It follows as shown in Chap. 2 that

$$K_{jb} = \frac{1}{\sum_{i=1}^{c} \alpha_{ji}x_{ji}} \tag{5-42}$$

and thus

$$y_{ji} = \frac{\alpha_{ji}l_{ji}}{\sum_{i=1}^{c} \alpha_{ji}l_{ji}} \tag{5-43}$$

where

$$x_{ji} = l_{ji} \Bigg/ \sum_{i=1}^{c} l_{ji} \tag{5-44}$$

The K value for the base component was taken to be an exponential function of temperature as follows

$$\ln K_{jb} = A_j - \frac{B_j}{T_j} \tag{5-45}$$

The vapor enthalpies were expressed in terms of the enthalpy departure function Ω (see Chap. 14) as follows

$$H = \sum_{i=1}^{c} y_i \bar{H}_i = \sum_{i=1}^{c} y_i(H_i^\circ + \Omega) = \sum_{i=1}^{c} y_i H_i^\circ + \Omega \tag{5-46}$$

where $\Omega = \Omega(P, T, \{y_i\})$ and H_i° is the enthalpy of component i in the perfect gas state at the temperature T. The enthalpy departure function was separated into a composition correction term and a temperature-dependent term by introducing new variables Φ_y and Φ_T as follows

$$\Omega(P, T, \{y_i\}) = \Phi_y + \Phi_T(T - T^*) \tag{5-47}$$

The variables Φ_T and Φ_y are defined by

$$\Phi_T = [\Omega(P, T, \{y_i^*\}) - \Omega(P, T^*, \{y_i^*\}]/(T - T^*) \tag{5-48}$$

$$\Phi_y = \Omega^* + \Omega(P, T, \{y_i\}) - \Omega(P, T, \{y_i^*\}) \tag{5-49}$$

where $\Omega^* = \Omega(P, T^*, \{y_i^*\})$
T^* = reference temperature
$\{y_i^*\}$ = reference set of mole fractions

Next the following set of variables was introduced

$$\Phi_i = (H_i^\circ - H_{bi}^\circ)/(T - T_b) \tag{5-50}$$

where H_{bi}° is the perfect gas enthalpy of component i evaluated at an arbitrary reference temperature T_b. Let the new variables Γ and Θ be defined as follows

$$\Theta = \sum_{i=1}^{c} y_i \Phi_i$$
$$\Gamma = T - T_b \tag{5-51}$$

The total vapor enthalpy may now be expressed as follows

$$H = \Gamma\Theta + \Phi_y + \Phi_T(\Gamma - \Gamma^*) + \sum_{i=1}^{c} y_i H_{bi}^\circ \tag{5-52}$$

The variable Θ was partitioned into two factors Θ_b and Θ_r as follows

$$\Theta = \Theta_b \Theta_r \tag{5-53}$$

where Θ_b is strongly dependent on vapor composition and Θ_r is a weak function of both vapor composition and temperature. These functions are defined as follows

$$\Theta_r = \sum_{i=1}^{c} y_{ri} \Phi_{ri}$$

$$\Theta_b = \sum_{i=1}^{c} y_i \Phi_{bi}$$

$$\Phi_{bi} = \lim_{T \to T_b} \Phi_i \tag{5-54}$$

$$\Phi_{ri} = \Phi_i/\Phi_{bi}$$

$$y_{ri} = y_i \Phi_{bi} \bigg/ \sum_{i=1}^{c} y_i \Phi_{bi}$$

The enthalpy of the liquid phase was treated in a manner similar to that of the vapor phase. In this case, the partial molar enthalpy $\bar{h}_i$ is related to the excess partial molar enthalpy as follows

$$\bar{h}_i = h_i + \bar{h}_i^E \tag{5-55}$$

where h_i is the enthalpy of a pure component evaluated at P and T of the mixture, and $\bar{h}_i^E$ is the excess partial molar enthalpy. Next a new set of variables was introduced. First, the variable ϕ_i is defined

$$\phi_i = (h_i - h_{bi}^\circ)/(T - T_b) \tag{5-56}$$

where h_{bi}° is the liquid enthalpy evaluated at the temperature T_b. Then the total enthalpy per mole of liquid mixture is given by

$$h = \Gamma\Theta + h^E + \sum_{i=1}^{c} x_i h_{bi}^\circ \tag{5-57}$$

where

$$\theta = \sum_{i=1}^{c} x_i \phi_i$$

$$h^E = \sum_{i=1}^{c} x_i h_i^E$$

The new variable θ was partitioned in the same manner as Θ, namely,

$$\theta = \theta_b \theta_r$$

where

$$\theta_r = \sum_{i=1}^{c} x_{ri} \phi_{ri}$$

$$\theta_b = \sum_{i=1}^{c} x_i \phi_{bi}$$

$$\phi_{bi} = \lim_{T \to T_b} \phi_i$$

$$\phi_{ri} = \phi_i / \phi_{bi}$$

$$x_{ri} = x_i \phi_{bi} \bigg/ \sum_{i=1}^{c} x_i \phi_{bi}$$

The energy balance enclosing any interior stage j other than the feed plate or plate above it is given by

$$\sum_{i=1}^{c} l_{j-1,i} \bar{h}_{j-1,i} + \sum_{i=1}^{c} v_{j+1,i} \bar{H}_{j+1,i} - \sum_{i=1}^{c} l_{ji} \bar{h}_{ji} - \sum_{i=1}^{c} v_{ji} \bar{H}_{ji} = 0 \tag{5-58}$$

The results obtained above for vapor and liquid enthalpies together with other equations including the total-material balance equations and the component-material balance equations were substituted into Eq. (5-58) to obtain the following form of the energy balance equations which were employed to calculate the liquid phase rates for any stage $j < f - 1$.

$$(a_{j-1} - a_j')L_{j-1} - (a_j - a_{j+1}')L_j = (a_{j+1} - a_{j-1}')D \tag{5-59}$$

where

$$a_j = \Gamma_j \theta_j + h_j^E - \lambda_{bj}^{\circ}$$

$$a_j' = \Gamma_j \theta_j + \Phi_{yj} + (\Gamma_j - \Gamma_j^{\circ})\Phi_{Tj}$$

$$\lambda_{bj}^{\circ} = \sum_{i=1}^{c} x_{ji}(H_{bi}^{\circ} - h_{bi}^{\circ})$$

The new variables Φ_{yj}, Φ_{Tj}, Θ_{rj}, θ_{rj} and h^E are referred to as the energy parameters. Then parameters and the volatility parameters α_{ji}'s become the principal successive approximation variables of the algorithm. The following calculational procedure was used.

Step 1. Assume values for the energy and volatility parameters, the S_R's $\{\alpha_{ji}\}$, $\{\Phi_{Tj}\}$, $\{\Phi_{yj}\}$.

Step 2. Solve the component material balances for the liquid rates $\{l_{ji}\}$.

Step 3. Calculate the liquid mole fractions by use of Eq. (5-45).

Step 4. Calculate K_b's by use of Eq. (5-42) and the corresponding temperatures by use of Eq. (5-45).

Step 5. Calculate the coefficients of the energy balance equations (4-59) and solve for the total liquid rates. Then compute the vapor rates by use of total material balances.

Step 6. Compute the S_R's from the defining equation given beneath Eq. (5-39). Compare these values with the last assumed values. If they do not agree within an acceptable tolerance, assume a new set of values and return to step 2. A quasi-Newton method was used to find the new set of S_R's.

Step 7. Calculate the vapor mole fractions by use of Eq. (5-43).

Step 8. Evaluate the equilibrium ratios and enthalpies.

Step 9. Update the energy and volatility parameters using the defining equations. If they do not agree within an acceptable tolerance, retain the updated values and return to step 2.

Numerical examples solved by Boston and Sullivan[3] demonstrated that the method is both fast and stable for all types of problems considered.

PROBLEMS

5-1 Show that the expression given by Eq. (5-21) for S_1 reduces to the one given by Eq. (5-12).

5-2 Show that the assumptions

$$\sum_{i=1}^{c} \left(\frac{\partial \gamma_{jk}^L}{\partial l_{ji}} \right)_n (l_{ji})_{n+1} = 0 \qquad (k = 1, 2, \ldots, c)$$

$$\sum_{i=1}^{c} \left(\frac{\partial \gamma_{jk}^V}{\partial v_{ji}} \right)_n (v_{ji})_{n+1} = 0 \qquad (k = 1, 2, \ldots, c)$$

amount to neglecting the dependency of the γ_{ji}^L's on the l_{ji}'s and the γ_{ji}^V's on the v_{ji}'s in the differentiation of the functions $\{f_{ji}\}$ in the Newton-Raphson method. That is, show that the above assumptions are equivalent to carrying out the differentiation of the functions $\{f_{ji}\}$ as if the mixture formed an ideal solution.

5-3 Show that the multiplications implied by Eq. (12) of App. 5-1 are represented by the algorithm given above Eq. (14) of App. 5-1.

5-4 By Broyden's method, the corrected jacobian $\mathbf{J}_{k+1}$ is given by

$$\mathbf{J}_{k+1} = \mathbf{J}_k + \mathbf{u}_{k+1} c_{k+1} \mathbf{p}_{k+1}^T$$

Use Householder's identity to show that

$$\mathbf{J}_{k+1}^{-1} = (\mathbf{I} + \alpha_{k+1} \mathbf{v}_{k+1} \mathbf{p}_{k+1}^T) \mathbf{J}_k^{-1}$$

where $\mathbf{v}_{k+1}$ and α_{k+1} are defined by Eqs. (5) and (6) of App. 5-1.

5-5 (a) To illustrate the variety of choices of independent variables which may be made in the Almost Band formulation of the Newton-Raphson equations, display the variables and functions for the following formulations for absorbers: $N(2c + 1)$, $N(c + 1)$, and $N(c + 2)$.

(b) For a conventional distillation column for which the two additional specifications are taken to be the reflux rate L_1 and the bottoms rate L_N, display the variables and functions for the following formulations: $[N(2c + 1) + 2]$, $N(2c + 2)$, $[N(c + 1) + 2]$, and $N(c + 2)$.

5-6 (a) Formulate the equations for the determination of the bubble-point temperature in terms of $c + 1$ independent functions and the $c + 1$ independent variables $\mathbf{x} = [y_1 \ y_2 \ \cdots \ y_c \ T]^T$.

(b) Repeat part (a) for the determination of the dew-point temperature where the independent variables are $\mathbf{x} = [x_1 \ x_2 \ \cdots \ x_c \ T]^T$.

5-7 (a) Formulate the equations required to determine the solution of an isothermal flash problem in terms of $2c$ independent functions and $2c$ independent variables $\mathbf{x} = [l_1 \ l_2 \ \cdots \ l_c \ v_1 \ v_2 \ \cdots \ v_c]^T$.

(b) Repeat part (a) for the case where the formulation is in terms c independent functions and the c independent variables $\mathbf{x} = [l_1 \ l_2 \ \cdots \ l_c]^T$.

5-8 (a) Formulate the equations for an adiabatic flash of a highly nonideal solution in terms of $2c + 1$ independent functions and the $2c + 1$ independent variables

$$\mathbf{x} = [v_1 \ v_2 \ \cdots \ v_c \ l_1 \ l_2 \ \cdots \ l_c \ T]^T.$$

(b) Show that if the approximations given by Eqs. (5-9) through (5-11) are made, the enthalpy balance functions may be treated in the same manner as those of an ideal solution in the differentiation process.

REFERENCES

1. J. M. Bennett: "Triangular Factors of Modified Matrices," *Numerische Mathematik*, **7**:217 (1965).
2. D. S. Billingsley: "On the Equations of Holland in the Solution of Multicomponent Distillation," *IBM J. Res. Develop.*, **14**:33 (1970).
3. J. F. Boston and S. L. Sullivan, Jr.: "A New Class of Solution Methods for Multicomponent, Multistage Separation Processes," **52**:52 (1974).
4. J. A. Bruno, J. L. Yanosik, and J. W. Tierney: "Distillation Calculation with Nonideal Mixtures," *Extractive and Azeotropic Distillation*, Advances in Chemistry Series 115, American Chemical Society, Washington, D.C., 1972.
5. C. G. Broyden: "A Class of Methods for Solving Nonlinear Simultaneous Equation," *Math. Comp.*, **19**:577 (1965).
6. S. E. Gallun: M.S. Thesis, Texas A&M University, 1975.
7. ———: F. E. Hess, G. W. Bentzen, and C. D. Holland: "Algorithms for Solving Problems Involving the Separation of Ideal and Nonideal Solutions by Equilibrium Stage Processes," *Proceedings of the 5th Symposium on Computers in Chemical Engineering*, Oct. 5–9, 1977, Vysoké, Tatry, Czechoslovakia.

8. ——— and C. D. Holland: "Solve More Distillation Problems, Part 5—For Highly Nonideal Solutions," *Hydrocarbon Process.*, **55**(1):137 (1976).

9. ——— and ———: "A Modification of Bryoden's Method for the Solution of Sparse Systems—With Application to Distillation Problems Described by Nonideal Thermodynamic Functions," *Comp. Chem. Eng.*, **4**:93 (1980).

10. ———: Ph.D. Dissertation, Texas A&M University, 1979.

11. F. E. Hess, S. E. Gallun, G. W. Bentzen, and C. D. Holland: "Solve More Distillation Problems, Part 8—Which Method to Use," *Hydrocarbon Process.*, **53**(8):181 (1977).

12. C. D. Holland: "Energy Balances for Systems Involving Nonideal Solutions," *Ind. Eng. Chem. Fundam.*, **16**(1):143 (1977).

13. A. S. Householder: *Principles of Numerical Analysis*, McGraw-Hill Book Company, New York, 1953.

14. H. P. Hutchison and C. F. Shewchuk: "A Computational Method for Multiple Distillation Towers," *Trans. Inst. Chem. Eng.*, **52**:325 (1974).

15. Y. Ishii and F. D. Otto: "A General Algorithm for Multistage Multicomponent Separation Calculations," *Can. J. Chem. Eng.*, **51**:601 (1973).

16. M. Kubíček, V. Hlaváček, and F. Procháska: "Global Modular Newton-Raphson Technique for Simulation of an Interconnected Applied to Complex Rectifying Columns," *Chem. Eng. Sci.*, **31**:277 (1976).

17. L. M. Naphtali and D. P. Sandholm: "Multicomponent Calculations by Linearization," *AIChE J.*, **17**:148 (1971).

18. J. M. Prausnitz, C. A. Eckert, R. V. Orye, and J. P. O'Connell: *Computer Calculations for Multicomponent Vapor-Liquid Equilibrium*, Prentice-Hall, Englewood Cliffs, N.J., 1967.

19. E. C. Roche and H. K. Staffin: "Rigorous Solution of Multicomponent Multistage Liquid-Liquid Extraction Problems, 61st Annual AIChE Meeting, Los Angeles, California, December 1968.

20. ———: "Rigorous Solution of Multicomponent, Multistage Liquid-Liquid Extraction Problems, *Brit. Chem. Eng.*, **14**:1393 (1969).

21. L. K. Schubert: "Modification of a Quasi-Newton Method for Nonlinear Equations with a Sparse Jacobian," *Math. Comp.*, **25**:27 (1970).

22. I. S. Sokolnikoff and E. S. Sokolnikoff: *Higher Mathematics for Engineers and Physicists*, McGraw-Hill Book Company, New York, 1941.

23. R. P. Tewarson: *Sparse Matrices*, Academic Press, New York, 1973.

APPENDIX 5-1 DEVELOPMENT OF THE BROYDEN-HOUSEHOLDER ALGORITHM FOR SPARSE MATRICES [*S. E. Gallun and C. D. Holland, Comput. Chem. Eng.*, **4**:93 (1980), *by courtesy Comput. Chem. Eng.*]

The Householder identity is given by

$$(A + WCZ^T)^{-1} = A^{-1} - A^{-1}W(C^{-1} + Z^T A^{-1}W)^{-1}Z^T A^{-1} \qquad (1)$$

where A is an $n \times n$ matrix, W and X are $n \times m$ matrices and C is an $m \times m$ matrix. For the case where W and Z are vectors and C is a nonzero scalar, Householder's identity reduces to

$$(A + wcz^T)^{-1} = A^{-1} + \alpha A^{-1}wz^T A^{-1} \qquad (2)$$

where A is a square matrix, w and z are conformable column vectors, and α is the scalar given by

$$\alpha = \frac{-1}{(1/c) + z^T A^{-1} w}$$

If the vector $\mathbf{v}$ is defined by

$$\mathbf{A}\mathbf{v} = \mathbf{w}$$

then Eq. (2) may be factored and written as

$$(\mathbf{A} + \mathbf{w}c\mathbf{z}^T)^{-1} = (\mathbf{I} + \alpha\mathbf{v}\mathbf{z}^T)\mathbf{A}^{-1} \tag{3}$$

where

$$\alpha = \frac{-1}{(1/c) + \mathbf{z}^T\mathbf{v}}$$

Equation (3) and the definition of α provide the basis for an algorithm to efficiently solve the series of linear equations that arise during the implementation of Broyden's method.

An expression for $\mathbf{J}_{k+1}^{-1}$, where $\mathbf{J}_{k+1}$ is defined by Eq. (5-29) may be developed through the use of Eq. (3) as follows

$$\mathbf{J}_{k+1}^{-1} = (\mathbf{I} + \alpha_{k+1}\mathbf{v}_{k+1}\mathbf{p}_{k+1}^T)\mathbf{J}_k^{-1} \tag{4}$$

where

$$\mathbf{J}_k\mathbf{v}_{k+1} = \mathbf{u}_{k+1} \tag{5}$$

$$\alpha_{k+1} = \frac{-1}{(1/c_{k+1}) + \mathbf{p}_{k+1}^T\mathbf{v}_{k+1}} \tag{6}$$

Since

$$\mathbf{J}_k = \mathbf{J}_{k-1} + \mathbf{u}_k c_k \mathbf{p}_k^T \tag{7}$$

a similar expression can be developed for $\mathbf{J}_k^{-1}$ by replacing the subscript $k + 1$ in Eqs. (4) and (5) by k. Substitution of the expression so obtained for $\mathbf{J}_k^{-1}$ into Eq. (4) gives

$$\mathbf{J}_{k+1}^{-1} = (\mathbf{I} + \alpha_{k+1}\mathbf{v}_{k+1}\mathbf{p}_{k+1}^T)(\mathbf{I} + \alpha_k\mathbf{v}_k\mathbf{p}_k^T)\mathbf{J}_{k-1}^{-1} \tag{8}$$

Clearly, the process may be repeated for $\mathbf{J}_{k-1}^{-1}$, $\mathbf{J}_{k-2}^{-1}$, ..., $\mathbf{J}_1^{-1}$ to give

$$\mathbf{J}_{k+1}^{-1} = (\mathbf{I} + \alpha_{k+1}\mathbf{v}_{k+1}\mathbf{p}_{k+1}^T)(\mathbf{I} + \alpha_k\mathbf{v}_k\mathbf{p}_k^T) \cdots (\mathbf{I} + \alpha_1\mathbf{v}_1\mathbf{p}_1^T)\mathbf{J}_0^{-1} \tag{9}$$

where $\mathbf{I}$ is the identity matrix and the definitions of $\alpha_1, \alpha_2, \ldots, \alpha_k$ and $\mathbf{v}_1, \mathbf{v}_2, \ldots,$ $\mathbf{v}_k$ are of the same form as the expressions given in Eqs. (5) and (6), respectively.

Thus, if $\mathbf{J}_0$ is sparse and a sparse factorization is available, Broyden's procedure can be implemented while effectively maintaining the sparsity of the jacobian. The algorithm for effecting these calculations in an efficient manner is developed as follows. Suppose that it is desired to solve

$$\mathbf{J}_{k+1}\,\Delta\mathbf{x}_{k+1} = -\mathbf{f}_{k+1} \tag{10}$$

Since

$$\Delta\mathbf{x}_{k+1} = \mathbf{J}_{k+1}^{-1}(-\mathbf{f}_{k+1}) \tag{11}$$

post multiplication of the members of Eq. (9) by $(-\mathbf{f}_{k+1})$ gives

$$\Delta\mathbf{x}_{k+1} = (\mathbf{I} + \alpha_{k+1}\mathbf{v}_{k+1}\mathbf{p}_{k+1}^T)(\mathbf{I} + \alpha_k\mathbf{v}_k\mathbf{p}_k^T)\cdots(\mathbf{I} + \alpha_1\mathbf{v}_1\mathbf{p}_1^T)\mathbf{J}_0^{-1}(-\mathbf{f}_{k+1}) \quad (12)$$

Since it is supposed that an **LU** factorization of $\mathbf{J}_0$ is available, the result indicated by multiplication of $\mathbf{J}_0^{-1}$ by $(-\mathbf{f}_{k+1})$ can be obtained by solving

$$\mathbf{J}_0\mathbf{w} = -\mathbf{f}_{k+1} \quad (13)$$

for $\mathbf{w}$. After $\mathbf{w}$ has been obtained, it is necessary to perform $k + 1$ matrix multiplications in order to obtain $\Delta\mathbf{x}_{k+1}$. The development of the algorithm is simplified by first supposing that α_j, $\mathbf{v}_j$, and $\mathbf{p}_j$. $(j = 1, 2. \ldots, k + 1)$ are known as well as the **LU** factorization of $\mathbf{J}_0$. Let $\leftarrow$ be the replacement operator. Then

(1) $j = 0$

    SOLVE $\mathbf{J}_0\mathbf{w} = -\mathbf{f}_{k+1}$

(2) DO $j = 1, 2, \ldots, k + 1$

    $\beta = \alpha_j\mathbf{p}_j^T\mathbf{w}$

    $\mathbf{w} \leftarrow \mathbf{w} + \beta\mathbf{v}_j$

(3) $\Delta\mathbf{x}_{k+1} = \mathbf{w}$

    Although it was assumed in the development of the above algorithm that α_j and $\mathbf{v}_j$ $(j = 1, 2, \ldots, k + 1)$ were available, the values of these variables may be developed in parallel with the solution of Eq. (10). First observe that when Eq. (4) is post multiplied by $(-\mathbf{f}_{k+1})$, one obtains

$$\Delta\mathbf{x}_{k+1} = (\mathbf{I} + \alpha_{k+1}\mathbf{v}_{k+1}\mathbf{p}_{k+1}^T)\mathbf{J}_k^{-1}(-\mathbf{f}_{k+1}) \quad (14)$$

Now begin Eq. (12) and apply the above algorithm for $j = 0, 1, \ldots, k$ to obtain

$$\Delta\mathbf{x}_{k+1} = (\mathbf{I} + \alpha_{k+1}\mathbf{v}_{k+1}\mathbf{p}_{k+1}^T)\mathbf{w} \quad (15)$$

Upon comparison of Eqs. (14) and (15), it is seen that

$$\mathbf{J}_k\mathbf{w} = -\mathbf{f}_{k+1} \quad (16)$$

    The development of a formula for $\mathbf{v}_{k+1}$ in terms of $\mathbf{J}_0^{-1}$ is initiated by first solving Eq. (5) for $\mathbf{v}_{k+1}$ to give

$$\mathbf{v}_{k+1} = \mathbf{J}_k^{-1}\mathbf{u}_{k+1} \quad (17)$$

Application of Householder's formula to Eq. (7) permits Eq. (17) to be written as follows

$$\mathbf{v}_{k+1} = (\mathbf{I} + \alpha_k\mathbf{v}_k\mathbf{p}_k^T)\mathbf{J}_{k-1}^{-1}\mathbf{u}_{k+1} \quad (18)$$

Clearly, the process may be continued to give

$$\mathbf{v}_{k+1} = (\mathbf{I} + \alpha_k\mathbf{v}_k\mathbf{p}_k^T)(\mathbf{I} + \alpha_{k-1}\mathbf{v}_{k-1}\mathbf{p}_{k-1}^T)\cdots(\mathbf{I} + \alpha_1\mathbf{v}_1\mathbf{p}_1^T)\mathbf{J}_0^{-1}\mathbf{u}_{k+1} \quad (19)$$

If $\mathbf{z}$ is defined by

$$\mathbf{J}_0\mathbf{z} = \mathbf{u}_{k+1} \quad (20)$$

it is seen by comparison of Eqs. (12) and (19) that an algorithm for the calculation of v_{k+1} of the same form as that shown above for Δx_{k+1} exists, and that z has the same role in the algorithm for the calculation of v_{k+1} as w has in the algorithm for the calculation of Δx_{k+1}. From the above equations, it is seen that the algorithm presented in the text is appropriate for the solution of Eq. (10) with simultaneous development of the vector v_{k+1} and the scalar α_{k+1}.

SIX

SYSTEMS OF AZEOTROPIC AND EXTRACTIVE DISTILLATION COLUMNS

Azeotropic and *extractive distillation* are old processes which have become widely used since about 1930. In 1908 Emile Guillaume patented an extractive distillation process for the removal of fusel oil from fermentation alcohol.[10] The name "extractive distillation" is said to have been introduced by Dunn et al.[6]

Azeotropic and extractive distillation columns are examples of columns in the service of separating highly nonideal solutions. Algorithms for solving problems involving highly nonideal solutions are described in Chap. 5. *Azeotropic* and *extractive distillation* are the names given to processes in which advantage is taken of the nonideal behavior exhibited by certain mixtures in the presence of selected solvents. In Sec. 6-1, the qualitative aspects of azeotropic and extractive distillation are presented. In Sec. 6-2, several topics are presented which include the quantitative behavior of solvents, three-phase mixtures, and the solution of systems of interconnected columns. In the first of two methods proposed for systems of columns, a "column modular" method is presented in which the sets of equations for each column are solved sequentially. In the second procedure, called the "system modular method," the complete set of equations for the system are solved simultaneously.

6-1 QUALITATIVE CHARACTERISTICS OF AZEOTROPIC AND EXTRACTIVE DISTILLATION PROCESSES

Applications of azeotropic and extractive distillation have continued to expand because many very close boiling mixtures may be separated economically by use of these techniques. The separation of such mixtures by conventional distillation methods is usually uneconomical because of the large number of stages which would be required to effect such separations.

Because of the tendency toward nonideal behavior of mixtures, it is generally possible to find some component which when added to a given mixture will increase the difference between the volatilities of the light and heavy key components to be separated. The component or material added to the mixture to be separated is called the *solvent*. When the solvent added to the mixture is withdrawn from the column, usually in the distillate, as an azeotrope with one or both of the key components, the separation process is called *azeotropic distillation*. The name azeotropic distillation has also sometimes been given to processes where no azeotrope is formed and the solvent is withdrawn almost exclusively in the distillate.

When the solvent added to the mixture is withdrawn almost exclusively in the bottom product without forming an azeotrope, the process is called *extractive distillation*.

Behavior of Solvents

An effective solvent for an extractive distillation is one which is attracted to one or more of the components. This attraction of the solvent for these components reduces the volatility of the solvent as well as the volatilities of the components to which it is attracted. It is desirable that the attraction occur in the natural direction, that is, that the solvent be attracted to the relatively heavy components. However, this is not a necessary condition for the behavior of the solvent. Many separations are carried out in which one of the relatively light components is attracted by the solvent and removed in the bottom product with the solvent.

A variety of theories have been advanced for the roles of the solvent in azeotropic and extractive distillation. In the case of extractive distillation, attraction of the solvent for the certain components of the mixture is commonly attributed to one or more or a combination of the following phenomenon: (1) hydrogen bonding, (2) polar characteristics of the solvent and members of the mixture, (3) the formation of weak unstable chemical complexes, (4) chemical reactions between the solvent and one or more of the components of the mixture. A more complete statement of theories has been summarized by Berg.[3]

In the case of azeotropic distillation, the solvent should have the capacity to reduce the tendency of attraction between molecules. For example, a nonpolar solvent may be added to a mixture of polar molecules in order to increase the volatilities of the more polar compounds relative to the less polar compounds.

Although any one theory does not sufficiently explain all applications of azeotropic and extractive distillation, the theories do provide qualitative rules for the selection of solvents. The role of polarity has been elucidated by Hopkins and Fritsch[17] who described the use of products obtained by oxidation of selected hydrocarbons. Because of the dissimilarities in molecular structure, the oxidation products can be arranged in the order of increasing polarity,[4, 17] namely, esters, oxides, aldehydes, ketones, acetals, and alcohols. In any class of compounds, the polarity is inversely proportional to the molecular weight, the polarity of straight-chain molecules is greater than that of branched-chain structures, and olefinic compounds are more polar than their corresponding paraffin derivatives.

Separation of Hydrocarbon Oxidation Products by Azeotropic and Extractive Distillation

The products obtained by the oxidation of hydrocarbons have small differences in boiling points. Typical oxidation products are shown in Table 6-1. Straight fractionation to recover a pure product from one of these groups is uneconomical because too many plates would be required.

One hydrocarbon reaction product produced by Celanese contains over forty components which are capable of forming more than fifty binary azeotropes with each other.[17] Several ternary azeotropes are also known to exist. Azeotropes of eleven of the oxidation products are listed in Table 6-2. An examination of Table 6-2 shows that the separation of any one product from the mixture of eleven components would be most difficult to effect by straight fractionation.

If the solvent is highly polar, the volatility of the more polar compound is lowered and it is withdrawn with the solvent from the bottom of the column. The less polar compound is recovered in the top product. If a nonpolar compound is added, the volatility of the more polar compounds may be increased in some instances enough to permit the component with the higher polarity to be removed as an overhead product.[17]

Separation of Acetone and Methanol

The use of a polar and a nonpolar solvent to separate acetone and methanol from a mixture of tetramethylene oxide and other oxides has been described by Hopkins and Fritsch.[17] A schematic drawing of this purification process is shown in Fig. 6-1. The ternary azeotrope of acetone, methanol, and tetramethylene, a cyclic ether, may be broken by an extractive distillation using the highly polar solvent, water. The volatility of the methanol is lowered by the water to such an extent that the azeotrope of acetone and tetramethylene oxide may be distilled overhead in the extractive distillation column, and the methanol is withdrawn with the water from the bottom of the column. A second column is used to separate the azeotropic mixture of acetone and tetramethylene oxides by use of the relative nonpolar solvent, pentane. An azeotrope of pentane and acetone boiling at 32°C, is removed from the top of the column. The azeotrope is broken by adding water which results in the formation of two phases, a pentane phase and an acetone-water phase.

Purification of Methyl Ethyl Ketone

Another example presented by Hopkins and Fritsch[17] consists of the use of azeotropic and extractive distillation to recover part of the methyl ethyl ketone

Table 6-1 Typical groups of oxidation products of hydrocarbons [*W. C. Hopkins and J. J. Fritsch, Chem. Eng. 51(8):361 (1955), by courtesy McGraw-Hill Book Company.*]

Compound	Normal boiling point °C
Propionaldehyde	48.8
Acrolein	52.5
Ethyl acetate	77.1
Methyl propionate	79.7
Allyl alcohol	96.6
n-Propyl alcohol	97.8
sec-butyl alcohol	99.5

Table 6-2 Boiling points of selected oxygenated chemicals and their binary azeotropes [*L. H. Horsley, Azeotropic Data, by courtesy American Chemical Society, Washington, D.C., 1952*]

Pure compound or binary azeotrope	Normal boiling point, °C
Acetone, methyl alcohol	55.5
Acetone	56.5
Methyl alcohol, methyl ethyl ketone	63.5
Methyl alcohol	64.7
Methyl ethyl ketone, water	73.4
Ethyl alcohol, methyl ethyl ketone	75.7
Methyl ethyl ketone, isopropyl alcohol	77.9
Ethyl alcohol, water	78.2
Ethyl alcohol	78.4
Methyl ethyl ketone, tert-butyl alcohol	78.7
Methyl ethyl ketone	79.6
tert-butyl alcohol, water	69.9
Isopropyl alcohol, water	80.3
Isopropyl alcohol	82.5
tert-butyl alcohol	82.9
n-propyl alcohol, water	87.0
sec-butyl alcohol, water	87.5
Isobutyl alcohol, water	89.8
Water, n-butyl alcohol	92.7
n-propyl alcohol	99.5
sec-butyl alcohol	99.5
Water	100.0
Isobutyl alcohol	108.0
n-butyl alcohol	117.5

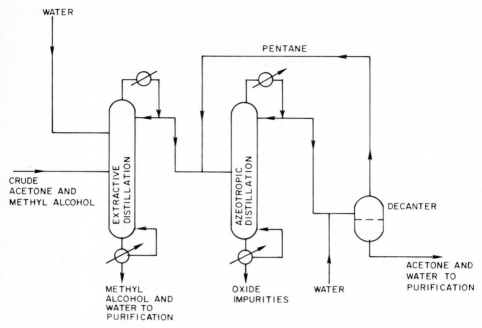

Figure 6-1 Separation of a mixture containing acetone and methyl alcohol by azeotropic and extractive distillation. [*W. C. Hopkins and J. J. Fritsch, Chem. Eng. 51(8):361 (1955), by courtesy McGraw-Hill Book Company.*]

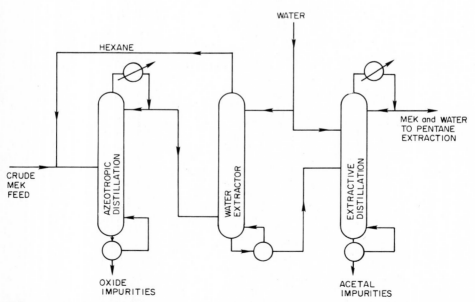

Figure 6-2 Purification of methyl ethyl keytone by azeotropic and extractive distillation. [*W. C. Hopkins and J. J. Fritsch, Chem. Eng. 51(8):361 (1955), by courtesy McGraw-Hill Book Company.*]

from a stream which also contained methyl-tetrahydrofuran, formals, acetals, and oxide impurities. The sequence of steps is shown in Fig. 6-2. All but a small portion of the oxide impurities are removed by azeotropic distillation. The azeotrope is taken overhead at 65°C. In the second column, the solvent is separated from the remaining components by water extraction. The remaining impurities are separated from the methyl ethyl ketone by a water extractive distillation with a water concentration on the trays of approximately 60 percent by weight. The polar solvent, water, reduces the volatilities of the acetals, and the azeotrope of methyl ketone and water is removed overhead.[26] The overhead, methyl ethyl ketone and water, is then dried by pentane extraction.

Separation of Ethanol and Water

An early application of azeotropic distillation was proposed by Guinot and Clark[11] for the separation of ethanol and water by the use of benzene as the solvent. This process is based on the fact that benzene forms a ternary azeotrope with ethanol and water, which has a higher ratio of water to ethanol than does the ethanol-water azeotrope. In the first column, shown in Fig. 6-3, an azeotropic distillation is carried out. A two-phase liquid separation at 20°C in the decanter is used to concentrate the benzene in the reflux to the first column. The solvent benzene is recovered in the second column and water is removed in the third column.

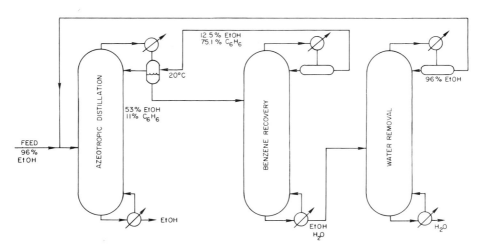

Figure 6-3 Azeotropic separation of ethanol and water by use of a benzene solvent. [*H. Guinot and F. W. Clark, Trans. Inst. Chem. Eng. (London) 16:187 (1938), by courtesy The Institute of Chemical Engineers (London).*]

Solvents for Hydrocarbon Separations

The high demand for relatively pure butadiene and toluene at the outbreak of World War II was met through the use of azeotropic distillation. Nitration grade toluene was needed for the production of explosives and butadiene was needed for the production of synthetic rubber. As a result of the need for these chemicals, azeotropic and extractive distillation became large-scale industrial processes.

An enumeration of some of the specifications which must be met by an efficient solvent follows. Obviously, the solvent should be noncorrosive to the equipment and should not react with the feed to form undesired products. It should produce a sufficient change in the volatilities of the components to be separated so that these components may be separated with a reasonable number of plates at an economical reflux ratio. The solvent should have an appropriate boiling point relative to the components of the feed to be separated. An azeotropic solvent should have a volatility near the major component desired in the overhead product and an extractive solvent should have a volatility lower than the major component to be withdrawn in the bottom product. The solvent should not be toxic, and it should be available in sufficient quantities at a reasonable price.

Azeotropic Separation of Butadiene from a Mixture of the C_4 Hydrocarbons by Use of Ammonia

One of the first processes employed to separate butadiene from a C_4 hydrocarbon stream was an azeotropic distillation which used liquid ammonia as the solvent. A description of this process has been presented by Poffenberger et al.,[24] who also gave a typical analysis of the C_4 stream together with the boiling points of hydrocarbons and their azeotropes. Other solvents such as furfural and acetonitrile are presently employed to effect this separation.[12]

Other Extractive and Azeotropic Separations

Many solvents have been investigated for the separation of toluene and other aromatics from paraffinic mixtures. Dunn et al.,[6] among others, have presented lists of possible solvents. The use of phenol for the extraction of toluene has been described by Dunn et al.[6] A solvent-to-feed ratio of approximately 3 to 1 was used. The first commercial plant for the recovery of nitration grade toluene by phenol extraction was constructed and put into operation in 1940 at the Houston Refinery of Shell Oil Company, Inc.[6] Because of the emergencies which existed at that time, it was necessary to go directly from the laboratory to the full-scale plant.

The production of butadiene from a butane feed generally requires a feed purification process which involves the separation of the butenes from the butanes. Atkins and Boyer[1] have described a process in which the separation of the butenes from the butanes was carried out by use of extractive distillation with a mixture of 85 percent acetone and 15 percent water as the solvent. A solvent-to-feed ratio of 0.85 was used.

Separation of Azeotropes by Fractionation

The formation of azeotropes in azeotropic distillation calls for a discussion of some of the techniques which may be employed in the separation of such mixtures. If the azeotropic composition is sensitive to the variation of the total pressure, homogeneous azeotropes may be separated by use of a two-column fractionation scheme as described by Van Winkle,[29] among others.

SEPARATION OF MINIMUM-BOILING HOMOGENEOUS AZEOTROPES

Suppose that the boiling-point diagrams for two different total pressures are as shown in Fig. 6-4. Such an azeotrope may be separated by use of two columns as shown in Fig. 6-5. The feed $(X_A = 0.3)$ is introduced to the second column at the higher pressure P_2. The bottom product contains a relatively pure component B and the top product consists of the minimum-boiling azeotrope $X_{D, A} = 0.8$ and $X_{D, B} = 0.2$. The distillate is fed to the first column which is operated at the total pressure of P_1. This column produces a bottom product which is relatively pure in A. The top product is essentially the azeotropic composition at P_1 $(X_{D, A} = 0.6$ and $X_{D, B} = 0.4)$, and thus it is added to the feed to the second column.

SEPARATION OF MAXIMUM-BOILING HOMOGENEOUS AZEOTROPES

The separation of these azeotropes may be effected, if they are sensitive to a change in the total pressure, in a manner similar to that described above for minimum-boiling azeotropes. Suppose that the boiling-point diagrams at the pressures P_1 and P_2 are as shown in Fig. 6-6. Then the separation may be effected by use of two columns as shown in Fig. 6-7.

SEPARATION OF HETEROGENEOUS AZEOTROPES

If the feed is in the two-phase region, the two phases may be separated and fed to each of two columns. Suppose that the boiling-point diagram for a mixture of components A and B is as shown in Fig. 6-8. The feed is introduced to the separator shown in Fig. 6-9 which operates at the temperature T_{sep}. Phase I is then fed to column 1 and phase II to column 2. The bottom products of columns 1 and 2 are relatively pure B and A, respectively. The distillate compositions are approximately those of the azeotrope, and consequently the distillates are fed to the separator.

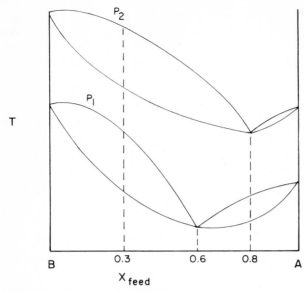

Figure 6-4 Boiling-point diagrams for a minimum-boiling homogeneous azeotrope which is sensitive to a change in the total pressure. (*M. Van Winkle, Distillation, 1967, by courtesy McGraw-Hill Book Company.*)

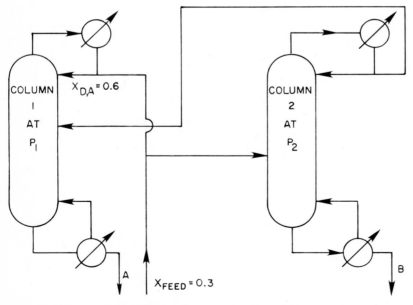

Figure 6-5 Use of two columns to separate a minimum-boiling homogeneous azeotrope which is sensitive to a change in total pressure. (*M. Van Winkle, Distillation, 1967, by courtesy McGraw-Hill Book Company.*)

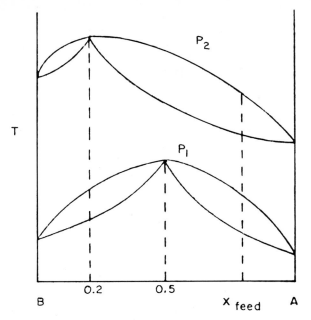

Figure 6-6 Boiling-point diagrams for a maximum-boiling azeotrope which is sensitive to a change in the total pressure. (*M. Van Winkle, Distillation, 1967, by courtesy McGraw-Hill Book Company.*)

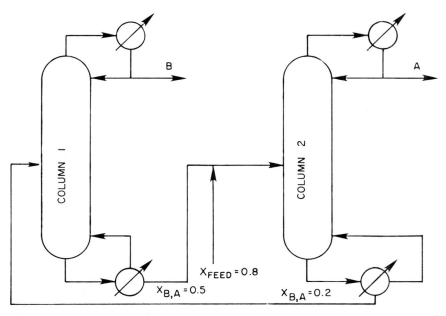

Figure 6-7 Use of two columns to separate a maximum-boiling homogeneous azeotrope which is pressure sensitive. (*M. Van Winkle, Distillation, 1967, by courtesy McGraw-Hill Book Company.*)

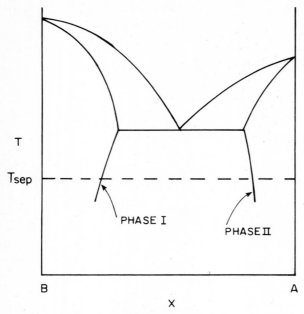

Figure 6-8 Boiling-point diagram for a heterogeneous minimum-boiling azeotrope. (*M. Van Winkle, Distillation, 1967, by courtesy McGraw-Hill Book Company.*)

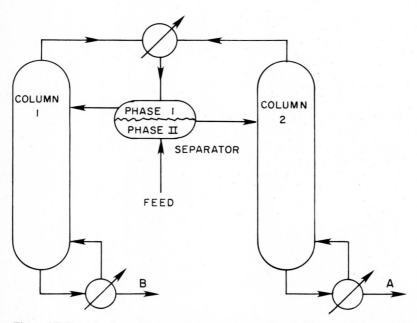

Figure 6-9 Use of a two-column system to separate a heterogeneous minimum-boiling azeotrope. (*M. Van Winkle, Distillation, 1967, by courtesy McGraw-Hill Book Company.*)

6-2 SOLUTION OF PROBLEMS INVOLVING SINGLE COLUMNS USED TO EFFECT AZEOTROPIC AND EXTRACTIVE DISTILLATIONS

Prior to the treatment of systems of azeotropic and extractive distillation columns, two other topics are considered, namely, the quantitative description of the behavior of solvents and the solution of problems involving columns in which three phases exist in the accumulator.

A Quantitative Description of the Behavior of Solvents

The behavior of solvents in the vapor and liquid phases which form nonideal solutions are described by the same set of expressions as those given by Eq. (5-1). Further insight into the behavior of solvents in azeotropic and extractive distillation may be gained by the reconsideration of the first expression of Eq. (5-1), which may be restated in the following form

$$y_i = \frac{\gamma_i^L K_i x_i}{\gamma_i^V} \tag{6-1}$$

where

$$K_i = \frac{f_i^L}{f_i^V}$$

If a mixture behaves as an ideal solution in the vapor phase $\gamma_i^V = 1$ for all i, and if it also behaves as an ideal solution in the liquid phase $\gamma_i^L = 1$ for all i, then Eq. (6-1) reduces to

$$y_i = K_i x_i \tag{6-2}$$

Azeotropic and extractive distillation are made possible by the unique variations of the activity coefficients when a solvent is added to the mixture. A measure of the effect of a given solvent on two components may be expressed in terms of their volatility ratio. Let the volatility of component i be defined by y_i/x_i. Then the volatility of component i relative to any component b (or the relative volatility α_i of component i relative to component b) is given by

$$\alpha_i = \frac{y_i/x_i}{y_b/x_b} = \frac{\gamma_i^L K_i/\gamma_i^V}{\gamma_b^L K_b/\gamma_b^V} \tag{6-3}$$

In many systems, the vapor phase exhibits ideal behavior and Eq. (6-3) reduces to

$$\alpha_i = \frac{y_i/x_i}{y_b/x_b} = \frac{\gamma_i^L K_i}{\gamma_b^L K_b} \tag{6-4}$$

If equations of state are available for estimating the effect of a solvent on the γ_i^L's, then Eq. (6-4) can be used in the screening of solvents for a given separation.[2, 6, 13, 23, 27, 29]

Almost Band Algorithms for Single Columns in the Service of Azeotropic and Extractive Distillations

The Almost Band Algorithm presented in Chap. 5 may be used to describe a single column in which either an azeotropic or extractive distillation is carried out provided that the accumulator contains only one liquid phase and one vapor phase. In many azeotropic distillation columns, the accumulator contains two liquid phases and one vapor phase. In order to describe a column whose accumulator contains three phases, the Almost Band Algorithms presented in Chap. 5 must be modified. To illustrate the modifications of the Almost Band Algorithm which are required in order to describe a column having three phases in the accumulator, the $[N(2c + 1) + 2]$ formulation of the Almost Band Algorithm for a column with a two-phase partial condenser is selected as the base case. Then the modifications required to describe a three-phase partial condenser are presented.

As shown in Chap. 5, for the case where L_1 and L_N are taken to be fixed (or specified), the $[N(2c + 1) + 2]$ functions of the Almost Band Algorithm are as follows

$$\mathbf{f} = [S_1 \ (f_{j, 1} \ f_{j, 2} \ \cdots \ f_{j, c} \ m_{j, 1} \ m_{j, 2} \ \cdots \ m_{j, c} \ G_j)_{j=1, N} \ S_N]^T \qquad (5\text{-}14)$$

The functions S_1 and S_N are given by Eqs. (5-12) and (5-13). The equilibrium functions and the material balance functions are of the same general form as those given by Eqs. (5-1) and (5-2). The enthalpy balance functions are given by Eqs. (5-4) and (5-17).

The $[N(2c + 1) + 2]$ independent variables are given by

$$\mathbf{x} = [Q_C \ (l_{j, 1} \ l_{j, 2} \ \cdots \ l_{j, c} \ v_{j, 1} \ v_{j, 2} \ \cdots \ v_{j, c} \ T_j)_{j=1, N} \ Q_R]^T \qquad (5\text{-}15)$$

In the above listing of the variables, v_{1i} has been used to denote d_i.

Columns Having Two Liquid Phases in the Accumulator

A sketch of the condenser-accumulator section for the general case where two liquid phases and one vapor phase are formed in the accumulator is shown in Fig. 6-10. The functions and variables selected depend upon the particular set of specifications made on the column. Three cases are considered. All of the specifications are the same for each case except for the specification of the type of condenser (partial or total) and the "product specifications" such as L_1 and L_N or Q_C and Q_R. In addition to these two specifications, the fractions α and β of the liquid streams withdrawn from the system may be specified. Generally, α and β are set equal to 0 or 1 such that one phase is returned to the column as reflux and the other phase is withdrawn as a product.

CASE 1. SPECIFICATION OF Q_C, Q_R, α, β, AND A PARTIAL CONDENSER

Except for the first stage, the equations are of the same general form presented above. The equilibrium relationships needed to describe the two

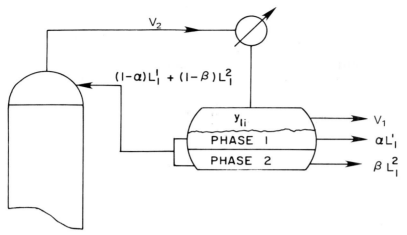

Figure 6-10 Sketch of a three-phase separator.

liquid phases are

$$f^1_{1i} = \frac{\gamma^{L1}_{1i} K^1_{1i} l^1_{1i}}{\sum\limits^c_{i=1} l^1_{1i}} - \frac{\gamma^V_{1i} v_{1i}}{\sum\limits^c_{i=1} v_{1i}} \tag{6-5}$$

$$f^2_{1i} = \frac{\gamma^{L2}_{1i} K^2_{1i} l^2_{1i}}{\sum\limits^c_{i=1} l^2_{1i}} - \frac{\gamma^V_{1i} v_{1i}}{\sum\limits^c_{i=1} v_{1i}} \tag{6-6}$$

where the superscripts 1 and 2 refer to the respective liquid phases. The component-material balance becomes

$$m_{1i} = v_{2i} - v_{1i} - l^1_{1i} - l^2_{1i} \tag{6-7}$$

and the enthalpy balance must be modified to account for all of the streams withdrawn from the accumulator. Two possible formulations follow. In the first formulation, the independent variables and functions are as follows

$$\mathbf{x} = [v_{1,1} \;\cdots\; v_{1,c} \; l^1_{1,1} \;\cdots\; l^1_{1,c} \; l^2_{1,1} \;\cdots\; l^2_{1,c} \; T_1$$

$$(v_{j,1} \;\cdots\; v_{j,c} \; l_{j,1} \;\cdots\; l_{j,c} \; T_j)_{j=2,\,N}]^T \tag{6-8}$$

$$\mathbf{f} = [m_{1,1} \;\cdots\; m_{1,c} \; f^1_1 \;\cdots\; f^1_{1,c} \; f^2_{1,1} \;\cdots\; f^2_{1,c} \; G_1$$

$$(m_{j,1} \;\cdots\; m_{j,c} \; f_{j,1} \;\cdots\; f_{j,c} \; G_j)_{j=2,\,N}]^T \tag{6-9}$$

and the number of variables is equal to the number of functions, namely, $[N(2c + 1) + c]$. In the second formulation, the independent variables and functions are as follows

$$\mathbf{x} = [L^2_1 \; v_{1,1} \;\cdots\; v_{1,c} \; l^1_{1,1} \;\cdots\; l^1_{1,c} \; l^2_{1,1} \;\cdots\; l^2_{1,c} \; T_1$$

$$(v_{j,1} \;\cdots\; v_{j,c} \; l_{j,1} \;\cdots\; l_{j,c} \; T_j)_{j=2,\,N} \; L_N]^T \tag{6-10}$$

$$\mathbf{f} = [S_1 \; m_{1,1} \;\cdots\; m_{1,c} \; f^1_{1,1} \;\cdots\; f^1_{1,c} \; f^2_{1,1} \;\cdots\; f^2_{1,c} \; G_1$$

$$(m_{j,1} \;\cdots\; m_{j,c} \; f_{j,1} \;\cdots\; f_{j,c} \; G_j)_{j=2,\,N} \; S_N]^T \tag{6-11}$$

The functions S_1 and S_2 are defined as follows

$$S_1 = \frac{\sum\limits_{i=1}^{c} l_{1,i}^2}{L_1^2} - 1 \tag{6-12}$$

$$S_N = \frac{\sum\limits_{i=1}^{c} l_{Ni}}{L_N} - 1 \tag{6-13}$$

The second formulation in terms of $[N(2c + 1) + 2 + c]$ variables and functions is seen to be symmetrical to the one given in Chap. 5 [see Eqs. (5-12) through (5-18)].

CASE 2. SPECIFICATION OF L_1^2, L_N, α, β, AND A PARTIAL CONDENSER

In this case Q_C and Q_R become variables, and the variables are given by

$$\mathbf{x} = [Q_C \ v_{1,1} \ \cdots \ v_{1,c} \ l_{1,1}^1 \ \cdots \ l_{1,c}^1 \ l_{1,1}^2 \ \cdots \ l_{1,c}^2 \ T_1$$

$$(v_{j,1} \ \cdots \ v_{j,c} \ l_{j,1} \ \cdots \ l_{j,c} \ T_j)_{j=2,N} \ Q_R]^T \tag{6-14}$$

and the functions by

$$\mathbf{f} = [S_1 \ m_{1,1} \ \cdots \ m_{1,c} \ f_{1,1}^1 \ \cdots \ f_{1,c}^1 \ f_{1,1}^2 \ \cdots \ f_{1,c}^2 \ G_1$$

$$(m_{j,1} \ \cdots \ m_{j,c} \ f_{j,1} \ \cdots \ f_{j,c} \ G_j)_{j=2,N} \ S_N]^T \tag{6-15}$$

for a total of $[N(2c + 1) + 2 + c]$ variables and an equal number of functions.

CASE 3. SPECIFICATION OF L_N, α, β, AND A TOTAL CONDENSER

Since no product is withdrawn as a vapor, the variables $v_{1,i}$ $(i = 1, 2, \ldots, c)$ are replaced by the mole fractions $y_{1,i}$ $(i = 1, 2, \ldots, c)$. The functions f_{1i}^1 and f_{1i}^2 are formulated by replacing the v_{1i}'s wherever they appear by the y_{1i}'s, and the material balance function m_{1i} becomes

$$m_{1i} = v_{2i} - l_{1i}^1 - l_{1i}^2 \tag{6-16}$$

The condition that the sum of the y_{1i}'s must be equal to unity is introduced as the function S_1 instead of Eq. (6-12) and S_N is given by Eq. (6-13). The complete sets of variables and functions are as follows

$$\mathbf{x} = [Q_C \ y_{1,1} \ \cdots \ y_{1,c} \ l_{1,1}^1 \ \cdots \ l_{1,c}^1 \ l_{1,1}^2 \ \cdots \ l_{1,c}^2 \ T_1$$

$$(v_{j,1} \ \cdots \ v_{j,c} \ l_{j,1} \ \cdots \ l_{j,c} \ T_j)_{j=2,N} \ Q_R]^T \tag{6-17}$$

and $\mathbf{f}$ is given by Eq. (6-15). The number of variables is equal to the number of functions which is equal to $[N(2c + 1) + 2 + c]$.

6-3 SYSTEMS OF COLUMNS IN THE SERVICE OF SEPARATING MIXTURES OF NONIDEAL SOLUTIONS

Two approaches may be used to treat systems of interconnected columns in which nonideal solutions are being separated. First, the complete set of Newton-Raphson equations may be solved simultaneously as suggested by Kubíček et al.[22] and Hutchinson and Shewchuk.[21] In such a formulation of the problem, the stages are numbered consecutively, beginning with the first stage of the first column and ending with the last stage of the last column. The Newton-Raphson equations for a system may be formulated in a manner similar to that described for the Almost Band Algorithm. In the discussions which follow, this method is called the *Almost Band Algorithm for Systems.*

Alternatively, a column modular approach may be used wherein the Almost Band Algorithm is applied successively to each column of the system. After one or more trials have been made on each column of the system, the capital Θ method of convergence is applied in a manner similar to that described for systems of interconnected columns.

FORMULATION OF THE CAPITAL Θ METHOD FOR SYSTEMS OF AZEOTROPIC AND EXTRACTIVE DISTILLATION COLUMNS

Consider the system shown in Fig. 6-11 and suppose that two of the specifications consist of the rates B_1 and B_2. Since $0.94\,B_2$ is recycled back to column 1 from column 2, and $0.06\,B_2$ is withdrawn from the system, the flow rates of all terminal streams may be computed where it is, of course, understood that the feed rates F_1, F_2, F_3, and their compositions have been specified. Let the positive multipliers Θ_1 and Θ_2 be defined as follows

$$\frac{b_{i,\,1}}{d_{i,\,1}} = \Theta_1 \left(\frac{b_{i,\,1}}{d_{i,\,1}}\right)_{ca} \tag{6-18}$$

$$\frac{b_{i,\,2}}{d_{i,\,2}} = \Theta_2 \left(\frac{b_{i,\,2}}{d_{i,\,2}}\right)_{ca} \tag{6-19}$$

where the subscript ca denotes the most recently calculated value of the variable, and $b_{i,\,1}/d_{i,\,1}$, and $b_{i,\,2}/d_{i,\,2}$ denote the corrected values which are to be found by the capital Θ method of convergence. The values of the Θ's are to be determined such that the corrected component-flow rates satisfy the component-material balances

$$F_1 X_{1i} + F_2 X_{2i} + 0.94 b_{i,\,2} - b_{i,\,1} - d_{i,\,1} = 0 \tag{6-20}$$

$$F_3 X_{3i} + b_{i,\,1} - b_{i,\,2} - d_{i,\,2} = 0 \tag{6-21}$$

and are in agreement with the specified values of B_1 and B_2. Equations (6-18) through (6-21) may be solved simultaneously for $d_{i,\,1}$ and $d_{i,\,2}$. However, in

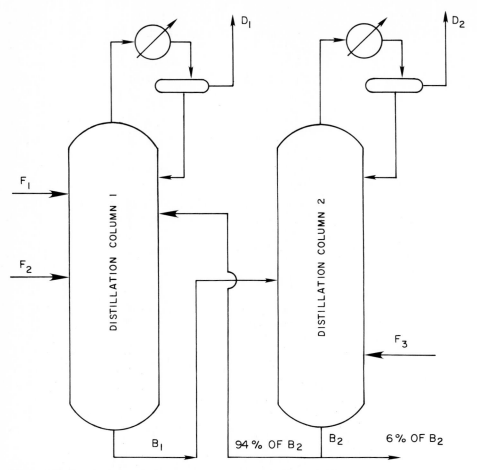

Figure 6-11 Sketch of system of columns used in Examples 6-1 and 6-2.

order to avoid possible numerical problems, it is better to solve for the ratios $p_{i,1}$ and $p_{i,2}$ which are defined by

$$p_{i,1} = d_{i,1}/(d_{i,1})_{ca} \tag{6-22}$$

$$p_{i,2} = d_{i,2}/(d_{i,2})_{ca} \tag{6-23}$$

To facilitate the solving for these ratios, let Eqs. (6-18) and (6-19) be restated as follows

$$b_{i,1} = r_{i,1} p_{i,1} \tag{6-24}$$

$$b_{i,2} = r_{i,2} p_{i,2} \tag{6-25}$$

where

$$r_{i,1} = \Theta_1(b_{i,1})_{ca}$$

$$r_{i,2} = \Theta_2(b_{i,2})_{ca}$$

Elimination of $b_{i,1}$ and $b_{i,2}$ from Eqs. (6-20) and (6-21) by use of Eqs. (6-24) and (6-25) followed by the restatement of the resulting equations in matrix notation yields

$$\begin{bmatrix} -R_{i,1} & 0.94r_{i,2} \\ r_{i,1} & -R_{i,2} \end{bmatrix} \begin{bmatrix} p_{i,1} \\ p_{i,2} \end{bmatrix} = - \begin{bmatrix} (F_1 X_{1i} + F_2 X_{2i}) \\ F_3 X_{3i} \end{bmatrix} \tag{6-26}$$

where

$$R_{i,1} = (d_{i,1})_{ca} + r_{i,1}$$

$$R_{i,2} = (d_{i,2})_{ca} + r_{i,2}$$

Equation (6-26) is readily solved for $p_{i,1}$ and $p_{i,2}$ to give

$$p_{i,1} = \frac{0.94F_2 X_{3i} r_{i,2} + (F_1 X_{1i} + F_2 X_{2i})R_{i,2}}{R_{i,1} R_{i,2} - 0.94r_{i,1} r_{i,2}} \tag{6-27}$$

$$p_{i,2} = \frac{(F_1 X_{1i} + F_2 X_{2i})r_{i,1} + R_{i,1} F_3 X_{3i}}{R_{i,1} R_{i,2} - 0.94r_{i,1} r_{i,2}} \tag{6-28}$$

The desired set of Θ's is that set of positive numbers which satisfy the component-material balances [Eq. (6-26)] and the specified values of the terminal flow rates. Thus, the desired set of Θ's is that set which makes $g_1 = g_2 = 0$, simultaneously, where

$$g_1(\Theta_1, \Theta_2) = \frac{1}{B_1} \left[\sum_{i=1}^{c} b_{i,1} \right] - 1 \tag{6-29}$$

$$g_2(\Theta_1, \Theta_2) = \frac{1}{B_2} \left[\sum_{i=1}^{c} b_{i,2} \right] - 1 \tag{6-30}$$

Formulas for $b_{i,1}$ and $b_{i,2}$ are given by Eqs. (6-24) and (6-25), respectively, and B_1 and B_2 denote the specified values. The desired values of Θ_1 and Θ_2 may be found by use of the Newton-Raphson method.[5] After the Θ's have been determined, the corrected b_i's and d_i's needed to initiate the next trial on the system are readily computed. Details pertaining to the sequence of calculations are presented below.

To demonstrate the application of the column modular method and system modular method (the Almost Band Algorithm for Systems), Examples 6-1 and 6-2 are presented. The statements of Examples 6-1 and 6-2 are presented in Table 6-3, and the solutions are given in Tables 6-4 through 6-6. Example 6-1 is a relatively easy problem to solve while Example 6-2 is more difficult. A flow diagram of the system involved in Examples 6-1 and 6-2 is shown in Fig. 6-11.

The data and curve fits used to solve Examples 6-1 and 6-2 are the same as those enumerated for Example 5-2. Likewise, the same assumptions and approximations were used.

In the solution of Example 6-1, the temperature profiles between the top and bottom for each column were assumed to be linear. The top and bottom temperatures for each column were taken equal to 100°F and 175°F, respectively. Initial

Table 6-3 Statement of Examples 6-1 and 6-2

I. Feeds for Examples 6-1 and 6-2 (The symbols used in the statement of these examples are identified in Fig. 6-11.)

Component	Flow rate (mol/h)		
	$F_1 X_{1i}$	$F_2 X_{2i}$	$F_3 X_{3i}$
Methanol	0	65	15
Acetone	0	25	0
Ethanol	0	5	0
Water	5	5	5

II. Other specifications for Examples 6-1 and 6-2

Variable	Example 6-1	Example 6-2
Number of stages		
Column 1	25	50
Column 2	20	40
Type of condenser	Total	Total
Reflux ratio		
Column 1	2.5	2.5
Column 2	1.5	1.5
Pressure	760 mmHg	760 mmHg
B_1	285 mol/h fed to stage 10 of column 2	275 mol/h fed to stage 21 of column 2
B_2	215 mol/h, 94% of B_2 is fed to stage 3 of column 1	215 mol/h, 94% of B_2 is fed to stage 6 of column 1
Feed plate location		
F_1	2	4
F_2	10	21
F_3	17	17
Thermal condition of		
F_1	Liquid of 100°F	Liquid at 100°F
F_2	Liquid of 170°F	Liquid at 100°F
F_3	Liquid at 100°F	Liquid at 100°F

profiles for the component-flow rates were also assumed to be linear, and they were estimated on the basis of assumed values for the b_i's and d_i's for each column.

The following procedure was used in the solution of Example 6-1 by the column modular method. After one complete trial had been made on column 1 by use of the Almost Band Algorithm, the component flow rates $\{b_{i, 1}\}$ so obtained were used as the feed to the second column. Then one trial on column 2 was made by use of the Almost Band Algorithm. Next the capital Θ method was applied to the most recent sets of terminal flow rates $\{d_{i, 1}\}$, $\{b_{i, 1}\}$, $\{d_{i, 2}\}$, and $\{b_{i, 2}\}$, and corrected sets were obtained which satisfied the component-material balances enclosing each column and the specified values of B_1 and B_2. To

Table 6-4 Solution values of the product

Example 6-1 Product flow rates (mol/h)

Component	$d_{i,1}$	$b_{i,1}$	$d_{i,2}$	$b_{i,2}$
Methanol	8.9252	98.9252	68.3395	45.5857
Acetone	8.4422	16.5579	16.5577	0.0001267
Ethanol	1.1318	12.0196	3.34823	8.6713
Water	3.6008	157.4972	1.75426	160.7429

Example 6-2 Product flow rates (mol/h)

Component	$d_{i,1}$	$b_{i,1}$	$d_{i,2}$	$b_{i,2}$
Methanol	5.0092	64.9705	74.6729	5.2976
Acetone	24.9999	0.1981×10^{-4}	0.1981×10^{-4}	0.0
Ethanol	0.3614	4.8876	4.6227	0.2648
Water	1.7297	205.1428	0.7042	209.4387

	Temperature (°F) of steam indicated			
Example	D_1	B_1	D_2	B_2
6-1	138.38	159.14	141.52	177.04
6-2	132.07	176.37	151.13	203.43

	Column modular method		Almost Band Algorithm	
Example	No. of trials	Time (s)	No. of trials	Time (s)*
6-1	14 (at $B_1 = 277$)	4	9 (at $B_1 = 277$)	3.7
6-2	70 (at $B_1 = 275$)	25.3	65 (at $B_1 = 275$)	27.3

* AMDAHL, FORTRAN H OPT 2.

initiate the second trial on column 1, the sets of $d_{i,1}$'s, $b_{i,1}$'s, and $b_{i,2}$'s found by use of the capital Θ method were used in the set of assumed values.

Example 6-1 was also solved by the Almost Band Algorithm for Systems (the system modular method) as well as the column modular method. Results obtained by each of these methods are presented in Tables 6-3 through 6-6.

No difficulties were encountered in the solution of Example 6-1 by either of the two methods. However, by making minor changes in the specifications, this system becomes more difficult to solve as demonstrated by Example 6-2.

Example 6-2 differs from Example 6-1 by the specification of more plates, the temperature of the feed F_2, and the flow rate B_1. With the additional stages specified, this problem becomes very sensitive to the value specified for B_1. For $B_1 = 277$ instead of the specified value $B_1 = 275$, Example 6-2 is relatively easy to solve by use of the combination of the column modular method. When B_1 was taken equal to 275, Example 6-2 could not be solved by either the column modular or the system modular method. However, Example 6-2 could be solved

Table 6-5 Final temperature and liquid rate profiles and other variables for column 1 of Example 6-2

I. Profiles

Plate	Temp (°F)	Liquid flow rate (lb mol/h)	Plate	Temp (°F)	Liquid flow rate (lb mol/h)
1	132.07	30.247	26	162.38	371.07
2	133.11	78.563	27	164.56	370.92
3	135.00	75.795	28	166.45	370.88
4	139.39	76.965	29	167.89	270.90
5	148.86	69.805	30	168.90	370.94
6	171.36	265.91	31	169.57	370.97
7	172.43	266.02	32	169.98	371.00
8	172.89	266.07	33	170.23	371.02
9	173.09	266.09	34	170.38	371.03
10	173.18	266.10	35	170.48	371.03
11	173.21	266.10	36	170.53	371.04
12	173.21	266.10	37	170.56	371.04
13	173.20	266.11	38	170.58	371.04
14	173.15	266.12	39	170.59	371.04
15	173.05	266.14	40	170.60	371.04
16	172.85	266.18	41	170.560	371.04
17	172.45	266.26	42	170.60	371.04
18	171.66	266.44	43	170.61	371.04
19	170.00	266.90	44	170.61	371.04
20	166.12	268.54	45	170.61	371.04
21	154.94	372.96	46	170.62	371.04
22	155.66	372.65	47	170.65	371.02
23	156.73	372.25	48	170.81	370.89
24	158.23	371.80	49	171.63	370.13
25	160.17	371.38	50	176.37	275.00

II. Final values of other variables

Component	y_{1i}†
Methanol	0.150667
Acetone	0.809183
Ethanol	0.0062675
Water	0.033923

$Q_C = 1.5515 \times 10^6$ Btu/h

$Q_R = 1.6023 \times 10^6$ Btu/h

† Mole fraction of each component in the vapor above the liquid in the accumulator.

Table 6-6 Final temperature and liquid rate profiles and other variables for column 2 of Example 6-2

				I. Profile	

Plate	Temp (°F)	Liquid flow rate (lb mol/h)	Plate	Temp (°F)	Liquid flow rate (lb mol/h)
1	151.13	120.00	21	174.48	385.37
2	152.05	119.85	22	174.47	385.37
3	152.93	119.69	23	174.45	385.37
4	153.73	119.54	24	174.43	385.37
5	154.44	119.37	25	174.41	385.37
6	155.06	119.20	26	174.40	385.37
7	155.60	119.02	27	174.38	385.37
8	156.07	118.82	28	174.36	385.37
9	156.65	118.60	29	174.34	385.37
10	156.91	118.35	30	174.32	385.37
11	157.31	118.07	31	174.29	385.38
12	157.75	117.73	32	174.22	385.42
13	158.25	117.33	33	174.08	385.53
14	158.85	116.84	34	173.73	385.82
15	159.66	116.22	35	172.79	406.87
16	160.74	115.44	36	173.47	406.33
17	162.20	114.46	37	175.31	404.86
18	164.24	113.25	38	180.22	402.04
19	167.02	111.85	39	190.81	400.25
20	170.58	110.45	40	203.41	215.00

II. Final values of other variables

Component	y_{1i}[†]
Methanol	0.960544
Acetone	0.625244×10^{-6}
Ethanol	0.0351114
Water	0.0043440

$Q_C = 3.132158 \times 10^7$ Btu/h

$Q_R = 3.272744 \times 10^7$

[†] Mole fraction of each component in the vapor above the liquid in the accumulator.

by the column modular method by using the following procedure. First, a modified version of Example 6-2 in which B_1 was taken equal to 277 was solved. The solution so obtained was then used as the starting values for Example 6-2. Selected values of the solution so obtained are presented in Tables 6-3 through 6-6. However, when the solution at $B_1 = 277$ was used as the starting values for Example 6-2, a solution could not be obtained by use of the Almost Band Algorithm for Systems (the system modular method).

In order to solve Example 6-2 by the system modular method, the following procedure was used. First, a solution to the following modified form of Example 6-2 was obtained. The recycle stream B_2 was replaced by an additional independent feed. This feed was taken to be liquid at 170°F and was assigned the following composition.

Component	mol/h
Methanol	0.25
Acetone	0.50
Ethanol	5.00
Water	189.50

The flow rate of B_1 was again set equal to 277.

After column 1 had been solved by use of the Almost Band Algorithm, the $b_{i,1}$'s so obtained were used to solve column 2 by use of the Almost Band Algorithm. The solutions so obtained for the respective columns were used as the starting values for the system modular method. The final solution values of selected variables for this example are presented in Tables 6-3 through 6-6.

The sensitivity of the system of columns of Example 6-2 to the specified value of B_1 is reflected by the fact that when the value of B_1 was changed from $B_1 = 277$ to $B_1 = 275$, the temperature profile and the mole fraction of acetone in B_1 changed markedly. For example, for $B_1 = 277$ and $B_1 = 275$, the corresponding temperatures of stage 30 were 168.90°F and 154.42°F, respectively, and the corresponding molar flow rates of acetone in B_1 were 1.0348 and 0.0000198, respectively. Example 6-2 was included in order to demonstrate that the selection of an initial set of values of the variables is difficult for some problems. Experience thus far suggests that it is generally easier to pick a suitable set of starting values of the variables for the column modular method (the combination of the capital Θ method and the Almost Band Algorithm) than it is for the system modular method (the Almost Band Algorithm for Systems). The effort required to pick the initial values of the variables is not reflected in the computational times listed in Table 6-4.

Although Example 6-2 was originally devised by Gallun[9] to test the proposed calculational procedures, it does serve to illustrate extractive distillation. Water is the extractive distillation solvent. The acetone is recovered in the first column and the alcohols are recovered by the second column. The purity of acetone in the distillate of the first column could be improved, perhaps, by changing the specifications on the total flow rates. A program of the type used to solve this problem may be utilized to find the optimal set of operating conditions needed to effect a given separation.

In conclusion, the techniques of azeotropic and extractive distillation have come of age. The advent of high-speed computers and the development of the mathematical techniques to solve problems involving these types of separations can only serve to increase and to quantify the application of azeotropic and extractive distillation.

Before concluding this chapter, a useful algorithm proposed by Kubíček[22] for solving the equations for the Almost Band Algorithm for Systems is presented.

Kubíček Algorithm for Systems[22]

This algorithm is useful for solving the Newton-Raphson equations for systems which contain several recycle streams. The elements lying in the submatrices along the principal diagonal are treated by gaussian elimination in the same manner as described for single columns while the nonzero elements lying above and below the submatrices along the principal diagonal may be treated by use of the Kubíček algorithm. Since the nonzero elements lying above the submatrices offer no particular difficulty in the gaussian elimination process, their treatment by the Kubíček algorithm is optional.

Like the Broyden-Householder method, Kubíček algorithm is based on Householder's identity.[19]

$$(A + WCZ^T)^{-1} = A^{-1} - A^{-1}W(C^{-1} + Z^TA^{-1}W)^{-1}Z^TA^{-1} \qquad (6\text{-}31)$$

where A is an n by n matrix, W and Z are n by m matrices, and C is an m by m matrix.

Suppose that a solution of the set of equations

$$Bx = b \qquad (6\text{-}32)$$

is desired where B is an n by n matrix and x and b are conformable column vectors. Then let

$$B = A + R = A + R_1 I_m R_2^T \qquad (6\text{-}33)$$

where R_1 and R_2 are of order $n \times m$ and I_m is an identity matrix of order $m \times m$. Thus

$$(A + R_1 I_m R_2^T)x = b \qquad (6\text{-}34)$$

Then

$$x = [A + R_1 I_m R_2^T]^{-1}b \qquad (6\text{-}35)$$

Application of Householder's identity [Eq. (6-31)] gives

$$x = [A^{-1} - A^{-1}R_1(I_m^{-1} + R_2^T A^{-1}R_1)^{-1}R_2^T A^{-1}]b$$
$$= A^{-1}b - A^{-1}R_1(I_m^{-1} + R_2^T A^{-1}R_1)^{-1}R_2^T A^{-1}b \qquad (6\text{-}36)$$

Let the vector **y** and the matrix **V** be defined as follows

$$Ay = b \qquad (6\text{-}37)$$

$$AV = R_1 \qquad (6\text{-}38)$$

Then Eq. (6-36) becomes

$$x = y - V(I_m^{-1} + R_2^T V)^{-1}R_2^T y \qquad (6\text{-}39)$$

Let

$$z = (I_m + R_2^T V)^{-1}R_2^T y \qquad (6\text{-}40)$$

Since $I_m^{-1} = I_m$, it is evident that Eq. (6-39) can be written as follows

$$x = y - Vz \qquad (6\text{-}41)$$

The order of calculations is as follows where it is supposed that the **LU** factorization of the matrix **A** has been obtained, and it is desired to solve the equation **Bx = b**.

Step 1. Find **R** by use of the defining equation **R = B − A**. Form the matrix R_1 from the columns of **R** containing nonzero elements and choose R_2 such that $R = R_1 I_m R_2^T$. (Alternatively, form the matrix R_2 from the rows of **R** containing nonzero elements and choose R_1 such that $R = R_1 I_m R_2^T$.)

Step 2. Solve Eq. (6-37) for **y**. Use the **LU** factorization of **A**.

Step 3. Solve Eq. (6-38) for **V**. Use the **LU** factorization of **A**.

Step 4. Compute $R_2^T V$ and $R_2^T y$.

Step 5. Solve Eq. (6-40) for **z**.

Step 6. Solve Eq. (6-41) for **x**.

Example 6-3 Use the Kubíček algorithm to solve the matrix equation, **Bx = b**, where

$$\begin{bmatrix} 1 & 0 & 0 & 0 \\ 0 & 2 & 0 & 0 \\ 0 & 0 & 3 & 0 \\ 0 & 3 & 2 & 4 \end{bmatrix} \begin{bmatrix} x_1 \\ x_2 \\ x_3 \\ x_4 \end{bmatrix} = \begin{bmatrix} 1 \\ 4 \\ 6 \\ 4 \end{bmatrix}$$

and the **LU** factorization of the matrix **A** is known

$$A = \begin{bmatrix} 1 & 0 & 0 & 0 \\ 0 & 2 & 0 & 0 \\ 0 & 0 & 3 & 0 \\ 0 & 0 & 0 & 4 \end{bmatrix} = \begin{bmatrix} 1 & 0 & 0 & 0 \\ 0 & 1 & 0 & 0 \\ 0 & 0 & 1 & 0 \\ 0 & 0 & 0 & 1 \end{bmatrix} \begin{bmatrix} 1 & 0 & 0 & 0 \\ 0 & 2 & 0 & 0 \\ 0 & 0 & 3 & 0 \\ 0 & 0 & 0 & 4 \end{bmatrix}$$

Form R_1 from the columns of **R** which contain nonzero elements.

SOLUTION

Step 1. In this case

$$\mathbf{B} = \mathbf{A} + \mathbf{R} = \begin{bmatrix} 1 & 0 & 0 & 0 \\ 0 & 2 & 0 & 0 \\ 0 & 0 & 3 & 0 \\ 0 & 0 & 0 & 4 \end{bmatrix} + \begin{bmatrix} 0 & 0 & 0 & 0 \\ 0 & 0 & 0 & 0 \\ 0 & 0 & 0 & 0 \\ 0 & 3 & 2 & 0 \end{bmatrix}$$

Thus,

$$\mathbf{R}_1 = \begin{bmatrix} 0 & 0 \\ 0 & 0 \\ 0 & 0 \\ 3 & 2 \end{bmatrix}$$

and

$$\mathbf{R}_1 \mathbf{I}_m \mathbf{R}_2^T = \begin{bmatrix} 0 & 0 \\ 0 & 0 \\ 0 & 0 \\ 3 & 2 \end{bmatrix} \begin{bmatrix} 1 & 0 \\ 0 & 1 \end{bmatrix} \begin{bmatrix} 0 & 1 & 0 & 0 \\ 0 & 0 & 1 & 0 \end{bmatrix} = \mathbf{R}$$

Step 2. Since $\mathbf{A}$ is a diagonal matrix, the equation $\mathbf{Ay} = \mathbf{b}$ may be solved directly for $\mathbf{x}$ by use of the multiplication rule, that is,

$$\begin{bmatrix} 1 & 0 & 0 & 0 \\ 0 & 2 & 0 & 0 \\ 0 & 0 & 3 & 0 \\ 0 & 0 & 0 & 4 \end{bmatrix} \begin{bmatrix} y_1 \\ y_2 \\ y_3 \\ y_4 \end{bmatrix} = \begin{bmatrix} 1 \\ 4 \\ 6 \\ 4 \end{bmatrix}$$

Thus

$$\mathbf{y} = \begin{bmatrix} 1 \\ 2 \\ 2 \\ 1 \end{bmatrix}$$

Step 3. Equation (6-38) becomes

$$\begin{bmatrix} 1 & 0 & 0 & 0 \\ 0 & 2 & 0 & 0 \\ 0 & 0 & 3 & 0 \\ 0 & 0 & 0 & 4 \end{bmatrix} \begin{bmatrix} v_{11} & v_{12} \\ v_{21} & v_{22} \\ v_{31} & v_{32} \\ v_{41} & v_{42} \end{bmatrix} = \begin{bmatrix} 0 & 0 \\ 0 & 0 \\ 0 & 0 \\ 3 & 2 \end{bmatrix}$$

and thus

$$\mathbf{V} = \begin{bmatrix} 0 & 0 \\ 0 & 0 \\ 0 & 0 \\ 3/4 & 1/2 \end{bmatrix}$$

Step 4.

$$\mathbf{R}_2^T \mathbf{V} = \begin{bmatrix} 0 & 1 & 0 & 0 \\ 0 & 0 & 1 & 0 \end{bmatrix} \begin{bmatrix} 0 & 0 \\ 0 & 0 \\ 0 & 0 \\ 3/4 & 1/2 \end{bmatrix} = \begin{bmatrix} 0 & 0 \\ 0 & 0 \end{bmatrix} = \mathbf{0}$$

and

$$\mathbf{R}_2^T \mathbf{y} = \begin{bmatrix} 0 & 1 & 0 & 0 \\ 0 & 0 & 1 & 0 \end{bmatrix} \begin{bmatrix} 1 \\ 2 \\ 2 \\ 1 \end{bmatrix} = \begin{bmatrix} 2 \\ 2 \end{bmatrix}$$

Step 5. When the results obtained above are substituted into the equation

$$(\mathbf{I}_m + \mathbf{R}_2^T \mathbf{V})\mathbf{z} = \mathbf{R}_2^T \mathbf{y}$$

one obtains

$$\begin{bmatrix} 1 & 0 \\ 0 & 1 \end{bmatrix} \begin{bmatrix} z_{11} \\ z_{21} \end{bmatrix} = \begin{bmatrix} 2 \\ 2 \end{bmatrix}$$

and thus

$$\mathbf{z} = \begin{bmatrix} 2 \\ 2 \end{bmatrix}$$

Step 6. Substitution of the results obtained above in the equation

$$\mathbf{x} = \mathbf{y} - \mathbf{V}\mathbf{z}$$

yields

$$\mathbf{x} = \begin{bmatrix} 1 \\ 2 \\ 2 \\ 1 \end{bmatrix} - \begin{bmatrix} 0 & 0 \\ 0 & 0 \\ 0 & 0 \\ 3/4 & 1/2 \end{bmatrix} \begin{bmatrix} 2 \\ 2 \end{bmatrix} = \begin{bmatrix} 1 \\ 2 \\ 2 \\ 1 \end{bmatrix} - \begin{bmatrix} 0 \\ 0 \\ 0 \\ 5/2 \end{bmatrix} = \begin{bmatrix} 1 \\ 2 \\ 2 \\ -3/2 \end{bmatrix}$$

NOTATION

b_i	molar flow rate at which component i leaves the accumulator
B	total molar flow rate at which the bottom product is withdrawn from a column
c	total number of components
d_i	molar flow rate at which component i leaves the column
D	total molar flow rate of the distillate
f_{ji}	equilibrium function for component i and plate j
$\mathbf{f}$	vector of functions; see Eq. (6-5)
$\hat{f}_i^V$	fugacity of component i in a vapor mixture; evaluated at the temperature, pressure, and composition of the vapor
$\hat{f}_i^L$	fugacity of component i in a liquid mixture; evaluated at the temperature, pressure, and composition of the liquid
f_i^V	fugacity of pure component i in the vapor phase at the temperature and pressure of the mixture
f_i^L	fugacity of pure component i in the liquid phase at the temperature and pressure of the mixture
g_1, g_2	functions; defined by Eqs. (6-29) and (6-30), respectively
G_j	enthalpy balance function; see Eq. (5-4)
K_i	equilibrium ratio; defined below Eq. (6-1)
l_{ji}	molar flow rate at which component i leaves plate j in the liquid phase
L_j	total molar flow rate at which the liquid leaves plate j
m_{ji}	material-balance function for component i and plate j
N	total number of stages
$p_{i,1}, p_{i,2}$	ratios; defined by Eqs. (6-24) and (6-25), respectively
Q_C	condenser duty
Q_R	reboiler duty
$r_{i,1}, r_{i,2}$	functions of Θ_1 and Θ_2, respectively; see Eqs. (6-24) and (6-25)
S_1	specification function; see Eq. (6-12)
S_N	specification function; see Eq. (6-13)
y_i	mole fraction of component i in the vapor phase
x_i	mole fraction of component i in the liquid phase
X_i	total mole fraction of component i in the feed to the column
$\mathbf{x}$	vector of variables; see Eq. (6-6)

Greek Letters

α, β	fractions of streams L_1^1 and L_1^2 withdrawn; see Fig. 6-10
γ_i^V	activity coefficient of component i in the vapor phase; see Eq. (6-1)
γ_i^L	activity coefficient of component i in the liquid phase; see Eq. (6-1)
Θ_1, Θ_2	multipliers; defined by Eqs. (6-18) and (6-19)

PROBLEMS

6-1 Use the Kubíček algorithm to solve the matrix equation

$$Bx = b$$

where

$$B = \begin{bmatrix} 1 & 0 & 0 & 0 \\ 0 & 2 & 0 & 0 \\ 0 & 2 & 3 & 0 \\ 0 & 3 & 2 & 4 \end{bmatrix}$$

$$b = [1 \quad 4 \quad 6 \quad 4]^T$$

$$x = [x_1 \quad x_2 \quad x_3 \quad x_4]^T$$

6-2 Repeat Prob. 6-1 for the case where the matrix **B** is given by

$$B = \begin{bmatrix} 1 & 0 & 0 & 0 \\ 2 & 2 & 0 & 0 \\ 0 & 0 & 3 & 0 \\ 0 & 3 & 2 & 4 \end{bmatrix}$$

and matrices **b** and **x** are the same as stated in Example 6-1.

6-3 Produce an $N[(2c + 1) + c + 2]$ Newton-Raphson formulation for case 1 (Q_C, Q_R, and a partial condenser) for a conventional distillation column having two liquid phases and one vapor phase in the partial condenser.

6-4 (a) If in Example 6-3, the matrix R_2^T is taken to be

$$R_2^T = [0 \quad 3 \quad 2 \quad 0]$$

find the R_1 needed in the Kubíček algorithm.

(b) Repeat Example 6-3 for the R_2^T and corresponding values of R_1 found in part (a).

6-5 (a) Show that the $2c + 3$ equilibrium relationships for a three-phase mixture may be stated in terms of the $\{y_i\}$ for the dew-point temperature as follows

$$F(x_1, x_2, T) = F_1^2(x_1, T) + F_2^2(x_2, T)$$

where

$$F_1(x_1, T) = \sum_{i=1}^{c} \frac{y_i}{K_{1i}} - 1$$

$$F_2(x_2, T) = \sum_{i=1}^{c} \frac{y_i}{K_{2i}} - 1$$

$x_1 = [x_{1,1} \quad x_{1,2} \quad \cdots \quad x_{1,c}]^T$, mole fractions for liquid phase 1.

$x_2 = [x_{2,1} \quad x_{2,2} \quad \cdots \quad x_{2,c}]^T$, mole fractions for liquid phase 2.

$K_{1i} = \dfrac{\gamma_{1i}^L}{\gamma_i^V} (K_{1i})_{ideal}$, where $(K_{1i})_{ideal}$ is the ideal solution value of K_{1i}.

$K_{2i} = \dfrac{\gamma_{2i}^L}{\gamma_i^V} (K_{2i})_{ideal}$, where $(K_{2i})_{ideal}$ is the ideal solution value of K_{2i}.

(b) For the case where P and $\{y_i\}$ are given and it is desired to find the dew-point temperature T and the compositions of the two liquid phases, devise a calculational procedure whereby it is possible to correct the values of x_1 and x_2 with each successive iteration.

6-6 Formulate the equations for the determination of the dew point of a three phase mixture in terms of $2c + 2$ independent functions and $2c + 2$ independent variables.

6-7 (a) State the $3c + 3$ equations required to describe a three-phase isothermal flash.

(b) For the case where the pressure P and temperature T of the three-phase flash are specified as well as the feed rate F and composition $\{X_i\}$, formulate the isothermal flash problem in terms of $3c$ independent functions and the $3c$ independent variables

$$\mathbf{x} = [l_{1,1} \cdots l_{1,c} \; l_{2,1} \cdots l_{2,c} \; v_1 \cdots v_c]^T$$

(c) Repeat part (b) for the case where the formulation is in terms of $2c$ independent functions and the $2c$ independent variables

$$\mathbf{x} = [l_{1,1} \cdots l_{1,c} \; l_{2,1} \cdots l_{2,c}]^T$$

6-8 (a) State the $3c + 4$ equations required to described a three-phase adiabatic flash.

(b) Formulate the adiabatic flash problem in which the composition, flow rate, and thermal condition of the feed is specified as well as the flash pressure, in terms of $3c + 1$ independent functions and the $3c + 1$ independent variables

$$\mathbf{x} = [l_{1,1} \cdots l_{1,c} \; l_{2,1} \cdots l_{2,c} \; v_1 \cdots v_c \; T]^T$$

6-9 (a) Formulate the three-phase isothermal flash in terms of the independent variables $\theta_1 = L_1/V$ and $\theta_2 = L_2/V$, where L_1 and L_2 are the flow rates of liquid phases 1 and 2, and V is the flow of the vapor formed by the flash. Begin with

$$0 = \frac{\sum\limits_{i=1}^{c} v_i}{V} - \frac{\sum\limits_{i=1}^{c} l_{1i}}{L_1}$$

$$0 = \frac{\sum\limits_{i=1}^{c} v_i}{V} - \frac{\sum\limits_{i=1}^{c} l_{2i}}{L_2}$$

and obtain the functions

$$F_1(\theta_1, \theta_2) = \frac{1}{V} \sum_{i=1}^{c} \left(\frac{1}{K_{1i}} - 1 \right) v_i$$

$$F_2(\theta_1, \theta_2) = \frac{1}{V} \sum_{i=1}^{c} \left(\frac{1}{K_{2i}} - 1 \right) v_i$$

where

$$V = \frac{F}{1 + \theta_1 + \theta_2}$$

$$v_i = \frac{F x_i}{1 + (\theta_1/K_{1i}) + (\theta_2/K_{2i})}$$

$K_{1i}, K_{2i} = K$ values for phases 1 and 2, respectively, defined in Prob. 6-5.

(b) Show that sufficient conditions for $F_1(\theta_1, \theta_2)$ to be monotonic in the $\theta_1 = 0$ and $\theta_2 = 0$ planes and that one and only one θ_1, and one and only one θ_2 exist such that $F_1(\theta_1, 0) = 0$ and $F_1(0, \theta_2) = 0$ are as follows:

$$(1) \quad \sum_{i=1}^{c} K_{1i} X_i > 1 \qquad\qquad (2) \quad \sum_{i=1}^{c} \frac{X_i}{K_{1i}} > 1$$

$$(3) \quad \sum_{i=1}^{c} \frac{Y_{2i}}{K_{1i}} < 1 \qquad\qquad (4) \quad \sum_{i=1}^{c} \frac{Y_{1i}}{K_{2i}} < 1$$

where

$$Y_{1i} = \frac{K_{1i} X_i}{\displaystyle\sum_{i=1}^{c} K_{1i} X_i} \qquad Y_{2i} = \frac{K_{2i} X_i}{\displaystyle\sum_{i=1}^{c} K_{2i} X_i}$$

(5) For each $K_{1i} < 1$, the corresponding value of $K_{2i} < 1$, and for each $K_{1i} > 1$, the corresponding values of $K_{2i} > 1$.

(c) Repeat part (b) for the function $F_2(\theta_1, \theta_2)$. In this case the corresponding set of five sufficient conditions are obtained by replacing the subscript 1 by the subscript 2 in part (b).

6-10 State the Newton-Raphson equations for the functions in Prob. 6-9, and formulate a calculational procedure for computing corrected values of x_1 and x_2 after improved values of θ_1 and θ_2 have been found on the basis of assumed sets of values for x_1 and x_2.

6-11 (a) In a manner analogous to the formulation of the flash function $p(\psi)$ of Chap. 1, show that the three-phase isothermal flash problem may be formulated as follows

$$P_1(\psi_1, \psi_2) = \sum_{i=1}^{c} \left| \frac{X_i}{\psi_1(1 - K_{1i}) + \psi_2[(K_{1i}/K_{2i}) - K_{1i}] + K_{1i}} \right| - 1$$

$$P_2(\psi_1, \psi_2) = \sum_{i=1}^{c} \left(\frac{K_{1i} X_i}{K_{2i}\{\psi_1(1 - K_{1i}) + \psi_2[(K_{1i}/K_{2i}) - K_{1i}]\} + K_{1i}} \right) - 1$$

where

$$\psi_1 = L_1/F \qquad \psi_2 = L_2/F$$

(b) Show that the functions P_1 and P_2 are not monotonic. (The functions P_1 and P_2 were suggested by Professor John H. Erbar, p. 62, *Proceedings 52 Annual Convention*, Natural Gas Processors Association, 1973.)

REFERENCES

1. G. T. Atkins and C. M. Boyer: "Application of the McCabe-Thiele Method to Extractive Distillation Calculations," *Chem. Eng. Prog.*, **45**: 553 (1949).
2. M. Benedict and L. C. Rubin: "Extractive and Azeotropic Distillation, Theoretical Aspects 1," *Trans. AIChE*, **45**: 353 (1945).
3. L. Berg: "Azeotropic and Extractive Distillation-Selecting the Agent for Distillation," *Chem. Eng. Proc.*, **65**(9): 52 (1969).
4. C. S. Carlson, P. V. Smith, Jr., and C. E. Morrell: "Separation Oxygenated Organic Compounds by Water Extractive Distillation," *Ind. Eng. Chem.*, **46**: 350 (1954).
5. B. Carnahan, H. Luther, and J. O. Wilkes: *Applied Numerical Methods*, John Wiley and Sons, New York, 1969.
6. C. L. Dunn, R. W. Millar, G. J. Pierotti, R. N. Shiras, and Mott Souders, Jr.: "Toluene Recovery of Extractive Distillation," *Trans. AIChE*, **41**: 631 (1945).
7. S. E. Gallun and C. D. Holland: "Solve More Distillation Problems," Part 5. *Hydrocarbon Process.*, **55**(1): 137 (1976).
8. ———: M.S. Thesis, Texas A&M University, 1975.
9. ———: Ph.D. Dissertation, Texas A&M University, 1977.

10. E. Guillaume: U.S. Patent 887,793 (May 19, 1908).
11. H. Guinot and F. W. Clark: "Azeotropic Distillation in Industry," *Trans. Inst. Chem. Eng.* (*London*), **16**: 187 (1938).
12. J. Happel, P. W. Cornell, Du B. Eastman, M. J. Fowle, C. A. Porter, and A. H. Schutte: "Extractive Distillation-Separation of C_4 Hydrocarbons Using Furfural," *Trans. Am. Inst. Chem. Eng.*, **42**: 189 (1946).
13. H. V. Hess, E. A. Naragon, and C. A. Coghlan: "Extractive Distillation and Separation of n-Butane from Butenes—2," *Chem. Eng. Prog. Symp. Ser.*, (2): 72 (1952).
14. ———, S. E. Gallun, G. W. Bentzen, and C. D. Holland: "Solve More Distillation Problems, Part 8—Which Method to Use," *Hydrocarbon Process.*, **56**(6): 181 (1977).
15. ———, ———, and ———: "Calculational Procedures for Systems of Distillation Columns," presented at the 83 National Meeting of the American Institute of Chemical Engineers, Mar. 20–24, 1977, Houston, Texas.
16. C. D. Holland: *Fundamentals and Modeling of Separation Processes, Absorption, Distillation, Evaporation, and Extraction*, Prentice-Hall, Inc., Englewood Cliffs, N.J., 1975.
17. W. C. Hopkins and J. J. Fritsch: "How Celanese Separates Complex Petrochemical Mixtures," *Chem. Eng.*, **51**(8): 361 (1955).
18. L. H. Horsley: *Azeotropic Data*, American Chemical Society, Washington, D.C., 1952.
19. A. S. Householder: *Principles of Numerical Analysis*, McGraw-Hill Book Company, New York, 1953.
20. M. J. Holmes and M. Van Winkle: "Prediction of Ternary Vapor-Liquid Equilibria from Binary Data," *Ind. Eng. Chem.*, **62**(1): 20 (1970).
21. H. P. Hutchison and C. F. Shewchuk: "A Computational Method for Multiple Distillation Towers," *Trans. Inst. Chem. Eng.*, **52**: 325 (1974).
22. Kubíček, V. Hlaváček, and F. Procháska: "Global Modular Newton-Raphson Technique for Simulation of an Interconnected Plant Applied to Complex Rectifying Columns," *Chem. Eng. Sci.*, **31**: 277 (1976).
23. H. R. Null and D. A. Palmer: "Azeotropic and Extractive Distillation-Predicting Phase Equilibria," *Chem. Eng. Prog.*, **66**(9): 47 (1969).
24. N. Poffenberger, L. H. Horsley, H. S. Nutting, and C. E. Britton: "Separation of Butadiene by Azeotropic Distillation with Ammonia," *Trans. AIChE.*, **42**: 815 (1946).
25. J. M. Prausnitz, C. A. Echert, R. V. Orye, and J. P. O'Connell: *Computer Calculations for Multi-Component Vapor-Liquid Equilibria*, Prentice-Hall, Englewood Cliffs, N.J., 1967.
26. N. C. Robertson: (to Celanese Corporation of America), U.S. Patent 2,477,087 (July 16, 1949).
27. D. Tassios: "Choosing Solvents for Extractive Distillation," *Chem. Eng.*, 118 (Feb. 10, 1969).
28. R. P. Tewarson: *Sparse Matrices*, Academic Press, New York, 1973.
29. M. Van Winkle: *Distillation*, McGraw-Hill Book Company, New York, 1967.

SEVEN

SOLUTION OF SYSTEMS OF COLUMNS WITH ENERGY EXCHANGE BETWEEN RECYCLE STREAMS

Three types of energy exchange between columns are considered. In the first type, one stream is used to condense another. In the particular example considered, the condenser of one column is used as the reboiler for another column. In the second type of energy exchange, energy is transferred from one recycle stream to another through the use of a heat exchanger. In the third type of energy exchange, an intercooler or interheater is used to adjust the heat content or temperature of a given stream.

A column modular method which makes use of the capital Θ method of convergence is presented for solving problems involving systems containing both mass and energy recycle streams. On the basis of an assumed set of values of the variables for the recycle streams, one complete trial is made on each column or unit of the system by use of the most appropriate calculational procedure for the column or unit. Then the capital Θ method is used to place all terminal streams in material and energy balance. The values of the variables so obtained for the recycle streams are used to make the next trial on each column or unit of the system.[6, 7]

The capital Θ method of convergence has been applied extensively to systems of all types of columns interconnected by recycle streams in which mass is transferred from one column to another.[5] Recycle streams of this type are referred to as *mass recycle streams*. The decomposition of a column or a system of columns into any number of columns interconnected by mass recycle streams has been effected by use of the capital Θ method of convergence.[2] For certain types of systems in which the columns are at the operating condition of total reflux, the capital Θ method has been shown to constitute an exact solution to problems involving such systems.[4]

The techniques developed in the application of the capital Θ method to the types of problems considered in Secs. 7-1, 7-2, and 7-3 may be used with the θ method to solve certain types of problems involving single columns as shown in Sec. 7-4.

7-1 COLUMNS HAVING A COMMON CONDENSER AND REBOILER

To illustrate this type of energy exchange, suppose that it is desired to use the condenser of column 2 of Fig. 7-1 as the reboiler for column 1. In order for such an operation to be feasible, it is of course necessary for the dew-point temperature $T_{2,2}$ of the overhead $V_{2,2}$ from column 2 to be greater than the bubble-point temperature $T_{N-1,1}$ of the liquid $L_{N-1,1}$ entering the reboiler of column 1. Obviously, many pairs of operating pressures $P_2 > P_1$ exist for which

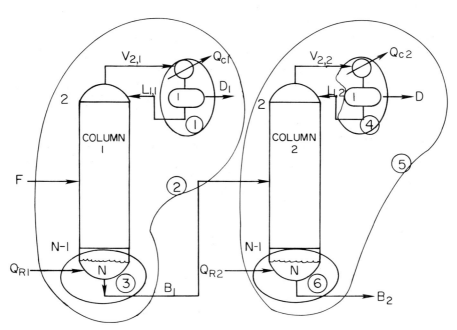

Figure 7-1 A system of two columns with energy recycle; the condenser of column 2 is the reboiler of column 1.

$T_{2,2} > T_{N-1,1}$, where P_1 and P_2 are the operating pressures of columns 1 and 2, respectively. There are also obviously many choices of $P_2 > P_1$ for which $T_{2,2} \leq T_{N-1,1}$. For any choice of pressures, it is possible, however, to find a solution for which the condenser duty of column 2 is equal to the reboiler duty of column 1 ($Q_{C2} = Q_{R1}$). Even though an equality of the duties may exist, it will be impossible to transfer the energy from $V_{1,2}$ to $L_{N-1,1}$ unless $T_{2,2} > T_{N-1,1}$. After a solution for which $Q_{C2} = Q_{R1}$ has been obtained for a given set of column pressures, the approach temperatures $T_{N-1,1}$ and $T_{2,2}$ may be examined to determine whether or not the heat can be transferred with a reasonable size heat exchanger. If the area required is too large, the difference in the approach temperatures may be increased by increasing the pressure of column 2 relative to the pressure of column 1. The balance between heat transfer area and operating pressures is an economic one.

There follows a description of the use of the capital Θ method in the determination of a solution for the system of columns shown in Fig. 7-1 such that the reboiler duty of column 1 is equal to the condenser duty of column 2, that is, such that $Q_{R1} = Q_{C2}$. For the system of columns shown in Fig. 7-1, suppose that column 1 has a partial condenser and that the flow rate, composition, and thermal condition of the feed are fixed as well as the pressure, the number of stages, and the feed-plate location. Likewise, suppose that the number of plates, the feed-plate location, and the pressure of column 2 are fixed. Two additional specifications may be made on each column. Since the precise choice of these additional specifications determines the form of the calculational procedures for each column as well as the system, only these additional specifications are listed in the problems considered. In each instance it is supposed, however, that the usual specifications of the type listed above have been made.

To demonstrate the use of the capital Θ method in the solution of problems involving such a system, suppose that the following set of additional specifications are made.

Set 1 Additional specifications

Column specifications		System specifications	
Column 1. $\dfrac{L_{1,1}}{D_1}$, $\left(\dfrac{V_{N1}}{B_1}\right)_n$		$\dfrac{L_{1,1}}{D_1}$, $Q_{R1} = Q_{C2}$	
Column 2. $\dfrac{L_{2,2}}{D_2}$, $\dfrac{V_{N2}}{B_2}$		$\dfrac{L_{2,2}}{D_2}$, $\dfrac{B_2}{V_{N2}}$	

The last subscript refers to the number of the column. The boilup ratio $(V_{N1}/B_1)_n$ of column 1 is to be adjusted from one system trial to the next until a solution has been found such that $Q_{C1} = Q_{R1}$.

The solution may be effected by use of the following calculational procedure.

On the basis of an assumed value for V_{N1}/B_1, denoted by $(V_{N1}/B_1)_n$, one complete trial is made on column 1. The results so obtained are used to make one complete trial on column 2. Then the capital Θ method of convergence is applied in order to find a new value $(V_{N1}/B_1)_{n+1}$ such that all of the energy and material balances enclosing the terminal streams as well as the equilibrium relationships for these streams are satisfied at the condition $Q_{R1} = Q_{C2}$. The value so obtained for V_{N1}/B_1 is used to make the next column trial on column 1.

In the application of the capital Θ method to the system of two columns shown in Fig. 7-1, two multipliers are defined

$$\frac{b_{i,1}}{d_{i,1}} = \Theta_1 \left(\frac{b_{i,1}}{d_{i,1}}\right)_{ca} \tag{7-1}$$

$$\frac{b_{i,2}}{d_{i,2}} = \Theta_2 \left(\frac{b_{i,2}}{d_{i,2}}\right)_{ca} \tag{7-2}$$

where the subscript ca denotes the values of the variables obtained by the most recent trial calculation for columns 1 and 2. The quantities on the left-hand side denote the corrected values of the variables which are to be found by use of the capital Θ method. To avoid numerical problems, it is recommended that variables $p_{i,1}$ and $p_{i,2}$ be used instead of $d_{i,1}$ and $d_{i,2}$, where

$$p_{i,1} = \frac{d_{i,1}}{(d_{i,1})_{ca}} \tag{7-3}$$

$$p_{i,2} = \frac{d_{i,2}}{(d_{i,2})_{ca}} \tag{7-4}$$

Restatement of Eqs. (7-1) and (7-2) in terms of these new variables gives

$$b_{i,1} = r_{i,1} p_{i,1} \tag{7-5}$$

$$b_{i,2} = r_{i,2} p_{i,2} \tag{7-6}$$

where

$$r_{i,1} = \Theta_1 (b_{i,1})_{ca}$$

$$r_{i,2} = \Theta_2 (b_{i,2})_{ca}$$

The Θ's are to be picked such that the corrected component-flow rates satisfy the component-material balances enclosing each column as well as the energy balances and equilibrium relationships stated below. The component-material balances to be satisfied are

$$FX_i - d_{i,1} - b_{i,1} = 0 \tag{7-7}$$

$$b_{i,1} - d_{i,2} - b_{i,2} = 0 \tag{7-8}$$

Equations (7-3) through (7-6) may be used to restate Eqs. (7-7) and (7-8) in terms of $p_{i, 1}$ and $p_{i, 2}$, namely,

$$(d_{i, 1})_{ca} p_{i, 1} + r_{i, 1} p_{i, 1} = FX_i \tag{7-9}$$

$$r_{i, 1} p_{i, 1} - (d_{i, 2})_{ca} p_{i, 2} - r_{i, 2} p_{i, 2} = 0 \tag{7-10}$$

Equations (7-9) and (7-10) are readily solved for $p_{i, 1}$ and $p_{i, 2}$ to give

$$p_{i, 1} = \frac{FX_i}{(d_{i, 1})_{ca} + r_{i, 1}} \tag{7-11}$$

$$p_{i, 2} = \frac{r_{i, 1} FX_i}{[(d_{i, 1})_{ca} + r_{i, 1}][(d_{i, 2})_{ca} + r_{i, 2}]} \tag{7-12}$$

To pick the set of Θ's such that the component-material balances and the energy balances enclosing the two columns are satisfied simultaneously requires the introduction of additional variables which appear in the energy balances. When all of the reflux and boilup ratios are regarded as fixed, these equations contain the following variables: $T_{1, 1}$, $T_{2, 1}$, $T_{N-1, 1}$, $T_{N, 1}$, $T_{1, 2}$, $T_{2, 2}$, $T_{N-1, 2}$, $T_{N, 2}$; and the reboiler and condenser duties: Q_{C1}, Q_{R1}, Q_{C2}, and Q_{R2}. Corresponding to each of the eight unknown temperatures, a vapor-liquid equilibrium function may be formulated. Six energy-balance functions may be formulated by use of the six energy balances corresponding to the enclosures shown in Fig. 7-1, for a total of 14 functions. (These results are summarized in Table 7-1 and variables and functions for other sets of specifications are summarized in Table 7-2.) When these equations are solved subject to the condition that $Q_{C2} = Q_{R1}$, one of the reflux or boilup ratios must be taken as an independent variable as discussed below.

Formulation of the Vapor-Liquid Equilibrium and the Enthalpy Functions

For the case where a partial condenser is used for column 1, the dew-point expression

$$\sum_{i=1}^{c} \frac{X_{Di, 1}}{K_i(T_{1, 1})} - 1 = 0 \tag{7-13}$$

may be restated in functional form in terms of the component-flow rates as follows

$$g_1 = \sum_{i=1}^{c} \frac{d_{i, 1}/D_1}{K_i(T_{1, 1})} - \frac{\sum\limits_{i=1}^{c} d_{i, 1}}{D_1} \tag{}$$

or

$$g_1 = \frac{1}{D_1} \left[\sum_{i=1}^{c} \left(\frac{1}{K_i(T_{1, 1})} - 1 \right) d_{i, 1} \right] \tag{7-14}$$

where

$$D_1 = \sum_{i=1}^{c} d_{i, 1}$$

In the calculational procedure, $d_{i, 1}$ is restated in terms of $p_{i, 1}$ as given by Eqs. (7-3) and (7-11), namely,

$$d_{i, 1} = (d_{i, 1})_{ca} p_{i, 1} = \frac{(d_{i, 1})_{ca} F X_i}{(d_{i, 1})_{ca} + r_{i, 1}} \tag{7-15}$$

The functions corresponding to the remaining temperatures are developed in an analogous manner. The results so obtained are listed in Table 7-1. Although these functions are stated in terms of the variables $d_{i, 1}$, $d_{i, 2}$, $b_{i, 1}$, and $b_{i, 2}$, these variables are readily restated in terms of the r_i's by use of Eqs. (7-3) through (7-6), and (7-11) and (7-12).

To illustrate the formulation of the enthalpy balance functions from the energy balances, the function corresponding to the enclosure of the condenser-accumulator of column 1 is developed. The enthalpy balance enclosing this section of column 1 is given by

$$\sum_{i=1}^{c} [v_{2i, 1} H_i(T_{2, 1}) - d_{i, 1} H_i(T_{1, 1}) - l_{1i, 1} h_i(T_{1, 1})] - Q_{C1} = 0 \tag{7-16}$$

Elimination of $l_{1i, 1}$ by use of the component-material balance

$$v_{2i, 1} - d_{i, 1} - l_{1i, 1} = 0 \tag{7-17}$$

yields

$$\sum_{i=1}^{c} \{v_{2i, 1}[H_i(T_{2, 1}) - h_i(T_{1, 1})] - d_{i, 1}[H_i(T_{1, 1}) - h_i(T_{1, 1})]\} - Q_{C1} = 0 \tag{7-18}$$

Then, let

$$g_9 = \frac{\displaystyle\sum_{i=1}^{c} d_{i, 1}[H_i(T_{1, 1}) - h_i(T_{1, 1})] + Q_{C1}}{\displaystyle\sum_{i=1}^{c} \{(d_{i, 1} + l_{1i, 1})[H_i(T_{2, 1}) - h_i(T_{1, 1})]\}} - 1 \tag{7-19}$$

Since the vapor stream $d_{i, 1}$ and the liquid stream $l_{1i, 1}$ are in equilibrium, $l_{1i, 1}$ appearing in Eq. (7-19) may be stated in terms of $d_{i, 1}$ as follows

$$l_{1i, 1} = \left(\frac{L_{1, 1}}{D_1 K_i(T_{1, 1})}\right) d_{i, 1} \tag{7-20}$$

Table 7-1 Functions for the capital Θ method for the system shown in Fig. 7-1

I. Vapor-liquid equilibrium functions

1. Dew-point function for the vapor stream D_1

$$g_1 = \frac{1}{D_1} \sum_{i=1}^{c} \left| \frac{1}{K_i(T_{1,1})} - 1 \right| d_{i,1}$$

where

$$D_1 = \sum_{i=1}^{c} d_{i,1}$$

2. Dew-point function for the vapor stream $V_{2,1}$

$$g_2 = \frac{1}{V_{2,1}} \sum_{i=1}^{c} \left| \frac{1}{K_i(T_{2,1})} - 1 \right| v_{2i,1}$$

$$v_{2i,1} = \left| \frac{L_{1,1}}{D_1} \frac{1}{K_i(T_{1,1})} + 1 \right| d_{i,1}$$

$$V_{2,1} = \sum_{i=1}^{c} v_{2i,1}$$

3. Bubble-point function for the liquid stream $L_{N-1,1}$

$$g_3 = \frac{1}{L_{N-1,1}} \sum_{i=1}^{c} [K_i(T_{N-1,1}) - 1]l_{N-1,i,1}$$

where

$$l_{N-1,i,1} = \left| \frac{V_{N,1} K_i(T_{N,1})}{B_1} + 1 \right| b_{i,1}$$

$$L_{N-1,1} = \sum_{i=1}^{c} l_{N-1,i,1}$$

4. Bubble-point function for the liquid stream B_1

$$g_4 = \frac{1}{B_1} \sum_{i=1}^{c} [K_i(T_{N,1}) - 1]b_{i,1}$$

where

$$B_1 = \sum_{i=1}^{c} b_{i,1}$$

5. Dew-point function for the vapor stream D_2

$$g_5 = \frac{1}{D_2} \sum_{i=1}^{c} \left| \frac{1}{K_i(T_{1,2})} - 1 \right| d_{i,2}$$

where

$$D_2 = \sum_{i=1}^{c} d_{i,2}$$

6. Dew-point function for the vapor stream $V_{2,2}$

$$g_6 = \frac{1}{V_{2,2}} \sum_{i=1}^{c} \left| \frac{1}{K_i(T_{2,2})} - 1 \right| v_{2i,2}$$

where

$$v_{2i,2} = \left| \frac{L_{1,2}}{D_2 K_i(T_{1,2})} + 1 \right| d_{i,2}$$

$$V_{2,2} = \sum_{i=1}^{c} v_{2i,2}$$

7. Bubble-point function for the liquid stream $L_{N-1,2}$

$$g_7 = \frac{1}{L_{N-1,2}} \sum_{i=1}^{c} [K_i(T_{N-1,2}) - 1] l_{N-1,i,2}$$

where

$$l_{N-1,i,2} = \left| \left(\frac{V_{N,2} K_i(T_{N,2})}{B_2} + 1 \right) b_{i,2} \right| \qquad L_{N-1,2} = \sum_{i=1}^{c} l_{N-1,i,2}$$

8. Bubble-point function for the liquid stream B_2

$$g_8 = \frac{1}{B_2} \sum_{i=1}^{c} [K_i(T_{N,2}) - 1] b_{i,2}$$

where

$$B_2 = \sum_{i=1}^{c} b_{i,2}$$

II. Energy balance functions

1. Enthalpy-balance function for the enclosure of the condenser-accumulator of column 1

$$g_9 = \frac{1}{Q_{C1}} \sum_{i=1}^{c} \left| \left[H_i(T_{2,1}) - H_i(T_{1,1}) \right] \right.$$
$$\left. + [H_i(T_{2,1}) - h_i(T_{1,1})] \frac{L_{1,1}}{D_1 K_i} \right| d_{i,1} - 1$$

2. Enthalpy-balance function for the enclosure of column 1

$$g_{10} = \frac{1}{Q_{C1}} \sum_{i=1}^{c} \{ FX_i [H_i(T_{feed}) - h_i(T_{N,1})] \}$$
$$- d_{i,1}[H_i(T_{1,1} - h_i(T_{N,1})] + \frac{Q_{R1}}{Q_{C1}} - 1$$

3. Enthalpy-balance function for the reboiler of column 1

$$g_{11} = \frac{1}{Q_{R1}} \sum_{i=1}^{c} \left| [h_i(T_{N,1}) - h_i(Y_{N-1,1})] \right.$$
$$+ \frac{V_{N,1} K_i(T_{N,1})}{B_1} [H_i(T_{N,1}) - h_i(T_{N-1,1})] \Big| b_{i,1} - 1$$

4. Enthalpy-balance function for the enclosure of the condenser-accumulator section of column 2

$$g_{12} = \frac{1}{Q_{C2}} \sum_{i=1}^{c} \left| [H_i(T_{2,2}) - H_i(T_{1,2})] \right.$$
$$\left. + [H_i(T_{2,2}) - h_i(T_{1,2})] \frac{L_{1,2}}{D_2 K_i(T_{1,2})} \right| d_{i,2} - 1$$

5. Enthalpy-balance function for the enclosure of column 2

$$g_{13} = \frac{1}{Q_{C2}} \sum_{i=1}^{c} \{ [h_i(T_{N,1}) - H_i(T_{1,2})] d_{i,2}$$
$$+ [h_i(T_{N,1}) - h_i(T_{N,2})] b_{i,2} \} + \frac{Q_{R2}}{Q_{C2}} - 1$$

6. Enthalpy-balance function for the enclosure of the reboiler of column 2

$$g_{14} = \frac{1}{Q_{R2}} \sum_{i=1}^{c} \left| [h_i(T_{N,2}) - h_i(T_{N-1,2})] \right.$$
$$+ [H_i(T_{N,2}) - h_i(T_{N-1,2})] \frac{V_{N,2} K_i(T_{N,2})}{B_2} \Big| b_{i,2} - 1$$

Table 7-2 Sets of specifications, variables, and functions for the capital Θ method for systems

Set	System	Column specifications	System specifications	System variables and functions
1	Fig. 7-1	$\dfrac{L_{1,1}}{D_1}$, $\left(\dfrac{V_{N1}}{B_1}\right)_n$	$\dfrac{L_{1,1}}{D_1}$, $Q_{R1} = Q_{C2}$	$x = [\Theta_1 \ \Theta_2 \ T_{1,1} \ T_{2,1} \ T_{N-1,1} \ T_{N1} \ T_{1,2} \ T_{2,2} \ T_{N-1,2} \ T_{N2} \ Q_{C1} \ Q_{R1} \ Q_{R2} \ V_{N1}/B_1]^T$
		$\dfrac{L_{1,2}}{D_2}$, $\dfrac{V_{N2}}{B_2}$	$\dfrac{L_{1,2}}{D_2}$, $\dfrac{V_{N2}}{B_2}$	$g = [g_1 \ g_2 \cdots g_{14}]^T$; replace Q_{C2} by Q_{R1} in Table 7-1
2	Fig. 7-2	$\dfrac{L_{1,1}}{D_1}$, $\dfrac{V_{N1}}{B_1}$	$\dfrac{L_{1,1}}{D_1}$, $\dfrac{V_{N1}}{B_1}$	$x = [\Theta_1 \ \Theta_2 \ T_{1,1} \ T_{2,1} \ T_{N-1,1} \ T_{N1} \ T_{1,2} \ T_{2,2} \ T_{N-1,2} \ T_{N2} \ Q_{C1} \ Q_{R1} \ Q_{C2} \ Q_{R2}]^T$
		$\dfrac{L_{1,2}}{D_2}$, $\dfrac{V_{N2}}{B_2}$	$\dfrac{L_{1,2}}{D_2}$, $\dfrac{V_{N2}}{B_2}$	$g = [g_1 \ g_2 \cdots g_{14}]^T$
3	Fig. 7-2	$L_{1,1}, D_1$	D_1, B_2	$x = [\Theta_1 \ \Theta_2]^T$
		$L_{1,2}, B_2$		$g = [g_1 \ g_2]^T$
4	Fig. 7-3	$\dfrac{L_{1,1}}{D_1}$, $\dfrac{V_{N1}}{B_1}$	Same as column specifications	$x = [\Theta_1 \ \Theta_2 \ T_{1,1} \ T_{2,1} \ T_{N-1,1} \ T_{N1} \ T_{1,2} \ T_{2,2}$
		$\dfrac{L_{1,2}}{D_2}$, $\dfrac{V_{N2}}{B_2}$		$T_{N-1,2} \ T_{N2} \ T_{N1,0} \ T_{N2,0} \ Q_{C1} \ Q_{C2} \ Q_{R1} \ Q_{R2}]^T$
		U and A of a double-pipe heat exchanger with counter flow		$g = [g_1 \ g_2 \cdots g_{15} \ g_{16}]^T$
5	Fig. 7-4	$V_{N1}/B_1, D_2, L_{1,2}$, W_1, W_2, B_2	B_2, D_2 W_1, W_2	$x = [\Theta_1 \ \Theta_2 \ \Theta_3 \ \Theta_4]^T$ $g = [g_1 \ g_2 \ g_3 \ g_4]^T$
6	Fig. 7-4	$V_{N1}/B_1, B_2, D_2$, $L_{1,2}, W_1, W_2$	B_2, D_2 W_1, W_2	$x = [\Theta_1 \ \Theta_2 \ \Theta_3 \ \Theta_4]^T$ $g = [g_1 \ g_2 \ g_3 \ g_4]^T$
		U and A of a double-pipe heat exchanger with counter-flow		

The remaining enthalpy balance functions g_{10} through g_{14} are developed in a manner analogous to that demonstrated for g_9.

Instead of regarding all of the reflux and boilup ratios as fixed, let one of them be varied, say V_{N1}/B_1, as required to satisfy the condition that $Q_{R1} = Q_{C2}$. The new variable V_{N1}/B_1 is added to the set of variables and the variable Q_{C2} is removed from the set of variables and functions by replacing Q_{C2} by Q_{R1}. Thus, to solve this problem, the variables are taken to be

$$\mathbf{x} = [\Theta_1 \ \Theta_2 \ T_{1,1} \ T_{2,1} \ T_{N-1,1} \ T_{N,1} \ T_{1,2} \ T_{2,2} \ T_{N-1,2} \ T_{N,2}$$
$$Q_{C1} \ Q_{R1} \ Q_{R2} \ V_{N1}/B_1]^T \qquad (7\text{-}21)$$

and the 14 functions are as shown in Table 7-1 with Q_{C2} replaced by Q_{R1}. In the solution of the g functions shown in Table 7-1 for the desired set of values of the variables, the terminal flow rates $d_{i,1}$, $b_{i,1}$, $d_{i,2}$, and $b_{i,2}$ are of course restated in terms of $p_{i,1}$ and $p_{i,2}$ by use of Eqs. (7-3) through (7-6).

Calculational Procedure

On the basis of an assumed value for V_{N1}/B_1, one complete trial is made on column 1. The set of bottom flow rates $\{b_{i,1}\}$ so obtained is used as the feed to column 2. The reboiler duty Q_{R1} found for column 1 is taken to be the assumed value Q_{C2} for making the trial on column 2. Then the capital Θ method is applied and a new value of V_{N1}/B_1 is obtained. This value of the boilup ratio is used in making the second trial on column 1. Also, in the application of the $2N$ Newton-Raphson method to column 1, the assumed values of Q_{C1} and Q_{R1} are set equal to the most recent set of values found by use of the capital Θ method. In the application of the $2N$ Newton-Raphson method to columns 1 and 2, the independent variables are taken to be Q_{C1}, Q_{R1}, $\{T_j\}$, $\{L_j/V_j\}$, for column 1 and Q_{C2}, Q_{R2}, $\{T_j\}$, $\{L_j/V_j\}$ for column 2.

When the θ method is used to solve the equations for each column where the reflux ratio and boilup ratio are specified rather than the reflux rate and the distillate rate, the appropriate equations needed to apply the θ method are developed in Sec. 7-4.

Example 7-1 In this example, it is intended for the overhead condenser of column 2 of Fig. 7-1 to serve as the reboiler for column 1. The problem is to be solved subject to the condition that $Q_{R1} = Q_{C2}$. The statement of the example is given in Table 7-3, the initial values in Table 7-4, the final temperatures, flow rates, and product flow rates are given in Table 7-5. The convergence characteristics appear in Table 7-6, and the number of trials and computer time required to solve this example are presented in Table 7-11. A comparison of $T_{2,2}$ and $T_{N-1,1}$ in Tables 7-5 and 7-6 shows that it is possible to transfer the heat from $V_{2,2}$ to $L_{N-1,1}$. This example and those which follow were taken from the results of Hass et al.[6]

Table 7-3 Specifications for Example 7-1 (See Fig. 7-1), Example 7-2 (See Fig. 7-2) and Example 7-3 (See Fig. 7-3)

---

1. Statement of Examples 7-1

Component	Component number	Feed FX_i (mol/h)
Ethane	1	15.0
Propylene	2	35.0
Propane	3	30.0
Iso-butene	4	20.0
		$\overline{20.0}$
		100.0

Other specifications

Column 1. $N = 20$ and $f = 11$. The column has a partial condenser, and is to be operated at a reflux ratio $L_{1,1}/D_1 = 4.0$ at a pressure of 250 lb/in² abs. The pressure drop across each plate is negligible. The feed enters the column as a liquid at its bubble point (551.56°R) at the column pressure. The boilup ratio of column 1 is to be selected such that the reboiler duty Q_{R1} of column 1 is equal to the condenser duty Q_{C2} of column 2. Use the vapor-liquid equilibrium and enthalpy data given in Tables B-1 and B-2. Since the K values in Table B-1 are at the base pressure of 300 lb/in² abs, approximate the K values at 250 lb/in² abs as follows

$$K_i|_{250} = (300/250)K_i|_{300}$$

Column 2. $N = 12$ and $f = 6$. The column has a partial condenser, and is to be operated at a reflux ratio $L_{1,2}/D_2 = 8.0$ at a pressure of 350 lb/in² abs. A boilup ratio V_{N2}/B_2 of 38.095 is to be used. Use the vapor-liquid equilibrium data and the enthalpy data given in Tables B-1 and B-2, approximate the K values in a manner analogous to that shown above for column 1.

---

2. Statement of Example 7-2 (see Fig. 7-2)

The specifications for this example are the same as those given for Example 7-1 except that the bottoms B_2 of column 2 is to be returned to stage $j = 15$ of column 1.

---

3. Statement of Example 7-3 (see Fig. 7-3)

The specifications for this example are the same as those given for Example 7-1 except that the bottoms B_2 of column 2 is to be returned to stage $j = 15$ of column 1 and the boilup ratio V_{N1}/B_1 is 5.6948. Also the reboiler of column 1 is to be operated independently of the condenser of column 2. For the heat exchanger (unit 3 of Fig. 7-3), the overall-heat-transfer coefficient U is 50 Btu/h ft² and the area is 10 square feet.

Table 7-4 Initial values of variables for Examples 7-1, 7-2, and 7-3

1. Initial values for Example 7-1 (see Fig. 7-1)

Column 1	Column 2
$T_{1,1} = 575°R$	$T_{1,2} = 620.00°R$
$T_{N,1} = 590°R$	$T_{N,2} = 700.0°R$
$T_{j,1} = T_{j-1,1} + 0.79$ ($j = 2, 3, \ldots, 19$)	$T_{j,2} = T_{j-1,2} + 7.275$ ($j = 2, 3, \ldots, 11$)
$V_{1,1} = D = 20$ mol/h	$V_{1,2} = D = 34$ mol/h
$B_1 = 80$ mol/h	$B_2 = 39.9$ mol/h
$L_{1,1} = 80$ mol/h	$L_{1,2} = 272$ mol/h
$V_{N,1} = 200$ mol/h	$V_{N,2} = 1520$ mol/h
$V_{j,1} = 100$ mol/h ($j = 2, 3, \ldots, 19$)	$V_{j,2} = 306$ mol/h ($j = 2, 3, \ldots, 11$)
$V_{N1}/B_1 = 2.5$	$Q_{C2} = 3.0 \times 10^6$ Btu/h
$Q_{C1} = 8.00 \times 10^5$ Btu/h	
$Q_{R1} = 1.00 \times 10^6$ Btu/h	

2. Initial values for Example 7-2 (see Fig. 7-2)

Same as Example 7-1 except for the assumed values for the recycle stream which were as follows:

Component	$b_{i,2}$ (mol/h)
Ethane	0.0
Propylene	3.3×10^{-3}
Propane	2.0×10^{-2}
Isobutene	9.5

3. Initial values for Example 7-3 (see Fig. 7-3)

The initial composition for the recycle stream is the same used in Example 7-2. The initial profiles for the columns are the same as in Example 7-1 except

$$Q_{C2} = 1.00 \times 10^6 \text{ Btu/h}$$

and for the double-pipe heat exchanger

$$T_{N1,0} = 600°R$$

$$T_{N2,0} = 680°R$$

Table 7-5 Solution values of the temperatures and vapor rates for Example 7-1 (see Fig. 7-1)

I. Temperature and vapor rate profiles

	Column 1			Column 2	
Stage	V_j (mol/h)	T_j (°R)	Stage	V_j (mol/h)	T_j (°R)
1	56.36	547.47	1	34.51	650.65
2	281.79	557.52	2	310.55	667.14
3	285.80	560.92	3	310.97	679.51
4	287.13	562.17	4	314.83	687.45
5	288.44	562.73	5	318.76	692.09
6	287.40	563.09	6	321.57	694.67
7	287.20	563.42	7	336.16	700.37
8	286.80	563.86	8	341.33	703.62
9	285.97	564.65	9	344.60	705.40
10	284.15	566.26	10	346.49	706.37
11	280.34	569.65	11	347.55	706.88
12	286.20	573.01	12	348.12	707.16
13	288.67	574.53			
14	289.51	575.39			
15	289.54	576.13			
16	288.91	577.15			
17	287.19	579.07			
18	283.22	583.13			
19	275.26	591.51			
20	263.19	606.26			

II. Product flow rates

Component	$d_{i, 1}$	$b_{i, 1}$	$d_{i, 2}$	$b_{i, 2}$
Ethane	15.0	1.076×10^{-4}	1.076×10^{-4}	2.74×10^{-10}
Propylene	28.276	6.725	6.722	3.0×10^{-3}
Propane	13.083	16.92	16.90	1.85×10^{-2}
Isobutene	1.637×10^{-4}	20.00	10.88	9.12×10^{2}
	56.36	43.64	34.51	9.139

Table 7-6 Convergence characteristics of the Θ method of convergence for Example 7-1 (see Fig. 7-1)

Trial	Θ_1	Θ_2	$T_{1, 1}$	$T_{2, 1}$	$T_{N-1, 1}$	$T_{N, 1}$	$T_{1, 2}$
Initial Values	....	....	575	575	589	590	620
1	0.28192	0.62362	547.94	558.07	590.47	605.52	650.11
2	2.2383	1.3552	547.51	557.56	591.42	606.19	650.62
3	0.80697	0.86111	547.50	557.55	591.44	606.21	650.61
4	1.0009	0.99947	547.47	557.52	591.51	606.26	650.65
5	1.0000	1.0000	547.47	557.52	591.51	606.26	650.65

A System of Two Distillation Columns in Which Mass is Recycled and the Condenser of Column 2 is to Serve as the Reboiler of Column 1

The system shown in Fig. 7-2 may be formed from the one shown in Fig. 7-1 by returning the bottoms B_2 from column 2 to column 1. The addition of the recycle mass transfer stream does not change the general form of the equations for the capital Θ method where the reflux ratios and boilup ratios are specified. The heat balance enclosing column 1 must be modified, however, in order to account for the recycle stream B_2. The resulting expression for the enthalpy function g_{10} is

$$g_{10} = \frac{\sum\limits_{i=1}^{c} [d_{i,\,1}\, H_i(T_{1,\,1}) + b_{i,\,1} h_i(T_{N,\,1})] + Q_{C1}}{\sum\limits_{i=1}^{c} [FX_i\, H_i(T_{\text{feed}}) + b_{i,\,2}\, h_i(T_{N,\,2})] + Q_{R1}} - 1 \qquad (7\text{-}22)$$

Also, the component-material balance enclosing column 1 must be altered in order to account for the return of B_2 to column 1. Thus, instead of Eq. (7-9), one obtains

$$(d_{i,\,1})_{ca} p_{i,\,1} + r_{i,\,1} p_{i,\,1} - r_{i,\,2} p_{i,\,2} = FX_i \qquad (7\text{-}23)$$

When the condition that $Q_{R1} = Q_{C2}$ is imposed on the system, the capital Θ method again consists of 14 functions in 14 unknowns; see specification set 1 of Table 7-2.

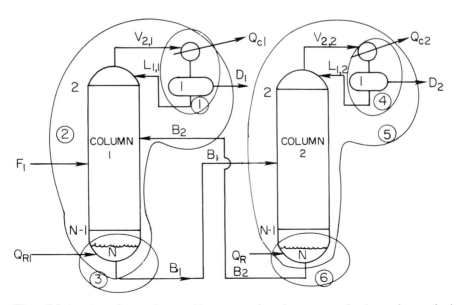

Figure 7-2 A system of two columns with mass recycle and energy recycle; the condenser of column 2 is the reboiler of column 1.

Example 7-2 This example is a minor variation of Example 7-1 in that the bottoms from column 2 are recycled to column 1 as shown in Fig. 7-2. Again the overhead condenser of column 2 is to serve as the reboiler for column 1, and the problem is to be solved subject to the condition that $Q_{R1} = Q_{C2}$. The statement of the example is given in Table 7-3, the initial values in Table 7-4, the final temperatures, the total flow rates, and the final product flow rates are given in Table 7-7, and the convergence characteristics are given in Table 7-8.

Table 7-7 Solution values of the temperatures and vapor rates for Example 7-2 (see Fig. 7-2)

I. Temperature and vapor rate profiles

	Column 1			Column 2	
Stage	V_j (mol/h)	T_j (°R)	Stage	V_j (mol/h)	T_j (°R)
1	62.23	549.84	1	37.77	671.55
2	311.16	558.99	2	339.95	684.32
3	315.28	562.07	3	345.48	692.04
4	316.60	563.20	4	350.36	696.31
5	316.89	563.72	5	353.56	698.56
6	316.83	564.07	6	355.39	699.72
7	316.62	564.39	7	372.16	703.24
8	316.19	564.82	8	375.98	705.19
9	315.30	565.58	9	378.22	706.25
10	313.38	567.12	10	379.48	706.82
11	309.37	570.35	11	380.16	707.12
12	315.17	573.69	12	380.53	707.28
13	317.15	575.51			
14	316.79	577.28			
15	314.11	580.25			
16	306.93	582.20			
17	303.21	585.99			
18	296.15	593.31			
19	285.63	605.76			
20	275.30	622.62			

II. Product flow rates

Component	$d_{i, 1}$	$b_{i, 1}$	$d_{i, 2}$	$b_{i, 2}$
Ethane	15.0	6.99×10^{-5}	6.99×10^{-5}	1.675×10^{-10}
Propylene	30.695	4.307	4.306	1.659×10^{-3}
Propane	16.54	13.48	13.46	1.224×10^{-2}
Isobutene	2.067×10^{-4}	29.97	20.00	9.975
	62.23	47.76	37.77	9.988

Table 7-8 Convergence characteristics of the Θ method of convergence for Example 7-2 (see Fig. 7-2)

Trial	Θ_1	Θ_1	$T_{1,1}$	$T_{2,1}$	$T_{N-1,1}$	$T_{N,1}$	$T_{1,2}$
Initial Values	....	....	575	579	589	590	620
1	0.30229	2.9160	550.11	559.30	604.95	622.15	671.34
2	1.6416	2.2738	549.88	559.03	605.64	622.55	671.52
3	0.91071	0.67345	549.84	558.99	605.76	622.63	671.55
4	0.99895	0.99390	549.84	558.99	605.76	622.62	671.55
5	0.99997	1.0024	549.84	558.99	605.76	622.62	671.55
6	1.0000	1.0001	549.84	558.99	605.76	622.62	671.55

Trial	$T_{2,2}$	$T_{N-1,2}$	$T_{N,2}$	Q_{C1}	Q_{R1}	Q_{R2}	V_{N1}/B_1
Initial Values	627	692	700	8×10^5	1×10^6	3×10^6	2.5
1	684.30	707.02	707.23	1.3912×10^6	1.7616×10^6	2.0578×10^6	5.7630
2	684.32	707.05	707.24	1.3899×10^6	1.7606×10^6	2.0566×10^6	5.7646
3	684.31	707.18	707.31	1.3897×10^6	1.7604×10^6	2.0564×10^6	5.7630
4	684.32	707.13	707.28	1.3897×10^6	1.7604×10^6	2.0564×10^6	5.7644
5	684.32	707.12	707.28	1.3897×10^6	1.7604×10^6	2.0564×10^6	5.7644
6	684.32	707.12	707.28	1.3897×10^6	1.7604×10^6	2.0564×10^6	5.7644

In the specifications given by set 2 of Table 7-2, the reflux ratios and boilup ratios are fixed and the capital Θ method again consists of 14 functions in 14 independent variables. If the reflux rates $L_{1,1}$, $L_{1,2}$ and the total flow rates D_1 and B_2 are specified instead of the reflux ratios and boilup ratios, the capital Θ method reduces to two functions in two independent variables Θ_1 and Θ_2; see set 3 of Table 7-2. In this case the g functions for the capital Θ method are given by

$$g_1 = \frac{1}{D_1} \sum_{i=1}^{c} d_{i,1} - 1 \tag{7-24}$$

$$g_2 = \frac{1}{B_2} \sum_{i=1}^{c} b_{i,2} - 1 \tag{7-25}$$

7-2 ENERGY EXCHANGE BETWEEN RECYCLE STREAMS OF SYSTEMS OF DISTILLATION COLUMNS

In many systems, energy is exchanged between recycle streams such as the streams B_1 and B_2 shown in Fig. 7-3. Consider the case where energy is transferred from stream B_2 to B_1 in a double-pipe heat exchanger with countercurrent flow.

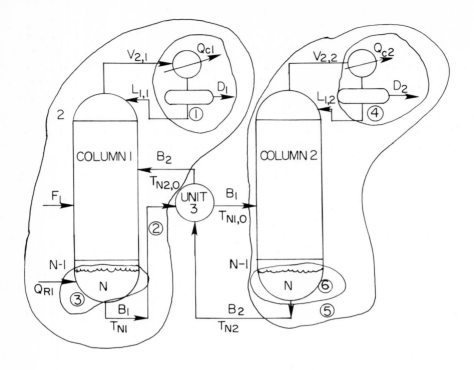

UNIT 3: DOUBLE-PIPE HEAT EXCHANGER

$$T_{N2,0} \xleftarrow{\quad B_2 \quad} T_{N2}$$

$$T_{N1} \xrightarrow{\quad B_1 \quad} T_{N1,0}$$

Figure 7-3 A system of two columns with heat exchange between recycle streams.

First, suppose that the specifications for the columns and the exchanger are given by set 4 in Table 7-2. The exchanger is regarded as an additional unit of the system, and it is treated as another module in the calculational procedure. The modular and system equations for this unit are identical. The heat exchanger may be described by three independent equations, an energy balance on the stream B_1, an energy balance on the stream B_2, and the rate equation, $Q = UA \, \Delta T_{lm}$. These equations may be stated as two functions with Q a dependent variable as follows

$$g_{15} = \frac{1}{Q} \sum_{i=1}^{c} b_{i,1} [h_i(T_{N1}) - h_i(T_{N1,0})] + 1 \qquad (7\text{-}26)$$

$$g_{16} = \frac{1}{Q} \sum_{i=1}^{c} b_{i,2} [h_i(T_{N2}) - h_i(T_{N2,0})] - 1 \qquad (7\text{-}27)$$

where

$$Q = \frac{UA[(T_{N2,0} - T_{N1}) - (T_{N2} - T_{N1,0})]}{\ln [(T_{N2,0} - T_{N1})/(T_{N2} - T_{N1,0})]}$$

The functions g_{10} and g_{13} must also reflect the fact that streams B_1 and B_2 have passed through the heat exchanger. In this case

$$g_{10} = \frac{\sum_{i=1}^{c} [d_{i,1} H_i(T_{1,1}) + b_{i,1} h_i(T_{N1})] + Q_{C1}}{\sum_{i=1}^{c} [FX_i H_i(T_{feed}) + b_{i,2} h_i(T_{N2,0})] + Q_{R1}} - 1 \qquad (7\text{-}28)$$

$$g_{13} = \frac{1}{Q_{C2}} \sum_{i=1}^{c} \{d_{i,2}[h_i(T_{N1,0}) - H_i(T_{1,2})]$$

$$+ b_{i,2}[h_i(T_{N1,0}) - h_i(T_{N2})]\} + \frac{Q_{R2}}{Q_{C2}} - 1 \quad (7\text{-}29)$$

When the specifications are taken to be those given as set 4 of Table 7-2, the capital Θ method consists of finding the set of 16 variables which satisfy the 16 g functions simultaneously.

Calculational Procedure

The first trial through the system is initiated as follows. First one trial is made on column 1. To make this trial, values are assumed for B_2, the composition of B_2, and the temperature $T_{N2,0}$. On the basis of the results so obtained (T_{N1}, B_2, and the composition of B_2) and an assumed value for T_{N2}, the equations for the double-pipe heat-exchanger [Eqs. (7-26) and (7-27)] are solved for $T_{N1,0}$ and $T_{N2,0}$. Next, on the basis of $T_{N1,0}$ and the values found for B_1 and its composition by the trial on column 1, one trial is made on column 2. Then the capital Θ method is applied by finding the 16 variables [the elements of x listed as set 4 of Table 7-2] which satisfy the 16 functions. The values of the variables found by the capital Θ method for the feedback streams [B_2, $\{x_{B2,i}\}$, $T_{N2,0}$, and T_{N2}] are used in making the next trial on each unit of the system.

Example 7-3 This example is a minor variation of Example 7-2 in that the streams B_1 and B_2 are passed through a heat exchanger as shown in Fig. 7-3, and columns 1 and 2 are to be provided with independent reboilers and condensers. The statement of this example is given in Table 7-3, the initial values in Table 7-4, the final temperatures, vapor rates, and product flow rates are given in Table 7-9. The convergence characteristics are given in Tables 7-10 and 7-11.

Table 7-9 Solution values of the temperatures and vapor rates for Example 7-3 (see Fig. 7-3)

			I. Temperature and vapor rate profiles			
	Column 1			Column 2		
Stage	V_j (mol/h)	T_j (°R)	Stage	V_j (mol/h)	T_j (°R)
1	61.54	549.59	1	38.46	670.70
2	307.72	558.83	2	346.13	683.70
3	311.83	561.94	3	351.69	691.61
4	313.15	563.08	4	356.71	696.00
5	313.44	563.61	5	360.03	698.33
6	313.39	563.95	6	361.95	699.54
7	313.17	564.28	7	374.76	703.14
8	312.75	564.71	8	378.69	705.14
9	311.86	565.47	9	381.69	706.22
10	309.96	567.03	10	382.29	706.80
11	305.97	570.28	11	381.00	707.11
12	311.77	573.63	12	383.38	707.28
13	313.77	575.44	13	383.38	707.28
14	313.48	577.18			
15	310.91	580.08			
16	307.88	581.95			
17	304.30	585.60			
18	297.40	592.70			
19	286.91	604.92			
20	276.31	621.75			

		II. Product rates		
Component	$d_{i,1}$	$b_{i,1}$	$d_{i,2}$	$b_{i,2}$
Ethane	15.00	7.464×10^{-5}	7.464×10^{-5}	0.0
Propylene	30.44	4.563	4.561	1.747×10^{-3}
Propane	16.105	13.910	13.897	1.258×10^{-2}
Isobutene	2.016×10^{-4}	30.049	20.001	10.049
	61.545	48.521	38.459	10.064

The final heat exchanger duty was 2.422×10^4 Btu/h

Table 7-10 Convergence characteristics of the Θ method of convergence for Example 7-3 (see Fig. 7-3)

Trial	Θ_1	Θ_2	$T_{1,1}$	$T_{2,1}$	$T_{N-1,1}$	$T_{N,1}$	$T_{1,2}$	$T_{2,2}$
Initial values	....	....	575	575	589.21	590	627.73	692.73
1	0.65512	176.57	549.87	559.17	603.74	620.90	670.38	683.61
2	2.5161	121.47	549.52	558.52	604.77	621.52	670.67	683.66
3	0.83911	58.991	549.59	558.84	604.64	621.47	670.63	683.66
4	0.99533	8.5640	549.58	558.82	604.83	621.65	670.67	683.69
5	0.99856	6.6618	549.58	558.83	604.88	621.70	670.69	683.69
6	0.99942	2.7844	549.83	558.83	604.90	621.72	670.69	683.69
7	0.99905	0.80965	549.58	558.83	604.93	621.75	670.70	683.70
8	1.0001	1.0145	549.58	558.83	604.83	621.75	670.70	683.70
9	0.9998	1.0006	549.58	558.83	604.92	621.75	670.70	683.70
10	1.0000	1.000	549.58	558.83	604.92	621.75	670.70	683.70

Trial	$T_{N-1,2}$	$T_{N,2}$	Q_{C1}	Q_{R1}	Q_{C2}	Q_{R2}	$T_{N1,0}$	$T_{N2,0}$
Initial values	692	700	8×10^5	1×10^6	1×10^6	3×10^6	620	680
1	704.51	705.91	1.3751×10^6	1.7641×10^6	1.7982×10^6	2.0765×10^6	634.36	648.35
2	704.75	706.02	1.3733×10^6	1.7626×10^6	1.7963×10^6	2.0745×10^6	635.87	648.79
3	705.29	706.31	1.3739×10^6	1.7632×10^6	1.7951×10^6	2.0742×10^6	634.90	648.93
4	706.40	706.90	1.3743×10^6	1.7648×10^6	1.7949×10^6	2.0729×10^6	635.15	649.36
5	706.74	707.08	1.3745×10^6	1.7640×10^6	1.7945×10^6	2.0723×10^6	635.23	649.51
6	706.89	707.16	1.3746×10^6	1.7641×10^6	1.7943×10^6	2.0721×10^6	635.26	649.57
7	707.13	707.28	1.3747×10^6	1.7642×10^6	1.7941×10^6	2.0719×10^6	635.31	649.67
8	707.11	707.27	1.3747×10^6	1.7642×10^6	1.7940×10^6	2.0719×10^6	635.31	649.65
9	707.11	707.28	1.3747×10^6	1.7642×10^6	1.7941×10^6	2.0719×10^6	635.31	649.65
10	707.11	707.28	1.3747×10^6	1.7642×10^6	1.7941×10^6	2.0719×10^6	635.31	649.65

Table 7-11 Execution time and number of trials† required to solve Examples 7-1, 7-2, and 7-3

Example	No. of system trials	Execution time (s)	Compiler
1	5	8.32	FORTRAN H Extended
2	6	11.72	FORTRAN H Extended
3	10	16.67	FORTRAN H Extended

† For Examples 7-1 through 7-3, the execution time and number of trials are those obtained when the equations for the individual columns are solved by use of the $2N$ Newton-Raphson method (4, 5).

7-3 ENERGY TRANSFER BY USE OF TRIM HEAT EXCHANGERS

Consider the system shown in Fig. 7-4 in which a "trim" heat exchanger is used to cool the stream B_2 to the temperature T_0 before it is returned to the reboiled absorber. This example was formulated and solved as Example 4-11. Specification of L_{N1}, D_2, W_1, and W_2 fixes V_1. Thus, in the capital Θ method there are four g functions as shown in Chap. 4.

In the application of the $2N$ Newton-Raphson method, V_1 is required as specified and the temperature T_0 of B_2 entering column 1 or the trim heat exchanger duty Q_C is taken as the new independent variable. In this case the $2N + 1$ variables $\mathbf{x}$ and the functions $\mathbf{f}$ are as follows

$$\mathbf{x} = [\theta_1 \; \theta_2 \; \cdots \; \theta_N \; T_0 \; T_1 \; T_2 \; \cdots \; T_N]^T$$

$$\mathbf{f} = [f_0 \; f_1 \; f_2 \; \cdots \; f_N \; G_1 \; G_2 \; \cdots \; G_N]^T$$

where $L_j/V_j = \theta_j(L_j/V_j)_a$. The new function f_0 corresponding to the new independent variable T_0 for the reboiled absorber is introduced to express the condition that the specified values $(V_1)_{sp}$ must be equal to the calculated value of V_1 at convergence, that is,

$$f_0 = \frac{V_1}{(V_1)_{sp}} - 1 \tag{7-30}$$

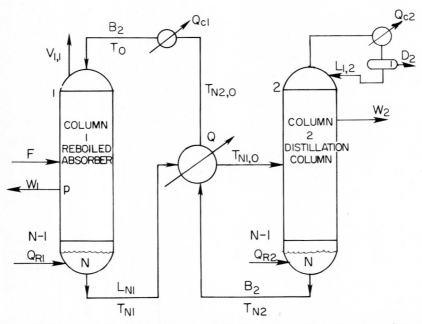

Figure 7-4 A system of an absorber and a distillation column with heat exchange between recycle streams and a trim heat exchanger.

The remaining functions $f_1, f_2, \ldots, f_N$ consist of the enthalpy balance functions; see Chap. 4.

The proposed calculational procedure consists of first making one trial calculation on the reboiled absorber by use of the $[2N + 1]$ formulation of the Newton-Raphson method and then one trial on the distillation column by use of the θ method. Then the capital Θ method is applied to the system in order to place it in agreement with the specified values of the terminal flow rates L_{N1}, W_1, W_2, and D_2.

After a solution to the problem has been found, the duty Q_{C1} of the trim heat exchanger is found by means of an energy balance enclosing it.

A Combination of Energy Exchange Between Recycle Streams and a Trim Heat Exchanger

A combination of the second and third types of energy exchange is illustrated by the system shown in Fig. 7-4. In this system a heat exchanger is used to remove heat from the bottoms B_2 of the reboiled absorber before returning it to the absorber. A "trim" heat exchanger is used to remove additional heat before the stream B_2 is fed to column 1.

When the specifications are given by set 6 of Table 7-2, the capital Θ method consists of 4 variables and 4 functions, namely,

$$\mathbf{x} = [\Theta_1 \ \Theta_2 \ \Theta_3 \ \Theta_4]^T$$

$$\mathbf{g} = [g_1 \ g_2 \ g_3 \ g_4]^T$$

where

$$g_1 = \frac{1}{W_1} \sum_{i=1}^{c} w_{i,1} - 1$$

$$g_2 = \frac{1}{W_2} \sum_{i=1}^{c} w_{i,2} - 1$$

$$g_3 = \frac{1}{D_2} \sum_{i=1}^{c} d_{i,2} - 1$$

$$g_4 = \frac{1}{B_2} \sum_{i=1}^{c} b_{i,2} - 1$$

The calculational procedure is as follows. On the basis of an assumed set of compositions for the stream B_2, one trial calculation is made on column 1 by use of the $2N + 1$ formulation of the Newton-Raphson method. Then the equations for the heat exchanger are solved for $T_{N1,0}$ and $T_{N2,0}$ on the basis of the most recently calculated values for T_{N1} and T_{N2}. For the first trial through the system, an assumed value for T_{N2} is used in Eq. (7-27) in the calculations for the heat exchanger. Then the value so obtained for $T_{N2,0}$ is used in making one trial on column 2 by use of the θ method. Next, the capital Θ method is applied.

The set of $b_{i,2}$'s so obtained are used to make the next trial calculation on column 1. After a solution to the problem has been obtained, the required value of the cooler duty Q_{C1} is readily determined by an enthalpy balance enclosing the trim heat exchanger.

7-4 OTHER APPLICATIONS OF THE THETA METHODS

In the same manner that the capital Θ method was applied above to solve problems involving specifications other than the total flow rates, the θ method for single columns may be applied to solve problems involving specifications other than the total-flow rates. To demonstrate this application of the θ method, it is applied to conventional distillation columns for which the reflux ratio L_1/D and the boilup ratio V_N/B are specified instead of the reflux rate L_1 and the distillate rate D. The remaining specifications for the column are the same as those enumerated in Chap. 2 in the application of the θ method of convergence to conventional distillation columns.

The specification of L_1/D and V_N/B gives rise to a θ method which consists of seven variables and seven functions.[8] The variables $\mathbf{x}$ are as follows

$$\mathbf{x} = [\theta \ T_1 \ T_2 \ T_{N-1} \ T_N \ Q_C \ Q_R]^T$$

The multiplier θ is defined by Eq. (2-25), and the formula for the corrected distillate rate is given by Eq. (2-57). The seven functions are based on the following relationships, a dew point on D (or a bubble point where a total condenser is used), a dew point on V_2, a bubble point on L_{N-1}, a bubble point on B, an energy balance enclosing the condenser, an energy balance enclosing the entire column, and an energy balance enclosing the reboiler. Expressions for these seven functions may be obtained by removing the column subscripts of the following functions which are presented in Table 7-1

$$\mathbf{f} = [g_1 \ g_2 \ g_3 \ g_4 \ g_9 \ g_{10} \ g_{11}]$$

The following calculational procedure may be used to solve problems of this type.

Step 1. Assume a set of temperatures $\{T_j\}$ and vapor rates $\{V_j\}$. Find the corresponding liquid rates $\{L_j\}$ by use of total material balances.

Step 2. Solve Eq. (2-18) for the vapor rates $\{v_{ji}\}$ for each component i and compute the corresponding liquid rates $\{l_{ji}\}$.

Step 3. Apply the θ method. Use the values so obtained $(T_1, T_2, T_{N-1}, T_N, D, V_2, L_{N-1}, V_N, Q_C, Q_R)$ in the energy balances and in the next trial.

Step 4. Compute the remaining temperatures $(T_3, T_4, ..., T_{N-2})$ by use of the K_b method on the basis of the compositions obtained by use of Eq. (2-59).

Step 5. Compute the remaining vapor and liquid rates $(V_3, V_4, ..., V_{N-1}, L_2, L_3, ..., L_{N-2})$ by use of enthalpy and total material balances [see Eqs. (2-34), (2-35), (2-37), and (2-38)]. If the convergence criteria are satisfied, cease calculations; otherwise return to step 1.

Table 7-12 Convergence characteristics of the θ method for Example 2-7 when the additional specifications are: $L_1/D = 2$ and $V_N/B = 1.80585$

Trial	θ	T_1	T_2	T_{N-1}	T_N	Q_C	Q_R
Initial values	....	610.0	635.0	885.0	910.0	3.0×10^5	2.0×10^6
1	3.5990	582.82	619.01	715.16	790.96	3.5863×10^5	1.1689×10^6
2	2.0270	566.62	595.20	773.53	827.08	3.8713×10^5	1.3081×10^6
3	0.77160	567.13	593.51	768.14	826.54	3.9443×10^5	1.3194×10^6
4	1.0820	567.29	593.89	767.93	826.48	3.9512×10^5	1.3199×10^6
5	0.98006	567.36	594.04	767.87	826.45	3.9542×10^5	1.3201×10^6
6	1.0034	567.36	594.03	767.88	826.45	3.9541×10^5	1.3201×10^6
7	0.99926	567.36	594.04	767.87	826.45	3.9543×10^5	1.3201×10^6
8	1.0002	567.36	594.04	767.87	826.45	3.9543×10^5	1.3201×10^6
9	0.99991	567.36	594.04	767.87	826.45	3.9543×10^5	1.3201×10^6

Example 7-4 Instead of specifying L_1 and D for Example 2-7, modify this example by taking the two additional specifications to be the reflux ratio $L_1/D = 2.0$ and the boilup ratio $V_N/B = 1.80585$. When these particular values for the reflux ratio and the boilup ratio are selected, the corresponding final solution is the same as the one shown in Tables 2-3 through 2-5. For this pair of additional specifications, nine iterations were required and 1.26 seconds of computer time (AMDAHL 470 V/6 computer, FORTRAN H EXTENDED). The convergence characteristics exhibited by this example for this version of the θ method are shown in Table 7-12.

The θ method may be used to solve absorber and stripper problems in a manner similar to that described above for distillation columns. Several examples were solved by Haas.[7] Test examples such as the one proposed by Boyum[1] were solved without difficulty. Convergence problems were encountered, however, in the solution of other examples. These problems were eliminated, however, by subdividing the column into an appropriate number of subunits and treating the subunits as columns of a system by use of the capital Θ method; see Problem 7-5. Also, fewer subunits were required to achieve convergence when about four trials were made on each column per system trial.

NOTATION

$b_{i,k}$	molar flow rate of component i in the bottoms of column k
B_k	total molar flow rate of the bottoms of column k
$d_{i,k}$	molar flow rate of component i in the distillate of column k
D_k	total molar flow rate of the distillate of column k
F	total molar flow rate of the feed
g_l	the lth function of the capital Θ method of convergence

$h_i(T_{j,k})$ molar enthalpy of component i in the liquid phase at the temperature $T_{j,k}$

$H_i(T_{j,k})$ molar enthalpy of component i in the vapor phase at the temperature $T_{j,k}$

$K_i(T_{j,k})$ K value for component i at the temperature $T_{j,k}$ and the pressure P_k of column k

$l_{ji,k}$ molar flow rate of component i in the liquid phase leaving plate j of column k

$L_{j,k}$ total molar flow rate of the liquid having plate j of column k

$p_{i,k}$ ratio of the corrected to the calculated distillate rates for component i and column k

$Q_{C,k}$ condenser duty of column k

$Q_{R,k}$ reboiler duty of column k

$r_{i,1}, r_{i,2}$ functions of Θ_1, Θ_2 and component i, defined by Eqs. (7-5) and (7-6)

T_{feed} temperature of the feed F

$T_{j,k}$ temperature of plate j of column k

$v_{ji,k}$ molar flow rate at which component i in the vapor phase leaves plate j of column k

$V_{j,k}$ total molar flow rate at which the vapor leaves plate j of column k

$w_{i,1}, w_{i,2}$ molar flow rate at which component i is withdrawn from plate p of columns 1 and 2, respectively

W_1, W_2 total molar flow rate at which component i is withdrawn from plate p columns 1 and 2, respectively

X_i total mole fraction of component i in the feed F

$\mathbf{x}$ column vector of variables

Mathematical symbols

$[x_1 \ x_2]^T = \begin{bmatrix} x_1 \\ x_2 \end{bmatrix}$, the transpose of a row vector is equal to a column vector

$\{x_i\}$ set of all values $x_1, x_2, \ldots, x_c$.

PROBLEMS

7-1 Develop the equations necessary to describe a column in which each stage is regarded as a unit of a system. Apply the capital Θ method to each stage and solve the equations by use of the Newton-Raphson method.

7-2 Describe the differences in the method described in Prob. 7-1 and the $2N$ Newton-Raphson method described in Chap. 4.

7-3 For the case where the reflux ratio L_1/D and the boilup ratio V_N/B are specified in lieu of the reflux rate L_1 and the distillate rate D for a conventional distillation column having a partial condenser, formulate the g functions of the θ method where the variables $\mathbf{x}$ are taken to be

$$\mathbf{x} = [\theta T_1 \ T_2 \ T_{N-1} \ T_N \ Q_C \ Q_R]^T$$

7-4 (*a*) Formulate the functions needed to solve a typical absorber problem by the θ method when the variables of the θ method are taken to be

$$\mathbf{x} = [\theta \ T_1 \ T_N]^T$$

(*b*) Formulate the equations necessary to compute the temperatures $T_2, T_3, \ldots, T_{N-1}$ by the K_b method and the flow rates $L_2, L_3, \ldots, L_{N-1}$ by the constant-composition method.

7-5 Formulate the functions for the capital Θ method for the case where the absorber is subdivided into two columns[8] as shown in Fig. P7-5. The independent variables of the capital Θ method are taken to be

$$\mathbf{x} = [\Theta_1 \ \Theta_2 \ T_1 \ T_M \ T_{M+1} \ T_N]^T$$

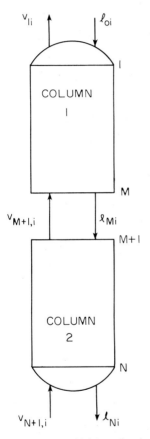

Figure P7-5 Division of a single column into a system of two columns.

7-6 Show that the function g_1 given in Table 7-1 for a conventional distillation column is a monotonic decreasing function of θ and has one zero lying between $\theta = 0$ and $\theta = \infty$.

REFERENCES

1. A. A. Boyum, Ph.D.: Dissertation Polytechnic Institute of Brooklyn, Brooklyn, N.Y. (1966).
2. F. E. Hess, J. R. Haas, and C. D. Holland: "Solution of a Single Column or a System of Columns as a System of Any Arbitrary Number of Columns," accepted for publication, *Hydrocarbon Process.*
3. C. D. Holland: *Fundamentals and Modeling of Separation Processes—Absorption, Extraction Distillation, Evaporation, and Extraction,* Prentice-Hall, Englewood Cliffs, N.J. (1975).
4. C. D. Holland and G. P. Pendon: "Solve More Distillation Problems: Part I. Improvements Give Exact Answers," *Hydrocarbon Process.,* **33**: 148 (1974).
5. T. A. Nartker, J. M. Srygley, and C. D. Holland: "Solution of Problems Involving Systems of Distillation Columns," *Can. J. Chem. Eng.,* **44**, (1966).
6. J. R. Haas, C. D. Holland, F. Dominguez, S., and A. Gomez, M.: "Solution of Systems of Columns Containing Mass and Energy Recycle Streams," in press, *Computers & Chem. Eng.*
7. J. R. Haas, M. S. Thesis, Texas A&M University, 1980.
8. J. R. Haas, Alejandro Gomez, M., and C. D. Holland: "Generalization of the Theta Methods of Convergence for Solving Distillation and Absorber type Problems," accepted for publication by *Separation Science* and *Technology.*

EIGHT

DISTILLATION ACCOMPANIED BY CHEMICAL REACTION

Many separations which would be difficult to achieve by conventional distillation processes may be effected by a distillation process in which a solvent is introduced which reacts chemically with one or more of the components to be separated. Three methods are presented for solving problems of this type. In Sec. 8-1, the θ method of convergence is applied to conventional and complex distillation columns. In Sec. 8-2, the $2N$ Newton-Raphson method is applied to absorbers and distillation columns in which one or more chemical reactions occur per stage. The first two methods are recommended for mixtures which do not deviate too widely from ideal solutions. For mixtures which form highly nonideal solutions and one or more chemical reactions occur per stage, a formulation of the Almost Band Algorithm such as the one presented in Sec. 8-3 is recommended.

8-1 APPLICATION OF THE THETA METHOD OF CONVERGENCE TO DISTILLATION COLUMNS IN WHICH CHEMICAL REACTIONS OCCUR

The θ method of convergence, which was originally formulated for the solution of distillation problems in which chemical reactions did not occur, is extended to include the case where one or more chemical reactions occur on each stage of a distillation column. The computational time of the θ method, which is one of the fastest known methods for solving distillation problems, is not significantly increased by the inclusion of one or more chemical reactions per stage. In the development which follows, the θ method of convergence is applied to distillation columns in which one chemical reaction occurs on each plate. The formulation of the algorithm is followed by the solution of several numerical examples.

In recent years, several calculational procedures have been proposed for solving distillation problems in which a chemical reaction occurs on each plate. Suzuki, et al.[24] used a procedure based on Muller's method; whereas, Jelinek and Hlaváček[14] used a relaxation method. Komatsu and Holland[17] and Nelson[19] proposed different formulations of the Newton-Raphson method.

Formulation of the θ Method of Convergence

Consider the case of an existing column such as the one shown in Fig. 2-3. Suppose that the specifications are taken to be: (1) the total number of stages and the location of the feed plate, (2) the complete definition of the feed (the total-flow rate, composition, and thermal condition), (3) the reflux rate, (4) the distillate rate, (5) the column pressure, and (6) the type of condenser (total or partial). On the basis of this set of specifications, it is desired to find the resulting compositions of the distillate and bottom products.

The equations required to describe this column are developed in the order in which they are solved sequentially in the proposed calculational procedure. On the basis of assumed temperature and L/V profiles, the material balances, the physical equilibrium relationships, and the chemical rate expressions (or chemical equilibrium expressions) are solved for the moles of each component which reacts per stage per unit time and for the component-flow rates. A formulation of the Newton-Raphson method is used.

The remainder of the calculational procedure is analogous to that proposed for distillation columns without chemical reactions. After the component-material balances have been solved for the moles reacted and the component-flow rates, a θ multiplier is found that places the column in overall material balance and in agreement with the specified value of the distillate rate D. Next, new sets of compositions are computed, and these are used to find a new set of temperatures by the K_b method. On the basis of these temperatures and the most recent sets of compositions, a new set of total-flow rates is found by use of the enthalpy balances and the total material balances. The enthalpy balances are stated in the constant-composition form.

Vapor-Liquid Equilibrium Relationships

The equilibrium relationship for stage j and component i [the first expression of Eq. (5-1)] is readily restated in the following form

$$l_{ji} = A_{ji} v_{ji} \tag{8-1}$$

where the absorption factor A_{ji} is defined in the usual way

$$A_{ji} = \frac{\gamma_{ji}^L L_j}{\gamma_{ji}^V K_{ji} V_j} \tag{8-2}$$

Material Balances

The equations are formulated first for the case where a single reaction

$$aA + bB \rightleftarrows cC + dD \tag{8-3}$$

occurs in the liquid phase on each stage. In the analysis which follows, this reaction is restated in the following equivalent form

$$v_A A + v_B B + v_C C + v_D D = 0 \tag{8-4}$$

where

$$v_A = -1 \qquad v_B = -b/a \qquad v_C = c/a \qquad v_D = d/a$$

The contents on each plate are assumed to be perfectly mixed.

Let η_j denote the moles of A which disappear by reaction per unit time for any plate j. Then the mass balance for component i for a typical interior stage j $(j \neq 1, f - 1, f, N)$ is given by

$$M_i v_{j+1, i} + M_i l_{j-1, i} - M_i v_{ji} - M_i l_{ji} + M_i \delta_{ji} = 0 \tag{8-5}$$

where

$$M_i = \text{molecular weight of component } i$$

$$\delta_{ji} = v_i \eta_j$$

Since M_i is common to each term of Eq. (8-5), the material balance for component i on stage j reduces to

$$v_{j+1, i} + l_{j-1, i} - v_{ji} - l_{ji} + \delta_{ji} = 0 \tag{8-6}$$

Use of the vapor-liquid equilibrium relationship $l_{ji} = A_{ji} v_{ji}$ to eliminate the liquid flow rates from Eq. (8-6) yields

$$A_{j-1, i} v_{j-1, i} - (1 + A_{ji}) v_{ji} + v_{j+1, i} = -\delta_{ji} \tag{8-7}$$

The complete set of component-material balances for the conventional column may be represented by the following matrix equation

$$\mathbf{A}_i \mathbf{v}_i = -f_i - v_i \mathbf{\eta} \tag{8-8}$$

where the tridiagonal matrix $\mathbf{A}_i$ and the column vectors $\mathbf{f}_i$ and $\mathbf{v}_i$ contain the elements shown beneath Eq. (2-18). The column vector $\mathbf{\eta}$ is of the form

$$\mathbf{\eta} = [\eta_1 \ \eta_2 \ \cdots \ \eta_N]^T$$

The total material balance enclosing a typical plate j is found by summing each member of Eq. (8-6) over all components i to give

$$V_{j+1} + L_{j-1} - V_j - L_j + \Delta_j = 0 \tag{8-9}$$

where

$$\Delta_j = \sum_{i=1}^{c} \delta_{ji} = \eta_j \sum_{i=1}^{c} v_i = \eta_j \sigma$$

$$\sigma = \sum_{i=1}^{c} v_i$$

To restate Eq. (8-9) in terms of an assumed set of L_j/V_j's, let b_j denote the assumed value of L_j/V_j for plate j. Then Eq. (8-9) may be restated in terms of the V_j's and b_j's as follows

$$b_{j-1}V_{j-1} - (1 + b_j)V_j + V_{j+1} = -\sigma\eta_j \tag{8-10}$$

and the complete set of total material balances for the column may be represented the matrix equation

$$\mathbf{BV} = -\mathscr{F} - \sigma\eta \tag{8-11}$$

where $\mathbf{B}$ is a tridiagonal matrix having the form

$$\mathbf{B} = \begin{bmatrix} -(1+b_1) & 1 & 0 & 0 & & 0 \\ b_1 & -(1+b_2) & 1 & 0 & & 0 \\ \cdots\cdots\cdots\cdots\cdots\cdots\cdots\cdots\cdots\cdots\cdots\cdots \\ 0 & & \cdots & 0 & b_{N-1} & -(1+b_N) \end{bmatrix}$$

and $\mathscr{F}$ and $\mathbf{V}$ are column vectors of the form

$$\mathscr{F} = [0 \;\cdots\; 0 \; V_F \; L_F \; 0 \;\cdots\; 0]^T$$
$$\mathbf{V} = [D \; V_2 \;\cdots\; V_N]^T$$

Observe that for a column having a partial condenser $D = V_1$ and for a column having a total condenser $V_1 = 0$.

Formulation of the Chemical Reaction Function

Consider first the rate of reaction for the second-order reversible reaction

$$r_{jA} = k_j C_{jA} C_{jB} - k'_j C_{jC} C_{jD} \tag{8-12}$$

where C_{ji} is the concentration of component i in the liquid phase in moles per unit volume, r_{jA} is the rate of disappearance of A in moles per unit time per unit volume, and k_j and k'_j are the forward and reverse rate constants in the consistent units. The concentrations may be expressed in terms of the liquid flow rates as follows

$$C_{ji} = \rho_j^L \left(\frac{l_{ji}}{L_j}\right) \tag{8-13}$$

where ρ_j^L is the molar density of the liquid phase. Thus, Eq. (8-12) becomes

$$r_{jA} = \frac{(\rho_j^L)^2}{L_j^2} [k_j l_{jA} l_{jB} - k'_j l_{jC} l_{jD}] \tag{8-14}$$

Let the volume of liquid on stage j be denoted by U_j^L. Then the moles of A reacting per unit time on stage j is given by

$$\eta_j = r_{jA} U_j^L \tag{8-15}$$

Multiplication of each member of Eq. (8-14) by U_j^L followed by rearrangement yields the following form of the reaction function, namely,

$$R_j = \frac{U_j^L (\rho_j^L)^2}{L_j^2} [k_j l_{jA} l_{jB} - k'_j l_{jC} l_{jD}] - \eta_j \tag{8-16}$$

Table 8-1 Expressions for the reaction function R_j for various types of reactions

1. Vapor phase reaction

$$r_{jA} = k_j C_{jA} C_{jB} - k'_j C_{jC} C_{jD}$$

where the concentrations are for the vapor phase. Then

$$R_j = \frac{U_j^V (\rho_j^V)^2}{V_j^2} [k_j v_{jA} v_{jB} - k'_j v_{jC} v_{jD}] - \eta_j$$

where ρ_j^V is the molar density of the vapor phase.

2. Chemical equilibrium in the vapor phase

$$K_{Rj}^V = \frac{(\hat{f}_{jC}^V)^{v_C}(\hat{f}_{jD}^V)^{v_D}}{(\hat{f}_{jA}^V)^{-v_A}(\hat{f}_{jB}^V)^{-v_B}}$$

where K_{Rj} is the chemical equilibrium constant for stage j. The expression for R_j is readily obtained through the use of the above expression for K_{Rj}^V and the expression for $\hat{f}_{ji}^V$, namely,

$$\hat{f}_{ji}^V = \frac{\gamma_{ji}^V f_{ji}^V v_{ji}}{V_j}$$

Thus

$$R_j = \frac{K_{\gamma j}^V K_{fj}^V}{K_{Rj}^V V_j^\sigma} \left[\frac{(v_{jC})^{v_C}(v_{jD})^{v_D}}{(v_{jA})^{-v_A}(v_{jB})^{-v_B}} \right] - 1$$

where

$$K_{\gamma j}^V = \frac{(\gamma_{jC}^V)^{v_C}(\gamma_{jD}^V)^{v_D}}{(\gamma_{jA}^V)^{-v_A}(\gamma_{jB}^V)^{-v_B}}$$

$$K_{fj}^V = \frac{(f_{jC}^V)^{v_C}(f_{jD}^V)^{v_D}}{(f_{jA}^V)^{-v_A}(f_{jB}^V)^{-\mu_B}}$$

$$\sigma = v_A + v_B + v_C + v_D$$

3. Chemical equilibrium in the liquid phase

$$K_{Rj}^L = \frac{(\hat{f}_{jC}^L)^{v_C}(\hat{f}_{jD}^L)^{v_D}}{(\hat{f}_{jA}^L)^{-v_A}(\hat{f}_{jB}^L)^{-v_B}}$$

By use of

$$\hat{f}_{ji}^L = \frac{\gamma_{ji}^L f_{ji}^L l_{ji}}{L_j}$$

and the definition of the ideal solution K value

$$K_{ji} = \frac{f_{ji}^L}{f_{ji}^V}$$

the following functional form of R_j is readily obtained

$$R_j = \frac{K_{\gamma j}^L K_{fj}^V K_{Kj}}{K_{Rj}^L L_j^\sigma} \left[\frac{(l_{jC})^{v_C}(l_{jD})^{v_D}}{(l_{jA})^{-v_A}(l_{jB})^{-v_B}} \right] - 1$$

where the expression for $K_{\gamma j}^L$ is obtained from the one given above for $K_{\gamma j}^V$ by replacing γ_{ji}^V by γ_{ji}^L. Likewise, the expression for K_{Kj} is obtained from the one given above for $K_{\gamma j}$ by replacing γ_{ji}^V by K_{ji}.

Observe that if all of the components which appear in both phases are in physical and chemical equilibrium, it follows then that

$$K_{Rj}^L = K_{Rj}^V$$

since $\hat{f}_i^V = \hat{f}_i^L$.

The functional forms of R_j for the case where the reaction occurs in the gas phase (at the rate r_{jA}) and for the cases where a state of dynamic equilibrium exists in the vapor and liquid phases are presented in Table 8-1.

Determination of the Moles Reacted per Stage

For a given set of T_j's and L_j/V_j's, it is desired to find the set of η_j's which satisfy the material balances [Eqs. (8-8) and (8-11)] and which make each $R_j = 0$. The solution set of η_j's may be found by use of the Newton-Raphson method which consists of the repeated solution of the Newton-Raphson equations

$$\mathbf{J}_R \, \Delta \boldsymbol{\eta} = -\mathbf{R} \tag{8-17}$$

where the jacobian matrix $\mathbf{J}_R$ has the form

$$\mathbf{J}_R = \begin{bmatrix} \dfrac{\partial R_1}{\partial \eta_1} & \cdots & \dfrac{\partial R_1}{\partial \eta_N} \\ \vdots & & \vdots \\ \dfrac{\partial R_N}{\partial \eta_1} & \cdots & \dfrac{\partial R_N}{\partial \eta_N} \end{bmatrix}$$

and

$$\Delta \boldsymbol{\eta} = [\Delta \eta_1 \ \Delta \eta_2 \ \cdots \ \Delta \eta_N]^T$$
$$\mathbf{R} = [R_1 \ R_2 \ \cdots \ R_N]^T$$

The solution may be effected by use of analytical expressions for the partial derivatives of the R_j's or by use of Broyden's method, which makes use of numerical approximations of the partial derivatives;[5] see Chaps. 4 and 15.

The θ Method of Convergence

The θ method of convergence is used to find a corrected set of compositions which are in turn used to find a new set of temperatures. After the above material balances have been solved, the set of calculated d_i's may not be in agreement with the specified value of D. The θ method of convergence is used to find a set of $(d_i)_{co}$'s and $(b_i)_{co}$'s which are in overall component-material balance

$$FX_i + v_i \sum_{j=1}^{N} \eta_j = (d_i)_{co} + (b_i)_{co} \tag{8-18}$$

and in agreement with the specified value of D. Again the θ multiplier can be defined by

$$\left(\frac{b_i}{d_i}\right)_{co} = \theta \left(\frac{b_i}{d_i}\right)_{ca} \tag{8-19}$$

where the subscript ca denotes the most recent values of $(b_i/d_i)_{ca}$ found by the solution of the material balance equations. Use of Eq. (8-19) permits Eq. (8-18) to be solved for $(d_i)_{co}$ in terms of the θ and the most recently calculated values of the terminal rates, namely,

$$(d_i)_{co} = (d_i)_{ca} p_i \tag{8-20}$$

where

$$p_i = \frac{FX_i + v_i \phi}{(d_i)_{ca} + \theta(b_i)_{ca}}$$

$$\phi = \sum_{j=1}^{N} n_j$$

The desired value of θ is that $\theta > 0$ that makes $g(\theta) = 0$, where

$$g(\theta) = \sum_{i=1}^{c} (d_i)_{co} - D \tag{8-21}$$

and the formula for $(d_i)_{co}$ is given by Eq. (8-20). Equation (8-21) may be solved for θ by use of Newton's method which consists of the repeated use of the relationship

$$\theta_{k+1} = \theta_k - \frac{g(\theta_k)}{g'(\theta_k)} \tag{8-22}$$

where

$$g'(\theta_k) = -\sum_{i=1}^{c} \frac{(d_i)_{ca}(b_i)_{ca} p_i^2}{FX_i + v_i \phi} \tag{8-23}$$

After the desired value of θ has been found, the corrected compositions are computed as follows

$$x_{ji} = \frac{(l_{ji})_{ca} p_i}{\sum_{i=1}^{c} (l_{ji})_{ca} p_i} \tag{8-24}$$

$$y_{ji} = \frac{(v_{ji})_{ca} p_i}{\sum_{i=1}^{c} (v_{ji})_{ca} p_i} \tag{8-25}$$

On the basis of the mole fractions given by Eqs. (8-24) and (8-25), the most recently assumed temperature profile, and the activity coefficients used in the A_{ji}'s [see Eq. (8-2)], the new temperature profile is found by use of the K_b method; see Eqs. (2-30) and (2-31).

Enthalpy Balances

When simultaneous mass transfer and chemical reaction occur, it is convenient to define the enthalpy of a component as the enthalpy it has at a given temperature T in the perfect gas state at one atmosphere above its elements in their standard states at some arbitrary datum temperature. For example, the enthalpy $H_i^\circ(T)$ of component i in the perfect gas state at the temperature T is defined as follows

$$H_i^\circ(T) = \Delta H_{f0, i}^\circ + (H_T - H_0^\circ)_i \tag{8-26}$$

Values for the heats of formation $\Delta H_{f0, i}^\circ$ at 0 K and the heat content of an element at temperature T above its enthalpy at 0 K, $(H_T - H_0^\circ)_i$, may be found in standard thermodynamic tables; see for example Ref. 22.

To account for the deviation of an actual mixture from a perfect gas mixture, it has been demonstrated that the virtual values of the partial molar enthalpies $\hat{H}_i$ may be used in lieu of the partial molar enthalpies $\bar{H}_i$ to obtain the correct enthalpy of the mixture,[13] that is,

$$n_T H = \sum_{i=1}^{c} n_i \hat{H}_i = \sum_{i=1}^{c} n_i \bar{H}_i \tag{8-27}$$

where H = enthalpy per mole of mixture
$\bar{H}_i$ = partial molar enthalpy
$\hat{H}_i$ = virtual value of the partial molar enthalpy
n_i, n_T = number of moles of component i and the total moles of mixture, respectively

As shown in Chap. 14, the virtual value of the partial molar enthalpy is defined by

$$\hat{H}_i = H_i^\circ + \Omega \tag{8-28}$$

where Ω is the departure of the enthalpy of one mole of an actual gas mixture from one mole of a perfect gas mixture. Expressions for computing Ω for several equations of state are also presented in Chap. 14. For the case of a liquid phase which forms an ideal solution which is independent of pressure, the enthalpy of the mixture is computed as follows

$$n_T H = \sum_{i=1}^{c} n_i \bar{h}_i = \sum_{i=1}^{c} n_i h_i^\circ = \sum_{i=1}^{c} n_i (H_i^\circ - \lambda_i) \tag{8-29}$$

where λ_i is the latent heat of vaporization at a given temperature T. Since

$$n_T H = \sum_{i=1}^{c} n_i \hat{h}_i = \sum_{i=1}^{c} n_i (H_i^\circ + \Omega^L) \tag{8-30}$$

it follows by comparison of this expression with Eq. (8-29) that

$$n_T \Omega^L = - \sum_{i=1}^{c} n_i \lambda_i \tag{8-31}$$

Let the virtual values of the enthalpies of component i in the vapor and liquid phases leaving plate j be denoted by $\hat{H}_{ji}$ and $\hat{h}_{ji}$, respectively. Then the enthalpy balance enclosing stage j $(j \neq 1, N)$ is given by

$$\sum_{i=1}^{c} [v_{j+1, i} \hat{H}_{j+1, i} + l_{j-1, i} \hat{h}_{j-1, i} - v_{ji} \hat{H}_{ji} - l_{ji} \hat{h}_{ji}] = 0 \tag{8-32}$$

Use of the component-material balance [Eq. (8-6)] to eliminate one of the component-flow rates, say l_{ji}, in Eq. (8-32) yields

$$\sum_{i=1}^{c} [v_{j+1,i}(\hat{H}_{j+1,i} - \hat{h}_{ji}) + l_{j-1,i}(\hat{h}_{j-1,i} - \hat{h}_{ji}) - v_{ji}(\hat{H}_{ji} - \hat{h}_{ji}) - \delta_{ji}\hat{h}_{ji}] = 0 \quad (8\text{-}33)$$

Since $v_{j+1,i} = V_{j+1} y_{j+1,i}$, Eq. (8-33) may be solved for V_{j+1}.

$$V_{j+1} = \frac{V_j \sum_{i=1}^{c} (\hat{H}_{ji} - \hat{h}_{ji})y_{ji} - L_{j-1} \sum_{i=1}^{c} (\hat{h}_{j-1,i} - \hat{h}_{ji})x_{j-1,i} + \sum_{i=1}^{c} \delta_{ji}\hat{h}_{ji}}{\sum_{i=1}^{c} (\hat{H}_{j+1,i} - \hat{h}_{ji})y_{j+1,i}} \quad (8\text{-}34)$$

For $j = f - 1$ and f, the numerator of Eq. (8-34) contains the following additive terms, namely, for $j = f - 1$,

$$V_F \sum_{i=1}^{c} (\hat{H}_{Fi} - \hat{h}_{f-1,i})y_{Fi}$$

and for $j = f$,

$$L_F \sum_{i=1}^{c} (\hat{h}_{Fi} - \hat{h}_{fi})x_{Fi}$$

For $j = 1$ and N, the following expressions are obtained for Q_C and Q_R

$$Q_C = V_2 \sum_{i=1}^{c} (\hat{H}_{2i} - \hat{h}_{1i})y_{2i} - D \sum_{i=1}^{c} (\hat{H}_{Di} - \hat{h}_{1i})X_{Di} - \sum_{i=1}^{c} \delta_{1i}\hat{h}_{1i}$$

$$Q_R = V_N \sum_{i=1}^{c} (\hat{H}_{Ni} - \hat{h}_{Ni})y_{Ni} - L_{N-1} \sum_{i=1}^{c} (\hat{h}_{N-1,i} - \hat{h}_{Ni})x_{n-1,i} + \sum_{i=1}^{c} \delta_{Ni}\hat{h}_{Ni} \quad (8\text{-}35)$$

It is of interest to note that the familiar heat of reaction term is included in the sum of the $\delta_{ji} h_{ji}$'s. Evaluation of this term by use of Eq. (8-28) gives

$$\sum_{i=1}^{c} \delta_{ji}\hat{h}_{ji} = \eta_j \left[\sum_{i=1}^{c} v_i(H_{ji}^\circ + \Omega_j^I) \right] = \eta_j[\Delta H_j^\circ + \sigma\Omega_j^I] = \eta_j(\Delta H_j)_{\text{liq}} \quad (8\text{-}36)$$

where $\Delta H_j^\circ = \sum_{i=1}^{c} v_i H_i^\circ$, the heat of reaction in the perfect gas state at the temperature T.

For the case where the liquid phase behaves ideally and is also independent of pressure, Eq. (8-31) may be used to further reduce Eq. (8-36) to

$$\sum_{i=1}^{c} \delta_{ji}\hat{h}_{ji} = \eta_j \left[\sum_{i=1}^{c} v_i(H_{ji}^\circ - \lambda_{ji}) \right] = \eta_j \sum_{i=1}^{c} v_i h_{ji}^\circ = \eta_j(\Delta H_j^\circ)_{\text{liq}} \quad (8\text{-}37)$$

Calculational Procedure

1. Assume sets of values $\{T_j\}$, $\{(L_j/V_j)\}$.
2. On the basis of the most recent sets $\{T_{jn}\}$, $\{(L_j/V_j)_n\}$ find the set of η_j's which satisfy the material balances and the reaction functions.
3. Find the θ that makes $g(\theta) = 0$.
4. Compute a new set of compositions by use of Eqs. (8-24) and (8-25).
5. Compute a new set of temperatures by use of Eqs. (2-30) and (2-31).
6. Compute a new set of total-flow rates by use of the energy balances and the total material balances. Then return to step 1.
7. Repeat the process until the convergence criterion has been satisfied.

Numerical example To demonstrate the characteristics of the proposed θ method of convergence for solving distillation problems in which a chemical reaction occurs on each stage, two numerical examples are presented. Example 8-1 is based on an esterification reaction

$$CH_3COOH + C_2H_5OH \rightleftarrows C_2H_5COOH + H_2O$$

which occurs in the liquid phase. The reaction is second order in both directions. A statement of this example appears in Table 8-2 and the solution is presented in Table 8-3. The K values, activity coefficients, vapor and liquid enthalpies, and the rate data used to solve this problem are presented in Tables B-19 through B-21. Although, these mixtures deviate significantly from ideal solutions, satisfactory convergence is achieved as shown in the last entry of Table 8-3. The θ method gives satisfactory convergence for all systems which do not deviate too widely from ideal solutions.[4]

Table 8-2 Statement of Example 8-1

	1. Column specifications and operating conditions		
Component	Component no.	X_i	
Acetic acid	1	0.4963	
Ethanol	2	0.4808	
Water	3	0.0229	
Ethyl acetate	4	0.0	

Specifications

Total condenser, $P = 1$ atm, boiling-point liquid feed, $N = 13$, $f = 6$, $F = 0.1078$ g mol/min, $D = 0.0208$ g mol/min. Use a reflux ratio of 10. The holdup is taken as 1.0 liters for the reboiler and 0.3 liter for each plate and the condenser. The molar densities of the liquids acetic acid, ethyl alcohol, water, and ethyl acetate were: 17.470, 17.129, 55.49, and 10.22 moles per liter respectively. The mole fraction average of these was used as the molar density of the mixture. Use the vapor-liquid equilibrium, enthalpy, and reaction rate data given in Tables B-19 through B-21.

Table 8-3 Solution of Example 8-1

1. Initial and final profiles

Stage	Temperature (K)		L/V	
	Initial	Final	Initial	Final
1	345.0	337.00	. . .	. . . .
2	346.5	336.91	1.0	0·9081
3	348.0	337.15	1.0	0.9017
4	349.5	338.05	1.0	0.8897
5	351.0	339.87	1.0	0.8798
6	352.5	347.71	1.0	1.3897
7	354.0	348.16	1.0	1.4010
8	355.5	348.53	1.0	1.4038
9	357.0	348.89	1.0	1.4061
10	358.5	349.37	1.0	1.4085
11	360.0	350.37	1.0	1.4156
12	361.5	353.68	1.0	1.4551
13	363.0	361.57	0.4	0.3844

Stage	L_j Final	V_j Final	η, gm mol/min	
			Initial	Final
1	0·2080	. . . .	0.0	-0.1986×10^{-3}
2	0.2077	0·2288	0.0	-0.1530×10^{-3}
3	0.2061	0.2285	0.0	-0.1144×10^{-3}
4	0.2019	0.2269	0.0	-0.0851×10^{-3}
5	0.1959	0.2227	0.0	-0.0408×10^{-3}
6	0.3012	0.2167	0.3	0.6029×10^{-3}
7	0.3004	0.2144	0.3	0.6490×10^{-3}
8	0.2999	0.2136	0.3	0.6748×10^{-3}
9	0.2996	0.2131	0.3	0.6848×10^{-3}
10	0.2998	0.2128	0.3	0.6898×10^{-3}
11	0.3015	0.2130	0.3	0.7426×10^{-3}
12	0.3125	0.2147	0.3	1.0620×10^{-3}
13	0.0867	0.2257	0.3	6.0336×10^{-3}

Execution Time = 3.01 s. (FORTRAN H)

2. Final product distribution

Component	b_i	d_i	FX_i	$v_i \sum \eta_j$†
1	0.04278	0.0002	0.05350	-0.01060
2	0.03781	0.00332	0.05183	-0.01060
3	0.00496	0.00811	0.00247	0.01060
4	0.00125	0.00935	0.0	0.01060
	0.08680	0.0208	0.1078	

† $\sum \eta_j = \eta_1 + \eta_2 + \cdots + \eta_{13}$

(Continued on page 286.)

Table 8-3 (*continued*)   3. Convergence characteristics of the θ method

Iteration	θ	No. of trials required to find η_j's
Initial value	5.1191	1
1	3.4557	3
2	1.0146	3
3	1.3308	3
4	0.9527	2
5	0.9566	3
6	1.0287	3
7	0.9763	3
8	1.2134	2
9	0.9781	2
10	1.0192	2
11	0.9696	2
12	1.0163	1
13	0.9975	1
14	1.0135	1
15	0.9957	1
16	1.0000	1

Generalization of the Proposed Calculational Procedure for the Case Where Any Number of Independent Reactions Occur on Each Stage

To demonstrate the technique for generalization of the proposed procedure, consider the case where two reactions such as

$$A + B \rightleftarrows C + D \tag{I}$$

$$A \rightleftarrows R + S \tag{II}$$

occur on each stage. The equations presented previously must now be modified to account for the fact that η_{Ij} moles of A react on stage j by reaction I and η_{IIj} moles of A react on stage j by Reaction II. Thus, the component-material balances [Eq. (8-5)] now become

$$A_i v_i = -f_i - v_{Ii}\eta_I - v_{IIi}\eta_{II} \tag{8-38}$$

and the total-material balances [Eq. (8-11)] become

$$BV = -\mathcal{F} - \sigma_I \eta_I - \sigma_{II}\eta_{II} \tag{8-39}$$

In the enthalpy balances [see Eq. (8-34)], the term $\sum_{i=1}^{c} \delta_{ji} \hat{h}_{ji}$ is replaced by

$$\left[\sum_{i=1}^{c} \delta_{Iji} \hat{h}_{ji} + \sum_{i=1}^{c} \delta_{IIji} \hat{h}_{ji} \right]$$

Also, there are now $2N$ reaction functions $\{R_{Ij}\}$ and $\{R_{IIj}\}$ instead of only N.

The equations for the θ method of convergence are the same form as those shown except for the fact that p_i which appears in Eq. (8-20) is now given by

$$p_i = \frac{FX_i + v_{Ii}\phi_I + v_{IIi}\phi_{II}}{(d_i)_{ca} + \theta(b_i)_{ca}} \tag{8-40}$$

In the general case of any number of reactions, care must be exercised in definition of the v_i's in order to avoid divisions by zero. In conclusion, the θ

method of convergence may be used to solve distillation problems involving conventional distillation columns in which any number of chemical reactions occur on each stage by making the appropriate modifications described above.

8-2 FORMULATION OF THE $N(r + 2)$ NEWTON-RAPHSON METHOD

In this formulation of the Newton-Raphson method, it is supposed that r independent chemical reactions occur on each stage of a column. The Newton-Raphson formulation presented is analogous to the one recommended previously for absorber-type columns in which chemical reactions did not occur.

The Newton-Raphson method is formulated first for an absorber in which one chemical reaction occurs per plate. Then, the method is modified as required to describe distillation columns in which chemical reactions occur. Although the resulting algorithm is readily applied to systems which are characterized by nonideal solution behavior, it is an exact application of the Newton-Raphson method for those systems in which ideal or near ideal solution behavior exists throughout the column. The algorithm presented is recommended for absorption-type columns which exhibit ideal or near ideal solution behavior.

Formulation of the $N(r + 2)$ Newton-Raphson Method for Absorbers

Consider first the case where the reaction given by Eq. (8-3) occurs in the liquid phase on each stage. The rate expression for this reaction is given by Eq. (8-12) and the moles of A reacted per unit time per unit volume on stage j is denoted by η_j. In this case $r = 1$ and $N(r + 2) = 3N$. The $3N$ independent variables are the $\{L_j/V_j\}$, $\{T_j\}$, and $\{\eta_j\}$. The variable L_j/V_j is normalized by introduction of the new variable θ_j which is defined as follows

$$\frac{L_j}{V_j} = \theta_j \left(\frac{L_j}{V_j}\right)_a = b_j \tag{8-41}$$

By taking $(L_j/V_j)_a$ to be equal to the most recent value of L_j/V_j found by the calculational procedure, it is evident that θ_j approaches unity as the solution set of values of the variables is approached.

Corresponding to these $3N$ independent variables, $3N$ independent functions must be selected. These are taken to be the dew-point functions $\{F_j\}$, the enthalpy balance functions $\{G_j\}$, and the reaction functions $\{R_j\}$. The dew-point functions are given by Eq. (4-3), and the reaction functions are given by Eq. (8-16). The enthalpy balance functions are formulated by stating Eq. (8-22) in functional form to give

$$G_j = \frac{\sum_{i=1}^{c} [v_{ji}\hat{H}_{ji} + l_{ji}\hat{h}_{ji}]}{\sum_{i=1}^{c} [v_{j+1,i}\hat{H}_{j+1,i} + l_{j-1,i}\hat{h}_{j-1,i}]} - 1 \tag{8-42}$$

As discussed below, $G_j = 0$ when the solution set of values of the variables has been found.

THE CONSTRAINING EQUATIONS

For any choice of values for the independent variables, the corresponding values of the dependent variables, the $\{v_{ji}\}$, $\{l_{ji}\}$, $\{V_j\}$, and $\{L_j\}$ may be found by use of the component-material balances, the equilibrium relationships, and the total-material balances.

As before in Sec. 8-1, the equilibrium relationships [Eq. (8-2)] are used to eliminate the l_{ji}'s from the component-material balances. The complete set of equations may be stated in matrix form as indicated by Eq. (8-8). For an absorber, the vectors $\mathbf{v}_i$ and $\mathbf{f}_i$ have the following meanings rather than those given beneath Eq. (2-18)

$$\mathbf{f}_i = [l_{0i} \ 0 \ \cdots \ 0 \ v_{N+1,\,i}]^T$$

and

$$\mathbf{v}_i = [v_{1i} \ v_{2i} \ \cdots \ v_{Ni}]^T$$

The total-material balances may be stated in terms of the independent variables $\{L_j/V_j\}$ and solved for the vapor rates $\mathbf{V}$ as indicated by Eq. (8-11). For an absorber, the vectors $\mathscr{F}$ and $\mathbf{V}$ have the following meanings rather than those given beneath Eq. (8-11)

$$\mathscr{F} = [L_0 \ 0 \ \cdots \ 0 \ V_{N+1}]^T$$
$$\mathbf{V} = [V_1 \ V_2 \ \cdots \ V_N]^T$$

Formulation of the Newton-Raphson Equations

It is desired to find the solution set of values of the independent variables

$$\mathbf{X} = [\eta_1 \ \eta_2 \ \cdots \ \eta_N \ \theta_1 \ \theta_2 \ \cdots \ \theta_N \ T_1 \ T_2 \ \cdots \ T_N]^T \tag{8-43}$$

such that each of the functions

$$\mathbf{f} = [R_1 \ R_2 \ \cdots \ R_N \ F_1 \ F_2 \ \cdots \ F_N \ G_1 \ G_2 \ \cdots \ G_N]^T \tag{8-44}$$

is equal to zero. This may be achieved by use of the Newton-Raphson method which consists of the repeated solution of the equation

$$\mathbf{J} \ \Delta \mathbf{X} = -\mathbf{f} \tag{8-45}$$

where

$$\mathbf{J} = \begin{bmatrix} \dfrac{\partial R_1}{\partial \eta_1} & \cdots & \dfrac{\partial R_1}{\partial T_N} \\ \vdots & & \vdots \\ \dfrac{\partial F_1}{\partial \eta_1} & \cdots & \dfrac{\partial F_1}{\partial T_N} \\ \vdots & & \vdots \\ \dfrac{\partial G_N}{\partial \eta_1} & \cdots & \dfrac{\partial G_N}{\partial T_N} \end{bmatrix}$$

$$\Delta \mathbf{X} = \mathbf{X}_{k+1} - \mathbf{X}_k = [\Delta \eta_1 \ \Delta \eta_2 \ \cdots \ \Delta \eta_N \ \cdots \ \Delta \theta_N \ \cdots \ \Delta T_N]^T$$

This conventional form of the Newton-Raphson method was very sensitive to the set of starting values selected for the η_j's, and it was difficult to find a set for which the calculational procedure would converge to the solution set. In order to eliminate the sensitivity of the Newton-Raphson method to the starting values selected for the η_j's, it was necessary to modify the procedure in the following manner.

For each choice of T_j's and θ_j's, the component-material balances and total-material balances were solved for the set of η_j's that made each of the R_j's equal to zero. The desired set of η_j's were found by use of the Newton-Raphson method. In this case, the Newton-Raphson equations are of the form

$$\mathbf{J}_R \, \Delta\boldsymbol{\eta} = -\mathbf{R} \tag{8-46}$$

where

$$\mathbf{J}_R = \begin{bmatrix} \dfrac{\partial R_1}{\partial \eta_1} & \cdots & \dfrac{\partial R_1}{\partial \eta_N} \\ \vdots & & \vdots \\ \dfrac{\partial R_N}{\partial \eta_1} & \cdots & \dfrac{\partial R_N}{\partial \eta_N} \end{bmatrix}$$

$$\Delta\boldsymbol{\eta} = \boldsymbol{\eta}_{n+1} - \boldsymbol{\eta}_n = [\Delta\eta_1 \; \Delta\eta_2 \; \cdots \; \Delta\eta_N]^T$$

$$\mathbf{R} = [R_1 \; R_2 \; \cdots \; R_N]^T$$

After the η_j's that made each $R_j = 0$ had been found, they were used together with the most recent set of T_j's and θ_j's to evaluate the elements of $\mathbf{J}$ and $\mathbf{f}$ of Eq. (8-45). This sequence of calculations in the application of the Newton-Raphson method is called procedure 1.

Also, in order to achieve convergence, it was necessary to perform column scaling followed by row scaling (see Chap. 15) of the jacobian matrix. Convergence could not be achieved, however, by use of variable scaling followed by row scaling (see Chap. 15).

Procedure 1: Newton-Raphson (with Analytical Expressions for the Partial Derivatives)

Step 1. Assume a set of values for $(L_j/V_j)_{a, k}$. Set each $\theta_{j, k} = 1$. Assume a set of values for $T_{j, k}$ (denoted by $\mathbf{T}_k$) and for $\eta_{j, n}$ (denoted by $\boldsymbol{\eta}_n$).

Step 2. On the basis of the values of the variables assumed in step 1, solve Eq. (8-49) repeatedly until a set of η's, denoted by $\boldsymbol{\eta}_k$, is found for which $\mathbf{R} = \mathbf{0}$.

Step 3. On the basis of the most recent set of values of the variables, $\boldsymbol{\eta}_k$, $(L_j/V_j)_k = \theta_{j, k}(L_j/V_j)_{a, k}$, and $\mathbf{T}_k$, evaluate $\mathbf{f}_k$. If each element of $\mathbf{f}_k - \mathbf{f}_{k-1}$ satisfies the convergence criterion, cease calculations; otherwise, evaluate the elements of $\mathbf{J}_k$ and solve Eq. (8-45) for $\Delta\mathbf{X}$ and compute $\mathbf{X}_{k+1}$. If each $T_{j, k+1}$ is within the range of the curve fits, and each $\theta_{j, k+1} > 0$, compute

$$\left(\frac{L_j}{V_j}\right)_{a, k+1} = \theta_{j, k+1}\left(\frac{L_j}{V_j}\right)_{a, k}$$

and return to step 1. Otherwise reduce the correction ΔX repeatedly by the factor β ($\beta = 1/2, 1/4, \ldots$)

$$X_{k+1} = X_k + \beta \, \Delta X$$

until an X_{k+1} is found which satisfies the above criterion. Then compute $(L_j/V_j)_{a,\, k+1}$. Replace the subscript k by $k+1$. The partial derivatives appearing in J and J_R may be evaluated by use of the analytical expressions appearing in Table 4-3 and Probs. 8-2 and 8-3.

Another significantly faster procedure for solving the Newton-Raphson equations, called the Broyden-Bennett method, was proposed by Hess et al.[10] The Broyden-Bennett algorithm may be used to solve the equations for an absorber-type column accompanied by a chemical reaction in the following manner.

Procedure 2: Broyden-Bennett Algorithm

Step 1. For the first trial, assume X_0, and use the Broyden-Bennett algorithm (see Chap. 4) to find an improved set of η_j's which satisfy Eqs. (8-8) and (8-11), and which make $R_0 = 0$.

Step 2. Evaluate the remaining partial derivatives of J_0 numerically, as shown in step 2 of procedure 2 of Chap. 4, and then find the factors L_0 and U_0 such that

$$J_0 = L_0 U_0$$

Step 3. Same as step 3 of procedure 2 of Chap. 4.

Step 4. Find the s_k such that the euclidean norm of $f(X_k + s_k P_k)$ is less than the norm of $f(X_k)$. For each choice of s_k, use the Broyden-Bennett algorithm (see Chap. 4) to find an improved η (called η_{k+1}) which satisfies Eqs. (8-8) and (8-11) and which makes

$$R_{k+1} = 0$$

The remainder of step 4 is the same as step 4 of procedure 2 of Chap. 4.

Step 5. Same as step 5 of procedure 2 of Chap. 4.

Step 6. Same as step 6 of procedure 2 of Chap. 4.

Formulation of the $N(r + 2)$ Newton-Raphson Method for Distillation Columns

The equations presented for the formulation of the $N(r + 2)$ Newton-Raphson method for absorbers may be applied to distillation columns in a manner similar to that demonstrated in Chap. 4 for the $2N$ Newton-Raphson method for absorbers and distillation columns.

Numerical examples Several mechanisms have been proposed for the reaction of monoethanolamine and carbon dioxide.[2, 7, 11, 16] A mechanism consisting of six reactions was proposed by Kent and Eisenberg,[15] and a mechanism consisting of two reactions was recently proposed by Hikita et al.[11] To demonstrate the proposed calculational procedures for absorbers, the mechanism proposed by Hikita et al. is used. The two-step mechanism consists of the reactions

$$CO_2 + HOC_2H_4NH_2 \rightleftarrows HOC_2H_4NHCOO^- + H^+ \qquad (A)$$

$$H^+ + HOC_2H_4NH_2 \rightleftarrows HOC_2H_4NH_3^+ \qquad (B)$$

The equilibrium constants of reactions (A) and (B) are of the order of 10^{-5} and 10^{10}, respectively.[11] Reaction (B) is ionic and virtually instantaneous, whereas reaction (A) is second order and is rate-controlling. Thus, the rate of reaction is given by reaction (A) and the stoichiometry is the sum of these reactions; that is, every time reaction (A) goes one time, reaction (B) then occurs instantaneously. Thus the overall reaction is

$$CO_2 + 2HOC_2H_4NH_2 \rightleftarrows HOC_2H_4NH_3^+ + HOC_2H_4NHCOO^-$$

Vapor-liquid equilibrium data and enthalpy data needed to describe this system were taken from several sources.[16, 18, 20]

A statement of the monoethanolamine example, Example 8-2, is given in Table 8-4. The solution is given in Table 8-5, and convergence characteristics of procedures 1 and 2 are given in Table 8-6. To demonstrate the use of procedures 1 and 2 in the solution of problems involving conventional distillation columns, Example 8-1 was solved. A statement of this example is

Table 8-4 Statement of Example 8-2

Component	Component no.	l_{0i} (lb mol/min)	$v_{N+1, i}$ (lb mol/min)
Monoethanolamine	1	9.88	0.0
Water	2	190.12	0.0
Carbon dioxide	3	0.0	4.0
Amine carbamate	4	0.0	0.0
Methane	5	0.0	96.0
		200.00	100.0

Other specifications

The absorber is to operate at 100 lb/in² abs, $N = 16$, $L_0 = 200$ lb mol/min and enters the absorber as a liquid at 100°F. The rich gas V_{N+1} enters at the rate of 100 lb mol/min at 90°F. The liquid holdup on each stage is taken to be equal to 60 lb mol. To initiate the calculational procedure, take $(L_j/V_j)_a = 1$, $\theta_j = 1$, $\eta_j = 0.05$, and $T_j = 100°F$ for all j. Use the vapor-liquid equilibrium, enthalpy, and reaction rate data given in Tables B-22 through B-23.

Table 8-5 Solution of Example 8-2

1. T_j, V_j, L_j, and η_j profiles

Stage	Temperature (°F)	L_j lb mol/min	V_j lb mol/min	η_j lb mol/min
1	100.08	200.03	96.86	0.0039
2	100.22	200.02	96.89	0.0059
3	100.37	200.02	96.90	0.0089
4	100.59	200.02	96.92	0.0135
5	100.93	201.01	96.94	0.0204
6	101.44	201.00	96.98	0.0308
7	102.23	199.99	97.03	0.0469
8	103.45	199.98	97.11	0.0716
9	105.39	199.97	97.24	0.1103
10	108.45	199.97	97.45	0.1719
11	113.11	199.97	97.79	0.2709
12	119.69	199.94	98.34	0.4259
13	127.51	199.72	99.17	0.6352
14	134.03	199.00	100.21	0.8128
15	135.10	197.61	101.12	0.7973
16	124.59	195.16	101.32	0.5626

2. Final product distribution

Component	l_{Ni}	v_{1i}	$v_i \sum \eta_j$
1	1.901	0.001	-7.978
2	189.250	0.872	0
3	0.003	0.008	-3.989
4	3.989	0	3.989
5	0.022	95.978	0
	195.165	96.859	

presented in Table 8-7. The convergence characteristics of procedures 1 and 2 for Example 8-1 are presented in Table 8-8. Procedure 2 (the Broyden-Bennett Algorithm) is seen to require more trials, but less computer time, than procedure 1 (the Newton-Raphson method) with analytical expressions for the partial derivatives. These same characteristics were exhibited for columns in which chemical reactions did not occur.[10] The speed advantage of procedure 2 over procedure 1 can be expected to become more pronounced as the number of components and stages are increased.

For distillation columns, the θ method presented in Sec. 8-1 required about the same amount of computer time as did procedures 1 or 2 for Example 8-1. As was found for distillation columns without reaction, the computer times required by the θ method and procedures 1 and 2 do not differ significantly. However, as the number of plates and components are increased, the speed advantages of the θ method over procedures 1 and 2 can be expected.

Table 8-6 Convergence characteristics of procedures 1 and 2 for Example 8-2

| | Procedure 1 | | | Procedure 2 | |
Iteration	No. of trials required to find $\{\eta_j\}$	Norm†	Iteration	No. of trials required to find $\{\eta_j\}$	Norm‡
1	5	0.444	1	5	0.444
2	3	0.346	2	3	0.346
3	2	0.272	3	3	0.201
4	2	0.215	4	3	0.834×10^{-1}
5	2	0.140	5	3	0.137
6	3	0.814×10^{-1}	6	2	0.151×10^{-1}
7	2	0.578×10^{-1}	7	3	0.981×10^{-1}
8	2	0.258×10^{-1}	8	1	0.151×10^{-1}
9	3	0.141×10^{-2}	9	2	0.749×10^{-2}
10	2	0.140×10^{-4}	10	2	0.985×10^{-2}
11	1	0.179×10^{-5}	11	2	0.749×10^{-2}
12	1	0.987×10^{-6}	12	2	0.136×10^{-3}
			13	2	0.259×10^{-4}
			14	2	0.140×10^{-4}
			15	2	0.186×10^{-4}
			16	2	0.140×10^{-4}
			17	1	0.111×10^{-5}
			18	3	0.672×10^{-4}
			19	1	0.106×10^{-5}
			20	1	0.663×10^{-6}

$$\dagger \text{ Norm} = \frac{1}{3N}\left[\sum_{j=1}^{N}(F_j^2 + R_j^2 + G_j^2)\right]^{1/2}$$

Execution time = 4.38 s (FORTRAN H)

‡ Norm = same as for procedure 1

Execution time = 4.16 s (FORTRAN H)

Table 8-7 Restatement and solution of Example 8-1, a conventional distillation column

1. Column specifications and operating conditions

Component	Component no.	X_i
Acetic acid	1	0.4963
Ethanol	2	0.4808
Water	3	0.0229
Ethyl acetate	4	0.0

Specifications

Total condenser, $P = 1$ atm, boiling point liquid feed, reflux ratio, $L_1/D = 10$, boilup ratio, $V_N/B = 2.588$. The holdup is taken as 1.0 liter for the reboiler and 0.3 liter for each plate and the condenser. The molar densities of the liquids acetic acid, ethyl alcohol, water, and ethyl acetate are 17.470, 17.129, 55.49, and 10.22 moles per liter, respectively. The mole fraction average of these was used as the molar density of the mixture.

2. Vapor-liquid equilibrium data, activity coefficients, and enthalpies; see Tables B-19 through B-21.

3. Solution; see Table 8-3.

Table 8-8 Convergence characteristics of procedures 1 and 2 for Example 8-1 by use of the $N(r + 2)$ Newton-Raphson method

	Procedure 1			Procedure 2	
Iteration	Number of trials required to find $\{\eta_j\}$	Norm†	Iteration	Number of trials required to find $\{\eta_j\}$	Norm‡
1	2	0.149×10^{-1}	1	3	0.258×10^{-1}
2	3	0.186×10^{-1}	2	3	0.192×10^{-1}
3	3	0.167×10^{-1}	3	3	0.117×10^{-1}
4	3	0.684×10^{-2}	4	3	0.854×10^{-2}
5	2	0.315×10^{-2}	5	3	0.571×10^{-2}
6	2	0.177×10^{-2}	6	2	0.568×10^{-2}
7	2	0.714×10^{-3}	7	2	0.373×10^{-2}
8	1	0.210×10^{-3}	8	2	0.125×10^{-2}
9	1	0.440×10^{-4}	9	2	0.783×10^{-3}
10	1	0.576×10^{-5}	10	2	0.486×10^{-3}
11	1	0.260×10^{-6}	11	1	0.292×10^{-3}
			12	1	0.825×10^{-4}
			13	1	0.559×10^{-4}
			14	1	0.647×10^{-5}
			15	1	0.735×10^{-6}

† Norm $= \dfrac{1}{3N} \left[\sum_{j=1}^{N} (F_j^2 + R_j^2 + G_j^2) \right]^{1/2}$ ‡ Norm = same as procedure 1

Execution time = 3.32 s (FORTRAN H compiler)   Execution time = 2.41 s, AMDAHL (FORTRAN H compiler)

8-3 FORMULATION OF AN ALMOST BAND ALGORITHM

For the case where one or more reactions occur on each stage of an absorber or distillation column and the vapor and liquid phases form highly nonideal mixtures, a formulation of the Almost Band Algorithm is recommended. In the present formulation for the case where one or more chemical reactions occur on each stage of an absorber, the following choice of $N(2c + 1 + r)$ independent variables and $N(2c + 1 + r)$ independent functions are made. In particular, for the case of one chemical reaction per stage, the independent variables and functions are taken to be

$$\mathbf{X} = [(l_{j1} \ l_{j2} \ \cdots \ l_{jc} \ v_{j1} \ v_{j2} \ \cdots \ v_{jc} \ T_j \ \eta_j)_{j=1, N}]^T \tag{8-47}$$

$$\mathbf{f} = [(f_{j1} \ f_{j2} \ \cdots \ f_{jc} \ m_{j1} \ m_{j2} \ \cdots \ m_{jc} \ G_j \ R_j)_{j=1, N}]^T \tag{8-48}$$

where the subscript $j = 1, N$ means that the arguments enclosed by parentheses are to be repeated for $j = 1, 2, \ldots, N$. The equilibrium relationships are denoted by the functions $f_{j1}, f_{j2}, \ldots, f_{jc}$, the component-material balance functions by

m_{j1}, m_{j2}, ..., m_{jc}, the enthalpy balance functions by G_j, and the chemical reaction functions by R_j.

For any stage j, the equilibrium functions are given by

$$f_{ji} = \frac{\gamma^L_{ji} K_{ji} l_{ji}}{\displaystyle\sum_{i=1}^{c} l_{ji}} - \frac{\gamma^V_{ji} v_{ji}}{\displaystyle\sum_{i=1}^{c} v_{ji}} \qquad (8\text{-}49)$$

The component-material balance function for any stage j

$$m_{ji} = v_{j+1, i} + l_{j-1, i} - v_{ji} - l_{ji} + \delta_{ji} \qquad (8\text{-}50)$$

The enthalpy balance functions are given by Eq. (8-42). Use of the characteristics of homogeneous functions of degree zero and the approximations presented in Chap. 5 make it possible to regard the mixtures as ideal solutions in the partial differentiation of the G_j's. This amounts to neglecting the dependencies of the enthalpies $\{\hat{H}_{ji}, \hat{h}_{ji}\}$ on compositions. Column scaling followed by row scaling is recommended. It is anticipated that an internal loop would be required. This loop would be similar to the one used above in the $N(r + 2)$ formulation in which the solution set $\{n_j\}$ is found on the basis of the assumed sets $\{l_{ji}\}$, $\{v_{ji}\}$, and $\{T_j\}$.

NOTATION

A_{ji}	$\gamma^V_{ji} L_j / \gamma^L_{ji} K_{ji} V_j$, absorption factor for component i and plate j
A_i	tridiagonal matrix; defined beneath Eq. (2-18)
b_j	assumed value of L_j/V_j
B	tridiagonal matrix; defined beneath Eq. (8-11)
b_i	molar flow rate of component i in the bottom product
B	total molar flow rate of the bottom product
C_{ji}	concentration of component i in the liquid or vapor phase (as indicated in the text) leaving plate j
d_i	molar flow rate of component i in the distillate
D	total molar flow rate of the distillate
f_i	feed vector for component i; defined beneath Eq. (2-18)
F	total molar flow rate of the feed
$g(\theta)$	a function of θ; defined by Eq. (8-21)
$\hat{h}_{ji}$	virtual value of the partial molar enthalpy of component i in the liquid phase leaving plate j
h°_{ji}	$H^\circ_{ji} - \lambda_{ji}$, where H°_{ji} is the enthalpy of pure component i in the perfect gas state at the temperature T_j and one atmosphere pressure above its elements in their standard states at some arbitrary datum temperature [see Eq. (8-26)], and λ_{ji} is the latent heat of vaporization at the temperature T_j as defined below
$\hat{H}_{ji}$	virtual value of the partial molar enthalpy; defined by Eq. (8-28)
H	total enthalpy of one mole of a mixture

H_{ji} partial molar enthalpy of component i in the vapor state; evaluated at the temperature, pressure, and composition of the vapor leaving plate j

$(\Delta H_r)_{liq}$ heat of reaction per mole of the base component A reacted; defined by Eq. (8-36)

$(\Delta H_r^\circ)_{liq}$ heat of reaction per mole of the base component A reacted; defined by Eq. (8-37)

k_j rate constant for the second-order forward reaction; see Eq. (8-12)

k_j' rate constant for the second-order reverse reaction; see Eq. (8-12)

l_{ji} molar flow rate of component i in the liquid phase leaving plate j

l_{Fi} molar flow rate of component i in the liquid part of the feed, $FX_i = v_{Fi} + l_{Fi}$

L_j total molar flow rate at which the liquid phase leaves plate j

L_F total molar flow rate of the liquid part of the feed, $F = V_F + L_F$

n_i molar flow rate of component i

n_T total molar flow rate

p_i ratio of the corrected to the calculated values of d_i, computed by use of the expression given beneath Eq. (8-20)

P total pressure

r_{jA} rate of reaction of the base component A; moles A reacted per unit volume at the conditions of plate j

R_j reaction function for plate j; defined by Eq. (8-16); see also Table 8-1

$\mathbf{R}$ column vector of the reaction functions; defined beneath Eq. (8-17)

v_{ji} molar flow rate at which component i leaves plate j in the vapor phase

$\mathbf{v}_i$ column vector of component flow rates; defined beneath Eq. (2-18)

v_{Fi} molar flow rate of component i in the vapor part of the feed

V_j total molar flow rate at which the vapor phase leaves plate j

$\mathbf{V}$ column vector of total vapor flow rates; defined beneath Eq. (8-11)

x_{ji} mole fraction of component i in the liquid phase leaving plate j

X_i total mole fraction of component i in the feed

y_{ji} mole fraction of component i in the vapor phase leaving plate j

Greek letters

α_{ji} relative volatility of component i at the temperature of plate j; $\alpha_{ji} = K_{ji}/K_{jb}$

γ_{ji}^V activity coefficient of component i in the vapor phase leaving plate j; $\gamma_{ji}^V = \gamma_{ji}^V(P_j, T_j, \{y_{ji}\})$

γ_{ji}^L activity coefficient of component i in the liquid phase leaving plate j; $\gamma_{ji} = \gamma_{ji}(P_j, T_j, \{x_{ji}\})$

δ_{ji} $v_i \eta_j$

Δ_j $\sum\limits_{i=1}^{c} \delta_{ji}$

η_j moles of the base component A reacted per unit time on stage j

$\boldsymbol{\eta}$ column vector of η_j's; defined beneath Eq. (8-8)

ϕ $\sum\limits_{j=1}^{n} \eta_j$

θ a multiplier; defined by Eq. (8-19)

ρ_j^L molar density of the liquid phase leaving plate j

ρ_j^V molar density of the vapor phase leaving plate j

λ_{ji} molar latent heat of vaporization for component i at the temperature of plate j

Ω a function of temperature, pressure and composition, see Eqs. (8-27) and (8-28)

Subscripts

ca calculated value

co corrected value

b base component for computing relative volatilities

i component number

j stage number

Superscripts

L liquid phase

V vapor phase

PROBLEMS

8-1 (a) Obtain the expression for $\partial F_j/\partial \eta_k$

(b) Develop the formulas needed to evaluate $\{\partial v_i/\partial \eta_k\}$ and $\{\partial l_i/\partial \eta_k\}$ by use of the calculus of matrices.

(c) Develop the formulas needed to evaluate $\{\partial V_j/\partial \eta_k\}$ and $\{\partial L_j/\partial \eta_k\}$ by use of the calculus of matrices.

8-2 For the first-order unidirectional reaction which occurs in the liquid phase

$$A \rightarrow R + S$$

develop the expressions for R_j and $\partial R_j/\partial \eta_k$.

8-3 Develop the complete set of equations for the application of the $N(r + 2)$ Newton-Raphson method to conventional distillation columns. In particular obtain the expressions for

(a) component-material balances

(b) total material balances

(c) enthalpy balance of functions

(d) vapor-liquid equilibrium functions

8-4 Develop the expressions given in Table 8-1 for R_j.

REFERENCES

1. H. J. Arnikar, T. S. Rao, and A. A. Bodne: "A Gas Chromatographic Study of the Kinetics of the Uncatalysed Esterification of Acetic Acid by Ethanol," *J. Chromatog.*, **47**:265 (1970).
2. G. Astarita, G. Marrucci, and F. Gioa, "The Influence of Carbonation Ratio and Total Amine Concentration on Carbon Dioxide Absorption in Aqueous Monoethanol-amine Solutions," *Chem. Eng. Sci.*, **19**:95 (1964).

3. J. M. Bennett: "Triangular Factors of Modified Matrices," *Numerische Mathematik*, **1**:217 (1965).
4. G. W. Bentzen, A. Izarraraz, R. G. Anthony, and C. D. Holland: "Algorithms for Simultaneous Distillation and Reaction, 86 National A.I.Ch.E. Meeting, April 1–5, Houston, 1979.
5. C. G. Broyden: "A Class of Methods for Solving Nonlinear Simultaneous Equations," *Math. Comput.* **19**:577 (1965).
6. S. D. Conte and Carl de Boor: *Elementary Numerical Analysis, An Algorithmic Approach*, 2d ed., McGraw-Hill Book Company, New York, 1972.
7. P. V. Danckwerts and K. M. McNeil: "The Absorption of Carbon Dioxide into Aqueous Amine Solutions and the Effects of Catalysis," *Trans. Inst. Chem. Eng.*, **45**:T32 (1967).
8. S. E. Gallun and C. D. Holland: "Solve More Distillation Problems Part 5—For Highly Non-ideal Mixtures," *Hydrocarbon Process.* **55**(1):137 (1976).
9. F. E. Hess and C. D. Holland: "Solve More Distillation Problems, Part 7—Absorber-type Pipestills," *Hydrocarbon Process.* **56**(5):241 (1977).
10. ———, S. E. Gallun, G. W. Bentzen, and C. D. Holland: "Solve More Distillation Problems, Part 8—Which Method to Use," *Hydrocarbon Process.* **56**(6):181 (1977).
11. H. Hikita, S. Asai, H. Ishikawa, and M. Honda: "The Kinetics of Reactions of Carbon Dioxide with Monoethanol-amine, Diethanolamine, and Triethanol-amine by a Rapid Mixing Method," *Chem. Eng. J.* (Printed in the Netherlands), **13**:7 (1977).
12. M. Hirata and H. Komatsu: "Vapor-Liquid Equilibrium Relations Accompanied with Chemical Reaction," *Kagaku Kogaku* (abridged edition), **5**(1):143 (1967).
13. C. D. Holland: *Fundamentals and Modeling of Separation Processes: Absorption, Distillation, Evaporation, and Extraction*, Prentice-Hall, Englewood Cliffs, N.J., 1975.
14. J. Jelinek and V. Hlaváĉek, "Steady-State Countercurrent Equilibrium Stage Separation with Chemical Reaction by Relaxation Method," *Chem. Eng. Sci.*, **2**:79 (1976).
15. R. L. Kent and B. Eisenberg: "Better Data for Amine Treating," *Hydrocarbon Process.* **55**(2):87 (1976).
16. A. L. Kohl and F. C. Riesenfeld, *Gas Purification*, McGraw-Hill Book Company, New York, 1960.
17. H. Komatsu and C. D. Holland: "A New Method of Convergence for Solving Reacting Distillation Problems," *J. Chem. Eng. Jpn.*, **10**(4):292 (1977).
18. Muhlbauer and P. R. Monaghan: "New Equilibrium Data on Sweeting Natural Gas with Ethanolamine Solutions," *Oil Gas. J.*, p. 145, Apr. 29, 1957.
19. P. A. Nelson: "Countercurrent Equilibrium Stage Separation with Reaction," *AIChE J.*, **17**(5):1043 (1971).
20. R. C. Reid, J. M. Prausnitz, and T. K. Sherwood: *The Properties of Gases and Liquids*, McGraw-Hill Book Company, New York, 1977.
21. ——— and T. K. Sherwood: *The Properties of Gases and Liquids*, McGraw-Hill Book Company, New York, 1966.
22. Selected Values of Properties of Hydrocarbon and Related Compounds, API Research Project 44, Thermodynamics Research Center, Texas A&M University.
23. I. Suzuki, H. Komatsu, and M. Hirata, "Formulation and Prediction of Quaternary Vapor-Liquid Equilibria Accompanied by Esterification," *J. Chem. Eng. Jpn.*, **3**(2):152 (1970).
24. ———, H. Yagi, H. Komatsu, and M. Hirata: "Calculation of Multicomponent Distillation Accompanied by a Chemical Reaction," *J. Chem. Eng. Jpn.*, **4**(3):26 (1971).
25. D. S. Viswanath and Gowd-Den Su, "Generalized Thermodynamic Properties of Real Gases," *AIChE J.*, **11**(2):202 (1965).

OPTIMUM DESIGN AND OPERATION OF CONVENTIONAL AND COMPLEX DISTILLATION COLUMNS

Procedures for solving several types of optimization problems commonly encountered in the design and operation of conventional and complex distillation columns are presented. The continued increase in energy costs for operating distillation columns has created the need for rapid calculational procedures both for the design of new distillation columns and for the selection of optimum operating conditions for existing columns. Problems of the following types are solved: (1) determination of the minimum number of stages required to effect a specified separation at a given reflux ratio, (2) optimum economic design of a distillation column, and (3) minimization of the reflux ratio for an existing column by determination of the optimum feed plate location. These problems are solved by use of a modified form of the search procedure called the *complex method*, which was proposed by Box.[2] The primary modification consists of including the constraints in the objective function which reduces the time required to solve a problem by a factor of 1/10 or less. Furthermore, the difficulty of finding feasible solutions (solutions which satisfy the constraints) throughout the domain of the search variables is thereby eliminated. The solution for which the objective function is a minimum is either a feasible solution or is in the near neighborhood of a feasible solution.

After the near neighborhood of the optimum solution has been located, the equations describing the distillation column are solved exactly. These exact solutions may be found by use of any algorithm which one might have available. Thus, this adaptation of the complex method makes it possible to solve a large variety of optimization problems by use of the calculational procedure which is the most efficient one for the particular system under consideration.

One of the first efforts to solve design problems by use of an optimization procedure was made by Srygley and Holland[19] who made use of the Hooke and Jeeves[13] search procedure. The proposed procedure was limited to the determination of the minimum number of stages required to effect a specified separation at a given reflux ratio. Sargent and Gaminibandara[16] considered the more general problem of the optimum configuration of columns needed to effect a specified separation. More recently, an iterative procedure based on the Naphtali and Sandholm[14] formulation of the Newton-Raphson method was proposed by Ricker and Grens[15] for the minimization of the number of stages required to effect a specified separation at a given reflux ratio.

The calculational procedures are presented first for conventional distillation columns and then for complex distillation columns. The conventional distillation column is completely determined by fixing the following variables: (1) the complete definition of the feed (total flow rate, composition, and thermal condition), (2) the column pressure (or the pressure at one point in the column, say in the accumulator), (3) the type of condenser, (4) k_1, the number of plates above and including the feed plate, (5) k_2, the total number of plates, and (6) two other specifications which are usually taken to be the reflux ratio and the distillate rate $\{L_1/D, D\}$ or two product specifications such as $\{b_l/d_l, b_h/d_h\}$, $\{X_{Dl}, x_{Bh}\}$, $\{T_D, T_B\}$, or combinations of these. The subscript l is used to denote the light key and the subscript h is used to denote the heavy key. In all of the optimization problems considered herein, the variables listed in items (1), (2), and (3) are always fixed. For convenience these variables are referred to collectively as the "usual specifications." The remaining four variables, k_1, k_2, and two other specifications such as L_1/D and D are called "additional specifications."

Since all of the variables included in the usual specifications are commonly fixed, only the four variables classified as additional specifications need to be fixed in order to completely determine the column. These four variables may be picked from the set $\{k_1, k_2, L_1/D, D, b_l/d_l, b_h/d_h\}$. Instead of b_l/d_l and b_h/d_h, any other pair of specifications such as those enumerated above may be selected. Since integral numbers of plates are to be used, the specified values for both b_l/d_l and b_h/d_h cannot necessarily be made exactly. However, better separations but not poorer separations of these components would be acceptable, which is expressed in the form of constraints

$$\frac{b_l}{d_l} \leq \left(\frac{b_l}{d_l}\right)_U \qquad \frac{b_h}{d_h} \geq \left(\frac{b_h}{d_h}\right)_L$$

where the subscripts U and L denote the upper and lower bounds, that is, the largest and smallest values of b_l/d_l and b_h/d_h, respectively, which are acceptable.

The light- and heavy-key components are usually but not necessarily adjacent in volatility. The feed may contain components having volatilities lying between those of the keys. The specification of the ratios b_l/d_l and b_h/d_h is equivalent to specifying the molar flow rates or fractional recoveries of the keys in the distillate D and the bottoms B. A material balance on the light key gives

$$FX_l = d_l + b_l = d_l[1 + b_l/d_l] \tag{9-1}$$

Thus

$$d_l = \frac{FX_l}{1 + b_l/d_l} \tag{9-2}$$

and the fractional recovery is given by

$$\frac{d_l}{FX_l} = \frac{1}{1 + b_l/d_l} \tag{9-3}$$

Expressions of the same form as those shown above are obtained for the heavy key by replacing the subscript l by the subscript h. For definiteness in the formulation of the optimization problem, the two product specifications are taken to be b_l/d_l and b_h/d_h (or d_l and d_h).

Let the total number of stages be denoted by N which includes the reboiler and the condenser, partial or total. The number of plates is then equal to $N - 2$ which is also equal to k_2 for a conventional column. The upper and lower bounds on k_1 are denoted by k_{1U} and k_{1L}, respectively. Similarly, the upper and lower bounds on the total number of plates are denoted by k_{2U} and k_{2L}, respectively. For example, values of k_2 greater than 50 or 100 are seldom encountered, and the designer might well take $k_{1U} = k_{2U} = 100$ (or 200). The lower bounds might well be taken to be $k_{1L} = k_{2L} = 1$.

The proposed application of the complex method makes it possible to use any existing calculational procedure for solving the equations for a distillation column exactly. However, since repetitive calculations are involved in the search technique, computer time can be conserved by use of the calculational procedure which is most efficient for a given type of column. In a series of papers,[4, 5, 8, 9] comparisons of the computer times required to solve a wide variety of numerical examples by use of the θ method and formulations of the Newton-Raphson method are presented. The following conclusions were reached. (1) For distillation-type columns in the service of separating ideal or near ideal solutions, the θ method is more efficient than the Newton-Raphson formulations; (2) for absorber-type columns in the service of separating ideal or near ideal solutions, the formulation of the Newton-Raphson method in terms of $2N$ independent variables (where N is equal to the number of stages) is recommended; (3) for columns of all types in the service of separated highly nonideal solutions, the Almost Band Algorithm presented in Chap. 5 is recommended.[8]

9-1 PROCEDURE 1. DETERMINATION OF THE MINIMUM NUMBER OF STAGES REQUIRED TO EFFECT A SPECIFIED SEPARATION AT A GIVEN REFLUX RATIO FOR CONVENTIONAL AND COMPLEX DISTILLATION COLUMNS

The development is presented first for conventional distillation columns and then for complex columns.

Conventional Distillation Columns

In the formulation of this problem it is supposed that the variables listed as usual specifications are fixed. Of the four remaining variables required to completely define the column, three are fixed, the purity specifications $\{b_l/d_l, b_h/d_h\}$ and the reflux ratio L_1/D. Thus, one additional variable remains to be fixed in order to determine the column. Consequently, the problem to be solved consists of finding the feed plate location k_1 which minimizes the total number of plates k_2 (see Fig. 9-1) required to achieve the purity specification at the specified value of the reflux ratio. A concise statement of the problem follows:

1. Specifications: $(b_l/d_l)_U$, $(b_h/d_h)_L$, and L_1/D
2. Constraints: $b_l/d_l \le (b_l/d_l)_U$, $b_h/d_h \ge (b_h/d_h)_L$

Thus, the problem is to find the values of k_1 and k_2 (where $k_2 = N - 2$) which

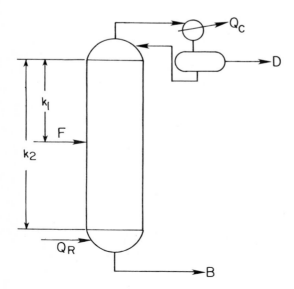

Figure 9-1 A conventional distillation column.

minimize the total number of stages N. The objective function O to be minimized is then

$$O = N - 2 \tag{9-4}$$

subject to the constraints

$$b_l/d_l \leq (b_l/d_l)_U \qquad b_h/d_h \geq (b_h/d_h)_L \tag{9-5}$$

and the limits

$$k_{1L} \leq k_1 \leq k_{1U} \qquad k_{2L} \leq k_2 \leq k_{2U} \qquad D_L \leq D \leq D_U \tag{9-6}$$

The lower bound on the distillate rate D is

$$D_L = d_l + d_h \tag{9-7}$$

where d_l and d_h are computed by use of the specified values of b_l/d_l and b_h/d_h and the component-material balances; see Eq. (9-2). The upper bound on D is, of course, equal to the total feed rate F, that is,

$$D_U = F \tag{9-8}$$

For definiteness, take $k_{1U} = k_{2U} = 200$, and $k_{1L} = k_{2L} = 2$.

The problem may be solved by use of the complex method of Box[2] which is based on the reduction of a simplex of solutions which satisfy the constraints. Finding such solutions can prove to be difficult, thereby requiring an excessive amount of computer time. To avoid this difficulty, the well-known technique of including the constraints in the objective function was employed. Several forms of the objective function including the one proposed by Srygley and Holland[19] were investigated. The best of these consisted of including the constraints as an additive term as follows

$$O = \frac{1}{f_1}[N - 2]^2 + \frac{1}{f_2}(s_l^2 + s_h^2) \tag{9-9}$$

where

$$s_l = (b_l/d_l)/(b_l/d_l)_U \qquad s_h = (b_h/d_h)_L/(b_h/d_h)$$

The factors f_1 and f_2 make it possible to place different weights on the number of plates $(N - 2)$ and the constraints. In the initial search, f_1 and f_2 were taken equal to a fraction or a multiple of the average of the simplex values of their respective numerators, and in the final search, f_1 was taken equal to a fraction or a multiple of the average of the simplex values of its respective numerator and f_2 was set equal to unity. This choice of f_2 tended to drive the search either into the neighborhood or the region of feasible solutions.

Use of the square of the terms shown in Eq. (9-9) tended to reduce the number of trials required to minimize the function O. In the initial search, Eq. (9-9) was used as stated, and in the final search it was used as stated for $s_l > 1$ and $s_h > 1$. For either $s_l < 1$ or $s_h < 1$, the term s_l or s_h was replaced by either $\ln s_l$ or $\ln s_h$. Use of this variable form of the function places a penalty on making separations which are better than those specified and tends to drive the objective function toward equality constraints.

In summary then, the function O given by Eq. (9-9) is to be searched for its minimum value over the variables $\{k_1, k_2, D\}$. The search is carried out by use of the following procedure.

Initial Search

In the initial search, approximate solutions are obtained to the equations describing a distillation column; see App. 9-2. This procedure makes use of component-material balances and equilibrium relationships, and it reflects fairly accurately the effect of varying k_1 and k_2 on the separations $\{b_l/d_l, b_h/d_h\}$. In the initial search, the minimum value of O is determined for the specified value of the reflux ratio. Also in the initial search, the total distillate rate D is taken to be dependent on N, and it is estimated as described in App. 9-2. Thus, the function O is searched over only two variables, k_1 and k_2, as follows:

Step 1. Determine the feasibility of the purity specifications (the constraints) at $N = N_U$ at total reflux. Use the Fenske equation as described in App. 9-1. If it is impossible to satisfy the constraints at total reflux, calculations are ceased. If the specified separations cannot be made at total reflux, they cannot be made, of course, at any operating reflux less than total. If the separations are feasible (all constraints are satisfied) at $N = N^L$ at total reflux, go to step 2.

Step 2. Search the objective function O over k_1 and k_2 by use of a modified version of the complex method by Box[2] which is described in App. 9-3. Commence by finding five solutions which define the initial simplex. To find the initial solution, pick arbitrary values of k_1 and k_2 lying between the upper and lower bounds as follows

$$k_1 = r_1 k_2 \tag{9-10}$$

$$k_2 = k_{2,L} + r_2(k_{2U} - k_{2L}) \tag{9-11}$$

where r_1 and r_2 are random numbers lying between 0 and 1. First k_2 is found by use of Eq. (9-10), and then k_1 is found by use of Eq. (9-10). For this set of values of k_1 and k_2, take D_1 to be equal to the values predicted by the total reflux models as described in App. 9-2. For each of the remaining solutions, the values of k_1 and k_2 are selected by use of random numbers as described above and the corresponding values of D are selected as described in App. 9-2.

Step 3. Search the function O by use of the complex method (see App. 9-3) over the variables k_1 and k_2 with the value of D being determined by the choice of k_1 and k_2 as described in App. 9-2.

Final Search

In the final search, exact solutions of the equations describing the distillation column are used. The objective function is searched over the variables which are most conveniently fixed in the particular calculational procedure used to solve the equations for the distillation column. If the θ method is used, the search variables are taken to be D, k_1, and k_2. On the other hand, if the $2N$ Newton-Raphson method is used, the search variables are taken to be V_N/B, k_1, and k_2.

In the final search, the function O is searched over the variables D, k_1, and k_2. The final search is initiated by finding a solution by an exact calculational procedure for each vertex D, k_1, and k_2 determined by the initial search. After the function O has been evaluated at each vertex, the complex method of Box is employed as described in App. 9-3.

In order to demonstrate the use of this procedure, Example 9-1 is presented. The statement of this example is given in Tables 9-1 and 9-2. Results of the initial and final searches are presented in Table 9-3, and the final solution appears in Table 9-4. The term "iteration" which appears in Table 9-3 is used to

Table 9-1 Common data and specifications for Examples 9-1 through 9-3

Component no.	Component	Feed composition X_i
1	Toluene	0.008
2	Ethyl benzene	0.5104
3	Styrene	0.4777
4	Isopropyl benzene	0.0005
5	1-methyl-3 ethyl benzene	0.0001
6	2-methyl styrene	0.0013
7	cis-1 propeneyl benzene	0.0020

Other specifications

The column has a total condenser which is to be operated at 40 mmHg. The pressure drop per stage may be taken to be 4.69 mmHg. The flow rate of the feed is 220.55 lb mol/h, and before entering the column, it is a liquid at its bubble-point temperature of 572°R at a pressure of 270 mmHg. The vapor-liquid equilibrium data and the enthalpy data are given in Tables B-3 and B-4.

Table 9-2 Statement of Example 9-1

The feed given in Table 9-1 is to be separated into two fractions by use of a conventional distillation column. In particular, it is desired to find the smallest number of plates required to effect the following separation between ethyl benzene (the light key) and styrene (the heavy key) at a reflux ratio $L_1/D = 5.44$.

$$b_l/d_l \leq (b_l/d_l)_U = 0.1248$$
$$b_h/d_h \geq (b_h/d_h)_L = 7.201$$

All other specifications pertaining to the feed, the column, and the data are given in Table 9-1.

Table 9-3 Initial and final simplexes for Example 9-1

Initial search

Initial simplex

Vertices

Variables	1	2	3	4	5
k_1	64	76	2	47	23
k_2	183	114	12	160	30
L_1/D	5.44	5.44	5.44	5.44	5.44
D	114	115	115	114	115
O [Eq. (9-9)]	2.24×10^{50}	4.58×10^{42}	5.12×10^3	1.35×10^{32}	9.08×10^2
b_l/d_l	1.87×10^{24}	2.67×10^{20}	3.62×10^{-3}	1.45×10^{15}	1.83×10^{-2}
b_h/d_h	4.39×10^{34}	1.24×10^{31}	0.102	1.08×10^{22}	3.54

Final simplex

Vertices

Variables	1	2	3	4	5
k_1	20	19	19	19	19
k_2	31	31	31	31	31
L_1/D	5.44	5.44	5.44	5.44	5.44
D	115	115	115	115	115
O [Eq. (9-9)]	1.97	2.01	2.01	2.01	2.01
b_l/d_l	1.77×10^{-2}	1.84×10^{-2}	1.84×10^{-2}	1.84×10^{-2}	1.84×10^{-2}
b_h/d_h	14.3	17.8	17.8	17.8	17.8

64 iterations were required to obtain the final simplex.

Final search

Initial simplex

Vertices

Variables	1	2	3	4	5
k_1	20	19	20	20	20
k_2	31	31	31	31	31
L_1/D	5.44	5.44	5.44	5.44	5.44
D	115	107	121	118	112
O [Eq. (9-9)]	0.246	2.50	1.93	1.49	1.53
b_l/d_l	0.116	0.185	8.43×10^{-2}	0.101	0.140
b_h/d_h	7.76	9.36	5.86	6.63	8.21

Table 9-3 (*continued*)

| | Final simplex | | | | |
| | Vertices | | | | |
Variables	1	2	3	4	5
k_1	20	20	20	20	20
k_2	31	33	33	32	32
L_1/D	5.44	5.44	5.44	5.44	5.44
D	115	115	115	115	115
O [Eq. (9-9)]	0.242	0.267	0.266	1.30	1.30
b_l/d_l	0.116	0.116	0.117	0.124	0.124
b_h/d_h	7.76	7.27	7.30	7.00	7.00

21 iterations were required to obtain the final simplex, and 6.09 seconds of computer time, AMDAHL (FORTRAN H with OPT2) were required for the initial and final searches.

denote the number of objective function evaluations. In order to reduce the time requirements of the final search, the complete set of equations needed to obtain an exact solution by the θ method were not iterated to convergence for each set of operating conditions. The criteria used were $|1 - \theta| < 0.01$ and $|\theta_{n+1} - \theta_n| < 0.01$ where n is the trial number. Generally, fewer than five trials were required to satisfy this condition. After having satisfied these conditions, the objective function was evaluated. This trial procedure was repeated until a simplex was found for which the standard deviation of the objective function was ≤ 0.005. In the selection of new points for the simplex, the same value of the normalization factors was used in the evaluation of the objective functions. After the optimum set of vertices had been found (see set 2 of Table 9-3), the problem was converged to $|1 - \theta| \leq 10^{-5}$ and $|(V_{j, n+1} - V_{j, n})/V_{j, n}| < 10^{-3}$. In the event that the final solution did not satisfy the constraints, the number of plates was increased by increments of one until a solution satisfying the constraints was obtained. Seldom was it necessary to add more than one plate.

As shown in Table 9-3, a total of 6 seconds of computer time were required to solve Example 9-1. To solve this same problem when it was uniquely specified (the number of plates, the feed-plate location, the reflux, and distillate rates are fixed) required 0.5 seconds of computer time by the θ method. Thus, Example 9-1 required about 12 times as much computer time to solve as was required to solve a typical distillation problem in which the plates, configuration, the reflux, and distillate rates are fixed. Since the θ method is 10 to 20 times faster than Newton-Raphson procedures, the proposed optimization procedure requires about the same amount of computer time as is required to solve a problem involving a fixed column by a Newton-Raphson procedure.

Table 9-4 Solution of Examples 9-1, 9-2 and 9-3

I. Pressure, temperature, vapor, and liquid rate profiles

Stage	T_j (°F)	P_j (mmHg)	L_j (lb mol/h)	V_j (lb mol/h)
1 (condenser)	126.91	39.98	621.12	114.07 (distillate)
2	136.28	44.68	604.61	735.19
3	140.66	49.38	598.18	718.68
4	144.67	54.07	592.75	712.25
5	148.44	58.77	587.86	706.82
6	151.99	63.46	583.24	701.93
7	155.35	68.16	578.63	697.31
8	158.52	72.85	573.54	692.70
9	161.50	77.55	566.76	687.61
10 (feed)	164.32	82.24	749.52	680.83
11	166.84	86.94	744.20	643.04
12	169.23	91.63	739.28	637.72
13	171.52	96.33	734.66	632.80
14	173.73	101.02	730.29	628.18
15	175.87	105.72	726.14	623.81
16	177.96	110.41	722.17	619.65
17	179.99	115.11	718.39	615.69
18	181.99	119.80	714.78	611.91
19	183.95	124.50	711.32	608.30
20	185.88	129.19	708.01	604.84
21	187.79	133.89	704.84	601.53
22	189.69	138.59	701.80	598.36
23	191.58	143.28	698.88	595.32
24	193.46	147.97	696.07	592.40
25	195.34	152.67	693.36	589.59
26	197.21	157.37	690.71	586.87
27	199.07	162.06	688.09	584.23
28	200.92	166.76	685.38	581.61
29	202.76	171.45	682.31	578.90
30	204.59	176.15	678.18	575.83
31	206.39	180.84	671.08	571.70
32	208.19	185.54	655.62	564.59
33 (reboiler)	209.97	190.23	106.48 (bottoms)	549.40

II. Product distribution

Component	b_i (lb mol/h)	d_i (lb mol/h)	b_i/d_i
1	7.1×10^{-10}	1.76	4.022×10^{-10}
2 (light key)	12.4	100.0	0.1233
3 (heavy key)	93.3	12.1	7.7140
4	0.11	1.34×10^{-3}	81.2981
5	2.2×10^{-2}	9.26×10^{-6}	2.3835×10^3
6	0.28	2.05×10^{-5}	1.3938×10^4
7	0.44	4.26×10^{-7}	1.0335×10^6
	$B = 106.27$	$D = 113.86$	

Procedure 1. Determination of the Minimum Number of Stages Required to Effect a Specified Separation at a Given Reflux Ratio for a Complex Distillation Column

For definiteness, consider the complex column shown in Fig. 9-2 which has a sidestream W withdrawn in addition to the distillate and bottoms. For a complex column, k_1 is equal to the number of plates above and inclusive of the plate from which the sidestream W is withdrawn, k_2 is equal to the number of plates above and inclusive of the plate on which the feed F enters, and k_3 denotes the total number of plates in the column, $k_3 = N - 2$.

The withdrawal of this sidestream introduces two new variables which may be specified such as the stage number $k_1 + 1$ of the sidestream withdrawal, and its flow rate W. Instead of k_1 and W, any other two product specifications may be made such as $\{w_r/d_r, w_s/d_s\}$, $\{x_{Ws}, x_{Wr}\}$, and $\{T_W, x_{Wr}\}$ as shown in Table 9-5. Components r and s consist of any two components of the feed which are selected in the order of increasing volatility; s is more volatile than r. In the following application of Box's complex method, it will be supposed that the additional specifications for the stream W are $\{w_s/d_s, w_r/d_r\}$. The complete set of additional purity specifications are taken to be

$$\{b_l/d_l, b_h/d_h, w_s/d_s, w_r/d_r\}$$

as shown by set 2 of item I of Table 9-5. At the specified reflux ratio L_1/D, the total number of plates $N - 2$ is to be minimized subject to the purity constraints. When the purity constraints are included in the objective function, one obtains

$$O = \frac{1}{f_1}[N - 2]^2 + \frac{1}{f_2}[s_l^2 + s_h^2 + s_r^2 + s_s^2] \qquad (9\text{-}12)$$

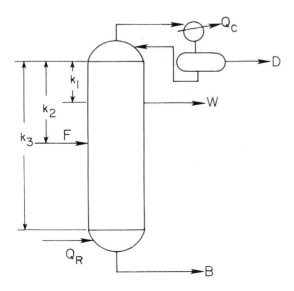

Figure 9-2 A complex distillation column with one sidestream.

Table 9-5 Various sets of specifications for complex columns

I. Complex column with one sidestream (Fig. 9-2)

Set	Additional specifications						Objective	Procedure
	1	2	3	4	5	6		
1	L_1/D	D	W	k_1	k_2	k_3	(Uniquely determined system)	(θ method)
2	L_1/D	b_l/d_l	b_h/d_h	w_s/d_s	w_r/d_r	...	min $(N-2)$	1
3	......	b_l/d_l	b_h/d_h	w_s/d_s	w_r/d_r	...	min (cost per mole of D)	2
4	N	b_l/d_l	b_h/d_h	w_s/d_s	w_r/d_r	...	min (L_1/D)	3
5	L_1/D	X_D	x_{Bh}	x_{Ws}	x_{Wr}	...	min $(N-2)$	1
6	L_1/D	T_D	T_B	T_W	x_{Wr}	...	min $(N-2)$	1

II. Complex column with two feeds (Fig. 9-3)

Set	Additional specifications					Objective	Procedure
	1	2	3	4	5		
1	L_1/D	D	k_1	k_2	k_3	(Uniquely determined system)	(θ method)
2	L_1/D	b_l/d_l	b_h/d_h	...	...	min $(N-2)$	1
3	......	b_l/d_l	b_h/d_h	...	...	min (cost per mole of D)	2
4	N	b_l/d_l	b_h/d_h	...	...	min (L_1/D)	3

Table 9-5 (*continued*)

III. Complex column with two sidestreams (Fig. 9-4)

Set	1	2	3	4	5	6	7	8	Objective	Procedure
1	L_1/D	D	W_1	W_2	k_1	k_2	k_3	k_4	(Uniquely determined system)	(θ method)
2	L_1/D	b_l/d_l	b_h/d_h	w_l/d_l	w_h/d_h	w_s/d_s	w_r/d_r	...	min $(N-2)$	1
3	......	b_l/d_l	b_h/d_h	w_l/d_l	w_h/d_h	w_s/d_s	w_r/d_r	...	min (cost per mole of D)	2
4	N	b_l/d_l	b_h/d_h	w_l/d_l	w_h/d_h	w_s/d_s	w_r/d_r	...	min (L_1/D)	3

IV. Complex column with two sidestreams and two feeds (Fig. 9-5)

Set	1	2	3	4	5	6	7	8	9	Objective	Procedure
1	L_1/D	D	W_1	W_2	k_1	k_2	k_3	k_4	k_5	(Uniquely determined system)	(θ method)
2	L_1/D	b_l/d_l	b_h/d_h	w_l/d_l	w_h/d_h	w_s/d_s	w_r/d_r	...	...	min $(N-2)$	1
3	......	b_l/d_l	b_h/d_h	w_l/d_l	w_h/d_h	w_s/d_s	w_r/d_r	...	...	min (cost per mole of D)	2
4	N	b_l/d_l	b_h/d_h	w_l/d_l	w_h/d_h	w_s/d_s	w_r/d_r	...	...	min (L_1/D)	3

where

$$s_l = (b_l/d_l)/(b_l/d_l)_U \qquad s_h = (b_h/d_h)_L/(b_h/d_h)$$

$$s_s = (w_s/d_s)/(w_s/d_s)_U \qquad s_r = (w_r/d_r)_L/(w_r/d_r)$$

which is subject to the limits

$$k_{1L} \le k_1 \le k_{1U} \qquad k_{2L} \le k_2 \le k_{2U} \qquad k_{3L} \le k_3 \le k_{3U}$$

$$D_L \le D \le D_U \qquad W_L \le W \le W_U \tag{9-13}$$

The following lower bounds were selected: $k_{1L} = k_{2L} = k_{3L} = 3$. The following upper bounds were selected: $k_{1U} = k_{2U} = k_{3U} = 200$.

The upper and lower bounds on D and W are established by use of the results obtained by finding the solutions at total reflux at the maximum and minimum number of plates as described in App. 9-1. The lower bounds on D and W were taken to be a fraction of the smaller values found at total reflux. Similarly, the upper bounds on D and W were taken to be a fraction larger than their corresponding values at total reflux. For the examples presented, the multipliers of 0.9 and 1.1 were used to compute the lower and upper bounds, respectively.

The calculational procedure employed is the same as the one described for conventional distillation columns except for the difference in the search variables and the objective function. In the initial search, the independent or search variables are $\{k_1, k_2, k_3\}$ and the dependent variables are $\{D, W\}$ as shown in Table 9-6. To find the initial vertex, pick an arbitrary value of k_3 lying between the upper and lower bounds as follows

$$k_3 = k_{3L} + r_3(k_{3U} - k_{3L})$$

where r_3 is a random number lying between 0 and 1. On the basis of this value of k_3, the values of k_1 and k_2 are obtained in the following manner

$$k_1 = r_1 k_3$$

$$k_2 = r_2 k_3$$

where r_1 and r_2 are random numbers lying between 0 and 1. The corresponding values of the dependent variables were selected as described in App. 9-2. The objective function given by Eq. (9-12) is used in both the initial and final searches.

In the final search, the function O is searched over $\{k_1, k_2, k_3, D, W\}$. The final search is initiated by finding a solution to the equations for the complex column by an exact calculational procedure for a set of vertices which are obtained by use of a 10 percent perturbation (by random numbers) about the final vertices $\{k_1, k_2, k_3, D, W\}$ of the initial search. The remainder of the calculational procedure is analogous to that described for conventional columns.

Procedure 1 is readily generalized for problems involving other types of

Table 9-6 Search variables

I. Procedure 1 and specification set 2 of items 1, 2, 3, 4 of Table 9-5

Column	Independent variables initial search	Dependent variables initial search	Independent variables final search
Fig. 9-2	k_1, k_2, k_3	D, W	k_1, k_2, k_3, D, W
Fig. 9-3	k_1, k_2, k_3	D	k_1, k_2, k_3, D
Fig. 9-4	k_1, k_2, k_3, k_4	D, W_1, W_2	$k_1, k_2, k_3, k_4, D, W_1, W_2$
Fig. 9-5	k_1, k_2, k_3, k_4, k_5	D, W_1, W_2	$k_1, k_2, k_3, k_4, k_5, D, W_1, W_2$

II. Procedure 2 and specification set 3 of items 1, 2, 3, 4 of Table 9-5

Column	Independent variables initial search	Dependent variables initial search	Independent variables final search
Fig. 9-2	$k_1, k_2, k_3^\dagger$	D, W	k_1, k_2, k_3, D, W
Fig. 9-3	$k_1, k_2, k_3^\dagger$	D	k_1, k_2, k_3, D
Fig. 9-4	$k_1, k_2, k_3, k_4^\dagger$	D, W_1, W_2	$k_1, k_2, k_3, k_4, D, W_1, W_2$
Fig. 9-5	$k_1, k_2, k_3, k_4, k_5^\dagger$	D, W_1, W_2	$k_1, k_2, k_3, k_4, k_5, D, W_1, W_2$

III. Procedure 3 and specification set 4 of items 1, 2, 3 and 4 of Table 9-5

Column	Independent variables initial search	Dependent variables initial search	Independent variables final search
Fig. 9-2	$k_1, k_2, L_1/D$	k_3, D, W	$k_1, k_2, D, W, L_1/D$ (k_3 is fixed, $k_3 = N - 2$)
Fig. 9-3	$k_1, k_2, L_1/D$	k_3, D	$k_1, k_2, D, L_1/D$ (k_3 is fixed)
Fig. 9-4	$k_1, k_2, k_3, L_1/D$	D, W_1, W_2, k_4	$k_1, k_2, k_3, D, W_1, W_2, L_1/D$ (k_4 is fixed)
Fig. 9-5	$k_1, k_2, k_3, k_4, L_1/D$	D, W_1, W_2, k_5	$k_1, k_2, k_3, k_4, D, W_1, W_2, L_1/D$ (k_5 is fixed)

† For the second part of the initial search L_1/D also is an independent variable

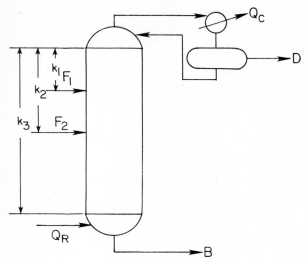

Figure 9-3 A complex distillation column with two feeds.

columns such as those shown in Figs. 9-3, 9-4, and 9-5. For specification set 2 of items II, III, and IV of Table 9-5, the generalized form of the objective function given by Eq. (9-12) becomes

$$O = \frac{1}{f_1}[N - 2]^2 + \frac{1}{f_2}\left[\sum_{i=1}^{2n+2} s_i^2\right] \tag{9-14}$$

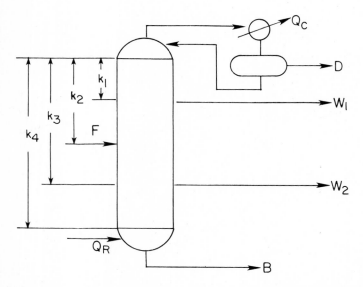

Figure 9-4 A complex distillation column with one feed and two sidestreams.

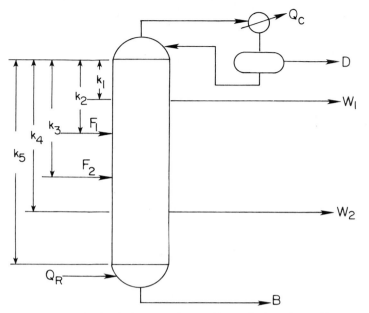

Figure 9-5 A complex distillation column with two feeds and two sidestreams.

where n is equal to the number of sidestreams withdrawn. The independent variables and dependent variables in the initial search and the independent variables of the final search are listed in Table 9-6.

9-2 PROCEDURE 2. OPTIMUM ECONOMIC DESIGN OF CONVENTIONAL AND COMPLEX DISTILLATION COLUMNS

Conventional Distillation Columns

In the formulation of this problem, it is supposed that in addition to the usual specifications, two purity specifications $(b_l/d_l$ and $b_h/d_h)$ are made, and that it is required to find the feed-plate location [plate number $(k_1 + 1)$], the total number of plates k_2, and the corresponding values of the operating variables L_1/D and D that minimize the operating costs and capital costs per mole of the most valuable product $(D$ or $B)$. Since only two of the four additional variables required to define the column are specified, many solutions may be obtained by making different choices for two of the four remaining variables. The best choice for the remaining variables has been made when the objective function is minimized.

When the purity constraints are included in the objective function, the problem to be solved may be stated as follows. The economic objective function

$$O = \frac{1}{f_1}[\{(1 + R)C_c + C_o\}/D]^2 + \frac{1}{f_2}[s_h^2 + s_l^2] \qquad (9\text{-}15)$$

is to be minimized subject to the limits

$$k_{1L} \leq k_1 \leq k_{1U} \qquad k_{2L} \leq k_2 \leq k_{2U}$$
$$D_L \leq D \leq D_U \qquad 0 < L_1/D < \infty \tag{9-16}$$

The capital and operating costs are denoted by C_c and C_o, respectively, and R is the fractional return on investment. If the bottom product B is more valuable than the top product D, then D is replaced by B in the objective function O given by Eq. (9-15). Alternatively, D could be replaced by the sum $c_1 D + c_2 B$, where c_1 and c_2 are the relative values of the distillate and bottoms. The formulation of the first term or the economic part of the objective function O is described in App. 9-4. Again the weight factors were selected as described in procedure 1. Also, the constraints s_i ($i = h, l$) were replaced by $(\ln s_i)$ for values of s_i less than unity. The objective function given by Eq. (9-15) was minimized by use of the following calculational procedure.

Initial Search

In the first initial search, the objective function, given by Eq. (9-1) is searched over the independent variables $\{k_1, k_2\}$ at some arbitrarily selected value of L_1/D in the same manner as described in procedure 1. In this first initial search, D is taken as a dependent variable. After the minimum number of stages N has been found by this search, a second initial search is made with the total number of stages fixed at this value of N. In the second initial search, the economic objective function given by Eq. (9-15) is searched over $\{k_1, L_1/D\}$ with k_2 and D being dependent. Note $k_2 = N - 2$, and D depends upon the fixed value of N.

The final search is initiated by finding exact solutions at perturbations of the vertices $\{k_1, k_2, D, L_1/D\}$ found by the second initial search. Then the economic objective function given by Eq. (9-15) is searched over these variables by use of the complex method of Box. To demonstrate the application of this procedure, Example 9-2 is presented in Table 9-7. The results of the initial and final searches are presented in Table 9-8.

Table 9-7 Statement of Example 9-2

Design a conventional distillation column to process the feed stream whose composition and thermal condition are stated in Table 9-1. The following separation is to be performed at the least possible cost per mole of distillate product.

$$b_l/d_l \leq (b_l/d_l)_U = 0.1248$$
$$b_h/d_h \geq (b_h/d_h)_L = 7.201$$

Use the design equations and cost data given in App. 9-4. Other design data and costs needed are as follows:

Overall heat transfer coefficient of condenser = 250 Btu/(h ft °F).
Cooling water at 55°F is available at a cost of $0.295/M gal.
Saturated steam at 490°F is available $2.50/MM Btu.

Table 9-8 Initial and final searches for Example 9-2

First initial search (at a fixed reflux ratio $L_1/D = 4$)

| | Initial simplex | | | | |
| | Vertices | | | | |
Variables	1	2	3	4	5
k_1	64	76	2	47	23
k_2	183	114	12	160	30
D	114	115	115	114	115
L_1/D	4.00	4.00	4.00	4.00	4.00
O [Eq. (9-9)]	2.66×10^{47}	1.63×10^{39}	3.14×10^3	1.04×10^{30}	9.2×10^2
b_l/d_l	6.44×10^{22}	5.04×10^{18}	4.88×10^{-3}	1.27×10^{14}	2.49×10^{-2}
b_h/d_h	1.33×10^{32}	2.00×10^{29}	0.132	8.30×10^{20}	1.74

| | Final simplex | | | | |
| | Vertices | | | | |
Variables	1	2	3	4	5
k_1	4	4	4	4	4
k_2	28	28	28	28	28
D	115	115	115	115	115
L_1/D	4.00	4.00	4.00	4.00	4.00
O [Eq. (9-9)]	2.00	2.00	2.00	2.00	2.00
b_l/d_l	9.99×10^{-2}	9.99×10^{-2}	9.99×10^{-2}	9.99×10^{-2}	9.99×10^{-2}
b_h/d_h	7.39	7.39	7.39	7.39	7.39

45 iterations were required to obtain the final simplex.

Second initial search (at a fixed $N = 30$)

| | Initial simplex | | | | |
| | Vertices | | | | |
Variables	1	2	3	4	5
k_1	4	5	17	18	24
k_2	28	28	28	28	28
D	115	115	115	115	115
L_1/D	4.00	4.25	4.50	4.75	5.00
Diameter (ft)	6	6	7	7	8
Cost, \$/mol D	8.95×10^{-2}	9.28×10^{-2}	9.09×10^{-2}	9.2×10^{-2}	8.58×10^{-2}
O [Eq. (9-15)]	80.2	86.5	92.1	97.6	190
b_l/d_l	9.99×10^{-2}	7.15×10^{-2}	1.18×10^{-2}	1.12×10^{-2}	2.84×10^{-2}
b_h/d_h	7.39	9.54	3.28	2.69	0.675

(Continued on page 318.)

Table 9-8 (*continued*)

Variables	1	2	3	4	5
			Final simplex		
			Vertices		
k_1	6	6	7	7	6
k_2	28	28	28	28	28
D	115	115	115	115	115
L_1/D	3.08	3.14	2.78	2.82	3.17
Diameter (ft)	6	5	5	5	5
Cost, \$/mol D	7.81×10^{-2}	7.39×10^{-2}	7.17×10^{-2}	7.22×10^{-2}	7.38×10^{-2}
O [Eq. (9-15)]	0.327	0.353	0.289	0.299	0.350
b_l/d_l	0.125	0.108	0.124	0.120	0.108
b_h/d_h	8.98	9.59	9.09	9.19	9.57

25 iterations were required to obtain the final simplex.

Final search

Variables	1	2	3	4	5
			Initial simplex		
			Vertices		
k_1	6	6	8	7	6
k_2	28	30	30	30	28
D	115	108	118	119	107
L_1/D	3.08	3.33	3.58	3.83	4.08
Diameter (ft)	5	5	5	5	5
Cost, \$/mol D	6.69×10^{-2}	7.24×10^{-2}	7.47×10^{-2}	7.80×10^{-2}	8.11×10^{-2}
O [Eq. (9-15)]	10.4	11.7	6.10	5.70	9.63
b_l/d_l	0.282	0.345	0.188	0.173	0.321
b_h/d_h	3.20	3.66	3.81	3.86	4.38

Variables	1	2	3	4	5
			Final simplex		
			Vertices		
k_1	10	10	10	10	10
k_2	33	33	33	33	33
D	114	114	114	114	114
L_1/D	5.44	5.45	5.43	5.45	5.46
Diameter (ft)	6	6	6	6	6
Cost, \$/mol D	0.106	0.106	0.106	0.106	0.106
O [Eq. (9-15)]	0.225	0.255	0.253	0.255	0.257
b_l/d_l	0.125	0.123	0.124	0.123	0.122
b_h/d_h	7.73	7.72	7.67	7.71	7.73

48 iterations were required to obtain the final solutions, and 14.6 seconds of computer time, AMDAHL (FORTRAN H, OPT 2) were required for the initial and final searches.

Procedure 2. Optimum Economic Design of a Complex Distillation Column

In the formulation of a typical problem of this type, consider again the complex column shown in Fig. 9-2, and suppose that the purity specifications or constraints are taken to be $(b_h/d_h)_L$, $(b_l/d_l)_U$, $(w_r/d_r)_L$, and $(w_s/d_s)_U$. Then the problem to be solved consists of finding the number of plates k_1, k_2, and k_3 and the corresponding values of the operating variables L_1/D, D, and W that minimize the operating costs and capital costs per mole of the most valuable product (D, B, or W). When the purity constraints are included in the objective function, the problem to be solved may be stated as follows. The economic objective function

$$O = \frac{1}{f_1}[\{(1 + R)C_c + C_o\}/D]^2 + \frac{1}{f_2}[s_l^2 + s_h^2 + s_r^2 + s_s^2] \tag{9-17}$$

is to be minimized subject to the limits

$$
\begin{array}{ccc}
k_{1L} \le k_1 \le k_{1U} & k_{2L} \le k_2 \le k_{2U} & k_{3L} \le k_3 \le k_{3U} \\
W_L \le W \le W_U & D_L \le D \le D_U & 0 < L_1/D < \infty
\end{array}
\tag{9-18}
$$

The initial and final searches are carried out in a manner analogous to that described for conventional distillation columns. In the first initial search, the objective function given by Eq. (9-12) is searched over the independent variables $\{k_1, k_2, k_3\}$ at an arbitrarily assumed reflux ratio L_1/D as described in procedure 1 for complex columns.

After the minimum number of stages has been found by this search, a second initial search is made with the total number of stages held fixed at this value of N. In the second initial search, the objective function given by Eq. (9-17) is searched over the variables $\{k_1, k_2, D, W, L_1/D\}$ with the value of the dependent variable k_3 being fixed by the value of N found in the initial search ($k_3 = N - 2$).

In the final search, the objective function given by Eq. (9-17) is searched over the variables $\{k_1, k_2, k_3, D, W, L_1/D\}$ in a manner analogous to that described in procedure 2 for conventional columns.

Procedure 2 is readily generalized for any of the complex columns shown in Figs. 9-3, 9-4, and 9-5. For the case where the additional specifications are given by set 3 of Table 9-5, the economic objective function may be restated in the following form

$$O = \frac{1}{f_1}[\{(1 + R)C_c + C_o\}/D]^2 + \frac{1}{f_2}\sum_{i=1}^{2n+2} s_i^2 \tag{9-19}$$

where n is equal to the number of sidestreams withdrawn. The independent and dependent variables in the initial search and the independent variables in the final search are listed in Table 9-6.

9-3 PROCEDURE 3. MINIMIZATION OF THE REFLUX RATIO FOR EXISTING CONVENTIONAL AND COMPLEX DISTILLATION COLUMNS

Conventional Columns

In this example, the reflux ratio is to be minimized by finding the optimum feed plate location for a fixed number of total stages N or plates $[k_2 = N - 2]$. This example was included in order to demonstrate the use of the optimization procedure for solving problems involving existing columns. Again the variables listed as usual specifications are fixed. Of the four remaining variables required to completely define the column, three are fixed, namely, $(b_l/d_l)_U$, $(b_h/d_h)_L$, and the total number of stages N. Thus, the column is completely determined by fixing one additional variable. This additional variable is taken to be the feed-plate location which minimizes the reflux ratio. When the constraints are included in the objective function, the problem to be solved consists of the minimization of

$$O = \frac{1}{f_1}[L_1/D]^2 + \frac{1}{f_2}[s_l^2 + s_h^2] \tag{9-20}$$

subject to the limits

$$D_L \le D \le D_U \tag{9-21}$$

The initial and final searches are carried out in a manner similar to that described in procedure 1. In the initial search, the objective function given by Eq. (9-20) is searched over the variables k_1 and L_1/D. Since the total number of stages is fixed, the value of D given by the calculational procedure in App. 9-2 is the same for all choices of L_1/D.

In the final search, the objective function given by Eq. (9-20) is searched over the variables L_1/D, k_1, and D. The results of the initial search $\{(L_1/D)_i, k_{1i}\}$ where $i = 1, 2, ..., 5$, are used as two of the three coordinates of each of the vertices of the initial simplex of the final search. A value of D is picked for each of the vertices of the initial simplex by choosing random values of D in the neighborhood of the value of D (say D_k) found in the initial search as follows

$$D_{k+1} = \frac{D_k}{1 + \gamma} + r_3\left[(1 + \gamma)D_k - \frac{D_k}{1 + \gamma}\right] \tag{9-22}$$

where r_3 is a random number lying between 0 and 1. The constant γ was taken equal to 0.2. In the event that the final solution did not satisfy the constraints, the reflux ratio was increased by increments of five percent until a solution was found which satisfied the constraints.

This procedure is demonstrated by the solution of Example 9-3 which is presented in Tables 9-1, 9-9, and 9-10. The results of this example demonstrate that the proposed optimization procedure may be used to solve problems involving existing columns as well as the design of new columns. Somewhat more

Table 9-9 Statement of Example 9-3

An existing column with 31 plates, a total condenser, and a reboiler is to be used to separate the feed given in Table 9-1 at the operating conditions enumerated in this table. The specified separations for the light and heavy keys are as follows

$$b_l/d_l \le (b_l/d_l)_U = 0.1248$$

$$b_h/d_h \ge (b_h/d_h)_L = 7.201$$

Find the feed-plate location which minimizes the reflux ratio required to effect this separation. Provisions may be made to introduce the feed on any plate between 2 and 28.

computer time was required to solve this example because the criterion on the standard deviation of the objective function in the final search was reduced from ≤ 0.005 to ≤ 0.001. Relative to other examples solved[1] such as the example presented in Table 9-11, the objective functions for Examples 9-1, 9-2, and 9-3 were relatively flat in the neighborhood of the minimum, and a relatively tight tolerance on the deviation of the objective function was required in order to force the optimization procedure to continue the search until the minimum value of the function had been found.

Procedure 3. Minimization of the Reflux Ratio for an Existing Complex Column

When the purity constraints are included as additive terms in the objective function, one obtains

$$O = \frac{1}{f_1}[L_1/D]^2 + \frac{1}{f_2}[s_l^2 + s_h^2 + s_r^2 + s_s^2]^2 \tag{9-23}$$

which is subject to the limits

$$D_L \le D \le D_U \qquad W_L \le W \le W_U \tag{9-24}$$

where N is fixed.

In the initial search, the function O given by Eq. (9-23) is searched over $\{k_1, k_2, L_1/D\}$ with $\{k_3, D, W\}$ being dependent ($k_3 = N - 2$). The flow rates D and W are found by the calculational procedure described in App. 9-2.

In the final search, the function O is searched over $\{k_1, k_2, L_1/D, W\}$ with the dependent variable k_3 being computed from $k_3 = N - 2$.

Discussion of the Results Obtained for Complex Columns

In order to demonstrate the application of the proposed procedures, problems were solved using each of the objective functions presented.[1] A statement of the specifications common to all examples is presented in Table 9-11. The results obtained for each of these various types of specifications (which are listed in

Table 9-10 Initial and final searches for Example 9-3

	Initial search				
	Initial simplex				
	Vertices				
Variables	1	2	3	4	5
k_1	19	25	18	17	23
k_2	31	31	31	31	31
D	115	115	115	115	115
L_1/D	4.0	4.5	5.0	5.5	6.0
O [Eq. (9-20)]	19.2	22.3	25.9	28.0	30.7
b_l/d_l	0.0482	0.0566	0.149	0.149	0.0629
b_h/d_h	32.4	55.8	67.0	89.0	84.6

	Final simplex				
	Vertices				
Variables	1	2	3	4	5
k_1	21	21	21	21	21
k_2	31	31	31	31	31
D	115	115	115	115	115
L_1/D	2.73	2.73	2.73	2.73	2.73
O [Eq. (9-20)]	0.740	0.739	0.739	0.739	0.738
b_l/d_l	0.0360	0.0620	0.062	0.062	0.062
b_h/d_h	7.22	7.21	7.22	7.20	7.21

73 iterations were required to obtain the final simplex.

	Final search				
	Initial simplex				
	Vertices				
Variables	1	2	3	4	5
k_1	21	22	21	22	20
k_2	31	31	31	31	31
D	115	119	108	114	126
L_1/D	2.73	3.23	3.73	4.23	4.73
O [Eq. (9-20)]	21.0	20.0	22.5	23.2	25.5
b_l/d_l	0.326	0.242	0.301	0.207	0.0895
b_h/d_h	2.76	2.99	4.35	4.51	4.17

Table 9-10 (*continued*)

Variables	Final simplex				
	Vertices				
	1	2	3	4	5
k_1	21	21	21	21	21
k_2	31	31	31	31	31
D	115	115	115	115	115
L_1/D	5.51	5.51	5.51	5.51	5.51
O [Eq. (9-20)]	0.250	0.250	0.250	0.250	0.250
b_l/d_l	0.125	0.125	0.125	0.125	0.125
b_h/d_h	7.21	7.21	7.21	7.21	7.21

100 iterations were required to obtain the final simplex, and 17.9 seconds of computer time, AMDAHL (FORTRAN H, OPT 2) were required for the initial and final searches.

Table 9-11 Specifications common to all examples for complex columns

Component no.	Component	Composition of feed F_1 X_{1i}	Composition of feed F_2 X_{2i}
1	Methane	0.050	0.020
2	Ethane	0.100	0.100
3	Propylene (*l*)	0.150	0.060
4	Propane (*s*)	0.150	0.125
5	Isobutane (*r*)	0.150	0.035
6	*n*-butane (*h*)	0.100	0.150
7	*n*-pentane	0.100	0.152
8	Hexane	0.100	0.113
9	Heptane	0.050	0.090
10	Octane	0.030	0.085
11	400 NBP	0.020	0.070
Bubble point at 300 lb/in² abs		63.018°F	164.442°F

Other specifications

The column has a partial condenser which is to be operated at 300 lb/in² abs. The distillate is removed as dewpoint vapor. The bottoms and the sidestream products are removed as bubble-point liquids. Feeds enter as liquids at their bubble point at the column pressure. Vapor-liquid equilibrium data and enthalpy data are given in Tables B-1 and B-2.

Table 9-12 Optimum values of k_1, k_2, k_3, k_r, k_5, L_1/D, D, W_1, W_2 for the types of sets of specifications listed in Table 9-5

Item	Set	k_1	k_2	k_3	k_4	k_5	L_1/D	D	W_1	W_2	Number of iterations: initial search	Number of iterations: final search	Total computer time (s)
I	1†	4	8	13	...	...	2.00	35.00	5.00	......	....	19	0.36
	2	4	5	11	...	...	2.00	34.00	4.61	......	72	40	4.15
	3	6	5	12	...	...	1.07	34.20	5.69	......	164	80	6.8
	4	7	5	13	...	...	1.01	33.40	4.30	......	38	73	6.86
	5	7	3	8	...	...	2.00	27.60	16.80	......	97	32	2.76
	6	3	3	9	...	...	2.00	23.70	20.9	......	72	29	3.16
II	1†	7	8	13	...	...	2.00	61.23	......	......	....	20	0.33
	2	7	10	12	...	...	2.00	62.8	......	......	96	59	5.14
	3	7	9	12	...	...	1.99	63.3	......	......	127	94	8.57
	4	7	11	13	...	...	1.68	61.4	......	......	39	100	6.01
III	1†	4	7	12	13	...	2.00	30.00	5.00	9.00	....	19	0.38
	2	7	6	12	13	...	2.00	30.6	5.45	7.12	97	66	7.35
	3	8	7	12	14	...	2.04	30.8	5.41	6.38	142	100	10.82
	4	8	6	12	13	...	1.49	30.9	5.37	5.67	50	88	7.25
IV	1†	4	5	6	12	13	2.00	65.00	10.00	25.00	....	18	0.37
	2	7	7	6	11	14	2.00	68.1	10.8	23.9	96	49	5.67
	3	6	5	16	13	17	1.99	66.7	10.3	23.1	201	61	8.17
	4	10	7	5	6	13	2.02	66.1	10.5	23.4	143	87	7.00

† These variables were specified for the first example of each set of specifications.

Table 9-4) are given in Table 9-12. All examples were solved by use of the θ method of convergence which is generally an order of magnitude (or more) faster than the Newton-Raphson formulations.[8]

In the first set of specifications for each type of problem (see, for example, item I, set 1 of Table 9-5), the specifications for an existing column are made, namely, L_1/D, D, k_1, k_2, and W_1. The results for this set as well as the other sets of specifications are given in Table 9-12. The computer time required to solve the optimization problems ranges from 10 to 20 times that required to solve a problem for an existing column.

The desirable characteristics of the initial search, which was originally proposed by Srygley and Holland[19] for minimizing the number of equilibrium stages at a fixed reflux ratio, are clearly demonstrated in the solutions of the examples considered. The initial search produced values for the minimum number of plates which were consistently close to the solution values obtained by the final search of procedure 1.

In the determination of the optimum economic design by procedure 2, the second part of the initial search is employed because the starting value for the reflux ratio L_1/D may be far from its optimum value. The improved value of L_1/D found by the second initial search reduces the region to be searched in the final search procedure, thereby reducing the computer time required to find the optimum economic solution.

The coordinates of the initial simplex of the final search were obtained by varying those of the final simplex of the initial search by ± 10 percent through the use of random numbers. This procedure promoted the expansion of the region of the search for the minimum. As in the case of conventional columns, the computer time requirement was reduced by the convergence criterion placed on θ at each vertex. At the end of the final search, one of the simplex vertices was picked as the optimum. Then additional trials were made as required in order to obtain a solution which met the convergence criteria. If the constraints were not satisfied by this solution, minor adjustments of the variables were made such as adding one plate or increasing the reflux ratio. In the solution of the problems presented, it was seldom necessary to make these adjustments.

NOTATION

A_{ji}	absorption faction for component i and stage j; $A_{ji} = L_j/K_{ji}/V_j$
b_i	molar flow rate of component i in the bottoms
B	total molar flow rate of the bottoms
C_c	capital cost per unit time
C_o	operating costs per unit time
d_i	molar flow rate of component i in the distillate
D	total molar flow rate of the distillate
F	total molar flow rate of the feed
$k_1, \ldots, k_5$	number of plates; see Figs. 9-1, 9-2, 9-3, 9-4, and 9-5
K_{ji}	ideal solution K value
l_{ji}	molar flow rate of component i leaving plate j
L_j	total molar flow rate of the liquid leaving plate j
N	total number of stages, which includes all plates, the reboiler and the condenser (partial or total)
O	objective function
p	plate number of a sidestream withdrawal
P	column pressure
q	plate number of a sidestream withdrawal
Q_C	condenser duty
Q_R	reboiler duty
r	random number lying between 0 and 1; also used to denote a component for which a specification is made; $K_r < K_s$.
R	expected return on investment
s	a component for which a specification is made

s_l, s_h	separation specifications on the light and heavy keys; defined beneath Eq. (9-9)
v_{ji}	molar flow rate of component i in the vapor leaving stage j
V_j	total molar flow rate of the vapor leaving stage j
V_F	total molar flow rate of the vapor part of a partially vaporized feed
W	total molar flow rate of a sidestream
x_{ji}	mole fraction of component i in the liquid leaving stage j
x_{Bi}	mole fraction of component i in the bottoms
X_i	mole fraction of component i in the feed

$$[x_1 \ x_2 \ x_3]^T \quad \begin{bmatrix} x_1 \\ x_2 \\ x_3 \end{bmatrix}$$

$\{x_i\}$ set of all numbers $x_1, x_2, \ldots, x_c$

PROBLEMS

9-1 A knowledge of the temperature of the cooling medium which may be used to produce liquid reflux and the knowledge of the approximate composition of the distillate are commonly used to set the column pressure. A distillation column having a total condenser and operating at a reflux ratio $L_1/D = 1$ is to produce 50 lb mol of distillate per hour. Cooling water is available to condense the overhead vapor in the condenser-accumulator section. The distillate and reflux are to leave the condenser-accumulator as a liquid at their boiling point of 150°F.

    Given:

Component	d_i (lb mol/h)	K_i
1	5.5	$0.01T/3P$†
2	18.0	$0.02T/3P$
3	26.5	$0.03T/3P$
	$D = 50.0$	

    † T is in °F and P in atm.

    (*a*) Compute the column pressure.
    (*b*) Compute the temperature of the overhead vapor.

9-2 Repeat Prob. 9-1 for the case of a partial condenser. In many instances where the distillate is to be used in the vapor state, it is economically advantageous to use a partial condenser. In general when a partial condenser is used, the column pressure can be reduced and/or a refrigerated coolant may be avoided. To illustrate the reduction in pressure, repeat Prob. 9-1 for the case where a partial condenser is used.

9-3 On the basis of the enthalpy data given in Example 1-5, compute the condenser duty required in Prob. 9-1.

9-4 On the basis of the enthalpy data given in Example 1-5, compute the condenser duty required for the column in Prob. 9-2.

(a) On the basis of the availability of cooling water at 80°F and an outlet temperature of 135°F, compute the flow rate of cooling water required in lb/h per 50 lb mol of distillate per hour.

(b) If the condenser has an overall heat transfer coefficient of 300 Btu/(h ft^2 °F), compute the condenser cooling area required per 50 lb mol of distillate per hour.

REFERENCES

1. N. S. Al-Haj-Ali: Ph.D. Dissertation, Texas A&M University, 1979.
2. M. J. Box: "A New Method of Constrained Optimization and a Comparison with Other Methods," *Comput. J.*, **8**:42 (1965).
3. M. R. Fenske: "Fractionation of Straight-Run Pennsylvania Gasoline," *Ind. Eng. Chem.*, **24**:482 (1932).
4. S. E. Gallun and C. D. Holland: "Solve More Distillation Problems, Part 5: For Highly Non-ideal Solutions," *Hydrocarbon Process.*, **55**(6):125 (1976).
5. ———, F. E. Hess, G. W. Bentzen, and C. D. Holland: "Algorithms for Solving Problems Involving the Separation of Ideal and Nonideal Solutions by Equilibrium Stage Processes," *Proc. 5th Symposium on Computers in Chemical Engineering*, Vysoke Tatry, Czechoslovakia, Oct. 5–9 (1977).
6. Glitsch, Inc.: "Ballast Tray Design Manual," Bulletin No. 4900, 3d ed., p. 17.
7. K. M. Guthrie: "Data and Techniques for Preliminary Capital Cost Estimating," *Chem. Eng.*, 142, Mar. 24, 1969.
8. F. E. Hess, S. E. Gallun, G. W. Bentzen, and C. D. Holland: "Solve More Distillation Problems, Part 8: Which Method to Use," *Hydrocarbon Process.*, **42**(6):181 (1977).
9. ——— and C. D. Holland: "Solve More Distillation Problems, Part 7: Absorber-Type Columns," *Hydrocarbon Process.*, **42**(5):241 (1977).
10. C. D. Holland: *Fundamentals and Modeling of Separation Processes, Absorption, Distillation, Evaporation, and Extraction*, Prentice-Hall, Englewood Cliffs, N.J., 1975.
11. ———: *Multicomponent Distillation*, Prentice-Hall, Englewood Cliffs, N.J., 1963.
12. ——— and G. P. Pendon: "Solve More Distillation Problems, Part 1: Improvements Give Exact Answers," *Hydrocarbon Process.*, **53**(7):148 (1974).
13. R. Hooke and T. A. Jeeves: "Direct Search Solution of Numerical and Statistical Problems," *J. Assoc. Comp. Mach.*, **8**(2):(1961).
14. L. M. Naphtali and D. P. Sandholm: "Multicomponent Separation Calculations by Linearization," *AIChE J.*, **17**(1):148 (1971).
15. N. L. Ricker and E. A. Grens: "A Calculational Procedure for Design Problems in Multicomponent Distillation," *AIChE J.*, **20**(2):238 (1974).
16. R. W. H. Sargent and K. Gaminibandara: "Optimum Design of Plate Distillation Columns," *Optimization in Action*, p. 266, L. C. W. Dixon (ed.), Academic Press, New York (1976).
17. R. C. Reid and T. K. Sherwood: *The Properties of Gases and Liquids*, 2d ed., McGraw-Hill Book Company, New York (1966).
18. *Selected Values of Properties of Hydrocarbon and Related Compounds*, API Research Project 44, Thermodynamics Research Center, Texas A&M University.
19. J. M. Srygley and C. D. Holland: "Optimum Design of Conventional and Complex Columns," *AIChE J.*, **11**(4):695 (1965).

APPENDIX 9-1   APPROXIMATE SOLUTIONS AT TOTAL REFLUX

The equations are presented first for conventional distillation columns and then for complex columns.

Conventional Columns

For a column having a partial condenser the Fenske equation[3] is given by

$$\frac{b_i}{d_i} = \frac{b_h}{d_h} \alpha_i^{-N} \tag{1}$$

where the relative volatility $\alpha_i = K_i/K_h$, and N is equal to the total number stages. Also, the heavy key is taken to be the base component and denoted by the subscript h. For a column having a total condenser, N in Eq. (1) is replaced by $N - 1$. Although Eq. (1) applies exactly only if the relative volatilities remain constant throughout the column, it is generally satisfactory for determining whether a realistic set of values have been specified for b_l/d_l and b_h/d_h, where the light key is denoted by the subscript l. Since the relative volatilities generally vary throughout the column, a set must be selected for use in Eq. (1). For definiteness, the α_i's are evaluated at the boiling-point temperature of the feed at the column pressure. For the case where b_l/d_l and b_h/d_h are specified, Eq. (1) may be solved for N to give

$$(N)_{\min} = \frac{\ln \ (b_h/d_h)_L/(b_l/d_l)_U}{\ln \ \alpha_l} \tag{2}$$

If

$$(N)_{\min} > N_U \tag{3}$$

where $N_U = k_{2,U} + 2$, then the specified separation is impossible. If

$$(N)_{\min} < N_U \tag{4}$$

the separation is feasible.

After finding N by use of Eq. (2), the complete set of b_i/d_i's are found through the use of Eq. (1). Then the total distillate rate at total reflux is found by solving the overall component-material balances for d_i, and summing all components to give

$$D = \sum_{i=1}^{c} d_i = \sum_{i=1}^{c} FX_i/[1 + (b_i/d_i)] \tag{5}$$

Other Specifications

Suppose that the purity specifications are taken to be $X_{D,l}$ and $x_{B,h}$. In general D and B are not determined by the specification of these two mole fractions except for the case of a binary mixture. To apply Fenske's equation to determine the feasibility of the specified separation, the following relationships may be used. First, observe that

$$x_{B,h} = \frac{b_h}{\displaystyle\sum_{i=1}^{c} b_i} \qquad X_{D,l} = \frac{d_l}{\displaystyle\sum_{i=1}^{c} d_i} \tag{6}$$

and

$$b_i = \frac{FX_i}{1 + d_i/b_i} = \frac{FX_i}{1 + (d_h/b_h)\alpha_i^N} \tag{7}$$

$$d_i = \frac{FX_i}{1 + b_i/d_i} = \frac{FX_i}{1 + (b_h/d_h)\alpha_i^{-N}} \tag{8}$$

The above expressions contain two unknowns, b_h/d_h and N. Use of the above expressions for d_i and b_i permits the two expressions given by Eq. (6) to be restated in functional form in terms of the two unknowns as follows

$$g_1(N, b_h/d_h) = \left[\frac{1}{(x_{Bh})_{\text{spec}}}\right] \frac{b_h}{\displaystyle\sum_{i=1}^{c} b_i} - 1 \tag{9}$$

$$g_2(N, b_h/d_h) = \left[\frac{1}{(X_{Dl})_{\text{spec}}}\right] \frac{d_l}{\displaystyle\sum_{i=1}^{c} d_i} - 1 \tag{10}$$

where b_i, b_h, d_i, and d_l are given by Eqs. (7) and (8). The desired value of the variables $(N > 0, b_h/d_h > 0)$ that make $g_1 = g_2 = 0$ may be found by use of the Newton-Raphson technique or other numerical methods.

Since the equilibrium relationship for the distillate and bottoms depends upon the respective sets of mole fractions $\{X_{Di}\}$ and $\{x_{Bi}\}$, the relationships may be stated in terms of the d_i's and b_i's. The determination of T_D and T_B at total reflux may be reduced to a trial-and-error problem in a manner analogous to that demonstrated above for the specification of a set of mole fractions.

Complex Distillation Columns

Modifications of the equations for conventional columns which are needed in order to obtain approximate solutions for complex columns are presented below for the special case of a column having one liquid sidestream withdrawn from stage number $(k_1 + 1)$.

The Fenske equation[3] given by Eq. (1) is valid for complex columns where N is equal to the total number of stages for a column having a partial condenser, and $\alpha_i = K_i/K_h$. The Fenske equation for the sidestream shown in Fig. 9-2 is

$$\frac{w_i}{d_i} = \frac{w_r}{d_r} \left(\frac{\alpha_i}{\alpha_r}\right)^{-(k_1+1)} \tag{11}$$

For the case where b_l/d_l, b_h/d_h, w_r/d_r, and w_s/d_s are specified, Eqs. (1) and (11) give

$$(N)_{min} = \frac{\ln\ (b_h/d_h)_L/(b_l/d_l)_U}{\ln\ \alpha_l} \tag{12}$$

$$(k_1)_{min} = \frac{\ln\ (w_r/d_r)_L/(w_s/d_s)_U}{\ln\ \alpha_s/\alpha_r} - 1 \tag{13}$$

If either

$$(N)_{min} > N_U$$

or

$$(k_1)_{min} > N_U$$

then the specified separation is impossible. If

$$(N)_{min} < N_U \quad\text{and}\quad (k_1)_{min} < N_U$$

then the specified separation is feasible. The generalization of the above equations for the case where any number of sidestreams are withdrawn is readily deduced.

For the case where specifications such as mole fractions rather than mole ratios are made, the development of the corresponding equations is analogous to that described for conventional distillation columns.

APPENDIX 9-2   APPROXIMATE SOLUTIONS AT OPERATING REFLUXES

The method is developed first for conventional distillation columns and then modified as required to describe complex columns.

Conventional Columns

The approximate method used consisted of solving the combined component-material balances and equilibrium relationships for the light key ($i = l$) and the heavy key ($i = h$) components on the basis of assumed temperature and total vapor rate (or total liquid rate) profiles. As shown in Chap. 2, the equations to be solved may be stated in the following matrix form

$$\mathbf{A}_i \mathbf{v}_i = -\mathbf{f}_i \tag{1}$$

The speed of the approximate method is enhanced by the fact that Eq. (1) is solved for only two components, the light and heavy keys ($i = l$, $i = h$). This equation is readily solved for d_i and the v_{ji}'s by use of the Thomas algorithm as demonstrated in Chap. 2. After v_{Ni} has been computed, b_i is found by use of the expressions

$$b_i = A_{Ni} v_{Ni} \tag{2}$$

This expression is readily developed by use of the equilibrium relationship of the reboiler

$$y_{Ni} = K_{Ni}x_{Ni} = K_{Bi}x_{Bi} \tag{3}$$

The calculation of b_i $(i = h, l)$ marks the end of one complete trial by the approximate method.

Temperature Profiles

Any number of methods could be used to pick adequate temperature profiles. The following relatively simple procedure was used to select these profiles. The temperatures of all intermediate stages were taken to be linear with stage number between the distillate temperature T_D and the bottoms temperature T_B. For different numbers of stages, the top and bottom temperatures were estimated as follows. First, the lowest overhead temperature $T_{D, L}$ and the highest bottoms temperature $T_{B, U}$ were found by use of a modified form of Fenske's equation which was deduced in the following manner. Since the value of $(N)_{min}$ computed by Eq. (2) on the basis of the specified values of b_l/d_l and b_h/d_h is less than N_U, a separation better than the specified one can be obtained. Thus, it is desirable to revise Eq. (1) such that each separation is better than the one which was specified. The result may be realized by introducing the factor μ as follows

$$N_U = \frac{\ln \dfrac{\mu(b_h/d_h)_L}{(1/\mu)(b_l/d_l)_U}}{\ln \alpha_l} \tag{4}$$

After Eq. (4) had been solved for μ, the corresponding set of b_i/d_i's is found by use of the following form of Eq. (1)

$$\frac{b_i}{d_i} = \mu\left(\frac{b_h}{d_h}\right)_L \alpha_i^{-N_U} \tag{5}$$

After the b_i/d_i's have been computed by use of Eq. (5), the d_i's are given by the following form of the overall material balance

$$d_i = \frac{FX_i}{1 + b_i/d_i} \tag{6}$$

Next the distillate rate D at N_U is found by summing the d_i's over all components i. Then the b_i's and the sets of mole fractions $\{X_{Di}\}$ and $\{x_{Bi}\}$ are computed. The mole fractions are used to compute the temperatures $T_{D, L}$ and $T_{B, U}$.

By replacing N_U by N_L (where $N_L = k_{2L} + 2$) in Eq. (5), the highest overhead temperature $T_{D, U}$ and the lowest bottoms temperature $T_{B, L}$ as well as the total distillate rate D at N_L may be obtained.

For any set of arbitrary values of k_1, k_2 lying between the upper and lower bounds of these respective variables, the overhead and bottoms temperatures $T_D(N)$ and $T_B(N)$ were found by assuming $T_D(N)$ to vary linearly between $T_{D, L}$ and $T_{D, U}$ with the total number of stages, and $T_B(N)$ to vary linearly between

$T_{B,L}$ and $T_{B,U}$ with the total number of stages. The temperatures of the intermediate stages were taken to be linear with stage number between $T_D(N)$ and $T_B(N)$.

For each choice of N lying between N_L and N_U, the distillate rate D was found by linear interpolation between the values of D at N_L and at N_U.

Total Vapor Rate (or Liquid Rate) Profiles

The total vapor rate was assumed to be constant throughout the rectifying section, and denoted by $V_j = V_r$. Then for a specified reflux ratio

$$V_r = L_1 + D = \left[\frac{L_1}{D} + 1\right]D \tag{7}$$

where the distillate rate D is computed as described above for each choice of N. The corresponding value of B was found by use of the overall material balance, $B = F - D$. Similarly, the total flow rates were assumed to be constant throughout the stripping section. These rates were computed as follows

$$L_s = L_r + L_F \tag{8}$$

and

$$V_s = L_s - B \tag{9}$$

Complex Distillation Columns

The method used to obtain approximate solutions to complex columns consists of an extension of the one proposed above for conventional columns. The equations used consist of a combination of the component-material balances and the equilibrium relationships. For the complex column with one feed shown in Fig. 9-2, the combination of the equilibrium relationships and the component-material balances are given by Eq. (1). For multiple feed plates and sidestream withdrawals, minor modifications of A_i and f_i are required.

The temperatures of all stages between the top temperature T_D of the overhead condenser and the temperature T_B of the reboiler were assumed to vary linearly with stage number. The temperatures T_D and T_B corresponding to the best possible separation are found by taking the total number of stages equal to the upper bound N_U. In order to find a set of specifications each of which is better than the specified value, let μ_1 and μ_2 be determined, respectively, by solving the following equations

$$N_U = \frac{\ln \dfrac{\mu_1(b_h/d_h)_L}{(1/\mu_1)(b_l/d_l)_U}}{\ln \alpha_l} \tag{10}$$

$$k_{1U} = \frac{\ln \dfrac{\mu_2(w_r/d_r)_L}{(1/\mu_2)(w_s/d_s)_U}}{\ln \alpha_s/\alpha_r} - 1 \tag{11}$$

After Eqs. (10) and (11) have been solved for μ_1 and μ_2, the new sets of b_i/d_i's and w_i/d_i's are found as follows

$$\frac{b_i}{d_i} = \mu_1 \left(\frac{b_h}{d_h}\right)_L \alpha_i^{-N_U} \qquad \frac{w_i}{d_i} = \mu_2 \left(\frac{w_r}{d_r}\right)\left(\frac{\alpha_i}{\alpha_r}\right)^{-(k_{1U}+1)} \tag{12}$$

After the complete sets of b_i/d_i's and w_i/d_i's have been computed by use of Eq. (12), the d_i's are found by use of the following form of an overall material balance

$$d_i = \frac{FX_i}{1 + b_i/d_i + w_i/d_i} \tag{13}$$

Next the b_i's and w_i's are found

$$b_i = \left(\frac{b_i}{d_i}\right)d_i \qquad w_i = \left(\frac{w_i}{d_i}\right)d_i \tag{14}$$

The total distillate rate D and the sidestream withdrawal rate W are found by summing the d_i's and the w_i's, respectively, over all components.

Next the mole fractions $\{X_{Di}\}$ and $\{x_{Bi}\}$ are computed. By use of these mole fractions and the equilibrium relationships, the lower bound on the distillate temperature T_D (denoted by $T_{D,L}$) and the upper bound on the bottoms temperature T_B (denoted by $T_{B,U}$) are determined.

By replacing N_U by N_L (where $N_L = k_{3L} + 2$), the upper bound $T_{D,U}$ of the distillate temperature and the lower bound $T_{B,L}$ on the bottoms temperature as well as the corresponding values of D and W are determined.

For any set of arbitrary values of k_1, k_2, and k_3 lying between the upper and lower bounds of these respective variables, the overhead and bottoms temperature $T_D(N)$ and $T_B(N)$ are found by assuming $T_D(N)$ to vary linearly between $T_D(N)$ and $T_B(N)$ with the total number of stages. Similarly, $T_B(N)$ is assumed to vary linearly between $T_{B,L}$ and $T_{B,U}$ with the total number of stages. The temperatures of the intermediate stages are assumed to vary linearly with stage number between $T_D(N)$ and $T_B(N)$.

For each choice of N lying between N_L and N_U, the distillate rate D and the sidestream withdrawal rate W are found by linear interpolation between the values of D and W, respectively, at N_L and at N_U.

The total flow rates are assumed to be constant throughout each section of the column, say k_1, k_2, k_3. Adjustments to the total flow rates are made, however, at each point where a stream is either introduced or withdrawn.

APPENDIX 9-3 A MODIFIED FORM OF THE COMPLEX METHOD OF BOX[2] FOR OBJECTIVE FUNCTIONS HAVING NO CONSTRAINTS

An abbreviated statement of this revised form of the Complex Method proposed by Box[2] follows. This method makes use of at least $n + 1$ vertices or solutions, where n is the number of variables over which the objective function is searched.

The total number of vertices is denoted by k, where $k \geq n + 1$. First an initial solution is found. Then an additional $k - 1$ vertices of the first simplex figure are found by choosing sets of independent variables lying between the upper and lower bounds, x_{iU} and x_{iL}, respectively, of the independent variables by use of random numbers as follows

$$x_{i, j} = x_{iL} + r_{i, j}(x_{iU} - x_{iL}) \qquad (i = 1, 2, \ldots, n) \tag{1}$$

where the $r_{i, j}$'s are random numbers generated in the range 0 to 1, and j denotes the sets of variables for the jth solution. The initial set is denoted by $j = 1$, and the remaining sets by $j = 2, 3, \ldots, k$.

The objective function is evaluated at each vertex, and the vertex yielding the poorest value of the objective function is rejected and replaced by the point N located at a distance $\beta(\beta \geq 1)$ times as far from the centroid of the remaining points as the distance of this rejected vertex, and in the direction of the vector which points from the rejected point to the centroid (see Fig. 9-6). The rejected point R is excluded in computing the centroid M

$$x_{i, M} = \frac{1}{k - 1} \left[\sum_{j = 1, \neq R}^{k} x_{i, j} \right] \tag{2}$$

The coordinates of point N are found by

$$x_{i, N} = \beta(x_{i, M} - x_{i, R}) + x_{i, M} \tag{3}$$

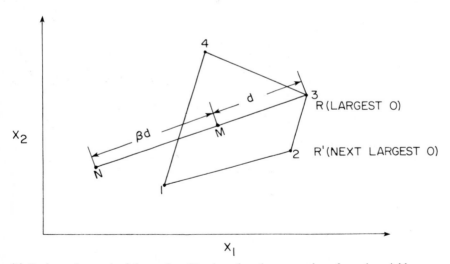

(1) Evaluate O at each of the vertices. Number of vertices > number of search variables.
(2) Compute centroid M based on points 1, 2, 4, and locate point N.
(3) If $O_N < O_R$, accept point N and reject R.
(4) If $O_N > O_R$, evaluate O_M; if $O_M < O_R$, accept point M and reject R.
(5) If $O_N > O_R$, $O_M > O_R$, repeat procedure for the next poorest point R'.

Figure 9-6 Graphical representation of Box's method.

If the value of O at x_N is less than its value at the rejected point R, then the new point N is accepted. If the new point N gives a poorer value for the objective function than that given by the rejected point R, then the value of the objective function at the centroid M is found. If the value of the objective function at the centroid M is less than its value at the rejected point R, then the centroid M is accepted as the new point. Otherwise, if neither N nor M give a smaller value for O than does the rejected point R, then the rejected point of the simplex is retained, and the next poorest point of the simplex is selected as the rejected point and the procedure described above is repeated.

Integral values for the number of plates were found by rounding the corresponding values of $x_{i,\,j}$ to the nearest integer. For example, if the value of $x_{i,\,j}$ corresponding to the number of plates in a given section is found to be 5.6, then the integral number of plates used for this section of the column in the evaluation of the objective function was taken equal to 6. If, however, a value of $x_{i,\,j} = 5.4$ is found, then the corresponding integral number of plates was taken equal to 5 in the evaluation of the objective function. The original values of $x_{i,\,j}$ (5.6 or 5.4) were used, however, in the subsequent generation of vertices.

APPENDIX 9-4 FORMULATION OF THE ECONOMIC OBJECTIVE FUNCTION

The economic objective function is formulated on the basis of (1) capital costs, (2) return on investment, and (3) operating costs. The objective function is formulated as an expression of the desire to minimize the total costs per mole of the most valuable product. The capital costs consist of the installed costs of the following items: the pressure vessel or shell of the distillation column, the trays, the condenser, and the reboiler.

After the column diameter had been estimated by use of the relationships recommended by Glitsch, Inc.,[6] the cost of the shell was estimated by use of the data presented by Guthrie.[7] Guthrie's data were also used to estimate the costs of the plates, the condenser, and the reboiler. Guthrie's data (originally presented in graphical form) were converted to the form of curve fits which are given below.

A straight-line depreciation of the capital investment over a 10-year period was assumed, and the corresponding annual capital investment, C_o, was computed. If R is used to denote the expected fractional return on investment, then the return on the investment per year is RC_c, and the total cost associated with the capital investment is $(1 + R)C_c$.

Operating costs, C_o, consist of the costs of supplying the required condenser and reboiler duties for one year. A summary of the equations used to evaluate the objective function follow.

Investment Costs

To compute the costs of the empty column (or the shell), the diameter of the column is needed. Consider the case where the column is to be operated at 80 percent of vapor flood, and two-pass ballast trays with 24-inch tray spacings are to be used. The column diameter may be estimated by use of the following relationship given by Glitsch.[6]

$$D = \frac{(L_{load} - 500)}{5.22(V_{load} - 5) + 461} + \frac{V_{load} - 5}{3.75} + 5.2 \tag{1}$$

where D = inside diameter of the column in feet
L_{load} = liquid load in gal/min
V_{load} = vapor load in ft^3/s

The liquid and vapor loads were approximated on the basis of the molar flow rates, composition, and temperature of the top plate as follows

$$L_{load} = \sum_{i=1}^{c} [(l_{1i}M_i)]\left(\frac{1\ h}{60\ min}\right)\left(\frac{ft^3}{\rho_L\ lb}\right)\left(\frac{7.48\ gal}{ft^3}\right) \tag{2}$$

where M_i is the molecular weight of component i. The mass density may be estimated as proposed by Reid and Sherwood[17] as follows

$$\rho_L = \frac{\sum_{i=1}^{c} l_{1i}M_i}{\sum_{i=1}^{c} l_{1i}(1/\rho_{Li})M_i} \tag{3}$$

The mass density of component i in the liquid phase ρ_{Li} may be approximated as suggested by Reid and Sherwood.[17]

$$\rho_{Li} = \frac{1}{C_i(5.3 + 3.0T/T_C)} \tag{4}$$

where ρ_{Li} = mass density
C_i = constant for each component i
T_C = critical temperature

The vapor load was computed as follows

$$V_{load} = \left(\sum_{i=1}^{c} v_{1i}M_i\right)\left(\frac{1\ h}{3600\ s}\right)\left(\frac{ft^3}{\rho_V\ lb}\right)\frac{\rho_V}{\rho_L - \rho_V} \tag{5}$$

The mass density of the vapor was estimated as follows

$$\rho_V = \left(\frac{P_1}{Z_m R T_1}\right)\left(\sum_{i=1}^{c} x_{1i}M_i\right) \tag{6}$$

where $Z_m = \sum_{i=1}^{c} Z_{1i}y_{1i}$, Z_{1i} is equal to the compressibility factor of pure component i at the temperature T_1 and pressure P_1 of plate 1, and y_{1i} is the mole fraction of component i in the vapor that leaves plate 1.

After the diameter D of the column has been computed, the cost of the empty column may be computed by use of the following curve fit which is based on the data of Guthrie[7]

$$C_{shell} = \left\{ [620 + 398(D - 2)] \left(\frac{b}{4} \right)^{0.8} \right\} F_M F_p I \tag{7}$$

where b = height of the pressure vessel in feet
$\quad C_{shell}$ = cost (FOB) of pressure vessel (or shell) in 1977 dollars
$\quad F_M$ = adjustment factor for the material of construction
$\quad F_p$ = adjustment factor for the column pressure
$\quad I$ = inflation index to correct the cost from mid-1968 to the end of 1977 dollars

The cost of the trays, C_{tray}, was computed by use of Guthrie's data in the form of the curve fit

$$C_{tray} = N_T \left[16 \left(\frac{D}{2} \right)^x (F_S + F_t + F_M) \right] I \tag{8}$$

where N_T = total number of trays
$\quad D$ = internal diameter of the column in feet
$\quad F_M$ = adjustment factor for the material of construction
$\quad F_S$ = adjustment factor for the tray spacing
$\quad F_t$ = adjustment factor for the type of trays
$\quad I$ = index of inflation factor to correct the cost from mid-1968 to the end of 1977
$\quad x$ = cost factor for capacity; see tabular material below
$\quad C_{tray}$ = cost of the trays (FOB) in 1977 dollars

The cost factor x as a function of the internal diameter is presented in the following table.

Variation of the cost factor x with the column diameter D (ft)

D (ft)	$0 \le D \le 3$	$3 < D \le 5$	$5 < D \le 7$	$7 < D \le 10$
x	1.0	1.16	1.36	1.47

The cost of installation, C_{inst}, was computed as follows by use of Guthrie's data.

$$C_{inst} = [C_{shell} + C_{tray}] F_I \tag{9}$$

where C_{inst} = cost of column with the trays installed in 1977 dollars
$\quad F_I$ = factor needed to convert FOB costs to installed costs

The cost of the condenser depends upon the surface area required to remove Q_C (Btu/h) from the overhead vapor. The heat transfer area A_C and the associated coolant flow rate w_C (of exit temperature $T_{C,\,out}$) may be found by solving the following equations simultaneously

$$Q_C = w_C C_{pC}(T_{C,\,out} - T_{C,\,in}) \qquad (10)$$

$$Q_C = U_C A_C \, \Delta T_{lm} \qquad (11)$$

where A_C = heat transfer area required, ft^2
$\quad U_C$ = overall heat transfer coefficient, Btu/(h ft^2 °F)
$\quad \Delta T_{lm}$ = log mean of the distillate temperature T_D, and the inlet and outlet
$\qquad$ temperatures of the cooling water, $T_{C,\,in}$ and $T_{C,\,out}$
$\quad w_C$ = flow rate of the coolant, lb/h

The value of A_C so obtained was used to compute the cost of the condenser, C_{cond}, by use of the following curve fit of Guthrie's data

$$C_{cond} = \left[2075\left(\frac{A_C}{100}\right)^{0.65}\right][(0.85 + F_P)F_M]I \qquad (12)$$

where C_{cond} = cost (FOB) of the condenser in 1977 dollars

The remaining symbols have the same meanings given previously.
The heat transfer area A_R of the reboiler and the flow rate w_R of the steam (or the condensing vapor) were computed by use of the following equations

$$Q_R = \lambda_s w_R \qquad (13)$$

$$Q_R = U_R A_R(T_s - T_B) \qquad (14)$$

where A_R = heat transfer area of the reboiler, ft^2
$\quad U_R$ = overall heat transfer coefficient of the reboiler, Btu/(h ft^2 °F)
$\quad T_s$ = saturation temperature of the condensing vapor
$\quad w_R$ = flow rate of the condensing vapor, lb/h
$\quad \lambda_s$ = latent heat of vaporization of the heat transfer medium

The cost of the reboiler was computed as follows

$$C_{reboil} = \left[2075\left(\frac{A_R}{100}\right)^{0.65}\right][(1.35 + F_p)F_M]I \qquad (15)$$

where C_{reboil} is the FOB cost of the reboiler in 1977 dollars, and the symbols F_p, F_M, and I have the same meanings given above.
The installed costs of the condenser and reboiler were computed by use of an equation of the same form as Eq. (9).
The operating costs were computed by use of the flow rates w_C and w_R found by solving Eqs. (10), (11), (13), and (14).

CHARACTERISTICS OF CONTINUOUS DISTILLATION COLUMNS AT VARIOUS MODES OF OPERATION

In this chapter the behavior of continuous distillation columns is examined over a wide range of operating conditions. Continuous distillation columns may be operated at a variety of conditions which range from total reflux to columns at a fixed, finite reflux rate with distillate rates ranging from zero to the feed rate. The characteristic behavior which can be expected of continuous columns at these operating conditions is presented in Secs. 10-1 through 10-3. Calculational procedures needed to predict the behavior of continuous distillation columns at these operating conditions are also presented. The operating conditions and geometrical considerations required to place a continuous and a batch distillation column in a one-to-one correspondence at total reflux are also presented.

The operating condition of total reflux is subject to several interpretations. Four general types of total reflux are considered in this chapter, namely,

Type 1: Total Reflux in Continuous Columns:

$$\lim_{V_{j+1} \to \infty} \frac{L_j}{V_{j+1}} = 1, \ B + D = F, \ B \text{ and } D \text{ are finite.}$$

Type 2: Total Reflux in Both Sections of a Column:

$$D = 0, \ B = 0, \ F = 0$$

Type 3: Total Reflux in the Rectifying Section:

$$D = 0, \ B = F$$

Type 4: Total Reboil in the Stripping Section:

$$B = 0, \ D = F$$

Columns operating at these four types of total reflux are considered in Secs. 10-1 through 10-3.

When the term total reflux is used, one customarily visualizes the type of operation classified as type 2 wherein $D = 0$, $B = 0$, $F = 0$. Such a column contains enough vapor and liquid to be operational at steady state but has neither feed nor product streams. In such a column, all of the overhead vapor leaving the top plate is condensed and returned to the column as reflux and all of the liquid leaving the bottom of the column is vaporized in the reboiler and returned to the column. This operating condition is easily achieved in either a laboratory or a commercial column.

10-1 CONTINUOUS DISTILLATION COLUMNS AT TOTAL REFLUX, TYPE 1: $(L_j/V_{j+1} = 1, F = D + B, D \neq 0, B \neq 0)$

Total reflux of type 1 may be approached in a continuous distillation column by approaching total reflux conditions in both the rectifying and stripping sections. The designer approaches this type of operation of a continuous distillation column as the reflux ratio is increased indefinitely at a fixed feed rate and nonzero product rates. In the limit, this type of operation is recognized as total reflux of type 1, continuous distillation columns at total reflux. The necessary conditions for an equivalence to exist between columns at the operation conditions of total reflux of types 1 and 2 are presented in Sec. 10-2. Total reflux of types 1 and 2 are of significant interest because columns at these types of operation produce the best possible separations.

In the analysis of continuous distillation columns at total reflux, three different calculational procedures are presented; namely, the θ method of convergence and two formulations of the Newton-Raphson method, one analogous to the $2N$ Newton-Raphson method and the other to the Almost Band Algorithm. The formulation of each of these methods for continuous distillation columns at total reflux is presented below.

The θ Method of Convergence

For definiteness, suppose that the following specifications are made on a column: (1) the feed rate F and its composition $\{X_i\}$, (2) the distillate rate D, (3) the column pressure, (4) the number of stages N, and (5) the type of condenser (total or partial).

Fewer equations are required to describe distillation columns at total reflux than are required to describe columns operating at finite reflux ratios. The condition that $L_j/V_{j+1} = 1$ (in the limit as $V_{j+1} \to \infty$) eliminates the necessity for the determination of the total flow rates, and consequently the energy balance for each stage may be omitted from the set of equations to be solved. This type of total reflux appears to have been first proposed by Robinson and Gilliland.[13]

As a consequence of the condition $L_j/V_{j+1} = 1$ (as $V_{j+1} \to \infty$), the component-material balance for each stage reduces to

$$y_{j+1, i} = x_{ji} \quad (j = 1, 2, \ldots, N - 1) \tag{10-1}$$

as shown in Chap. 1. Since F, D, and B are all finite and positive, the component-material balance enclosing the entire column is

$$FX_i = d_i + b_i \tag{10-2}$$

In the following development of the equations needed for the θ method of convergence and the Newton-Raphson method which follows, it is supposed that both the vapor and liquid phases form ideal solutions throughout the column. For any component i, the N equilibrium relationships for a column having a partial condenser ($y_{1i} = X_{Di}$)

$$y_{ji} = K_{ji} x_{ji} \qquad (j = 1, 2, \ldots, N) \tag{10-3}$$

may be solved simultaneously with the N component-material balances given by Eq. (10-1) to give

$$\frac{x_{1i}}{X_{Di}} = \frac{1}{K_{1i}}$$

$$\frac{x_{2i}}{X_{Di}} = \frac{1}{K_{2i} K_{1i}}$$

$$\frac{x_{ji}}{X_{Di}} = \frac{1}{K_{ji} K_{j-1,i} \cdots K_{2i} K_{1i}} \tag{10-4}$$

$$\frac{x_{Ni}}{X_{Di}} = \frac{x_{Bi}}{X_{Di}} = \frac{1}{K_{Ni} K_{N-1,i} \cdots K_{2i} K_{1i}}$$

Fenske[3] was among the first to state the equations in this form.

For the case where the column has a total condenser ($x_{1i} = X_{Di}$), Eqs. (10-1) and (10-3) may be solved to give

$$\frac{x_{1i}}{X_{Di}} = 1$$

$$\frac{x_{2i}}{X_{Di}} = \frac{1}{K_{2i}}$$

$$\vdots \tag{10-5}$$

$$\frac{x_{ji}}{X_{Di}} = \frac{1}{K_{ji} K_{j-1,i} \cdots K_{3i} K_{2i}}$$

$$\frac{x_{Ni}}{X_{Di}} = \frac{x_{Bi}}{X_{Di}} = \frac{1}{K_{Ni} K_{N-1,i} \cdots K_{3i} K_{2i}}$$

After $(x_{Bi}/X_{Di})_{ca}$ has been evaluated by use of Eq. (10-4) [or Eq. (10-5)] on the basis of an assumed temperature profile, the value of $(b_i/d_i)_{ca}$ is computed as follows

$$\left(\frac{b_i}{d_i}\right)_{ca} = \frac{B}{D}\left(\frac{x_{Bi}}{X_{Di}}\right)_{ca} \tag{10-6}$$

The θ method is developed in the same manner as described in Chap. 2. The resulting equations are as follows

$$\left(\frac{b_i}{d_i}\right)_{co} = \theta\left(\frac{b_i}{d_i}\right)_{ca}$$

$$FX_i = (d_i)_{co} + (b_i)_{co}$$

$$(d_i)_{co} = \frac{FX_i}{1 + \theta(b_i/d_i)_{ca}} \tag{10-7}$$

$$g(\theta) = \sum_{i=1}^{c} (d_i)_{co} - D$$

The corrected compositions $\{x_{ji}\}$ and $\{y_{yi}\}$ are computed as follows

$$x_{ji} = \frac{(x_{ji}/X_{Di})_{ca}(X_{Di})_{co}}{\sum\limits_{i=1}^{c} (x_{ji}/X_{Di})_{ca}(X_{Di})_{co}} \qquad y_{ji} = \frac{(y_{ji}/X_{Di})_{ca}(X_{Di})_{co}}{\sum\limits_{i=1}^{c} (x_{ji}/X_{Di})_{ca}(X_{Di})_{co}} \tag{10-8}$$

where

$$(X_{Di})_{co} = \frac{(d_i)_{co}}{D}$$

On the basis of these compositions (either the x_{ji}'s or the y_{ji}'s), a new set of temperatures may be computed by use of the K_b method. The temperatures so obtained are used for the next trial. This formulation of the θ method was first proposed by Lyster et al.[9] To demonstrate the convergence characteristics of this method, Example 10-1 stated in Table 10-1 was solved. The solution of this example is presented in Tables 10-2 and 10-3.

The N Newton-Raphson Method

The formulation of this method parallels the formulation presented in Chap. 4 for the $2N$ Newton-Raphson method for columns operating at finite reflux ratios. Again ideal solutions are assumed, and the condition that $L_j/V_{j+1} = 1$ (as $V_{j+1} \to \infty$) eliminates the necessity for making energy balances. Suppose that the specifications for the conventional distillation column are the same as those stated above in the formulation of the θ method.

In the following development, which is based on the work of Izarraraz,[8] the y_{ji}'s are taken to be the independent variables and the x_{ji}'s as the dependent variables. The y_{ji}'s and x_{ji}'s play the same roles in the following development as did the v_{ji}'s and l_{ji}'s, respectively, in the development of the $2N$ Newton-Raphson method in Chap. 4. When the x_{ji}'s are eliminated from the component-material balances [Eqs. (10-1) and (10-2)] by use of the equilibrium relationships

Table 10-1 Statement of Example 10-1

Component	FX	Specifications
C_3H_8	5.0	Boiling point liquid feed, partial condenser, column pressure =
$i\text{-}C_4H_{10}$	15.0	300 lb/in² abs, 3 rectifying plates, and 8 stripping plates
$n\text{-}C_4H_{10}$	25.0	plus a reboiler. The column is at total reflux in both the
$i\text{-}C_5H_{12}$	20.0	rectifying and stripping section. The feed rate $F = 100$,
$n\text{-}C_5H_{12}$	35.0	and $D = B = 50$ mol/h. Use the equilibrium and enthalpy data
		given in Tables B-1 and B-2 of App. B.

$F = 100.0$

Table 10-2 Convergence characteristics of the θ method of convergence for Example 10-1

		Temperature (°F), trial number				
Plate	Initial Profile	1	2†	3	4	5
1 (distillate†)	190	246.24	239.34	239.72	239.71	239.71†
2	200	266.88	254.08	255.27	255.17	255.17
3	210	286.67	269.41	271.59	271.59	271.62
4	220	310.89	284.29	288.09	287.49	287.57
5 (feed)	230	326.32	297.29	301.75	300.97	301.09
6	240	335.07	307.58	311.90	311.13	311.25
7	250	338.88	315.13	318.87	318.16	318.27
8	260	339.82	320.41	323.33	322.84	322.92
9	270	339.35	324.06	326.27	325.92	325.97
10	280	338.29	328.24	328.24	327.99	328.03
11	290	337.05	328.46	329.63	329.46	329.48
12	300	335.85	329.86	330.66	330.55	330.57
13 (reboiler)	310	334.79	330.97	331.49	331.42	331.43
D (calculated)	348.684	58.511	49.238	50.128	49.981	50.003
θ	3.0842×10^{-3}	3.54508	0.835601	1.02858	0.995848	1.00057

AMDAHL, Computer time $= 0.16$ s, WATFIV Compiler
† The temperature 239.71°F is the dew-point temperature of the distillate, and the corresponding bubble-point temperature of the distillate (in the liquid state) is 224.83°F.

Table 10-3 Product Distribution for Example 10-1

Component	b_i	d_i
C_3H_8	0.0	4.9999
$i\text{-}C_4H_{10}$	0.0079	14.992
$n\text{-}C_4H_{10}$	0.1341	24.866
$i\text{-}C_5H_{12}$	16.209	3.709
$n\text{-}C_5H_{12}$	33.651	1.3492

[Eq. (10-3)], the following set of equations is obtained for a column having a partial condenser

$$-y_{1i} - \frac{By_{Ni}}{DK_{Ni}} = \frac{-FX_i}{D}$$

$$\frac{y_{1i}}{K_{1i}} - y_{2i} = 0$$

$$\frac{y_{2i}}{K_{2i}} - y_{3i} = 0 \tag{10-9}$$

$$\vdots$$

$$\frac{y_{N-1, i}}{K_{N-1, i}} - y_{Ni} = 0$$

The first expression of Eq. (10-9) consists of a modified form of the overall component-material balance [Eq. (10-2)] in which X_{Di} has been replaced by its equivalent y_{1i} (for a partial condenser) and x_{Bi} has been expressed in terms of y_{Ni} by use of the equilibrium relationship $y_{Ni} = K_{Ni} x_{Ni} = K_{Ni} x_{Bi}$. Thus the combination of the component-material balances and equilibrium relationships may be represented by the matrix equation

$$A_i y_i = -f_i \tag{10-10}$$

where

$$A_i = \begin{bmatrix} -1 & 0 & 0 & 0 & \cdots & 0 & -\dfrac{B}{DK_{Ni}} \\ \dfrac{1}{K_{1i}} & -1 & 0 & 0 & \cdots & 0 & 0 \\ 0 & \dfrac{1}{K_{2i}} & -1 & 0 & \cdots & 0 & 0 \\ \hdotsfor{7} \\ 0 & \cdots & & 0 & & \dfrac{1}{K_{N-1, i}} & -1 \end{bmatrix}$$

$$f_i = [(-FX_i/D)\ 0\ 0\ \cdots\ 0]^T$$

$$y_i = [y_{1i}\ y_{2i}\ \cdots\ y_{Ni}]^T$$

This matrix equation is readily solved for the y_{ji}'s by use of the following algorithm which is developed in a manner analogous to the Thomas algorithm (see Prob. 10-1). For the set of equations

$$B_1 x_1 + C_N x_N = D_1$$

$$A_2 x_1 + B_2 x_2 = D_2$$

$$A_3 x_2 + B_3 x_3 = D_3 \tag{10-11}$$

$$\vdots \qquad \vdots \qquad \vdots$$

$$A_N x_{N-1} + B_N x_N = D_N$$

in which the A's, B's, C's and D's are constant, compute

$$e_1 = \frac{C_N}{B_1} \qquad e_k = -\frac{A_k}{B_k} e_{k-1} \qquad (k = 2, 3, \ldots, N)$$

$$g_1 = \frac{D_1}{B_1}$$

$$g_k = \frac{D_k - A_k g_{k-1}}{B_k} \qquad (k = 2, 3, \ldots, N - 1) \tag{10-12}$$

$$g_N = \frac{D_N - A_N g_{N-1}}{(1 - e_N)B_N}$$

Then

$$x_N = g_N \qquad (k = N, N - 1, \ldots, 2)$$
$$x_{k-1} = g_{k-1} - e_{k-1} x_N \qquad (k = 2, 3, \ldots, N) \tag{10-13}$$

The N functions based on the equilibrium relationships are stated in terms of dew-point functions, namely,

$$F_j = \sum_{i=1}^{c} \left(\frac{1}{K_{ji}} - 1 \right) y_{ji} \tag{10-14}$$

and the N Newton-Raphson equations may be represented by the matrix equation

$$\mathbf{J} \, \Delta \mathbf{T} = -\mathbf{f} \tag{10-15}$$

where the Jacobian $\mathbf{J}$ contains the partial derivatives $\partial F_j / \partial T_k$ $(j = 1, 2, \ldots, N;$ $k = 1, 2, \ldots, N)$, and

$$\Delta T = [\Delta T_1 \ \Delta T_2 \ \ldots \ \Delta T_N]^T$$
$$\mathbf{f} = [F_1 \ F_2 \ \ldots \ F_N]^T$$

The partial derivatives appearing in the jacobian $\mathbf{J}$ may be evaluated in a manner similar to that demonstrated for procedure 1 in Chap. 4. For any particular T_j, say T_k,

$$\frac{\partial F_j}{\partial T_k} = \sum_{i=1}^{c} \left(\frac{1}{K_{ji}} - 1 \right) \frac{\partial y_{ji}}{\partial T_k} - \sum_{i=1}^{c} \frac{1}{K_{ji}^2} \frac{\partial K_{ji}}{\partial T_k} \tag{10-16}$$

where $\partial K_{ji} / \partial T_k = 0$ for $j \neq k$. The partial derivatives of the y_{ji}'s with respect to each T_k may be found by use of the calculus of matrices in a manner analogous to that demonstrated in Chap. 4. In particular, after the component-material balances [Eq. (10-10)] have been solved for the y_{ji}'s, the partial derivatives may

Table 10-4 Convergence characteristics of the N Newton-Raphson method for Example 10-1

Plate	Initial profile	\multicolumn{5}{c}{Temperature (°F), trial number}				
		1	2	3	4	5†
1 (distillate†)	190	231.72	231.60	237.46	239.60	239.71†
2	200	229.50	244.01	249.56	254.56	255.16
3	210	230.69	258.86	262.80	269.88	271.59
4	220	235.09	279.61	280.20	285.44	287.50
5 (feed)	230	242.40	307.06	302.40	301.01	301.07
6	240	266.11	338.84	325.24	313.76	311.28
7	250	265.42	355.71	335.17	320.12	322.91
8	260	282.62	362.26	336.42	323.45	322.91
9	270	302.28	360.55	333.53	325.94	325.96
10	280	319.62	355.36	330.73	327.98	328.02
11	290	330.27	348.76	329.52	329.49	329.48
12	300	334.24	342.70	329.81	330.59	330.56
13 (reboiler)	310	334.81	338.47	330.77	331.44	331.43
D (calculated)	36.978	44.030	55.514	50.687	49.978	49.998

AMDAHL, Computer time = 1.17 s, WATFIV Compiler
† The temperature of 239.71°F is the dew-point temperature of the distillate, and its corresponding bubble-point temperature of the distillate (in the liquid state) is 224.83°F.

be computed by use of

$$A_i \frac{\partial y_i}{\partial T_k} = -\left(\frac{\partial A_i}{\partial T_k}\right) y_i \qquad (10\text{-}17)$$

in a manner similar to that shown in Chap. 4 for $\partial v_i / \partial T_k$.

Example 10-1 was also solved by use of the N Newton-Raphson method.[8] The convergence characteristics of this method for this example are shown in Table 10-4. Again as in Chap. 4, the corrections ΔT were reduced by the factors $1/2, 1/4, 1/8, \ldots$, until the T_j's so obtained were within the range of the curve fits. A comparison of the computer times given in Tables 10-2 and 10-4 shows that the θ method is several times faster than the N Newton-Raphson method.

The Almost Band Algorithm

For highly nonideal solutions, the Almost Band Algorithm is recommended. The following formulation of the Almost Band Algorithm follows closely the one recommended by Izarraraz et al.[7] Suppose that the column has a partial condenser and that the distillate rate V_1 is specified. The equations required to describe the column consist of the equilibrium relationships

$$\gamma_{ji}^V y_{ji} - \gamma_{ji}^L K_{ji} x_{ji} = 0 \qquad \left(\begin{array}{l} j = 1, 2, \ldots, N \\ i = 1, 2, \ldots, c \end{array}\right)$$

$$\sum_{i=1}^{c} x_{ji} = 1 \qquad (j = 1, 2, \ldots, N) \qquad (10\text{-}18)$$

$$\sum_{i=1}^{c} y_{ji} = 1 \qquad (j = 1, 2, \ldots, N)$$

and the material balances

$$y_{j+1, i} = x_{ji} \qquad \left(\begin{array}{l} j = 1, 2, \ldots, N - 1 \\ i = 1, 2, \ldots, c \end{array}\right) \qquad (10\text{-}19)$$

$$FX_i = v_{1i} + l_{Ni} \qquad (i = 1, 2, \ldots, c)$$

Let the following set of independent variables be selected.

$$\mathbf{x} = [v_{1, 1} \cdots v_{1, c} \, (x_{j, 1} \cdots x_{j, c} \, T_j)_{j=1, N-1} \, l_{N, 1} \cdots l_{N, c} \, T_N]^T \qquad (10\text{-}20)$$

The problem may be formulated in terms of the following $[N(c + 1) + c]$ independent functions

$$\mathbf{f} = [m_{1, 1} \, m_{1, 2} \cdots m_{1, c} \, S_1 \, (f_{j, 1} \, f_{j, 2} \cdots f_{j, c} \, F_j)_{j=1, N-1} \, f_{N, 1} \, f_{N, 2} \cdots f_{N, c}]^T \qquad (10\text{-}21)$$

where

$$S_1 = \frac{\sum_{i=1}^{c} v_{1i}}{V_1} - 1$$

$$m_{1i} = FX_i - v_{1i} - l_{Ni}$$

$$f_{1i} = \frac{\gamma_{1i}^V v_{1i}}{\sum_{i=1}^{c}} - \gamma_{1i}^L K_{1i} x_{1i} \qquad (j = 2, 3, \ldots, N - 1)$$

$$f_{ji} = \gamma_{ji}^V x_{j-1, i} - \gamma_{ji}^L K_{ji} x_{ji}$$

$$f_{Ni} = \gamma_{Ni}^V x_{N-1, i} - \frac{\gamma_{Ni}^L K_{Ni} l_{Ni}}{\sum_{i=1}^{c} l_{Ni}}$$

$$F_j = \sum_{i=1}^{c} x_{ji} - 1 \qquad (j = 1, 2, \ldots, N)$$

When the independent variables and functions are ordered as indicated by Eqs. (10-20) and (10-21), a banded jacobian matrix is obtained in which most of the elements lie along the principal diagonal as shown in Fig. 10-1.

Example 10-1 was also solved by use of the Almost Band Algorithm, and the results obtained are presented in Table 10-5. Again the computer time required by the Almost Band Algorithm is several times that required by the θ method.

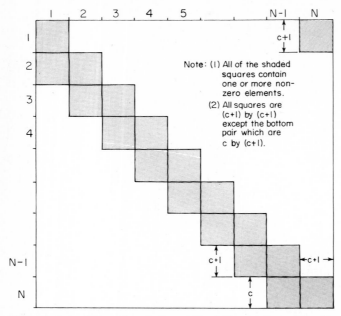

Figure 10-1 An $[N(c + 1) + c]$ Almost Band Formulation of the Newton-Raphson method for a distillation column at total reflux.

Table 10-5 Convergence characteristics of the Almost Band Algorithm for Example 10-1

| | Initial profile | Temperature (°F), trial number | | | | | | |
		1	2	3	4	5	6	7
1 (distillate†)	190	192.23	217.20	226.82	233.07	236.33	239.70	239.71
2	200	208.36	233.64	242.06	248.60	251.96	255.22	255.18
3	210	219.54	246.34	255.57	263.88	267.91	271.70	271.62
4	220	230.10	257.33	268.87	279.08	283.59	287.68	287.56
5 (feed)	230	240.13	267.32	281.64	292.82	297.28	301.21	301.08
6	240	248.82	274.19	291.53	303.04	307.44	311.35	311.24
7	250	257.75	281.86	300.58	311.05	314.89	318.34	318.26
8	260	266.13	287.30	306.18	315.75	319.47	322.95	322.91
9	270	275.34	294.83	312.11	319.94	323.04	325.99	325.97
10	280	284.61	301.99	316.77	322.00	325.56	328.04	328.23
11	290	293.79	308.42	320.27	325.29	327.40	329.48	329.48
12	300	302.92	314.36	323.43	327.19	328.88	330.56	330.56
13 (reboiler)	310	312.07	320.21	326.46	329.05	330.23	331.42	331.43
D (calculated)	59.5	58.313	54.156	52.078	51.039	50.519	50.0000	50.0000

AMDAHL, Computer time = 3.92 s, WATFIV Compiler

† The temperature of 239.71°F is the dew-point temperature of the distillate, and its corresponding bubble-point temperature of the distillate (in the liquid state) is 224.83°F.

10-2 CONTINUOUS AND BATCH DISTILLATION COLUMNS AT TOTAL REFLUX IN BOTH SECTIONS, TYPE 2: $(D = 0, B = 0, F = 0)$

This operating condition may be approached in a number of ways. A continuous distillation column may be placed on total reflux operation by setting the external rates equal to zero. This type of operation is commonly used in the startup of either a continuous or a batch distillation column. After a given quantity of feed has been introduced into the column, it may be allowed to come to a steady state in which no streams enter or leave the column. As will be demonstrated below, the precise steady state which is reached depends upon the holdup on each plate as well as the holdups in the accumulator and the reboiler.

The equations required to describe distillation columns at total reflux in both sections are formulated in a manner analogous to that shown in Sec. 10-1 for the θ method and the two Newton-Raphson methods. For purposes of illustration, the equations for the θ method are developed below.

The θ Method of Convergence

For a distillation column operating at steady state at total reflux of type 2 $(D = 0, B = 0, F = 0)$, the overhead vapor leaving the top of the column is condensed in a total condenser and returned to the column as liquid reflux. Similarly all of the liquid leaving the bottom plate enters the reboiler and after it has been vaporized reenters the bottom plate as vapor.

In the interest of simplicity, the vapor holdup throughout the column is regarded as negligible. Suppose that the remaining specifications are taken to be (1) the total number N of equilibrium stages, (2) the column pressure, (3) the composition $\{X_i\}$ and total moles U_F of feed introduced to the column, and (4) the moles of liquid holdup U_j ($j = 1, 2, \ldots, N$) on $N - 1$ of the N stages. The total moles of feed U_F introduced to the column is of course equal to the sum of the holdups U_j ($j = 1, 2, \ldots, N$).

The component-material balances may be combined with the equilibrium relationships for ideal solutions to give the following set of equations

$$\frac{u_{2i}}{u_{1i}} = \frac{1}{K_{2i}} \frac{U_2}{U_1}$$

$$\frac{u_{3i}}{u_{1i}} = \frac{1}{K_{2i} K_{3i}} \frac{U_3}{U_1}$$

$$\vdots$$

$$\frac{u_{Ni}}{u_{1i}} = \frac{1}{K_{2i} K_{3i} \cdots K_{Ni}} \frac{U_N}{U_1}$$

$$(10\text{-}22)$$

The set of $N - 1$ specified values of U_j and U_F serve to determine the complete set of U_j's since

$$\sum_{j=1}^{N} U_j = U_F \tag{10-23}$$

Thus on the basis of the known values of the U_j's and a set of assumed T_j's $(j = 1, 2, \ldots, N)$, the u_{ji}/u_{1i}'s are readily computed by use of Eq. (10-22). Of the N holdups, let $U_2, U_3, \ldots, U_N$ be selected as the independent set. Each of these $N - 1$ independent holdups gives rise to a θ multiplier defined by

$$\left(\frac{u_{ji}}{u_{1i}}\right)_{co} = \theta_j \left(\frac{u_{ji}}{u_{1i}}\right)_{ca} \qquad (j = 2, 3, \ldots, N) \tag{10-24}$$

as suggested by Barb and Holland[1,2] and Holland.[5] The θ_j's are to be determined such that the corrected u_{ji}'s are in overall component-material balance and in agreement with the specified values of the U_j's. The formula for $(u_{1i})_{co}$ is developed as follows. Since the corrected u_{ji}'s are required to be in overall component-material balance, it follows that

$$U_F X_i = \sum_{j=1}^{N} (u_{ji})_{co} \tag{10-25}$$

By use of the defining equation for the θ_j's [Eq. (10-24)], the following formula for $(u_{1i})_{co}$ is readily obtained from Eq. (10-25)

$$(u_{1i})_{co} = \frac{U_F X_i}{1 + \sum_{j=2}^{N} \theta_j (u_{ji}/u_{1i})_{ca}} \tag{10-26}$$

The desired set of θ's is that set of positive numbers that makes $g_2 = g_3 = \cdots = g_N = 0$, simultaneously,

$$g_j(\theta_2, \theta_3, \ldots, \theta_N) = \sum_{i=1}^{c} (u_{ji})_{co} - U_j \qquad (j = 2, 3, \ldots, N) \tag{10-27}$$

where $(u_{1i})_{co}$ is given by Eq. (10-26) and

$$(u_{ji})_{co} = \theta_j \left(\frac{u_{ji}}{u_{1i}}\right)_{ca} (u_{1i})_{co} \tag{10-28}$$

Instead of the set g_j's $(j = 2, 3, \ldots, N)$, the set of g_j's $(j = 1, 2, \ldots, N - 1)$ may be used to find the set of θ_j's $(j = 2, 3, \ldots, N)$. After the θ_j's have been computed by use of a numerical method such as the Newton-Raphson method, the corrected u_{1i}'s are found by use of Eq. (10-28). Then the mole fractions are readily computed by employing the next expression which follows from the definition of a mole fraction and the fact that the θ_j's have been found that make

$$\sum_{i=1}^{c} (u_{ji})_{co} = U_j$$

for each j. Thus

$$x_{ji} = \frac{(u_{ji})_{co}}{U_j} = \frac{(u_{ji}/u_{1i})_{ca}(u_{1i})_{co}}{\sum_{i=1}^{c} (u_{ji}/u_{1i})_{ca}(u_{1i})_{co}} \tag{10-29}$$

After the x_{ji}'s have been determined, the temperatures are found in the usual way by the K_b method. These temperatures are used to find the next set of u_{ji}/u_{1i}'s by Eq. (10-22).

The Modified θ Method of Convergence

In the interest of increasing the speed of performing the calculations, the following approximations are made. Let the θ's corresponding to the holdups be set equal to each other, namely,

$$\theta = \theta_j \qquad (j = 2, 3, \ldots, N) \tag{10-30}$$

This θ is selected such that the sum of the specified holdups

$$\sum_{j=2}^{N} U_j$$

is satisfied, that is, such that

$$g(\theta) = \sum_{j=2}^{N} \sum_{i=1}^{c} (u_{ji})_{co} - \sum_{j=2}^{N} U_j \tag{10-31}$$

where

$$(u_{1i})_{co} = \frac{U_F X_i}{1 + \theta \sum_{j=2}^{N} (u_{ji}/u_{1i})_{ca}}$$

and $(u_{ji})_{co}$ is given by Eq. (10-28) with θ_j replaced by θ. [Note that the function $g(\theta)$ is merely the sum of the g_j's given by Eq. (10-27).]

EXAMPLES To demonstrate the effect of the holdup specifications on the steady state solution of a batch distillation column at total reflux (a column operating at total reflux of type 2: $D = 0$, $B = 0$, $F = 0$), Examples 10-2 and 10-3 are presented in Table 10-6. The temperature profiles given in Table 10-7 were found by solving Examples 10-2 and 10-3 by use of the calculational procedure described above.

Table 10-6 Statement of Examples 10-2 and 10-3: effect of holdup on distillation columns at total reflux, type 1 ($D = 0$, $B = 0$, $F = 0$)

Examples 10-2 and 10-3		Stage	Holdups (moles)	
			Example 10-2	Example 10-3
Component	$U_F X_i$	1 (condenser)	4	3.57143
		2	1	3.57143
C_3H_8	2.5	3	1	3.57143
$i\text{-}C_4H_{10}$	7.5	4	1	3.57143
$n\text{-}C_4H_{10}$	12.5	5	1	3.57143
$i\text{-}C_5H_{12}$	10.0	6	1	3.57143
$n\text{-}C_5H_{12}$	17.5	7	1	3.57143
	50.0	8	1	3.57143
		9	1	3.57143
Other conditions		10	1	3.57143
		11	1	3.57143
$Q_R = 350,000$ Btu/h; column pressure		12	1	3.57143
$= 300$ lb/in² abs. The column has		13	1	3.57143
a total condenser, 12 plates, and a		14 (reboiler)	34	3.57143
reboiler. The K data and enthalpy			50	50.00000
data are given in Tables B-1 and B-2.				

Table 10-7 Solution of Examples 10-2 and 10-3: effect of holdup distributions on distillation columns at total reflux ($D = 0$, $B = 0$, $F = 0$)

Stage	Temperature profiles (°F) for Examples 10-2 and 10-3	
	Example 10-2	Example 10-3
1 (condenser)	175.24	175.67
2	190.55	191.05
3	203.61	208.89
4	213.24	225.04
5	219.91	240.95
6	224.67	258.31
7	228.57	276.78
8	232.58	294.00
9	237.80	307.79
10	245.51	317.54
11	256.82	323.92
12	271.59	327.95
13	289.73	330.49
14 (reboiler)	302.34	332.12

The Limiting Conditions of Equivalence for Total Reflux of Types 1 and 2

Consider first the equations for total reflux of type 2, and note that the U_j's appearing explicitly may be eliminated from the component-material balances by multiplying each expression of Eq. (10-22) by U_1/U_j. But the expressions so obtained still depend upon the holdups because the θ_j's, and hence the x_{ji}'s, depend upon the holdups. The component-material balances given by

Eq. (10-22) are of the same general form at those given by Eq. (10-4) for a continuous column.

For the particular set of specifications that follow, the steady state solution obtained for the batch distillation (a column at total reflux of type 2, $D = 0$, $B = 0$, $F = 0$) is the same as that obtained for a continuous distillation column with the same number of plates at total reflux.

1. $U_j = 0$ $(j = 2, 3, \ldots, N-1)$
2. $U_1/U_N = D/B$ and $U_F/U_1 = F/D$
3. The feeds F and U_F have the same composition

Note U_j $(j = 1, 2, \ldots, N)$ and U_F refer, of course, to the batch distillation column. By following the analysis offered by Barb and Holland,[1, 2] the equivalence of the solutions for the batch and continuous columns is readily shown as follows. Since $U_j = 0$ $(j = 2, 3, \ldots, N-1)$, the material balance given by Eq. (10-25) reduces to

$$U_F X_i = (u_{1i})_{co} + (u_{Ni})_{co} \tag{10-32}$$

and

$$(u_{1i})_{co} = \frac{U_F X_i}{1 + \theta_N(u_{Ni}/u_{1i})_{ca}} \tag{10-33}$$

Thus, Eq. (10-27) reduces to

$$g_1(\theta_N) = \sum_{i=1}^{c} (u_{1i})_{co} - U_1 \tag{10-34}$$

In view of conditions 2 and 3 given above and Eq. (10-32), Eq. (10-34) may be restated as follows

$$\frac{D}{U_1} g_1(\theta_N) = \sum_{i=1}^{c} \frac{F X_i}{1 + \theta_N(b_i/d_i)_{ca}} - D \tag{10-35}$$

Now recall that for a continuous column at total reflux of type 1, the g function given by Eq. (10-7) may be restated in the form

$$g(\theta) = \sum_{i=1}^{c} \frac{F X_i}{1 + \theta(b_i/d_i)_{ca}} - D$$

For a given set of $(b_i/d_i)_{ca}$'s, the $\theta_N > 0$ which makes $(D/U_1)g_1(\theta_N) = 0$ is obviously equal to the θ which makes the above function $g(\theta) = 0$. Also, it is readily shown that Eq. (10-22) reduces to the same set of x_{ji}/x_{1i}'s as those given by Eq. (10-5) for the continuous column. First, the expression for each stage j of Eq. (10-22) is multiplied by U_1/U_j, and then condition 1 ($U_j = 0$, $j = 2, 3, \ldots$, $N-1$) is imposed. The resulting set of equations are identical to those given by Eq. (10-5) for a continuous distillation with a total condenser.

Since all actual columns have finite holdups on each stage, the separations achieved in a column at total reflux of type 2 ($D = 0$, $B = 0$, $F = 0$) may differ considerably from those given by the continuous column at total reflux of type 1. The deviation of the separations achieved by the two types of operation is illustrated by the following example taken from Barb and Holland.[1, 2]

Table 10-8 Statement of Example 10-4

Conditions: All conditions are the same as those given for Examples 10-2 and 10-3 except the holdups which are specified as follows:

$$U_F = 50; \quad U_1 = U_N; \quad U_k = U_j \qquad j = 2, 3, \ldots, N - 1.$$

$$U_1 = \frac{50 - 12U_k}{2}$$

Obtain the solutions for a sequence of problems with U_k ranging from 2.0 to 0.001.

From the statement of Example 10-4 presented in Table 10-8, it is seen that the holdups U_j $(j = 2, 3, \ldots, N - 1)$ were decreased while the sum of the U_j's $(j = 1, 2, \ldots, N)$ was held fixed with $U_1 = U_N$. As the U_j's $(j = 2, 3, \ldots, N - 1)$ go to zero, Example 10-4 approaches Example 10-1. In the statement for the continuous column, $D/B = 1$ and $F/D = 2$; whereas, for Example 10-4, $U_1/U_N = 1$ and $U_F/U_1 = 2$. As is to be expected, the solution obtained is the same as the one obtained for Example 10-1, the continuous column. The approach of the solution for the batch column to that for the continuous column is displayed in Fig. 10-2 by showing the approach of the bubble-point temperature of the distillate (or reflux) of the batch column to 224.83°F, the bubble-point temperature of the distillate of the continuous column.

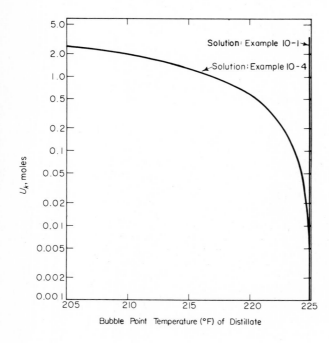

Figure 10-2 Approach of the steady state solutions for the start-up period of a batch column to the solution for a corresponding continuous column. [C. D. Holland, *Unsteady State Processes with Applications in Multicomponent Distillation*, 1966, by courtesy Prentice-Hall, Inc.]

10-3 OPERATING CHARACTERISTICS OF CONTINUOUS DISTILLATION COLUMNS OVER WIDE RANGES OF DISTILLATE AND REFLUX RATES

To determine the behavior of a distillation column over wide ranges of operating conditions, calculational procedures are needed for treating certain limiting conditions which arise such as $D = 0$ and $D = F$. Procedures are presented and discussed below.

One of the more interesting results obtained in this section is shown in Fig. 10-3. Examination of this figure shows that as the distillate rate is varied from $D = 0$ to $D = F$ at a fixed reflux rate L_1, there exists a particular D $(0 < D < F)$ at which the mole fraction of each component other than the lightest and the heaviest of the mixture passes through a maximum. The lightest component is seen to have its maximum X_D at $D = 0$, and the heaviest component has its maximum X_D (namely $X_D = X$) at $D = F$. The results shown in Fig. 10-3 were obtained by solving Example 10-1 at each of several distillate rates ranging from $D = 0$ to $D = F$ at $L_1 = 125$.

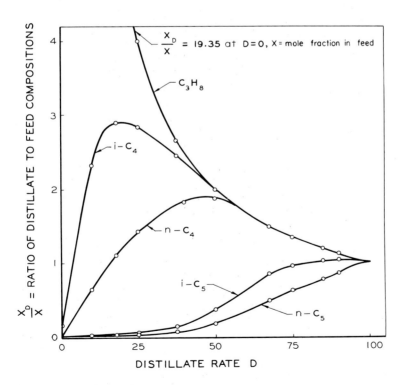

Figure 10-3 Variation of the distillate rate D at a fixed reflux rate, $L_1 = 125$.

A Continuous Distillation Column at Total Reflux in the Rectifying Section $D = 0$, $B = F$, and $L_1 = $ Finite Number

In addition to the above specifications, it is also supposed that the following variables and operating conditions are fixed: the number of stages N, the feed plate location f, the model for the feed plate behavior, the complete definition of the feed (the composition, flow rate, and thermal condition), and the column pressure. Since $D = 0$, the condenser behaves as a total condenser, and thus $x_{1i} = X_{Di}$.

Problems of this type may be solved by use of the θ method of convergence as demonstrated by Lyster et al.[9] However, since $(b_i)_{co}$ is known in advance to be equal to FX_i, the θ method reduces to direct iteration and convergence may become slow. Thus, the application of the $2N$ Newton-Raphson method to problems of this type is recommended.

The problem may be formulated in terms of either the mole fractions or the component-flow rates. When the component-flow rates are used, the resulting set of equations differ only in minor respects from those presented in Chap. 4 for conventional distillation columns. The problem is formulated in terms of the following set of independent variables

$$\mathbf{x} = [\theta_2 \ \theta_3 \ \cdots \ \theta_N \ Q_C \ T_1 \ T_2 \ \cdots \ T_N \ Q_R]^T \tag{10-36}$$

Except for the component-material balance enclosing the condenser-accumulator section

$$-l_{1i} + v_{2i} = 0 \tag{10-37}$$

the remaining component-material balances are the same as those shown for conventional distillation columns in Chap. 4. The combined set of component-material balances and equilibrium relationships may be represented again by the matrix equation

$$\mathbf{A}_i \mathbf{v}_i = -\mathbf{f}_i \tag{10-38}$$

where

$$\mathbf{A}_i = \begin{bmatrix} -1 & 1 & 0 & 0 & 0 & 0 \\ 1 & -(1 + A_{2i}) & 1 & 0 & 0 & 0 \\ 0 & A_{2i} & -(1 + A_{3i}) & 1 & 0 & 0 \\ \hdotsfor{6} \\ 0 & \cdots & 0 & A_{N-2, i} & -(1 + A_{N-1, i}) & 1 \\ 0 & \cdots & \cdots & 0 & A_{N-1, i} & -(1 + A_{Ni}) \end{bmatrix}$$

$$\mathbf{v}_i = [l_{1i} \ v_{2i} \ v_{3i} \ \cdots \ v_{Ni}]^T$$

$$\mathbf{f}_i = [0 \cdots 0 \ v_{Fi} \ l_{Fi} \ 0 \ \cdots \ 0]^T$$

Note that by commencing with the top row and adding each row to the one below it and then replacing the latter by the result so obtained, an upper triangular matrix which is bidiagonal is obtained (see Prob. 10-8).

Similarly, the total-material balances are the same as those presented for conventional distillation columns except for the condenser-accumulator section, and for this stage,

$$-L_1 + V_2 = 0 \tag{10-39}$$

The complete set of total-material balances may be represented in the usual way by

$$\mathbf{RV} = -\mathscr{F} \tag{10-40}$$

where the matrix $\mathbf{R}$ is readily formed from the matrix $\mathbf{A}_i$ of Eq. (10-38) by replacing A_{ji} by R_j, where

$$R_j = \theta_j (L_j/V_j)_a$$

The vectors $\mathbf{V}$ and $\mathscr{F}$ contain the elements

$$\mathbf{V} = [L_1 \ V_2 \ V_3 \ \cdots \ V_N]^T$$
$$\mathscr{F} = [0 \ \cdots \ 0 \ V_F \ L_F \ 0 \ \cdots \ 0]^T \tag{10-41}$$

The matrix equation $\mathbf{RV} = -\mathscr{F}$ is readily solved by converting $\mathbf{R}$ to an upper triangular matrix which is bidiagonal by the same procedure described above for $\mathbf{A}_i$.

Observe that in the solution of Eq. (10-40), L_1 is regarded as a dependent variable, that is, on the basis of a given set of θ_j's $(j = 2, 3, \ldots, N)$, L_1 and the complete set of V_j's $(j = 2, 3, \ldots, N)$ may be computed. When the solution set of θ_j's has been found, the value of L_1 given by Eq. (10-40) must be equal, of course, to the specified value of L_1. This condition is assured by taking one of the functions to be

$$S_1 = \frac{L_1}{(L_1)_{\text{spec}}} - 1 \tag{10-42}$$

This is only one of the $2N + 1$ independent functions which are needed to solve this problem in terms of the $2N + 1$ independent variables enumerated by Eq. (10-36). The remaining functions are taken to be the equilibrium functions F_j $(j = 1, 2, \ldots, N)$ and the energy balance functions G_j $(j = 1, 2, \ldots, N)$. Thus, the complete set $\mathbf{f}$ is

$$\mathbf{f} = [S_1 \ F_1 \ F_2 \ \cdots \ F_N \ G_1 \ G_2 \ \cdots \ G_N]^T \tag{10-43}$$

Except for the function F_1, the equilibrium functions are of the same form as those stated in Chap. 4. The bubble-point form of the function F_1 is used, namely,

$$F_1 = \frac{1}{L_1} \sum_{i=1}^{c} (K_{1i} - 1) l_{1i} \tag{10-44}$$

Enthalpy balance functions which are somewhat simpler in form than those presented in Chap. 4 may be obtained by taking advantage of the fact that

$d_i = 0$. In this formulation of the enthalpy balance functions, the enclosures for the energy balances include the condenser-accumulator and each stage below it. [It should be observed that the Newton-Raphson method could have been formulated in terms of $2N - 1$ independent variables by omitting Q_C, Q_R, and T_N.] The condenser duty could have been omitted by enclosing the top stage instead of the condenser section. The temperature T_N could have been omitted because the composition of the bottoms is known to be that of the feed, and the reboiler duty Q_R could have been omitted because it appears in only one function, G_N.

In Example 10-5, the specifications are taken to be $D = 0$, $B = F$ and $L_1 = 125$ lb mol/h for the example stated in Table 10-1. Composition profiles for a column at this operation are presented in Fig. 10-4.

When the distillate rate is fixed and the reflux rate is varied, the compositions obtained for the example stated in Table 10-1 are shown in Fig. 10-5. As V_2/L_1 approaches 1, a continuous distillation column at total reflux in both sections is obtained. The range of operation displayed in Fig. 10-5 extends from total reflux, $V_2/L_1 = 1$, to the limiting condition of no liquid reflux ($L_1 = 0$) for

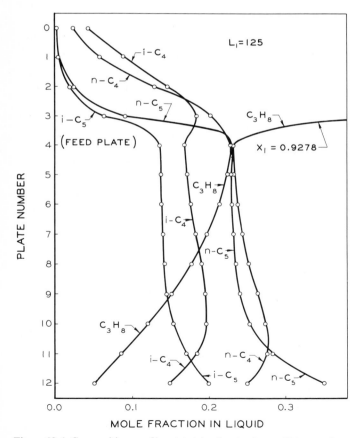

Figure 10-4 Composition profiles at total reflux in the rectifying section.

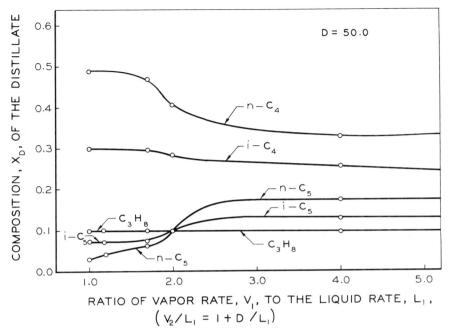

Figure 10-5 Variation of the reflux rate L_1 at a fixed distillate rate D.

the rectifying section. In the limit as L_1 approaches zero, X_{Di} approaches the mole fraction of component i in the vapor that enters the plate above the feed plate.

A Continuous Distillation Column at Total Reboil in the Stripping Section $(B = 0, D = F, \text{ and } L_1 = \text{Finite Number})$

When a finite value is specified for L_1 and D is taken equal to F, the condition of total reboil $(B = 0)$ in the stripping section is realized. The equations needed to apply the $2N$ Newton-Raphson method are formulated in a manner analogous to that demonstrated above for total reflux in the rectifying section. The condition that $B = 0$ is reflected by the equations for the Nth stage. The matrices of Eq. (10-38), the component-material balances, contain the following elements

$$\mathbf{A}_i = \begin{bmatrix} -(1 + A_{1i}) & 1 & 0 & 0 & & 0 & 0 \\ 1 & -(1 + A_{2i}) & 1 & 0 & & 0 & 0 \\ \cdots\cdots\cdots\cdots\cdots\cdots\cdots\cdots\cdots\cdots\cdots\cdots\cdots\cdots\cdots\cdots\cdots \\ 0 & & 0 & A_{N-2,i} & -(1 + A_{N-1,i}) & 1 \\ 0 & & \cdots & \cdots & 0 & A_{N-1,i} & -1 \end{bmatrix} \quad (10\text{-}45)$$

$$\mathbf{v}_i = [d_i \ v_{2i} \ v_{3i} \ \cdots \ v_{Ni}]^T$$

$$\mathbf{f}_i = [0 \ \cdots \ 0 \ v_{Fi} \ l_{Fi} \ 0 \ \cdots \ 0]^T$$

Table 10-9 Statement and solution of Example 10-6

1. Statement of Example 10-6

The statement of this example is the same as stated for Example 10-1 except that in this case $D = F$, $B = 0$, and $L_1 = 125$ lb mol/h.

2. Convergence characteristics of the temperatures and vapor rates

Temperature (°F), trial number

Plate	Initial profile	1	2	3	4	5	6
1	270	269.14	269.14	269.14	269.14	269.14	269.14
2	280	292.03	292.99	293.00	293.00	293.00	293.00
3	290	301.12	302.48	302.48	302.48	302.48	302.48
4	300	305.91	306.26	306.19	306.19	306.19	306.19
5	310	308.52	307.77	307.69	307.69	307.69	307.69
6	320	319.61	319.19	319.14	319.14	319.14	319.14
7	330	326.43	325.93	325.89	325.89	325.89	325.89
8	340	330.17	329.85	329.83	329.83	329.83	329.83
9	350	330.68	332.16	332.17	332.17	332.17	332.17
10	360	326.52	333.21	333.61	333.62	333.62	333.62
11	370	314.72	331.34	334.49	334.56	334.56	334.56
12	380	289.84	318.64	333.39	335.20	335.22	335.22
13	390	241.98	274.39	312.20	332.53	335.65	335.70

Vapor rates, lb mol/h

Plate	Initial profile	1	2	3	4	5	6
2	235.0	225.00	225.00	225.00	225.00	225.00	225.00
3	241.0	236.85	228.81	228.21	228.21	228.21	228.21
4	244.6	243.27	231.69	230.25	230.24	230.24	230.24
5	244.6	242.02	232.27	231.16	231.11	231.11	231.11
6	244.6	246.97	238.22	237.06	237.05	237.05	237.05
7	244.6	253.72	245.47	244.30	244.29	244.29	244.29
8	244.6	257.06	250.27	249.33	249.33	249.33	249.33
9	244.6	257.83	253.12	252.55	252.54	252.54	252.54
10	244.6	257.17	255.28	254.55	254.54	254.54	254.54
11	244.6	256.46	259.74	255.87	255.81	255.81	255.80
12	244.6	258.07	273.93	258.82	256.65	256.64	256.64
13	244.6	266.97	311.11	291.06	259.66	257.28	257.22

3. Final bottoms composition

Component	FX	x_B
C_3H_8	5.0	0.247965×10^{-5}
$i\text{-}C_4H_{10}$	15.0	0.331362×10^{-3}
$n\text{-}C_4H_{10}$	25.0	0.236626×10^{-2}
$i\text{-}C_5H_{12}$	20.0	0.145221
$n\text{-}C_5H_{12}$	35.0	0.852083

Matrices **R**, **V**, and $\mathscr{F}$ of Eq. (10-40) contain the following elements

$$\mathbf{R} = \begin{bmatrix} -(1+R_1) & 1 & 0 & 0 & 0 & 0 \\ R_1 & -(1+R_2) & 1 & 0 & 0 & 0 \\ \cdots & \cdots & \cdots & \cdots & \cdots & \cdots \\ 0 & \cdots & 0 & R_{N-2} & -(1+R_{N-1}) & 1 \\ 0 & \cdots & \cdots & 0 & R_{N-1} & -1 \end{bmatrix} \quad (10\text{-}46)$$

$$\mathbf{V} = [D_1 \ V_2 \ \cdots \ V_N]^T$$

$$\mathscr{F} = [0 \ \cdots \ 0 \ V_F \ L_F \ 0 \ \cdots \ 0]^T$$

By first adding row N to row $N-1$ and replacing row $N-1$ by the result so obtained and then continuing this operation for successive rows, it is possible to transform Eqs. (10-45) and (10-46) to bidiagonal form.

The independent variables are

$$\mathbf{x} = [\theta_1 \ \theta_2 \ \cdots \ \theta_{N-1} \ Q_C \ T_1 \ T_2 \ \cdots \ T_N Q_R]^T \quad (10\text{-}47)$$

and the corresponding functions **f** are

$$\mathbf{f} = [S_1 \ F_1 \ F_2 \ \cdots \ F_N \ G_1 \ G_2 \ \cdots \ G_N]^T \quad (10\text{-}48)$$

where S_1 is again given by Eq. (10-42) and the functions $\{F_j\}$ are the same as those stated in Chap. 4. Functions G_1 through G_N are formulated in a manner analogous to that described for the previous case where $D = 0$, $B = F$, and L_1 is nonzero and finite. To illustrate the application of these equations, Example 10-6 was solved. The statement and solution of this example appears in Table 10-9.

Calculational procedures for complex columns at total reflux are readily developed by extending the procedures for the θ method and the Newton-Raphson method in a manner analogous to that demonstrated in Chaps. 3 and 4, respectively.

The procedures presented in Secs. 10-1 through 10-3 are useful in the analysis of the column behavior over wide ranges of operating conditions as demonstrated above. Also, the procedures presented may be used in Chap. 9 in the design of distillation columns instead of the approximate method based on the Fenske equation.

NOTATION

u_{ji} molar holdup of component i in the liquid state on plate j.
U_j total molar holdup of the liquid on stage j.
U_F total moles of feed introduced to a batch distillation column.

PROBLEMS

10-1 Begin with the set of algebraic equations given by Eq. (10-11) and develop the recurrence formulas given by Eq. (10-12).

10-2 (*a*) For a batch distillation column operating in the product period ($D > 0$), show that the component-material balances and the total-material balance over the time period from t_n to $t_n + \Delta t$ are given by

$$\int_{t_n}^{t_n + \Delta t} [-d_i] \, dt = U_N x_{Ni}|_{t_n + \Delta t} - U_N x_{Ni}|_{t_n}$$

$$\int_{t_n}^{t_n + \Delta t} [-D] \, dt = U_N|_{t_n + \Delta t} - U_N|_{t_n}$$

where it is assumed that the holdups for stages $j = 1, 2, \ldots, N - 1$ are negligible relative to the reboiler or still ($j = N$).

(*b*) Use the mean value theorems (see App. A) and the appropriate limiting process to reduce the above integral-difference equations to the following differential equations

$$-DX_{Di} = \frac{d(U_N x_{Ni})}{dt}$$

$$-D = \frac{dU_N}{dt}$$

(*c*) From the results given in part (*b*), obtain the equation of Smoker and Rose,[14]

$$\ln \frac{U_N}{U_N^0} = \int_{x_{Ni}^0}^{x_{Ni}} \frac{dx_{Ni}}{X_{Di} - x_{Ni}}$$

where the superscript 0 denotes the values of the variables at time $t = 0$. Smoker and Rose[14] proposed that the integral appearing on the right-hand side of the above equation be evaluated by use of the graphical method of McCabe and Thiele.[10]

10-3 A batch distillation column with a single plate, the reboiler, is carried out at constant temperature and pressure by increasing the rate of flow of steam to the column to compensate for the decrease in the concentration of the lower-boiling components over the course of the distillation. Also, the unit is to be operated such that the partial pressure of steam in the vapor product is less than the saturation pressure of steam at the temperature of the reboiler. (This problem is based on the development given in Holland and Welch.[6])

(*a*) Beginning with the overall material balance on component i in a batch distillation column with a single stage and a withdrawal rate D

$$\int_{t_n}^{t_n + \Delta t} [-DX_{Di}] \, dt = u_i|_{t_n + \Delta t} - u_i|_{t_n}$$

show that

$$-DX_{Di} = \frac{du_i}{dt}$$

where the plate subscript which would normally be carried by u_i has been dropped, and where

$D = D_v + D_s$

D_v = the molar flow rate of two-phase (or volatile) components in the distillate

D_s = molar flow rate of steam in the distillate at any time t.

(b) Show

$$-D_v = \frac{dU}{dt}$$

(c) By use of the result obtained in part (a), show that

$$\alpha_i \left(\frac{u_i}{u_b}\right) = \frac{du_i}{du_b}$$

and that

$$u_i = u_i^0 \left(\frac{u_b}{u_b^0}\right)^{\alpha_i}$$

for all components ($i = 1$ through $i = c$) except steam ($i \neq s$). The superscript 0 refers to the values of the variables at $t = 0$. Also in the development of this expression, equilibrium between the vapor (X_{Di}) leaving the column and the liquid $\{x_i\}$ in the reboiler is assumed, that is,

$$X_{Di} = K_i x_i = K_i \frac{u_i}{U}$$

(d) Show that

$$U \frac{du_b}{u_b} = K_b[dU - dS]$$

where S = total moles of steam required. Next, show that the steam requirement S is given by

$$\sum_{i=1, \neq s}^{c} \frac{u_i^0}{\alpha_i} \left[1 - \left(\frac{u_b}{u_b^0}\right)^{\alpha_i}\right] = K_b[U^0 - U + S]$$

where s refers to steam.

10-4 (a) If the model for the vaporization efficiency (defined by $y_{ji} = E_{ji} K_{ji} x_{ji}$) is taken to be

$$E_{ji} = \beta_j \bar{E}_i$$

as originally suggested by Professor W. H. McAdams (according to Perry[11]), show that α_i in the final result should be replaced by

$$\frac{\bar{E}_i \alpha_i}{\bar{E}_b}$$

(b) Suppose that the initial batch in the reboiler contains a single-phase liquid component (a nonvolatile liquid) that is soluble in all proportions with the other components in the liquid phase. Show that, for this case, the following term should be added to the left-hand side of the expression given in part (d) of Prob. 10-3,

$$u_k \ln \frac{u_b^0}{u_b}$$

where u_k denotes the moles of inert or single-phase liquid in the initial charge to the reboiler.[6] Also, the summation that appears in this expression should be over all components except steam and the single-phase liquid ($i = 1$ to $i = c$, and $i \neq s, k$).

10-5 On the basis of the following definition of η_j, develop Eqs. (10-27) and (10-29) from first principles

$$(u_{ji})_{co} = \eta_j \left(\frac{u_{ji}}{u_{1i}}\right)_{ca} (u_{1i})_{co}$$

where the sum of the corrected values of u_{ji} is equal to the specified value of U_j,

$$\sum_{i=1}^{c} (u_{ji})_{co} = U_j$$

and the corrected values of u_{1i} is to be found by means of a component-material balance over the complete column.

10-6 Obtain the column vector corresponding to the right-hand side of Eq. (10-17).

10-7 Formulate the $2N + 2$ functions when the Newton-Raphson method is used to solve a problem involving a column operating at $D = 0$, $B = F$, and L is finite, and the choice of independent variables is given by Eq. (10-36).

10-8 (a) Begin with Eq. (10-38) and carry out the row operations of the gaussian elimination process which are required to transform the matrix $\mathbf{A}_i$ to bidiagonal form.

  (b) Apply the multiplication rule to the result found in part (a) to obtain an algorithm for solving such problems.

REFERENCES

1. D. K. Barb: "Solution of Problems Involving the Separation of Multi-Component Mixtures by Batch Distillation," Ph.D Dissertation, Texas A&M University, 1967.
2. D. K. Barb and C. D. Holland: "Batch Distillation," *Proceedings of the 7th World Petroleum Cong.* **4**:31 (1967).
3. M. R. Fenske: "Fractionation of Straight-Run Pennsylvania Gasoline," *Ind. Eng. Chem.* **24**:482 (1932).
4. A. V. Fiacco and G. P. McCormick: "Computational Minimization Technique for Nonlinear Programming, *Manage. Sci.*, **10**(4):601 (1964).
5. C. D. Holland: *Unsteady State Processes with Applications in Multicomponent Distillation*, Prentice-Hall, Englewood Cliffs, N.J., 1966.
6. ——— and N. W. Welch: "Steam Batch Distillation Calculations," *Hydrocarbon Process. Pet. Refiner*, **36**:251 (1957).
7. A. Izarraraz, J. R. Haas, and C. D. Holland: *Hydrocarbon Process.*, in press.
8. A. Izarraraz: M.S. Thesis, "Application of Theta Method to Distillation Columns Where One or More Reactions Occur per Stage A&M University, 1978.
9. W. N. Lyster, S. L. Sullivan, Jr., J. A. McDonough, and C. D. Holland: "Figure Distillation this New Way. Part 4—Explore Operating Characteristics of a Column," *Pet. Refiner*, **39**(8):121 (1960).
10. W. L. McCabe and E. W. Thiele: "Graphical Design of Fractionating Columns," *Ind. Eng. Chem.*, **17**:605 (1925).
11. J. H. Perry: Editor-in-Chief. *Chemical Engineers Handbook*, 2d ed., McGraw Hill Book Company, New York, 1941, 1938.
12. Lord Rayleigh (J. Strutt): "On the Distillation of Binary Mixtures," *Phil. Mag.* (6)**4**:527 (1902).
13. C. S. Robinson and E. R. Gilliland: *Elements of Fractional Distillation*, 4th ed. McGraw-Hill Book Company, New York, 1950.
14. E. H. Smoker and Arthur Rose: "Graphic Determination of Batch Distillation Curves for Binary Mixtures," *Trans. Am. Inst. Chem. Eng.* **36**:285 (1940).

ELEVEN

CONVENTIONAL AND COMPLEX DISTILLATION COLUMNS AT MINIMUM REFLUX

Two general classes of problems are solved in this chapter. These are perhaps best described by consideration of a conventional distillation column. The problems solved in Sec. 11-1 consist of those in which infinitely many stages are specified for the rectifying and stripping sections as well as the reflux and distillate rates (L_1 and D), and it is required to find the corresponding product distribution. In Sec. 11-2 two separations (say b_l/d_l and b_h/d_h) are specified and it is required to find the smallest reflux ratio required to effect these two separations. The smallest reflux ratio is found by employing infinitely many stages in the rectifying and stripping sections. An application of the θ method of convergence is used in Sec. 11-2 to determine the minimum reflux ratio. In Sec. 11-3, an application of the Newton-Raphson method (which is formulated in terms of two independent variables per stage) is used to solve problems requiring the determination of the product distribution as well as problems requiring the determination of the minimum reflux ratio. In Sec. 11-4 a formulation of the Newton-Raphson method, called an Almost Band Algorithm, is presented for solving problems requiring the determination of the product distribution as well as problems requiring the determination of the minimum reflux ratio. In Sec. 11-5, a collection of proofs dealing with the behavior columns with infinitely many stages is presented.

Unlike the earlier analytical approaches of Underwood,[15] Murdoch and Holland,[9, 10] and Acrivos and Amundson[1] that were based on constant-molar overflows and constant-relative volatilities between the rectifying and stripping section pinches, the numerical methods presented in this chapter permit variations in both the flow rates and the relative volatilities. Shiras, Hanson, and Gibson[12] as well as Bachelor[2] used the equations of Thiele and Geddes[14] in the treatment of columns at minimum reflux.

11-1 DETERMINATION OF PRODUCT DISTRIBUTIONS FOR COLUMNS WITH INFINITELY MANY STAGES BY USE OF THE θ METHOD

To initiate the development of this method, some qualitative ideas concerning the behavior of distillation columns with infinitely many stages are presented.

For a binary mixture, it is demonstrated graphically in Fig. 1-10 that if stepwise calculations are initiated at the top of the column and continued indefinitely, a condition is approached at the intersection of the operating line and the equilibrium curve at which the change in composition from one plate to the next approaches zero. An analogous situation exists in the case of a multicomponent mixture. That is, if calculations are initiated at the top of the column on the basis of a given distillate composition and flow rate, and calculations are continued down the column on the basis of a given reflux rate by solving the material balances, energy balances, and equilibrium relationships simultaneously, a limiting condition will be found in which the change in composition from one plate to the next is less than any small preassigned number. This limiting condition has come to be known as the rectifying pinch. Similarly, when calculations are initiated at the bottom of the column and continued indefinitely, a limiting condition in the stripping section is approached in which the change in composition from one plate to the next is less than any small preassigned number. This limiting condition has come to be known as the stripping pinch. For the case of a binary mixture, the stripping pinch is seen to occur at the intersection of the operating line for the stripping section and the equilibrium curve as demonstrated in Fig. 1-10. Also, for the case of a binary mixture, it is evident from Fig. 1-10 that the rectifying and stripping pinches occur at and adjacent to the feed plate. This may or may not be the situation for a multicomponent mixture. If all of the components of the feed are present in the distillate, then the rectifying pinch occurs at and adjacent to the feed plate. If one or more of the heavy components do not appear in the distillate, then the rectifying pinch occurs above the feed plate after the heavy component (or components) have been separated. Similarly, if one or more of the light components of the feed do not appear in the bottoms, then the stripping pinch occurs below the feed plate after the light component (or components) have been separated.

If all of the components of the feed are present in the distillate but not in the bottoms, then the rectifying pinch occurs at and adjacent to the feed plate while the stripping pinch occurs below the feed plate as the light components become separated. Conversely, if all of the components of the feed are present in the bottoms but not in the distillate, the stripping pinch occurs at and adjacent to the feed plate while the rectifying pinch occurs above the feed plate as the heavy components become separated.

Historically, Brown and Holcomb[3] were among the first to describe the behavior of multicomponent mixtures in columns in which infinitely many stages are used, and they prepared the graph shown in Fig. 11-1 from a set of calculations which were made manually by use of the calculational procedure known as the Lewis and Matheson method.[5] Demonstration of the fact that pinches do occur when infinitely many stages are employed is readily demonstrated by use of this method as shown below.

Use of the Lewis and Matheson Method[5] for the Demonstration of the Existence of a Pinch

Although this method has not been described herein for solving problems in which a finite number of plates are used, it may be used for conventional distillation columns in conjunction with the θ method of convergence as described by Holland.[4] However, when more than one feed plate is involved, the procedure is subject to problems of roundoff error which appear to be unsolvable with today's computer capabilities. In spite of this difficulty, the Lewis and Matheson method is ideally suited for aid in the visualization of the characteristics of a column with infinitely many plates.

In the Lewis and Matheson method, the product flow rates, the $\{d_i\}$ and the $\{b_i\}$, are taken to be the independent variables. On the basis of sets of assumed values for the product flow rates $\{d_i\}$ and $\{b_i\}$, calculations are initiated at the top and at the bottom of the column by solving in succession the material balances, equilibrium relationships, and energy balances for each stage as suggested by Fig. 11-2. If the compositions obtained at the feed plate by making calculations down from the top of the column are in agreement with those obtained by making calculations up from the bottom of the column, then the correct sets of product rates were assumed to initiate the calculations.

For the case of infinitely many plates in each section of the column, the product flow rates, the $\{d_i\}$, are assumed for all components in the distillate, the heavy key and all components lighter, and calculations are carried out to the rectifying pinch. Then small quantities of the components heavier than the heavy key are introduced by assigning to them values of d_i's, and calculations are continued to the feed plate. Similarly, calculations are initiated at the bottom of the column on the basis of a set of assumed rates $\{b_i\}$ for the light key and all components heavier, and calculations are continued to the stripping pinch. Then small amounts of the components lighter than the light key are added in the form of b_i's and calculations are continued to the feed plate. If the correct sets $\{d_i\}$ and $\{b_i\}$ were assumed to initiate the calculations and to calculate from the pinches to the feed plate, then both sets of calculations will terminate in the same set of compositions for the feed plate.

The composition profiles shown in Fig. 11-1 represent the result obtained after many sets of trial calculations had been made in search of the correct sets of product flow rates $\{b_i\}$ and $\{d_i\}$. In this example, the light key is $i\text{-}C_4H_{10}$ and the heavy key is $n\text{-}C_4H_{10}$. The separated lights are C_2H_6 and C_3H_8, and the separated heavies are $i\text{-}C_5H_{12}$, $n\text{-}C_5H_{12}$, and $n\text{-}C_6H_{14}$.

That rectifying and stripping pinches do exist when infinitely many plates are used in the rectifying and stripping sections is readily demonstrated by use of the following relatively simple numerical example. In the interest of simplicity, it

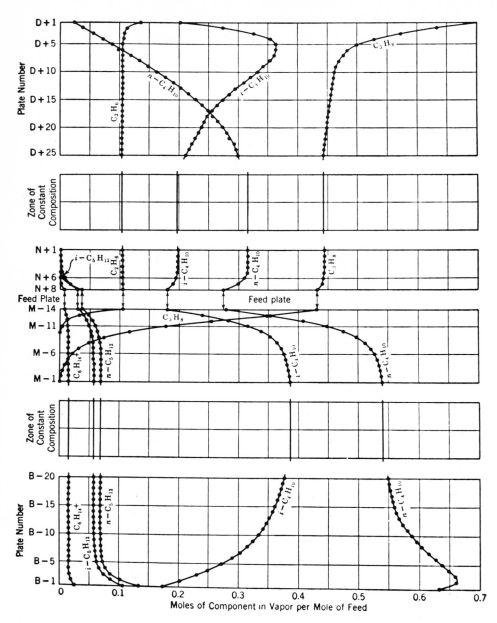

Figure 11-1 Composition of the vapor in a column containing infinitely many plates, moles of component per mole of feed. [G. G. Brown and D. E. Holcomb, *Pet. Eng.*, **11**, 23 (August 1940), *Courtesy Petroleum Engineer.*]

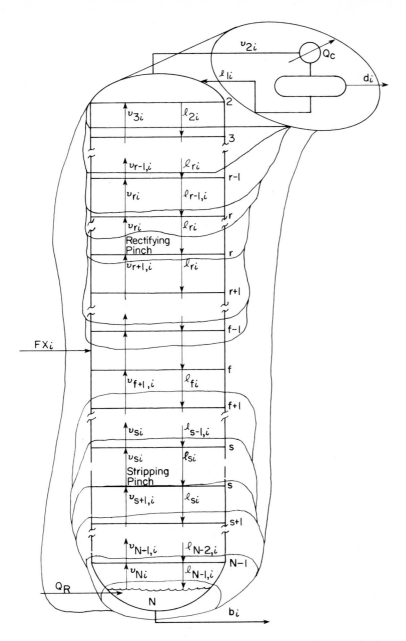

Figure 11-2 Material-balance enclosures for the Lewis and Matheson method.

is supposed that the molar overflows are constant throughout each section of the column and that the relative volatilities are independent of temperature. This example was taken from Murdoch and Holland.[9]

Example 11-1 Make calculations down from the top of the column until the change in the y_{ji}'s between successive stages is less than 10^{-7}.
Given:

Component	α_i	d_i	Other specifications
1	0.25	0	The column has a total condenser. The distillate
2	0.50	1.0714	rate is 66.429 and the reflux rate is 62.2705
3	1.00	42.8576	mol/h. The total flow rates remain constant in
4	2.00	22.5000	the rectifying section.

SOLUTION For the rectifying section, $V_r = L_r + D = 128.6995$, and $L/D = 0.937399$.

Component	$l_{1i} = \left(\dfrac{L_r}{D}\right)(d_i)$	$v_{2i} = l_{1i} + d_i$	y_{2i}	y_{2i}/α_i
2	1.004329	2.075727	0.0161285	0.03225696
3	40.17468	83.03227	0.6451616	0.6451616
4	21.09146	43.59145	0.3387071	0.1693535
		128.69994		$K_{2b} = 0.8467744$

Component	$x_{2i} = \left(\dfrac{1}{K_{2b}}\right)\left(\dfrac{y_{2i}}{\alpha_i}\right)$	$l_{2i} = L_r x_{2i}$	$v_{3i} = l_{2i} + d_i$	y_{3i}
2	0.0380939	2.372127	3.443535	0.026756
3	0.7619070	47.44435	90.30194	0.7016495
4	0.1999980	12.45400	34.95398	0.271593

Component	$\dfrac{y_{3i}}{\alpha_i}$	$x_{3i} = \left(\dfrac{1}{K_{3b}}\right)\left(\dfrac{y_{3i}}{\alpha_i}\right)$
2	0.053512	0.060061
3	0.7016495	0.787522
4	0.1357965	0.152416

$K_{3b} = 0.890958$

Continuation of this procedure yields
$$K_{4b} = 0.9130991 \qquad K_{5b} = 0.9276029, \ldots, \qquad K_{72b} = 1.027867$$
$$K_{73b} = 1.027867$$

The values of the y_i's and x_i's at the end of the 72d trial were as follows

Component	y_{72i}	$\dfrac{y_{72i}}{\alpha_i}$	$x_{72i} = \left(\dfrac{1}{K_{72b}}\right)\left(\dfrac{y_{72i}}{\alpha_i}\right)$
2	0.1421873	0.2843746	0.2766647
3	0.6291738	0.6291738	0.6121154
4	0.2286388	0.1143194	0.1112199

$$K_{72b} = 1.027867$$

Pinch Equations

From the results of the above example, it is evident that calculations may be continued until the values of the y_{ji}'s, x_{ji}'s, and K_{jb}'s for successive stages are less than any small preassigned number, that is,

$$\lim_{j\to\infty} y_{ji} = y_{ri} \qquad \lim_{j\to\infty} x_{ji} = x_{ri} \qquad \lim_{j\to\infty} K_{jb} = K_{rb} \tag{11-1}$$

where the subscript r denotes the pinch in the rectifying section. Consequently, the component-material balance

$$v_{j+1, i} = l_{ji} + d_i = A_{ji} v_{ji} + d_i \tag{11-2}$$

reduces to

$$v_{ri} = A_{ri} v_{ri} + d_i \tag{11-3}$$

for the pinch in the rectifying section. Thus, for the heavy key and all components lighter, the flow rates in the rectifying section are given by

$$v_{ri} = \frac{d_i}{1 - A_{ri}} \tag{11-4}$$

Similarly, for the light key and all components heavier, the component-material balance enclosing the bottom of the column and the stripping pinch reduces to

$$l_{si} = \frac{b_i}{1 - S_{si}} \tag{11-5}$$

Quantitative Definitions of the Various Types of Components

The *heavy key* component is the heaviest (the least volatile) component which appears in the distillate D. Thus, for this component $d_h > 0$, and since this component is in the distillate, it is also present in the rectifying pinch, and thus $v_{rh} > 0$. The heavy key is the heaviest component for which

$$A_{ri} < 1 \tag{11-6}$$

Similarly, the *light key* is the lightest component for which

$$S_{si} < 1 \tag{11-7}$$

The light key, the heavy key, and all components having K values between those of the keys are called *distributed components*. Components heavier than the heavy key are called *separated heavy components* because they are separated in the rectifying section and do not appear in the distillate. Components lighter than the light key are called *separated light components* because they are separated in the stripping section and do not appear in the bottom product B. More precisely, a separated heavy component is any component for which

$$A_{ri} \geq 1 \tag{11-8}$$

and a separated light component is any component for which

$$S_{si} \geq 1 \tag{11-9}$$

A further insight into the pinch equations is afforded by the following example.

Example 11-2 When Eq. (11-4) is summed over all components present in the rectifying pinch and restated in functional form, one obtains

$$\Omega_r(\phi_r) = \sum_{i=h}^{c} \frac{d_i}{1 - \phi_r/\alpha_i} - V_r \tag{A}$$

where the components are numbered in the order of increasing volatility, and

$$\phi_r = \frac{L_r}{V_r K_{rb}} \quad \left(\text{Note:} \quad A_{ri} = \frac{L_r}{V_r K_{ri}} = \frac{L_r}{V_r K_{rb}\alpha_i} = \frac{\phi_r}{\alpha_i} \right)$$

(a) Find the positive value of ϕ_r $(\alpha_{h-1} < \phi_r < \alpha_h)$ of Example 11-1 which is less than α_h that makes $\Omega_r(\phi_r) = 0$. (Note there are several other roots, but the one which is just less than α_h is the only one which has physical meaning.)

(b) Show that the $\{y_{ri}\}$ obtained by the pinch equations are the same as those found in Example 11-1.

SOLUTION (a) Assume that K_{rb} has the value given by the calculations shown in Example 11-1. Thus

$$\phi_r = \frac{L_r}{V_r K_{rb}} = \frac{62.2705}{(128.6995)(1.027867)} = 0.4707260$$

Then

Component	α_i	$\dfrac{\phi_r}{\alpha_i}$	$1 - \dfrac{\phi_r}{\alpha_i}$	d_i	$v_{ri} = \dfrac{d_i}{1 - \phi_r/\alpha_i}$
2	0.5	0.941452	0.058548	1.0714	18.299514
3	1.0	0.470726	0.5292740	42.8576	80.974371
4	2.0	0.235363	0.764637	22.5000	29.425727

$$V_r = 128.6996$$

and

$$\Omega_r(0.4707260) = 128.6996 - 128.6995 \cong 0$$

(b) By use of the above results compute $\{y_{ri}\}$ as follows:

Component	$y_{ri} = \dfrac{v_{ri}}{V_r}$
2	0.1421878
3	0.6291734
4	0.2286388

From the results of Example 11-2, it is evident that the compositions and temperature (or K_{rb}) at the rectifying pinch could have been obtained by solving Eq. (A) for $\phi_r(\alpha_{h-1} < \phi_r < \alpha_h)$. Similarly, the conditions at the stripping pinch are determined as shown in the following example.

Example 11-3 Since $l_{si} = A_{si} v_{si}$ and $S_{si} = 1/A_{si}$, Eq. (11-5) may be rearranged to give

$$v_{si} = \frac{b_i}{A_{si} - 1} \tag{A}$$

Summation over all components present in the stripping pinch followed by rearrangement and statement in functional form yields

$$\Omega_s(\phi_s) = \sum_{i=1}^{l} \frac{b_i}{1 - \phi_s/\alpha_i} + V_s \tag{B}$$

At the stripping section pinch

$$\phi_s = \frac{L_s}{V_s K_{sb}}$$

This is the positive root which is just greater than α_l and lies between α_l and α_{l+1}. Find the ϕ_s $(\alpha_l < \phi_s < \alpha_{l+1})$ for the following example.

Component	α_i	b_i	Other specifications
1	0.25	10.0000	Same as Example 11-1 plus
2	0.5	21.4286	the fact that $V_s = V_r$
3	1.0	2.1424	$= 128.6995$ and $L_s = L_r + F$
4	2.00		$F = 162.2705$

SOLUTION The root $\phi_s = 1.020432$ that makes $\Omega_s(\phi_s) = 0$ was found by use of the method of interpolation *regula falsi*.

Component	$\dfrac{\phi_s}{\alpha_i}$	$1 - \dfrac{\phi_s}{\alpha_i}$	$\dfrac{b_i}{1 - \phi_s/\alpha_i}$	y_{si}
1	4.081728	− 3.081728	− 3.2449327	0.0252157
2	2.040864	− 1.040864	− 20.587319	0.1599798
3	1.020432	− 0.020432	− 104.85512	0.8148074
			− 128.687	

Note: $v_{si} = -b_i/(1 - \phi_s/\alpha_i)$.

If one continued calculations from the pinches found in Examples 11-2 and 11-3 to the feed plate by introducing appropriate amounts of components 1 and 4 as described previously, a match in compositions would be obtained at the feed plate because the solution sets of $\{d_i\}$ and $\{b_i\}$ are given in the statement of Examples 11-1 and 11-3; see also Probs. 11-1 and 11-2.

The Lewis and Matheson approach in which the product flow rates $\{d_i\}$ and $\{b_i\}$ are selected as the independent variables is pursued no farther because of the problems of roundoff error encountered when this method is applied to complex columns with two or more feeds. The Lewis and Matheson method, however, has served the purpose of demonstrating the existence of the rectifying and stripping pinches for the case where the relative volatilities are constant and the total-flow rates of the vapor and liquid are constant within each section of the column. A proof of the existence of the pinches for the general case where the relative volatilities vary with temperature and the total-flow rates vary throughout each section of the column is presented in Sec. 11-5.

In the remainder of the discussion in the development of the θ method, the temperatures, and the total-flow rates are regarded as the independent variables, and the product rates $\{b_i\}$ and $\{d_i\}$ are dependent variables.

Development of the Component-Material Balances

To initiate the calculational procedure for the determination of the product distribution for specified reflux and distillate rates, a number of plates between the two pinches is selected. (As discussed in a subsequent section, too few plates but not too many plates may be selected.) Next L/V and temperature profiles for the plates between and including the two pinches as well as the distillate and bottoms temperatures are selected. Next the components of the feed are classified according to the above criteria. Since it is supposed that the complete definition of the feed, the reflux and distillate rates, as well as the column pressure and type of condenser are specified, the component-material balances can be solved for the component-flow rates throughout the column. The component-material balances may be simplified by taking advantage of the unique characteristics of the three classes of components, the distributed components, the separated lights, and the separated heavies.

MATERIAL BALANCES FOR THE DISTRIBUTED COMPONENTS
($A_{ri} < 1,\ S_{si} < 1$)

The enclosures used to make the material balances for the distributed components are shown in Fig. 11-3. The resulting equations follow.

$$
\begin{aligned}
v_{2i} - l_{1i} - d_i &= 0 \\
v_{ri} + l_{1i} - v_{2i} - l_{ri} &= 0 \\
v_{r+1,\,i} + l_{ri} - v_{ri} - l_{ri} &= 0 \\
v_{r+2,\,i} + l_{ri} - v_{r+1,\,i} - l_{r+1,\,i} &= 0 \\
\vdots \qquad\qquad \vdots \qquad\qquad \vdots & \\
v_{si} + l_{s-2,\,i} - v_{s-1,\,i} - l_{s-1,\,i} &= 0 \\
v_{si} + l_{s-1,\,i} - v_{si} - l_{si} &= 0 \\
l_{si} - v_{si} - b_i &= 0
\end{aligned}
\tag{11-10}
$$

Use of the equilibrium relationship, $l_{ji} = A_{ji} v_{ji}$, to eliminate the $\{l_{ji}\}$ permits the above equations to be stated in terms of the $\{v_{ji}\}$, $\{d_i\}$, and $\{b_i\}$. The resulting set of equations may be represented by the matrix equation

$$
\mathbf{A}_i \mathbf{v}_i = -\mathbf{\mathcal{f}}_i
\tag{11-11}
$$

where

$$
\mathbf{A}_i =
\begin{bmatrix}
-\rho_{1i} & 1 & 0 & 0 & \cdots & \cdots & 0 \\
A_{1i} & -1 & (1 - A_{ri}) & 0 & \cdots & & 0 \\
0 & 0 & -1 & 1 & 0 & \cdots & 0 \\
0 & 0 & A_{ri} & -\rho_{r+1,\,i} & 1 & 0 & 0 \\
\hline
0 & \cdots & 0 & A_{s-2,\,i} & -\rho_{s-1,\,i} & 1 & 0 \\
0 & \cdots & \cdots & 0 & A_{s-1,\,i} & -A_{si} & 0 \\
0 & \cdots & \cdots & \cdots & 0 & (A_{si} - 1) & -1
\end{bmatrix}
$$

$\mathbf{v}_i = [d_i \ v_{2i} \ v_{ri} \ v_{r+1,\,i} \ \cdots \ v_{s-2,\,i} \ v_{s-1,\,i} \ v_{si} \ b_i]^T$

$\mathbf{\mathcal{f}}_i = [0 \ \cdots \ 0 \ v_{Fi} \ l_{Fi} \ 0 \ \cdots \ 0]^T$, where v_{Fi} and l_{Fi} lie in rows $f - 1$ and f, respectively.

$A_{ji} = L_j/(K_{ji} V_j)$ for all j except $j = 1$. For $j = 1$, $A_{1i} = L_1/K_{1i}D$ for a partial condenser, and $A_{1i} = L_1/D$ for a total condenser.

$\rho_{ji} = 1 + A_{ji}$.

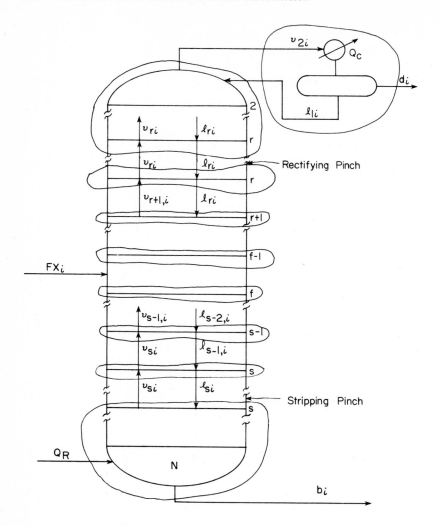

Figure 11-3 Material-balance enclosures for the distributed components—used in the θ method of convergence.

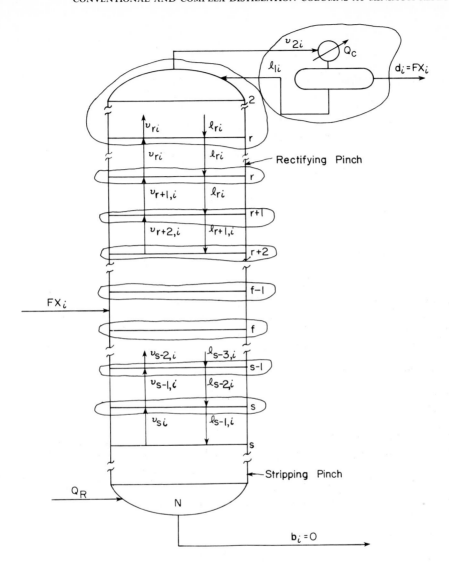

Figure 11-4 Material-balance enclosures for the separated lights—used in the θ method of convergence and the Newton-Raphson methods.

MATERIAL BALANCES FOR THE SEPARATED LIGHTS
$(S_{si} > 1, d_i = FX_i, b_i = 0)$

When the first balance encloses the top of the column and any stage r in the rectifying pinch and each of the other balances enclose the single stages, $r, r + 1$, $r + 2, \ldots, f - 1, f, \ldots, s - 2$, and $s - 1$ (see Fig. 11-4), the following equations are obtained.

$$
\begin{aligned}
v_{2i} - l_{1i} - FX_i &= 0 \\
v_{ri} + l_{1i} - l_{ri} - v_{2i} &= 0 \\
v_{r+1, i} + l_{ri} - v_{ri} - l_{ri} &= 0 \\
v_{r+2, i} + l_{ri} - v_{r+1, i} - l_{r+1, i} &= 0 \\
\vdots \qquad \vdots \qquad \vdots \qquad & \\
v_{s-1, i} + l_{s-3, i} - v_{s-2, i} - l_{s-2, i} &= 0 \\
v_{si} + l_{s-2, i} - v_{s-1, i} - l_{s-1, i} &= 0
\end{aligned}
\tag{11-12}
$$

Elimination of the liquid rates by use of the equilibrium relationship followed by rearrangement permits these equations to be stated in the form of matrix Eq. (11-11) where the elements of $\mathbf{A}_i$, $\mathbf{v}_i$, and f_i are as follows

$$
\mathbf{A}_i =
\begin{bmatrix}
1 & 0 & \cdots & \cdots & \cdots & 0 \\
-1 & (1 - A_{ri}) & 0 & \cdots & \cdots & 0 \\
0 & -1 & 1 & 0 & & \\
0 & A_{ri} & -\rho_{r+1, i} & 1 & 0 & 0 \\
\hdotsfor{6} \\
0 & 0 & A_{s-3, i} & -\rho_{s-2, i} & 1 & 0 \\
0 & \cdots & 0 & A_{s-2, i} & -\rho_{s-1, 1} & 1
\end{bmatrix}
\tag{11-13}
$$

$$
\mathbf{v}_i = \begin{bmatrix} v_{2i} & v_{ri} & v_{r+1, i} & \cdots & v_{fi} & v_{f+1, i} & \cdots & v_{s-2, i} & v_{s-1, i} & v_{si} \end{bmatrix}^T
$$

$$
f_i = \begin{bmatrix} (-\rho_{1i} FX_i), & (A_{1i} FX_i), & 0 \cdots 0, & v_{Fi}, & l_{Fi}, & 0 \cdots 0 \end{bmatrix}^T
$$

$$
\rho_{ji} = 1 + A_{ji}
$$

The solution set of values for the v_{ji}'s is readily obtained by application of the multiplication rule beginning with the first row. Originally, the component-material balances for the separated lights were formulated by enclosing the condenser-accumulator section and each plate in succession, $r, r + 1, r + 2, \ldots,$ $s - 3, s - 2,$ and $s - 1.$[4] The equations so obtained are equivalent to the set given by Eq. (11-12); see Prob. 11-7.

MATERIAL BALANCES FOR THE SEPARATED HEAVIES
$(A_{ri} > 1, b_i = FX_i, d_i = 0)$

The enclosures for the component-material balances shown in Fig. 11-5 yield the corresponding equations. Each stage between r and s are enclosed by a balance, and stage s and the bottom of the column are enclosed.

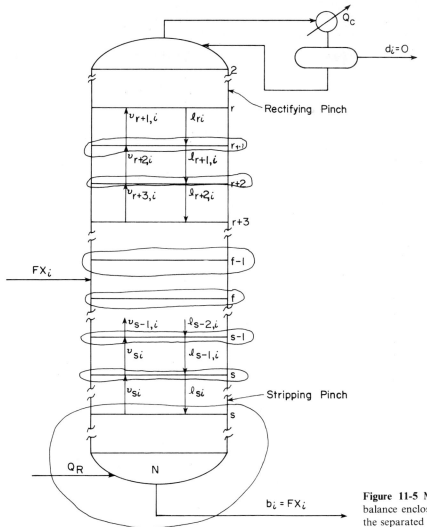

Figure 11-5 Material-balance enclosures for the separated heavies.

$$v_{r+2, i} + l_{ri} - v_{r+1, i} - l_{r+1, i} \quad = 0$$

$$v_{r+3, i} + l_{r+1, i} - v_{r+2, i} - l_{r+2, i} = 0$$

$$\vdots \qquad \vdots \qquad \vdots$$

$$v_{si} + l_{s-2, i} - v_{s-1, i} - l_{s-1, i} \quad = 0$$

$$v_{si} + l_{s-1, i} - v_{si} - l_{si} \quad = 0$$

$$l_{si} - v_{si} - FX_i \quad = 0$$

$$(11\text{-}14)$$

After the liquid rates have been eliminated, the resulting set of equations may be rearranged and stated in the form of the matrix equation given by Eq. (11-11) where the matrices A_i, v_i, and $\mathcal{f}_i$ contain the following elements.

$$
A_i = \begin{bmatrix}
A_{ri} & -\rho_{r+1,i} & 1 & 0 & 0 & 0 \\
0 & A_{r+1,i} & -\rho_{r+2,i} & 1 & 0 & 0 \\
\cdots & \cdots & \cdots & \cdots & \cdots & \cdots \\
0 & \cdots & 0 & A_{s-2,i} & -\rho_{s-1,i} & 1 \\
0 & \cdots & \cdots & 0 & A_{s-1,i} & -A_{si} \\
0 & \cdots & \cdots & \cdots & 0 & (1-A_{si})
\end{bmatrix}
\tag{11-15}
$$

$$
v_i = \begin{bmatrix} v_{ri} & v_{r+1,i} & \cdots & v_{f-1,i} & v_{fi} & \cdots & v_{s-2,i} & v_{s-1,i} & v_{si} \end{bmatrix}^T
$$

$$
\mathcal{f}_i = \begin{bmatrix} 0 & \cdots & 0 & v_{Fi} & l_{Fi} & 0 & \cdots & 0 & FX_i \end{bmatrix}^T
$$

Equation (11-5) is readily transformed into a bidiagonal matrix by use of the following procedure. Beginning with the bottom row, each row is added to the one above it and then the one above is replaced by the result so obtained. Originally,[4] the component-material balances were formulated by enclosing the bottom of the column and stages s, $s-1$, $s-2$, ..., $r+3$, $r+2$, and $r+1$. The equations so obtained are equivalent to those given by Eq. (11-14); see Prob. 11-8.

Application of the θ method

The θ method was first used to solve a problem of this type by McDonough and Holland.[6] After the component-material balances have been solved for the component-flow rates, the θ method is applied. In order to avoid numerical problems resulting from the presence of separated components, the equations for the θ method should be stated in the following form

$$
g(\theta) = \sum_{i=1}^{c} (d_i)_{ca} p_i - D
\tag{11-16}
$$

where

$$
p_i = \frac{FX_i}{(d_i)_{ca} + \theta(b_i)_{ca}}
$$

Observe that for a separated light component, $(b_i)_{ca} = 0$, and $p_i = 1$. For a separated heavy component, $(d_i)_{ca} = 0$, and $p_i = 1/\theta$. After the $\theta > 0$ that makes $g(\theta) = 0$ has been found, the compositions are computed as follows

$$
x_{ji} = \frac{(l_{ji})_{ca} p_i}{\sum\limits_{i=1}^{c} (l_{ji})_{ca} p_i} \qquad y_{ji} = \frac{(v_{ji})_{ca} p_i}{\sum\limits_{i=1}^{c} (v_{ji})_{ca} p_i}
\tag{11-17}
$$

Determination of the Temperatures

On the basis of the compositions found by use of Eq. (11-17), the temperatures to be used for the next trial are found by use of the K_b method [see Eqs. (2-30) and (2-31)]. In addition to the temperatures for each of the stages, the temperatures of the distillate and bottoms are found by use of the corrected flow rates, $\{(d_i)_{co}\}$ and $\{(b_i)_{co}\}$, and the K_b method.

Enthalpy Balances

The enthalpy balances are stated in the form of the constant-composition method in a manner analogous to that demonstrated in Chap. 2 for conventional distillation columns with a finite number of stages. The balance enclosing the condenser-accumulator section may be stated in the form

$$Q_C = L_1 \left[\sum_{i=1}^{c} (H_{2i} - h_{1i}) x_{1i} \right] + D \left[\sum_{i=1}^{c} (H_{2i} - H_{Di}) X_{Di} \right] \qquad (11\text{-}18)$$

The balances enclosing the top of the column and stages $r, r+1, r+2, \ldots, f-2, f-1, \ldots, s-2$, and $s-1$ are as follows

$$L_j = \frac{D \sum_{i=1}^{c} (H_{Di} - H_{j+1,i}) X_{Di} + Q_C}{\sum_{i=1}^{c} (H_{j+1,i} - h_{ji}) x_{ji}} \qquad (j = r, r+1, \ldots f-3, f-2)$$

$$L_{f-1} = \frac{D \sum_{i=1}^{c} (H_{Di} - H_{fi}) X_{Di} + V_F \sum_{i=1}^{c} (H_{fi} - H_{Fi}) y_{Fi} + Q_C}{\sum_{i=1}^{c} (H_{fi} - h_{f-1,i}) x_{f-1,i}} \qquad (11\text{-}19)$$

$$L_j = \frac{D \sum_{i=1}^{c} (H_{Di} - H_{j+1,i}) X_{Di} - F \sum_{i=1}^{c} (H_i - H_{j+1,i}) X_i + Q_C}{\sum_{i=1}^{c} (H_{j+1,i} - h_{ji}) x_{ji}}$$

$$(j = f, f+1, \ldots, s-3, s-2, s-1)$$

Note: $L_{s-1} = L_s$

The reboiler duty is found by use of the energy balance enclosing the entire column namely,

$$Q_R = DH_D + Bh_B + Q_C - FH \qquad (11\text{-}20)$$

The corresponding total vapor rates are found by use of the following total-material balances.

$$V_2 = L_1 + D$$

$$V_r = L_r + D$$

$$V_{j+1} = L_j + D \qquad (j = r + 1, r + 2, \ldots, f - 3, f - 2)$$

$$\text{(Note: } V_r = V_{r+1}) \quad (11\text{-}21)$$

$$V_f + V_F = L_{f-1} + D$$

$$V_{j+1} = L_j - B \qquad (j = f, f + 1, \ldots, s - 3, s - 2)$$

$$V_s = L_s - B$$

Selection of a Suitable Finite Set of Plates Between the Two Pinches

This selection was made by use of a concept described by Bachelor.[2] When the keys are two distinct components other than the lightest and heaviest of the feed, the pinches are separated by infinitely many plates. The pinches occur in the limit as the "separated" components are separated in the rectifying and stripping sections. That is, for any small preassigned positive number, there exists a plate (going from the feed plate toward the top of the column) beyond which the mole fraction for the component just heavier than the heavy key is less than the preassigned number. Thus, too few plates may be selected but not too many. In view of this, some arbitrary number of plates may be selected for the first trial, say five in the rectifying section and five in the stripping section. At the end of each trial, the original estimate may be adjusted by the addition of one (or more) plates to each section of the column. (Of course, it is necessary from a practical point of view to set an upper bound on the total number of plates.) In the rectifying section, a plate is added just above the pinch for the previous trial and becomes the pinch plate r for the next trial. In the stripping section, it is added just below the previous pinch and becomes the pinch plate s for the next trial. The total molar flow rates of the added pinch plates are taken equal to those calculated for the respective pinches by the previous trial. The temperatures of the pinches (the two plates added) are taken equal to the corrected values obtained for the plates denoted by r and s in the previous trial. For all problems considered, the solutions were obtained with less than sixty plates above and sixty plates below the feed plate.

Special Cases

When all of the components are distributed, the pinches in the rectifying and stripping sections occur at and adjacent to the feed plate. A method which consists of a direct solution to this problem has been presented previously.[6] However, problems in which all of the components are distributed may be solved by use of the general procedure outlined above and demonstrated by Example 11-5. This example was solved by use of the general procedure described above and by the method presented by McDonough and Holland.[6] The same results (see Table 11-3) were obtained by both methods.

When all of the components are separated (a perfect split) or when only one component is distributed, a solution to the problem may be obtained by solving each of two pairs of equations simultaneously, one pair for the rectifying section and one pair for the stripping section. The equations for the rectifying pinch are developed as follows. Since $l_{ri} = A_{ri} v_{ri}$, it follows from Eq. (11-4) that

$$l_{ri} = \frac{A_{ri} d_i}{1 - A_{ri}} \tag{11-22}$$

which may be rewritten as follows

$$x_{ri} = \frac{X_{Di}}{1 - \Psi_r(1 - K_{ri})} \tag{11-23}$$

where

$$\Psi_r = V_r/D$$

After Eq. (11-23) has been summed over all components i and the result so obtained has been stated in functional form, the following material balance function is obtained

$$P_1(\Psi_r, T_r) = \sum_{i=1}^{c} \frac{X_{Di}}{1 - \Psi_r(1 - K_{ri})} - 1 \tag{11-24}$$

It is to be understood, of course, that $X_{Di} = 0$ for all of the separated heavy components in this and the other equations to follow for the rectifying pinch.

The enthalpy function is formulated by use of the enthalpy balance enclosing the top of the column and the rectifying pinch [see the first expression given by Eq. (11-19)]. This expression for L_r contains the condenser duty Q_C. However since $\{d_i\}$, D, and V_2 are known, the top of the column is uniquely determined at the outset (Note: $v_{2i} = l_{1i} + d_i$), and Q_C may be computed by use of Eq. (11-18). Use of Eq. (11-23) permits the enthalpy balance enclosing the rectifying pinch to be restated in the following functional form

$$P_2(\Psi_r, T_r) = \frac{D}{Q_C} \sum_{i=1}^{c} [(\Psi_r - 1)\lambda_{ri} x_{ri} + (H_{ri} - H_{Di})X_{Di}] - 1 \tag{11-25}$$

where x_{ri} is given by Eq. (11-23) and

$$\lambda_{ri} = H_{ri} - h_{ri}$$

Equations (11-24) and (11-25) represent two independent functions in the two unknowns Ψ_r and T_r. The desired solution is that pair of positive values Ψ_r and T_r that not only give $P_1 = P_2 = 0$ but also satisfy the conditions given by Eq. (11-26). For definiteness, consider a conventional distillation column in which D, L_1, and infinitely many stages are specified. Suppose that components $h, h + 1, h + 2, \ldots, c$ appear in the rectifying pinch and components $i = h, h - 1$,

$h - 2, \ldots, 1$ appear in the stripping pinch. For such a system the desired set of roots must satisfy the conditions

$$1 < \Psi_r < \frac{1}{1 - K_{rh}}$$

$$T_{DP} < T_r < T_h \tag{11-26}$$

where T_{DP} is that temperature obtained by making a dew-point calculation on the basis of the X_{Di}'s, and T_h is that temperature at which $K_{rh} = 1$.

The following material-balance and energy functions for the stripping section pinch are developed in a manner analogous to that demonstrated for the rectifying pinch. The material-balance function is

$$p_1(\Psi_s, T_s) = \sum_{i=1}^{c} \frac{x_{Bi}}{1 - \Psi_s(K_{si} - 1)} - 1 \tag{11-27}$$

where

$$\Psi_s = V_s/B$$

Also, it is to be understood that $x_{Bi} = 0$ for all of the separated light components in this and the other equations for the stripping pinch. The enthalpy-balance function for the stripping pinch is as follows

$$P_2(\Psi_s, T_s) = \frac{B}{Q_R} \sum_{i=1}^{c} [(\Psi_s + 1)\lambda_{si} x_{si} - (H_{si} - h_{Bi})x_{Bi}] - 1 \tag{11-28}$$

where

$$x_{si} = \frac{x_{Bi}}{1 - \Psi_s(K_{si} - 1)}$$

At the outset Q_R may be computed by an energy balance enclosing the entire column. Equations (11-27) and (11-28) may be solved for the positive values of Ψ_s and T_s that not only make $p_1 = p_2 = 0$, but also satisfy the conditions

$$0 < \Psi_s < \frac{1}{K_{sl} - 1}$$

$$T_l < T_s < T_{BP} \tag{11-29}$$

where component l is the lightest component present in the stripping pinch and T_{BP} is the temperature of the bottoms.

Illustrative examples A wide variety of problems have been solved successfully by McDonough et al.[8] by use of the procedures described above. In order to illustrate these methods, a fairly simple problem used by McDonough and Holland[6] was selected.

The operating conditions were varied as required in order to produce different arrays of separated components. In Example 11-4 three components were separated, in Example 11-5 all components were distributed, and in Example 11-6 all components were separated. The statements and solutions are presented in Tables 11-1 through 11-4.

Table 11-1 Statement and specifications of Examples 11-4, 11-5, 11-6, and 11-9

Component	Component no.	Examples 11-4, 11-5, and 11-6 FX_i	Example	L/D	D
C_3H_8	1	20	11-4	150	40
$i\text{-}C_4H_{10}$	2	20	11-5	55	40
$n\text{-}C_4H_{10}$	3	20	11-6	200	60
$i\text{-}C_5H_{12}$	4	20	11-9	$b_2/d_2 = 0.37200$	$b_3/d_3 = 2.687363$
$n\text{-}C_5H_{12}$	5	20			

Other specifications

For all examples, the column pressure is 400 lb/in² abs, a partial condenser is used, and the feed enters as a liquid at its bubble-point temperature. The initial temperature profiles are taken to be linear between the pinches. Assumed values for T_r and T_s were $T_r = T_{feed} - 15°F$ and $T_s = T_{feed} + 15°F$, and initial values for T_D and T_B were $T_D = T_{feed} - 40°F$ and $T_{feed} + 40°F$. Initially, take $V_j = V_r = V_2$ for the rectifying section and $V_j = V_s = V_r$ for the stripping section. Use the equilibrium and enthalpy data given in Tables B-2 and B-24 of App. B.

Table 11-2 Solution of Example 11-4 (the solution shows components 1, 4, and 5 are separated)

	Temperature profile and vapor rates		Product distribution		
Stage	$T(°F)$	V_j (lb-mol/h)	Component	d_i (lb mol/h)	b_i (lb mol/h)
1 (distillate)	217.64	. . . .	C_3H_8	20	0.0
r	244.88	151.388	$i\text{-}C_4H_{10}$	14.57588	5.425142
$r + 1$	244.88	151.388	$n\text{-}C_4H_{10}$	5.424149	14.57666
$r + 2$	244.88	151.387	$i\text{-}C_5H_{12}$	0.0	20
$r + 3$	244.88	151.385	$n\text{-}C_5H_{12}$	0.0	20
f (feed)	272.05	136.562			
$s - 2$	289.93	151.374			
$s - 1$	289.93	151.377			
s	289.94	151.380			
(Bottoms)	323.21				
D (calculated)	39.99997				

In these examples, the temperatures of the distillate and the bottoms were correct to within four or five digits of the final values at the end of five to ten trials. This accuracy in the terminal temperatures suggests a corresponding accuracy in the product distribution, the determination of which is the primary objective of the general procedure.

However, in order to obtain the same accuracy for the pinch temperatures, several additional trials were required. This was caused in part by an

Table 11-3 Solution of Example 11-5 (the solution shows all components are distributed)

Stage	Temperature profile and vapor rates		Product distribution		
	T (°F)	V_j (lb mol/h)	Component	d_i (lb mol/h)	b_i (lb mol/h)
1 (distillate)	253.33	. . . .	C_3H_8	17.16039	2.831960
r	267.34	54.706	$i\text{-}C_4H_{10}$	9.501945	10.498054
$r + 1$	267.34	54.706	$n\text{-}C_4H_{10}$	7.361634	12.638365
$r + 2$	267.34	54.706	$i\text{-}C_5H_{12}$	3.440505	16.559494
$r + 3$	267.34	54.706	$n\text{-}C_5H_{12}$	2.527875	17.472125
f (feed)	267.34	54.706			
$s - 2$	267.34	54.706			
$s - 1$	267.34	54.706			
s	267.34	54.706			
N (bottoms)	304.29				
D (calculated)	39.99995				

Table 11-4 Solution of Example 11-6 (the solution shows that all components are separated)

Stage	Temperature profiles (°F) Trial no.				Find vapor rates (lb mol/h) Trial no.	
	5	10	25	Solution by Eqs. (11-24), (11-25), (11-27), and (11-28)	25	Solution by Eqs. (11-24), (11-25), (11-27), and (11-28)
1 (distillate)	237.57	237.00	237.00	237.00		
r	306.14	284.25	254.87	254.36	203.42	206.40
$r + 1$	306.14	284.36	254.88	. . . .	203.42	
$r + 2$	306.14	284.57	254.89	. . . .	203.40	
$r + 3$	306.14	284.86	254.90	. . . .	203.38	
f (feed)	307.92	288.37	284.24	. . . .	182.39	
$s - 2$	319.82	313.97	341.53	. . . .	207.75	
$s - 1$	320.23	314.26	341.63	. . . .	205.18	
s	320.43	314.31	341.67	352.88	205.18	201.45
N (bottoms)	355.41	354.96	354.90	354.96		
D (calculated)	65.95	60.00000	60.00000			
θ	77.96	1.00000	1.00000			
Separated components	1, 2, 3	All	All			

insufficient number of plates between the pinches. When this situation exists, a sizable mole fraction of a separated heavy component, for example, is computed at the rectifying pinch by use of Eqs. (11-11) and (11-15). Thus, additional trials are required for such examples in order to provide a sufficient number of plates between the pinches. This need was most pronounced for cases where either all or all except one component were separated. However, for such examples the pinch temperatures may be found by means of a direct solution [see Eqs. (11-24), (11-25), (11-27), and (11-28)], which is of course recommended. Several trials should be carried out by use of the general procedure in order to make fairly certain that a system classified as a special case has been found. Then the general calculational procedure is abandoned, and Eqs. (11-24), (11-25), (11-27), and (11-28) are solved.

When all of the components are separated, θ must be set equal to unity in all of the equations for the general procedure, and the general method becomes relatively slow. Fortunately, this type of problem as well as the one in which only one component is distributed may be solved directly by use of the equations listed above because the rate of convergence of the θ method tends to decrease as the number of distributed components is decreased.

Complex Columns

Problems involving complex columns may be solved by use of the same type of calculational procedure described for conventional columns. To demonstrate the application of the method, a column having one feed plate and one sidestream withdrawn (in addition to the distillate and bottoms) is considered. Two possible locations for the sidestream W_1 are considered. In the first, the sidestream is withdrawn a finite number of plates below the top of the column; and in the second, it is withdrawn a finite number of plates above the feed plate.

In the following development, it is supposed that the specifications are taken to be L_1, D, W_1, and infinitely many plates in both the rectifying and stripping sections. In addition to these specifications, the type of condenser, the column pressure, the flow rate of the feed, its composition, and thermal condition, as well as the location of the sidestream withdrawal, are specified. For this set of specifications, the problem is to find the product distribution.

SIDESTREAM W_1 IS WITHDRAWN A FINITE NUMBER OF PLATES BELOW THE TOP OF THE COLUMN AND INFINITELY MANY PLATES ABOVE THE FEED PLATE

Except for the material balance enclosing plate $p + 2$ and any stage in the rectifying pinch, and the material balance enclosing the bottom of the column and any stage s in the stripping pinch, the remaining balances enclose only one stage (see Fig. 11-6). After the liquid flow rates have been eliminated by use of the equilibrium relationships, the resulting equations may be rearranged and

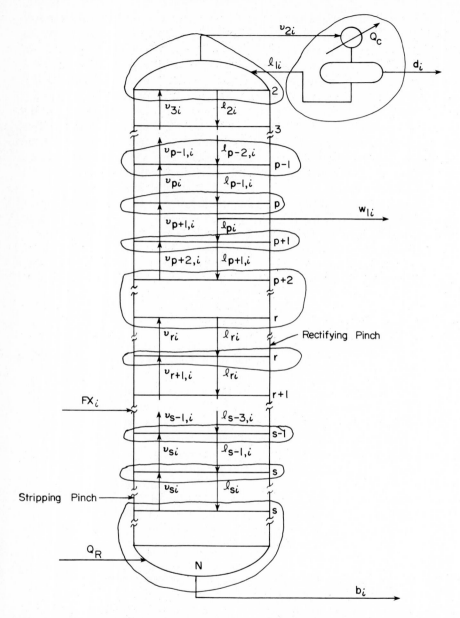

Figure 11-6 Material-balance enclosures for the distributed components. W_1 is withdrawn a finite number below the top of the column.

stated in the form of matrix Eq. (11-11). The component-material balances for the separated lights and heavies are formulated in a similar manner. The g functions are given by

$$g_0(\theta_0, \theta_1) = \frac{1}{D} \sum_{i=1}^{c} (d_i)_{ca} p_i - 1 \tag{11-30}$$

$$g_1(\theta_0, \theta_1) = \frac{1}{W_1} \sum_{i=1}^{c} \theta_1(w_{1i})_{ca} p_i - 1 \tag{11-31}$$

The variable p_i is defined by

$$p_i = \frac{FX_i}{(d_i)_{ca} + \theta_0(b_i)_{ca} + \theta_1(w_{1i})_{ca}} \tag{11-32}$$

and the multipliers θ_0 and θ_1 are defined by

$$\left(\frac{b_i}{d_i}\right)_{co} = \theta_0 \left(\frac{b_i}{d_i}\right)_{ca} \tag{11-33}$$

$$\left(\frac{w_{1i}}{d_i}\right)_{ca} = \theta_1 \left(\frac{w_{1i}}{d_i}\right)_{ca} \tag{11-34}$$

After θ_0 and θ_1 have been computed by use of Eqs. (11-30) and (11-31), the corrected mole fractions are computed by use of Eq. (11-17). Then a new set of temperatures is found by use of the K_b method. Improved sets of total-flow rates are found by use of the enthalpy balances (stated in constant-composition form).

SIDESTREAM W_1 IS WITHDRAWN A FINITE NUMBER OF PLATES ABOVE THE FEED PLATE

A sketch of this column is shown in Fig. 11-7. In this case each component-material balance also encloses only one stage, except for the material balance enclosing plate 2 and any plate r in the rectifying pinch, and the balance enclosing any plate s in the stripping pinch and the bottom of the column. Elimination of the liquid rates through the use of the equilibrium relationships followed by rearrangement permits the equations to be stated in the form of the matrix equation given by Eq. (11-11). The g functions are given by Eqs. (11-30) and (11-31).

Illustrative examples The calculation of the product distributions for complex columns having infinitely many stages is demonstrated by Examples 11-7 and 11-8. A statement of these examples is given in Table 11-5 and their solutions are presented in Tables 11-6 and 11-7. These examples were taken from McDonough and Holland.[7]

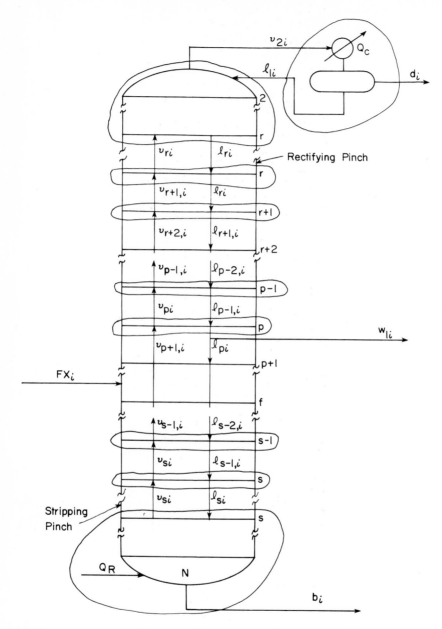

Figure 11-7 Material-balance enclosures for the distributed components. W_1 is withdrawn a finite number of plates above the feed plate.

Table 11-5 Statement and Specifications for Examples 11-7 and 11-8

Component	Component no.	All examples FX_i	Example	Specifications V_2	D	W_1
C_3H_8	1	20	11-7	150	30	20
$i\text{-}C_4H_{10}$	2	20	11-8	100	30	15
$n\text{-}C_4H_{10}$	3	20				
$i\text{-}C_5H_{12}$	4	20				
$n\text{-}C_5H_{12}$	5	20				

Other specifications

For all examples: column pressure = 400 lb/in² abs, partial condenser, thermal condition of the feed is boiling-point liquid. Initial temperature profile: linear between the pinches (252–282°F) for all examples. For Examples 11-7 and 11-8 the initial temperature profile was taken to be linear between 227 and 242°F for plates 1 through 5 and the temperature of the distillate was taken to be 217°F. Initial vapor rates: $V_j = V_2$ for all j. In Example 11-7 the liquid sidestream W_1 is withdrawn from plate 6 (the accumulator is assigned the number 1), and in Example 11-8, it is withdrawn five plates above the feed plate. Use the equilibrium and enthalpy data given in Tables B-2 and B-24 of the Appendix.

Table 11-6 Solution of Example 11-7 (W_1 is withdrawn from plate 6)

Plate	Temperature profile and vapor rates T (°F)	V_j (lb mol/h)	Product distribution Component	d_i	b_i	w_{1i}
1 (distillate)	211.21	....	C_3H_8	16.9733	0.0	3.0267
6 (plate p)	241.50	151.74	$i\text{-}C_4H_{10}$	9.9098	0.7052	9.3850
r	246.14	151.79	$n\text{-}C_4H_{10}$	3.1169	9.2948	7.5883
$r + 1$	246.14	151.79	$i\text{-}C_5H_{10}$	0.0	20.0000	0.0
$r + 2$	246.14	151.79	$n\text{-}C_5H_{10}$	0.0	20.0000	0.0
f (feed)	277.43	135.59				
$s - 2$	293.96	149.83				
$s - 1$	293.96	149.83				
s	293.96	149.83				
N (bottoms)	337.22					

D (calculated)	30.000
θ_0	0.99999
W_1 (calculated)	20.000
θ_1	0.99999

Table 11-7 Solution of Example 11-8 (W_1 is withdrawn from 5 plates above the feed plate)

Plate	Temperature profile and vapor rates		Product distribution			
	T (°F)	V_j (lb mol/h)	Component	d_i	b_i	w_{1i}
1 (distillate)	211.13	....	C_3H_8	17.344	0.0	2.6564
r	239.62	100.36	$i\text{-}C_4H_{10}$	8.9474	6.8042	4.2484
$r+1$	239.62	100.36	$n\text{-}C_4H_{10}$	3.7091	11.3418	4.9490
$r+2$	239.62	100.36	$i\text{-}C_5H_{12}$	0.0	17.9732	2.0267
p (sidestream)	255.45	94.17	$n\text{-}C_5H_{12}$	0.0	18.881	1.1193
f (feed)	271.96	89.23				
$s-2$	295.44	102.15				
$s-1$	295.44	102.16				
s	295.44	102.16				
N (bottoms)	366.02					

D (calculated)	30.000
θ_0	1.0000
W_1 (calculated)	15.000
θ_1	0.9999

11-2 DETERMINATION OF THE MINIMUM REFLUX RATIO BY THE θ METHOD

The θ method may be used to solve the classical problem in which the separations are specified for two components (the light and heavy keys), and it is required to find the smallest reflux ratio L_1/D required to effect the specifications. The smallest reflux ratio is achieved through the use of infinitely many stages in each section of the column.

Consider first the case where the specifications b_l/d_l and b_h/d_h are made for a conventional distillation column. In addition, the column pressure, the type of condenser, and the complete definition of the feed are specified. This type of problem may be solved by use of a combination of the procedure used to find the product distribution for a column with infinitely many plates and an optimization procedure such as the one described in Chap. 9.

Since it is desired to find the smallest reflux ratio L_1/D at a specified distillate rate which satisfies the specifications on the keys, the problem may be formulated in terms of the following objective function

$$O = \frac{1}{f_1}[L_1/D]^2 + \frac{1}{f_2}[s_l^2 + s_h^2] \tag{11-35}$$

where

$$s_l = (b_l/d_l)/(b_l/d_l)_{\text{spec}}$$

$$s_h = (b_h/d_h)_{\text{spec}}/(b_h/d_h)$$

The values of L_1/D in Eq. (11-35) are, of course, understood to be finite and positive. Also, the components lighter than the light key and heavier than the heavy key are separated. The quantities $1/f_1$ and $1/f_2$ are weight factors which were selected as described below.

For the case where the keys are adjacent in volatility and their b_i/d_i's are specified, the total distillate rate D is known in advance and there remains only one search variable L_1/D. For each of several initial values (say three) of L_1/D, the product distribution for a column with infinitely many plates is found. The results so obtained are used to evaluate the objective function [Eq. (11-35)]. An improved value for L_1/D is found by use of a search procedure such as the one proposed by Box (see Chap. 9).

The weight factor f_1 is taken equal to the average of the simplex values of its numerator, $[L_1/D]^2$, and f_2 is set equal to unity. For either $s_l < 1$ or $s_h < 1$, the term s_l or s_h is replaced by either $\ln s_l$ or $\ln s_h$. Use of this revised form of the function places a penalty on making separations which are better than those specified and tends to drive the objective function toward equality constraints.

11-3 A NEWTON-RAPHSON FORMULATION: TWO INDEPENDENT VARIABLES PER STAGE†

This formulation of the Newton-Raphson method for columns with infinitely many stages is analogous to the $2N$ Newton-Raphson method for a column with a finite number of stages. First the procedure is developed for a conventional distillation column with infinitely many stages for which the condenser duty Q_C (or the reflux ratio L_1/D) and the reboiler duty Q_R (or the boilup ratio V_N/B) are specified and it is required to find the product distribution. Then the procedure is modified as required to find the minimum reflux ratio required to effect the specified separation of two key components.

---

† Sections 11-3 and 11-4 are based on the work of An Feng who plans to use these results in partial fulfillment for the Ph.D. requirements.

Determination of the Product Distribution for a Conventional Distillation Column at Specified Condenser and Reboiler Duties Q_C and Q_R or Specified Values of L_1/D and V_N/B

The component-material balances are formulated in a manner analogous to that shown in Sec. 11-1 except that in this case, stages s and $N - 1$ are enclosed with material and energy balances as well as stage N. This minor variation makes it possible to include the boilup ratio V_N/B in the set of possible specifications. In effect the energy balance enclosing the entire column which was used to determine Q_R in the θ method is replaced by an energy balance which encloses the reboiler, stage N, alone.

The set of component-material balances is obtained by replacing the component-material balance given by Eq. (11-11) enclosing stage s and the reboiler by two component-material balances, one enclosing stage s and $N - 1$, and the second enclosing stage N as shown in Fig. 11-8.

$$v_{Ni} + l_{si} - v_{si} - l_{N-1,i} = 0$$

$$l_{N-1,i} - v_{Ni} - b_i = 0$$

(11-36)

With this modification, the matrices A_i, v_i and f_i of Eq. (11-11) for the distributed components take the following forms.

$$A_i = \begin{bmatrix} -\rho_{1i} & 1 & 0 & 0 & 0 & 0 & \cdots & 0 \\ A_{1i} & -1 & (1-A_{ri}) & 0 & 0 & 0 & \cdots & 0 \\ 0 & 0 & -1 & 1 & 0 & 0 & \cdots & 0 \\ 0 & 0 & A_{ri} & -\rho_{r+1,i} & 1 & 0 & \cdots & 0 \\ \hline 0 & \cdots & 0 & A_{s-2,i} & -\rho_{s-1,i} & 1 & 0 & 0 \\ 0 & \cdots & \cdots & 0 & A_{s-1,i} & -A_{si} & 0 & 0 \\ 0 & \cdots & \cdots & \cdots & 0 & (A_{si}-1) & -1 & 1 \\ 0 & \cdots & \cdots & \cdots & \cdots & 0 & 1 & -\rho_{Ni} \end{bmatrix}$$

(11-37)

$$v_i = [d_i \; v_{2i} \; v_{ri} \; v_{r+1,i} \; \cdots \; v_{s-2,i} \; v_{s-1,i} \; v_{si} \; l_{N-1,i} \; v_{Ni}]^T$$

$$f_i = [0 \; \cdots \; 0 \; v_{Fi} \; l_{Fi} \; 0 \; \cdots \; 0]$$

The liquid flow ratio $l_{N-1,i}$ is taken as one of the variables rather than $v_{N-1,i}$ because the liquid flow rate appears in the energy balances and the vapor flow rate does not. Furthermore, since the phase equilibrium function for stage $N - 1$ may be formulated in terms of the liquid rates, it is not necessary to compute the $v_{N-1,i}$'s.

The component-material balances for the separated heavies are obtained from the set given by Eq. (11-14) by replacing the component-material balance enclosing the reboiler and stage s by the two expressions given by Eq. (11-36); see Fig. 11-9. The component-material balances for the separated lights are the same as those given by Eq. (11-12) and the enclosures shown in Fig. 11-4.

The total-material balances are formulated in a manner analogous to that demonstrated for the component-material balances. The complete set of total-

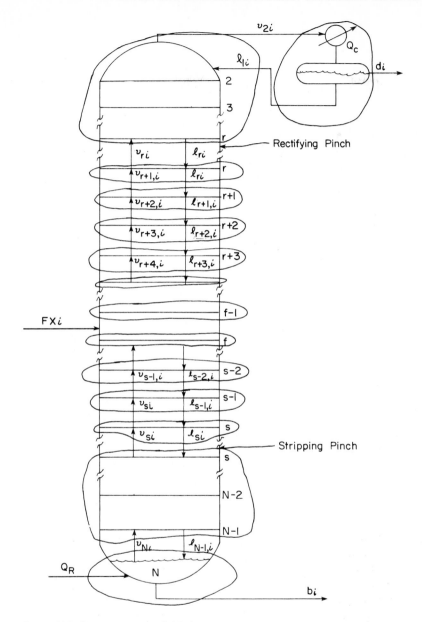

Figure 11-8 Component-material balance enclosures for the distributed components—used in the Newton-Raphson methods.

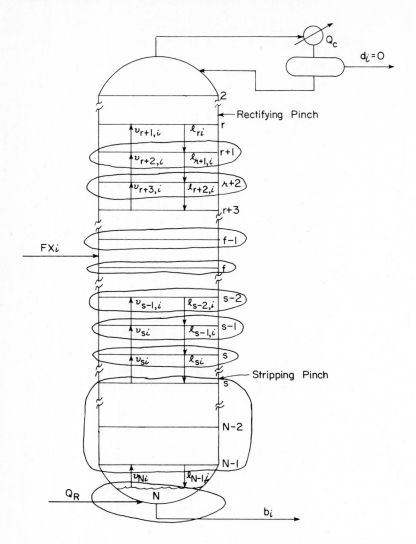

Figure 11-9 Component-material balance enclosures for the separated heavy components—used in the Newton-Raphson methods.

material balances are as follows

$$
\begin{aligned}
V_2 - L_1 - D &= 0 \\
V_r + L_1 - V_2 - L_r &= 0 \\
V_{r+1} + L_r - V_r - L_r &= 0 \\
V_{r+2} + L_r - V_{r+1} - L_{r+1} &= 0 \\
V_{r+3} + L_{r+1} - V_{r+2} - L_{r+2} &= 0 \\
&\cdots \cdots \cdots \cdots \cdots \cdots \cdots \cdots \cdots \\
V_{s-1} + L_{s-3} - V_{s-2} - L_{s-2} &= 0 \\
V_s + L_{s-2} - V_{s-1} - L_{s-1} &= 0 \\
V_s + L_{s-1} - V_s - L_s &= 0 \\
V_N + L_s - V_s - L_{N-1} &= 0 \\
L_{N-1} - V_N - B &= 0
\end{aligned}
\tag{11-38}
$$

Again as in the $2N$ Newton-Raphson method, the variable θ_j is introduced for the ratio L_j/V_j. More precisely, let

$$
\frac{L_j}{V_j} = \theta_j \left(\frac{L_j}{V_j}\right)_a
\tag{11-39}
$$

for $j = 4$, $r + 1$, $r + 2$, ..., $s - 2$, $s - 1$, s. For $j = 1$ and $j = N$, $L_1/D = \theta_1(L_1/D)_a$ and $B/V_N = \theta_N(B/V_N)_a$. For purposes of normalization, the assumed value, denoted by the subscript a, is taken to be the most recent value of the variable. Use of Eq. (11-39) permits Eq. (11-38) to be restated in the following matrix form

$$
\mathbf{RV} = -\mathscr{F}
\tag{11-40}
$$

where

$$
\mathbf{R} =
\begin{bmatrix}
-(R_1 + 1) & 1 & 0 & 0 & 0 & 0 & \cdots & 0 \\
R_1 & -1 & (1 - R_r) & 0 & 0 & 0 & \cdots & 0 \\
0 & 0 & -1 & 1 & 0 & 0 & \cdots & 0 \\
0 & 0 & R_r & -(1 + R_{r+1}) & 1 & 0 & \cdots & 0 \\
\hline
0 & \cdots & 0 & R_{s-2} & -(1 + R_{s-1}) & 1 & 0 & 0 \\
0 & \cdots & \cdots & 0 & R_{s-1} & -R_s & 0 & 0 \\
0 & \cdots & \cdots & \cdots & 0 & (R_s - 1) & -1 & 1 \\
0 & \cdots & \cdots & \cdots & \cdots & 0 & 1 & -(1 + R_N)
\end{bmatrix}
$$

$$
\mathbf{V} = [D \ V_2 \ V_r \ V_{r+1} \ V_{r+2} \ \cdots \ V_{s-2} \ V_{s-1} \ V_s \ L_{N-1} \ V_N]^T
$$

$$
\mathscr{F} = [0 \ \cdots \ 0 \ V_F \ L_F \ 0 \ \cdots \ 0]^T
$$

$$
R_j = \theta_j \, (L_j/V_j)_a, j = r, r + 1, r + 2, \ldots, s - 2, s - 1, s; \ R_1 = \theta_1(L_1/D)_a
$$

$$
R_N = \theta_N \, (B/V_N)_a
$$

The flow rate L_{N-1} rather than V_{N-1} is computed because the bubble-point form of the equilibrium relationship is used for stage $N - 1$ as discussed previously and shown below in Eq. (11-46).

In the choice of the independent variables and functions, care must be exercised in the treatment of the stages adjacent to the limiting conditions (the pinches r and s) in order to avoid singularities in the jacobian matrix of the Newton-Raphson equations. In particular, the following limiting conditions which appear in the material balances must be observed

$$\begin{matrix} v_{ri} = v_{r+1,\,i} & l_{s-1,\,i} = l_{si} \\ V_r = V_{r+1} & L_{s-1} = L_s \end{matrix} \qquad (11\text{-}41)$$

Since a dew-point calculation based on the $\{v_{ri}\}$ gives the same temperature as a dew-point calculation based on the $\{v_{r+1,\,i}\}$, it follows that

$$T_r = T_{r+1} \qquad (11\text{-}42)$$

Similarly, since a bubble-point calculation based on the $\{l_{s-1,\,i}\}$ gives the same temperature as a bubble-point calculation based on $\{l_{si}\}$, it follows that

$$T_{s-1} = T_s \qquad (11\text{-}43)$$

On the other hand, it cannot be inferred that L_r is equal L_{r+1}. (Note: $V_r = L_r + D$, $V_{r+1} = L_r + D$, but $V_{r+2} = L_{r+1} + D$). Likewise it cannot be inferred that V_s is equal to V_{s-1}. (Note: $L_s = V_s + B$, $L_{s-1} = V_s + B$, but $L_{s-2} = V_{s-1} + B$).

SPECIFICATION OF Q_C AND Q_R

It is of course understood that the remaining usual specifications are made such as the type of condenser, the column pressure, and the complete definition of the feed. The choice of independent variables $\mathbf{x}$ is as follows

$$\mathbf{x} = [\theta_1\ \theta_r\ \theta_{r+1}\ \cdots\ \theta_{s-2}\ \theta_{s-1}\ \theta_s\ \theta_N\ T_1\ T_2\ T_r\ T_{r+2}\ T_{r+3}\ \cdots$$
$$T_{s-3}\ T_{s-2}\ T_s\ T_{N-1}\ T_N]^T \qquad (11\text{-}44)$$

and the corresponding set of independent functions is as follows:

$$\mathbf{f} = [F_1\ F_2\ F_r\ F_{r+2}\ F_{r+3}\ \cdots\ F_{s-3}\ F_{s-2}\ F_s\ F_{N-1}\ F_N$$
$$G_1\ G_2\ G_{r+1}\ G_{r+2}\ \cdots\ G_{s-2}\ G_{s-1}\ G_{N-1}\ G_N] \qquad (11\text{-}45)$$

In the choice of the equilibrium functions F_j, functions corresponding to stages $r+1$ and $s-1$ are excluded because $T_r = T_{r+1}$ and $T_{s-1} = T_s$. The functions $\{F_j\}$ corresponding to the phase-equilibrium relationships are as follows:

$$F_1 = \frac{1}{D} \sum_{i=1}^{c} \left(\frac{1}{K_{1i}} - 1 \right) d_i \qquad\qquad F_{s-2} = \frac{1}{V_{s-2}} \sum_{i=1}^{c} \left(\frac{1}{K_{s-2,\,i}} - 1 \right) v_{s-2,\,i}$$

$$F_2 = \frac{1}{V_2} \sum_{i=1}^{c} \left(\frac{1}{K_{2i}} - 1 \right) v_{2i} \qquad\qquad F_s = \frac{1}{V_s} \sum_{i=1}^{c} \left(\frac{1}{K_{si}} - 1 \right) v_{si}$$

$$(11\text{-}46)$$

$$F_r = \frac{1}{V_r} \sum_{i=1}^{c} \left(\frac{1}{K_{ri}} - 1 \right) v_{ri} \qquad\qquad F_{N-1} = \frac{1}{L_{N-1}} \sum_{i=1}^{c} (K_{N-1,\,i} - 1) l_{N-1,\,i}$$

$$F_{r+2} = \frac{1}{V_{r+2}} \sum_{i=1}^{c} \left(\frac{1}{K_{r+2,\,i}} - 1 \right) v_{r+2,\,i} \qquad\qquad F_N = \frac{1}{V_N} \sum_{i=1}^{c} \left(\frac{1}{K_{Ni}} - 1 \right) v_{Ni}$$

For the case where a total condenser is used, the above expression for F_1 is replaced by

$$F_1 = \frac{1}{D} \sum_{i=1}^{c} (K_{1i} - 1)d_i \tag{11-47}$$

The enthalpy-balance functions which are based on the enclosures shown in Fig. 11-8 are as follows

$$G_1 = \frac{\sum\limits_{i=h}^{c} [l_{1i}h_{1i} + d_i H_{Di}] + Q_C}{\sum\limits_{i=h}^{c} v_{2i}H_{2i}} - 1$$

$$G_2 = \frac{\sum\limits_{i=1}^{c} [v_{2i}H_{2i} + l_{ri}h_{ri}]}{\sum\limits_{i=h}^{c} [v_{ri}H_{ri} + l_{1i}h_{1i}]} - 1$$

$$G_{r+1} = \frac{\sum\limits_{i=1}^{c} [v_{r+1,i}H_{r+1,i} + l_{r+1,i}h_{r+1,i}]}{\sum\limits_{i=1}^{c} [v_{r+2,i}H_{r+2,i} + l_{ri}h_{ri}]} - 1$$

$$G_{r+2} = \frac{\sum\limits_{i=1}^{c} [v_{r+2,i}H_{r+2,i} + l_{r+2,i}h_{r+2,i}]}{\sum\limits_{i=1}^{c} [v_{r+3,i}H_{r+3,i} + l_{r+1,i}h_{r+1,i}]} - 1$$

$$\vdots \qquad\qquad \vdots \qquad\qquad \vdots \tag{11-48}$$

$$G_{s-2} = \frac{\sum\limits_{i=1}^{c} [v_{s-2,i}H_{s-2,i} + l_{s-2,i}h_{s-2,i}]}{\sum\limits_{i=1}^{c} [v_{s-1,i}H_{s-1,i} + l_{s-3,i}h_{s-3,i}]} - 1$$

$$G_{s-1} = \frac{\sum\limits_{i=1}^{c} [v_{s-1,i}H_{s-1,i} + l_{s-1,i}h_{s-1,i}]}{\sum\limits_{i=1}^{c} [v_{si}H_{si} + l_{s-2,i}h_{s-2,i}]} - 1$$

$$G_{N-1} = \frac{\sum\limits_{i=1}^{l} [v_{si}H_{si} + l_{N-1,i}h_{N-1,i}]}{\sum\limits_{i=1}^{l} [v_{Ni}H_{Ni} + l_{si}h_{si}]} - 1$$

$$G_N = \frac{\sum\limits_{i=1}^{l} [v_{Ni}H_{Ni} + b_i h_{Ni}]}{\sum\limits_{i=1}^{l} [l_{N-1,i}h_{N-1,i}] + Q_R} - 1$$

The functions G_{f-1} and G_f differ in form for the above expressions by the additive terms

$$\sum_{i=1}^{c} H_{Fi} v_{Fi} \quad \text{and} \quad \sum_{i=1}^{c} h_{Fi} l_{Fi}$$

The first of these appears in the denominator of G_{f-1} and the second appears in the denominator of G_f.

In the above expressions for the rectifying section, $v_{r+1,i}$, $h_{r+1,i}$, and $H_{r+1,i}$ may be replaced by their equivalents, v_{ri}, h_{ri}, and H_{ri}, respectively. Similarly, in the expressions for the stripping section, $l_{s-1,i}$, $h_{s-1,i}$, and $H_{s-1,i}$ may be replaced by their equivalents l_{si}, h_{si}, and H_{si}, respectively. If desired, the enthalpy balances may be stated in terms of one less flow rate by using the component-material balance, $v_{j+1,i} + l_{j-1,i} - v_{ji} - l_{ji} = 0$ to eliminate one of the flow rates. For example, the resulting G function for stage $r + 2$ is given by

$$G_{r+2} = \frac{\sum_{i=1}^{c} [v_{r+3,i}(H_{r+3,i} - h_{r+2,i}) + l_{r+1,i}(h_{r+1,i} - h_{r+2,i})]}{\sum_{i=1}^{c} [v_{r+2,i}(H_{r+2,i} - h_{r+2,i})]} - 1$$

This problem may be solved by use of any of the three formulations of the $2N$ Newton-Raphson method which were presented in Chap. 4. The classification of the components and the selection of the appropriate stages between the pinches are performed in precisely the same manner as described in Sec. 11-1. It should be noted, however, that if analytical expressions for the partial derivatives are used, one must not overlook the fact that in the case of the separated lights, the vector f_i will have derivatives since it contains A_{1i}. Because of the slight differences in the equations for the distributed and separated components, Broyden's method, or the Broyden-Bennett algorithm is recommended.

For the case where the reflux ratio L_1/D and the boilup ratio V_N/B are specified rather than Q_C and Q_R, the variables θ_1 and θ_N in x are replaced by Q_C and Q_R to give

$$x = [Q_C \ \theta_r \ \theta_{r+1} \ \theta_{r+2} \ \cdots \ \theta_{s-2} \ \theta_{s-1} \ \theta_s \ Q_R$$
$$T_1 \ T_2 \ T_r \ T_{r+2} \ T_{r+3} \ \cdots \ T_{s-3} \ T_{s-2} \ T_s \ T_{N-1} \ T_N]^T \quad (11\text{-}49)$$

The functions are the same as those enumerated above.

SPECIFICATION OF THE SEPARATIONS FOR THE LIGHT AND HEAVY KEYS AND INFINITELY MANY STAGES

This is the classical problem in which it is desired to find the minimum reflux ratio required to effect the specified separations (say b_l/d_l and b_h/d_h) of the light and heavy key components. The material balances, F functions, and G functions are the same as those given above. However, in this case, L_1/D, V_N/B, Q_C, and

Q_R are all unknown. Thus, this problem contains two more variables than did the one considered above, namely,

$$\mathbf{x} = [Q_C \ \theta_1 \ \theta_r \ \theta_{r+1} \ \theta_{r+2} \ \cdots \ \theta_{s-2} \ \theta_{s-1} \ \theta_s \ \theta_N \ Q_R$$
$$T_1 \ T_2 \ T_r \ T_{r+1} \ T_{r+2} \ T_{r+3} \ \cdots \ T_{s-3} \ T_{s-2} \ T_s \ T_{N-1} \ T_N]^T \quad (11\text{-}50)$$

Corresponding to these two additional variables, two additional independent functions are required. These functions are formulated on the basis of the specifications, namely,

$$S_1 = \frac{d_l}{(d_l)_{\text{spec}}} - 1 \tag{11-51}$$

and

$$S_N = \frac{d_h}{(d_h)_{\text{spec}}} - 1 \tag{11-52}$$

Thus, the set of functions $\mathbf{f}$ is given by

$$\mathbf{f} = [S_1 \ F_1 \ F_2 \ F_r \ F_{r+2} \ F_{r+3} \ \cdots \ F_{s-3} \ F_{s-2} \ F_s \ F_{N-1} \ F_N$$
$$G_1 \ G_2 \ G_{r+1} \ G_{r+2} \ \cdots \ G_{s-2} \ G_{s-1} \ G_{N-1} \ G_N \ S_N]^T \quad (11\text{-}53)$$

Again, problems of this type may be solved by use of any of the three variations of the $2N$ Newton-Raphson method described in Chap. 4.

In order to demonstrate the use of this method, Example 11-9 was solved. A statement of Example 11-9 is given in Table 11-1. The specified values of b_l/d_l and b_h/d_h were taken to be the solution values of Example 11-4. The number of trials and computer times required to solve Example 11-4 are given in Table 11-8. An examination of these results shows that the θ method is significantly faster than the $2N$ formulations of the Newton-Raphson method.

11-4 AN ALMOST BAND ALGORITHM FORMULATION OF THE NEWTON-RAPHSON METHOD

For columns in the service of separating mixtures which form highly nonideal solutions, the following formulation of the Newton-Raphson method is recommended. For columns having infinitely many stages, the formulation of the Almost Band Algorithm is carried out in a manner similar to that demonstrated in Chap. 5 for columns having a finite number of stages. The formulation presented appears to be similar to the one recently described by Tavana and Hanson.[13]

Table 11-8 Number of trials required to solve Examples 11-4 and 11-9

Method	Example	Number of trials	Computer time (s)[†]	Compiler
θ method	11-4	17	1.46	WATFIV
Newton-Raphson with Broyden's method (one trial for each choice of plates)	11-4	20	184.93	WATFIV
θ method	11-4	19	0.62	FORTRAN H EXTENDED
Newton-Raphson with Broyden's method (one trial for each choice of plates)	11-4	20	36.48	FORTRAN H EXTENDED
Newton-Raphson with Broyden's method (two trials for each choice of plates)	11-4	20	36.82	FORTRAN H EXTENDED
Newton-Raphson with Broyden's method (one trial for each choice of plates)	11-4	20	38.82	FORTRAN H EXTENDED

[†] AMDAHL 470 V/6.

Consider first the case where the conventional distillation column has a partial condenser, and the condenser and reboiler duties Q_C and Q_R are specified. The material and energy balance enclosures are the same as those shown in Figs. 11-4, 11-8, and 11-9. Except for stages 2 and $N - 1$, the phase equilibrium functions are of the general form

$$f_{ji} = \frac{\gamma_{ji}^L K_{ji} l_{ji}}{\sum\limits_{i=1}^{c} l_{ji}} - \frac{\gamma_{ji}^V v_{ji}}{\sum\limits_{i=1}^{c} v_{ji}} \tag{11-54}$$

However, in the particular case of stage 2, the set $\{v_{2i}\}$ appears in the component-material balances, but not the set $\{l_{2i}\}$. Thus, the new set of variables $\{X_{2i}\}$ is introduced in order to obtain an appropriate phase-equilibrium function for stage 2. The variables $\{X_{2i}\}$ are related to the moles fractions $\{x_{2i}\}$ as follows

$$x_{2i} = \frac{X_{2i}}{\sum\limits_{i=1}^{c} X_{2i}} \tag{11-55}$$

The phase equilibrium function for stage 2 may be expressed as follows

$$f_{2i} = \frac{\gamma_{2i}^L K_{2i} X_{2i}}{\sum\limits_{i=h}^{c} X_{2i}} - \frac{\gamma_{2i}^V v_{2i}}{\sum\limits_{i=h}^{c} v_{2i}} \tag{11-56}$$

[Note that in the expressions for the activity coefficients $\{\gamma_{2i}^L\}$, x_{2i} is replaced wherever it appears by its equivalent as given by Eq. (11-55).]

Similarly, for stage $N - 1$, the set $\{l_{N-1, i}\}$ appears in the component-material balances but not the set $\{v_{N-1, i}\}$. Consequently, the new set of variables $\{Y_{N-1, i}\}$ is introduced to give the following equilibrium function

$$f_{N-1, i} = \frac{\gamma_{N-1, i}^L K_{N-1, i} l_{N-1, i}}{\sum\limits_{i=1}^{l} l_{N-1, i}} - \frac{\gamma_{N-1, i}^V Y_{N-1, i}}{\sum\limits_{i=1}^{l} Y_{N-1, i}} \tag{11-57}$$

where the variables $\{Y_{N-1, i}\}$ are related to the mole fractions $\{y_{N-1, i}\}$ as follows

$$y_{N-1, i} = \frac{Y_{N-1, i}}{\sum\limits_{i=1}^{l} Y_{N-1, i}} \tag{11-58}$$

Although the same enclosures are used for the component-material balances in the formulation of the Almost Band Algorithm as were used in the formulation of the $2N$ Newton-Raphson method, it is convenient in this case (because of the form of the phase equilibrium relationships) to include T_{r+1}, T_s, $\{v_{r+1, i}\}$, and $\{v_{si}\}$ in the set of independent variables. Thus, in the formulation of the Almost Band Algorithm, the following choice of independent variables is made.

$$\begin{aligned}
\mathbf{x} = [d_h &\cdots d_c \; l_{1, h} \cdots l_{1, c} \; T_1 \; v_{2, h} \cdots v_{2, c} \; X_{2, h} \cdots X_{2, c} \; T_2 \; v_{r, h} \cdots \\
v_{r, c} &\; l_{r, h} \cdots l_{r, c} \; T_r \; v_{r+1, 1} \cdots v_{r+1, c} \; l_{r+1, 1} \cdots \\
l_{r+1, c} &\; T_{r+1} \; v_{r+2, 1} \cdots v_{r+2, c} \; l_{r+2, 1} \cdots l_{r+2, c} \; T_{r+2} \cdots \\
v_{s-2, 1} &\cdots v_{s-2, c} \; l_{s-2, 1} \cdots l_{s-2, c} \; T_{s-2} \; v_{s-1, 1} \cdots v_{s-1, c} \; l_{s-1, 1} \cdots \\
l_{s-1, c} &\; T_{s-1} \; v_{s, 1} \cdots v_{s, l} \; l_{s, 1} \cdots l_{s, l} \; T_s \; Y_{N-1, 1} \cdots Y_{N-1, l} \; l_{N-1, 1} \\
&\cdots l_{N-1, l} \; T_{N-1} \; v_{N, 1} \cdots v_{N, l} \; b_1 \cdots b_l \; T_N]^T \tag{11-59}
\end{aligned}$$

The corresponding functions are given by

$$\begin{aligned}
\mathbf{f} = [m_{1, h} &\cdots m_{1, c} \; f_{1, h} \cdots f_{1, c} \; G_1 \; m_{2, h} \cdots m_{2, c} \; f_{2, h} \cdots f_{2, c} \; G_2 \; m_{r, h} \cdots \\
m_{r, c} &\; f_{r, h} \cdots f_{r, c} \; G_r \; m_{r+1, 1} \cdots m_{r+1, c} \; f_{r+1, 1} \cdots f_{r+1, c} \\
G_{r+1} &\; m_{r+2, 1} \cdots m_{r+2, c} \; f_{r+2, 1} \cdots f_{r+2, c} \; G_{r+2} \cdots m_{s-2, 1} \cdots \\
m_{s-2, c} &\; f_{s-2, 1} \cdots f_{s-2, c} \; G_{s-2} \; m_{s-1, 1} \cdots m_{s-1, c} \; f_{s-1, 1} \cdots \\
f_{s-1, c} &\; G_{s-1} \; m_{s, 1} \cdots m_{s, l} \; f_{s, 1} \cdots f_{s, l} \; G_s \; m_{N-1, 1} \cdots m_{N-1, l} \\
f_{N-1, 1} &\cdots f_{N-1, l} \; G_{N-1} \; m_{N, 1} \cdots m_{N, l} \; f_{N, 1} \cdots f_{N, l} \; G_N]^T \tag{11-60}
\end{aligned}$$

The component-material balance functions are as follows

$$m_{1i} = v_{2i} - l_{1i} - d_i \qquad\qquad (i = h, \ldots, c)$$

$$m_{2i} = v_{ri} + l_{1i} - v_{2i} - l_{ri} \qquad\qquad (i = h, \ldots, c)$$

$$m_{ri} = v_{r+1, i} + l_{ri} - v_{ri} - l_{ri} \qquad\qquad (i = h, \ldots, c)$$

$$m_{r+1, i} = v_{r+2, i} + l_{ri} - v_{r+1, i} - l_{r+1, i} \qquad\qquad (i = 1, \ldots, c)$$

$$m_{r+2, i} = v_{r+3, i} + l_{r+1, i} - v_{r+2, i} - l_{r+2, i} \qquad\qquad (i = 1, \ldots, c)$$

$$\cdots\cdots\cdots\cdots\cdots\cdots\cdots\cdots\cdots\cdots\cdots\cdots\cdots\cdots\cdots \qquad (11\text{-}61)$$

$$m_{s-2, i} = v_{s-1, i} + l_{s-3, i} - v_{s-2, i} - l_{s-2, i} \qquad\qquad (i = 1, \ldots, c)$$

$$m_{s-1, i} = v_{si} + l_{s-2, i} - v_{s-1, i} - l_{s-1, i} \qquad\qquad (i = 1, \ldots, c)$$

$$m_{si} = v_{si} + l_{s-1, i} - v_{si} - l_{si} \qquad\qquad (i = 1, \ldots, l)$$

$$m_{N-1, i} = v_{Ni} + l_{si} - v_{si} - l_{N-1, i} \qquad\qquad (i = 1, \ldots, l)$$

$$m_{Ni} = l_{N-1, i} - v_{Ni} - b_i \qquad\qquad (i = 1, \ldots, l)$$

For stages 2 and $N - 1$, the phase-equilibrium functions are given by Eqs. (11-56) and (11-57), and for the remaining stages Eq. (11-54) is applicable. The energy-balance functions are of the same general form as those given by Eq. (11-48) with the sets $\{H_{ji}\}$ and $\{h_{ji}\}$ replaced by the corresponding sets of virtual values of the partial molar values $\{\hat{H}_{ji}\}$ and $\{\hat{h}_{ji}\}$ (see Chap. 14) plus the following two functions for enclosures of stages r and s

$$\sum_{i=h}^{c} [v_{r+1, i}\hat{H}_{r+1, i} + l_{ri}\hat{h}_{ri} - v_{ri}\hat{H}_{ri} - l_{ri}\hat{h}_{ri}] = 0$$

or

$$G_r = \frac{\displaystyle\sum_{i=h}^{c} v_{ri} H_{ri}}{\displaystyle\sum_{i=h}^{c} v_{r+1, i} H_{r+1, i}} - 1 \qquad\qquad (11\text{-}62)$$

and

$$\sum_{i=1}^{c} [v_{si}\hat{H}_{si} + l_{s-1, i}\hat{h}_{s-1, i} - v_{si}\hat{H}_{si} - l_{si}\hat{h}_{si}] = 0$$

or

$$G_s = \frac{\displaystyle\sum_{i=1}^{l} l_{si}\hat{h}_{si}}{\displaystyle\sum_{i=1}^{l} l_{s-1, i}\hat{h}_{s-1, i}} - 1 \qquad\qquad (11\text{-}63)$$

SPECIFICATION OF THE REFLUX RATE L_1
AND THE BOTTOMS RATE L_N (OR B)

Consider again the case of a column with a partial condenser. When L_1 and L_N (or B) are specified, the condenser duty Q_C and the reboiler duty Q_R become variables and must be included in the vector $\mathbf{x}$ of independent variables. Let Q_C and Q_R be the first and last elements, respectively of $\mathbf{x}$. For the case of a column having a partial condenser, the remaining elements of $\mathbf{x}$ are the same as those given by Eq. (11-59). Corresponding to the two new variables, two new functions are required. These two functions are formulated such that the specified values L_1 and L_N are satisfied at convergence, namely,

$$S_1 = \frac{\sum\limits_{i=h}^{c} l_{1i}}{L_1} - 1 \tag{11-64}$$

$$S_N = \frac{\sum\limits_{i=1}^{l} l_{Ni}}{L_N} - 1 \tag{11-65}$$

The two new functions are taken to be the first and last elements of $\mathbf{f}$, and the remaining elements of $\mathbf{f}$ are the same as those given in Eq. (11-60). Again the components are classified and the stages between the pinches selected in the same manner as described in Sec. 11-1.

SPECIFICATION OF THE SEPARATIONS FOR THE LIGHT
AND HEAVY KEYS AND INFINITELY MANY STAGES

This is again the classical problem in which it is desired to find the minimum reflux ratio required to effect the specified separations of two key components.

Consider again the case of a conventional distillation column having a partial condenser. Again Q_C and Q_R are independent variables which are the first and last elements of $\mathbf{x}$.

Suppose that the two specifications made on the key components are $(b_l/d_l)_{\text{spec}}$ and $(b_h/d_h)_{\text{spec}}$. The two additional functions corresponding to the new variables Q_C and Q_R are based on the specified separations of the two key components, namely,

$$S_1 = \frac{d_l}{(d_l)_{\text{spec}}} - 1 \tag{11-66}$$

$$S_N = \frac{d_h}{(d_h)_{\text{spec}}} - 1 \tag{11-67}$$

where

$$(d_l)_{\text{spec}} = \frac{FX_l}{1 + (b_l/d_l)_{\text{spec}}}$$

$$(d_h)_{\text{spec}} = \frac{FX_h}{1 + (b_h/d_h)_{\text{spec}}}$$

Again the components are classified and the number of plates between the rectifying and stripping pinches are selected as described in Sec. 11-1.

After a solution has been obtained, the desired values for the reflux ratio and the boilup ratio are computed in the rather obvious manner

$$\frac{L_1}{D} = \frac{\sum\limits_{i=h}^{c} l_{1i}}{\sum\limits_{i=h}^{c} d_i} \tag{11-68}$$

and

$$\frac{V_N}{B} = \frac{\sum\limits_{i=1}^{l} v_{Ni}}{\sum\limits_{i=1}^{l} b_i}$$

For the case where the column has a total condenser, the above procedures are modified in a manner analogous to that shown in Chap. 5.

11-5 PROOF OF THE EXISTENCE OF THE RECTIFYING AND STRIPPING SECTION PINCHES

Since the existence of each of these pinches is established in the same manner, only the proof is presented for the rectifying pinch. This proof was first presented in Ref. 4. Use is made of the basic idea stated by Robinson and Gilliland[11] and demonstrated by Example 11-1, that if calculations are made by the Lewis and Matheson method from the top of the column down toward the feed plate on the basis of any given set of d_i's and a set of L's and V's which are constant (or computable by energy balances), a set of constant compositions will be obtained in the limit as the number of stepwise calculations are increased without bound.

The components in D are numbered in the order of decreasing volatility to give

$$K_{j1} > K_{j2} > K_{j3} > \cdots > K_{jc} \tag{11-69}$$

for any stage j. In the calculational procedure of Lewis and Matheson, a temperature is found for each plate j such that the dew-point equation

$$\sum_{i=1}^{c} \frac{y_{ji}}{K_{ji}} = 1 \tag{11-70}$$

is satisfied. Since the sum of the y_{ji}'s is by definition equal to unity, it follows that for at least one of the components $K_{ji} < 1$, and for at least one component, $K_{ji} > 1$. Thus, for the most volatile component, it can be said with certainty that $K_{ji} > 1$ for all j. Furthermore, this is the only component for which this statement can be made. By use of the Lewis and Matheson procedure, a temperature profile is found as well as a set of flow rates for each component. The temperature and total flow rates (assumed to be constant for the first part of the proof) may be used to represent the component-flow rates as follows

$$\frac{v_{2,1}}{d_1} = A_{1,1} + 1 \tag{11-71}$$

$$\frac{v_{j+1,1}}{d_1} = A_{j1}\left(\frac{v_{j1}}{d_1}\right) + 1 \tag{11-72}$$

Repeated substitution of the expression for v_{j1}/d_1 into the one for $v_{j+1,1}/d_1$ gives

$$\frac{v_{j+1,1}}{d_1} = 1 + A_{j1} + A_{j1}A_{j-1,1} + A_{j1}A_{j-1,1}A_{j-2,1} + \cdots$$

$$+ A_{j1}A_{j-1,1}\cdots A_{j-k,1} + \cdots + A_{ji}A_{j-1,1}\cdots A_{1,1} \tag{11-73}$$

Now consider the series obtained by setting $K_{j,1} = 1$ for all j

$$\frac{v_{j+1,1}}{d_1} < t_{j1} = 1 + \frac{L}{V} + \left(\frac{L}{V}\right)^2 + \cdots + \left(\frac{L}{V}\right)^j + \left(\frac{L}{V}\right)^j\left(\frac{L}{D}\right) \tag{11-74}$$

The first j terms of the series given by Eq. (11-74) are recognized as a geometric series whose sum is well known. Thus, the series t_{j1} may be restated in the form

$$t_{j1} = \frac{1 - (L/V)^{j+1}}{1 - (L/V)} + \left(\frac{L}{V}\right)^j\left(\frac{L}{D}\right) \tag{11-75}$$

Now let j take on all values to give an infinite series. Since $V > L$ and since L/D is finite, the infinite series converges to the sum given by

$$\lim_{j\to\infty} t_{j1} = \frac{1}{1 - (L/V)} \tag{11-76}$$

By use of the Weirstrass M test (Theorem A-5 of App. A), the series for $v_{j+1,1}/d_1$, Eq. (11-73) converges as j increases without bound because every term of the series given by Eq. (11-74) is larger than the corresponding term of the series given by Eq. (11-73). Also, the sum the series given by Eq. (11-73) as j approaches infinity (denoted by v_{r1}/d_1) is less than the sum given by Eq. (11-74).

Next it will be shown that the series for each of the remaining components has a limit as j is increased without bound. The approach used for component 1 cannot be followed for the remaining components because some of them may have values of $K_{ji} < 1$, and it may be that $A_{ji} > 1$, particularly for component c, for which $K_{jc} < 1$ for all j. The proof is continued as follows. First, observe that the definition of a mole fraction requires that

$$y_{j+1,1} = 1 - \sum_{i=1}^{c} y_{j+1,i} \tag{11-77}$$

Since both d_1 and V are finite, it follows that the limit of $y_{j+1,1}$ exists as j approaches infinity. For

$$y_{j+1,1} = \left(\frac{v_{j+1,1}}{d_1}\right)\left(\frac{d_1}{V}\right) \tag{11-78}$$

Hence the limit of the right-hand side of Eq. (11-77) exists and has the value given by

$$\lim_{j \to \infty} \sum_{i=1}^{c} \left(\frac{v_{j+1,i}}{d_i}\right)\left(\frac{d_i}{V}\right) = 1 - \left(\frac{v_{r1}}{d_1}\right)\left(\frac{d_1}{V}\right) \tag{11-79}$$

Consequently, since all of the d_i's as well as V are finite, it follows that

$$\lim_{j \to \infty} \sum_{i=2}^{c} \frac{v_{j+1,i}}{d_i}$$

exists. Thus, the infinite series (formed by adding the series for components 2, 3, ..., c)

$$(c-1) + \sum_{i=2}^{c} A_{ji} + \sum_{i=2}^{c} A_{ji} A_{j-1,i} + \sum_{i=2}^{c} A_{ji} A_{j-1,i} A_{j-2,i}$$

$$+ \cdots + \sum_{i=2}^{c} A_{ji} A_{j-1,i} \cdots A_{j-k,i} + \cdots$$

$$+ \sum_{i=2}^{c} A_{ji} A_{j-1,i} \cdots A_{1i} + \cdots \tag{11-80}$$

converges. Upon comparison of the series for component c with the one given by Eq. (11-80), the series for component c is seen to converge, since each term of the series for component c is less than the corresponding term of Eq. (11-80). Similarly, the series for each of the remaining components is shown to converge. Thus

$$\lim_{j \to \infty} \left(\frac{v_{j+1,i}}{d_i}\right) = \frac{v_{ri}}{d_i} \qquad \text{for all } i \tag{11-81}$$

Since each $y_{j+1,i}$ has the limit y_{ri}, the temperature also has a limit which follows from Eq. (11-70). Thus, A_{ji} has the limit A_{ri}. With this fact established, Eq. (11-4) is readily obtained as follows. In the limit as j approaches infinity, Eq. (11-84) (when stated for any component i) becomes

$$\frac{v_{ri}}{d_i} = A_{ri}\left(\frac{v_{ri}}{d_i}\right) + 1 \tag{11-82}$$

which may be solved for v_{ri} to give Eq. (11-4). Since each term in the series for $v_{j+1,i}/d_i$ is positive for all values of j, v_{ri}/d_i is positive and hence $A_{ri} < 1$ for all components in the distillate. Similarly, the existence of a pinch in the stripping section is established.

The above proof may be extended to include the variable L/V problem. First, observe that the above proof holds for any $L/V < 1$. To complete the proof, suppose that on the basis of specified values for L_1, D, and $\{d_i\}$, the component-material balances, the equilibrium relationships, and the energy balances are solved simultaneously for each stage j in succession. The results of such a sequence of calculations may be expressed in the series form given by Eq. (11-73).

To form the t_{j1} series, let the maximum of L_j/V_j for any stage be denoted simply by L/V and set $K_{j1} = 1$. Thus, Eq. (11-74) is again obtained. Under these conditions, it is again evident that each term of the series given by Eq. (11-73) is less than the corresponding term of Eq. (11-74). Thus, with changes outlined above, it is evident that the proof for constant L/V problems also holds for variable L/V problems.

NOTATION

p_1, p_2 functions of the stripping section pinch; defined by Eqs. (11-27) and (11-28)

P_1, P_2 functions of the rectifying section pinch; defined by Eqs. (11-24) and (11-25)

S_1, S_N specification functions; see Eqs. (11-51), (11-52), (11-64), and (11-65)

t_{j1} a series of constants; defined by Eq. (11-74)

T_h boiling-point temperature of pure component h at the column pressure

T_l boiling-point temperature of pure component l at the column pressure

T_N temperature of the reboiler of a column with infinitely many plates

Greek letters

ϕ_r root of $\Omega_r(\phi_r)$; see Example 11-2

ϕ_s root of $\Omega_s(\phi_s)$; see Example 11-3

Ψ_r, Ψ_r roots of the material and energy balance functions for the rectifying pinch; see Eqs. (11-24) and (11-25).

Ψ_s, Ψ_s roots of the material and energy balance functions for the stripping pinch; see Eqs. (11-27) and (11-28).

Ω_r, Ω_s material balance functions for the rectifying and stripping pinches, respectively; see Examples 11-2 and 11-3.

Subscripts

r rectifying pinch

s stripping pinch

N an arbitrary positive number assigned to the reboiler of a column with infinitely many plates, used for counting purposes. For example, the plate above the reboiler has the number $N - 1$.

PROBLEMS

11-1 (*a*) Locate the roots of $\Omega_r(\phi_r)$ and $\Omega_s(\phi_s)$ for the four-component examples considered in Examples 11-1, 11-2, and 11-3. Let these roots be numbered as follows

$$\alpha_{j-1} < \phi_{rj} < \alpha_j \tag{1}$$

$$\alpha_j < \phi_{sj} < \alpha_{j+1} \tag{2}$$

(*b*) Underwood[14] was the first to observe that at minimum reflux

$$\alpha_h < \phi_{rl} = \phi_{sh} < \alpha_l \tag{3}$$

Let this common root be denoted by ϕ_{fl}. For the case where $\phi_{rl} = \phi_{sh} = \phi_{fl}$, use the defining equations for Ω_r and Ω_s to obtain the new function

$$\Omega_f(\phi_f) = \sum_{i=1}^{c} \frac{X_i}{1 - \phi_f/\alpha_i} - (1-q) \tag{4}$$

where $V_r - V_s = 1 - q$. Equations (3) and (4), and the function Ω_r may be used to compute the minimum reflux ratio as outlined in the next problem.

11-2 Find the minimum reflux ratio (L_r/D) for the following example

Component	α_i	FX_i	b_i/d_i	Other specifications
1	0.25	10	....	The feed enters the column as
2	0.5	22.5	20.0	a liquid at its bubble-point
3	1.0	45	0.0499875	temperature at the column
4	2.0	22.5	....	pressure, and thus $q = 1$

Hint: Use Eq. (4) of Prob. 11-1 to compute ϕ_{fl}. Since $\Omega_r(\phi_{fl}) = 0$, compute V_r by use of Eq. (1) of Example 11-2.

Answer: $V_r = 128.69$ and $L_r/D = 0.9373$.

11-3 When the key components upon which the specifications are made are not adjacent in relative volatility, the total distillate rate D and the d_i's for the split keys (the components having α values lying between those of the keys) are unknown. To determine the minimum reflux ratio, it is necessary to make use of the fact that all of the roots of the function Ω_f lying between α_l and α_h ($\alpha_h < \phi_{f,h+1} < \alpha_{h+1} < \cdots < \phi_{fl} < \alpha_l$) satisfy the functions Ω_r and Ω_s. Murdoch and Holland[10] developed the following formulas for computing the minimum reflux ratio, the distillate rate D, and the d_i's of the split key components

$$\frac{L_r}{V_r} = v \frac{\sum\limits_{j=h,\,l,\,L} (\omega_j/\alpha_j)d_j}{\sum\limits_{j=h,\,l,\,L} \omega_j d_j} \tag{1}$$

where L denotes all components lighter than the light key and

$$v = \frac{\prod\limits_{i=h+1}^{l} \phi_{fi}}{\prod\limits_{i=h+1}^{l-1} \alpha_i} \qquad \omega_j = \frac{\prod\limits_{i=h+1}^{l-1} \alpha_j(\alpha_j - \alpha_i)}{\prod\limits_{i=h+1}^{l} (\alpha_j - \phi_{fi})}$$

$$D = \sum_{j=h,\,l,\,L} \omega_j d_j - v \sum_{j=h,\,l,\,L} \frac{\omega_j}{\alpha_j} d_j \tag{2}$$

The d_k of each split key component k $(k = h + 1, h + 2, ..., l - 1)$ in the distillate is computed as follows

$$d_k = -\frac{\alpha_k}{\omega'_k} \sum_{j=h, l, L} \frac{\omega'_j}{\alpha_j} d_j \tag{3}$$

where $k = h + 1, h + 2, ...,$ or $l - 1$

$$\frac{\omega'_j}{\alpha_j} = \frac{\displaystyle\prod_{i=h+1, \neq k}^{l-1} \alpha_j(\alpha_j - \alpha_i)}{\displaystyle\prod_{h+1}^{l} (\alpha_j - \phi_{fi})}$$

Compute the minimum reflux ratio (L_r/D), the distillate rate D, and the flow rate of d_k for each split key component for the following mixture

Component	FX_i	α_i	b_i/d_i	d_i	b_i
1	5	0.40	....	....	5
2	12	0.70	....	....	12
3	5	0.90	....	....	5
4	13	1.00	5	2.1667	10.8333
5	14	1.15			
6	8	1.25			
7	16	1.35	0.142857	14	2
8	14	1.50	....	14	
9	8	2.00	....	8	
10	5	3.00	,.......	5	

Thermal condition of feed $= 1 - q = 0.4$

Answer: $(L_r/D)_{min} = 2.6778$, $D = 55.37$, $d_5 = 6.79$, $d_6 = 5.42$

11-4 (a) For the following example, begin the stepwise calculations at the top of the column and make alternate use of the component-material balances and equilibrium relationships. Continue the calculations until the change in the y_{ji}'s is less than 10^{-7}.

Component	C_i	$K_i = C_i \exp(-E_i/T^*)$		
		E_i	X_{Di}	Specifications
1	$4.0 \times 10^3/P^{**}$	4.6447×10^3	1/6	The column has a total condenser and the
2	$8.0 \times 10^3/P$	4.6447×10^3	1/3	column pressure is 1 atm. $D = L_j = 50$
3	$12.0 \times 10^3/P$	4.6447×10^3	1/2	mol/h and $V_j = 100$ mol/h for all j

* T is in °R, P is in atm.

Show that at the rectifying pinch, $K_{rb} = 0.6371456$.
   (b) Find the $\phi_r < \alpha_l$ which makes $\Omega_r(\phi_r) = 0$; see Eq. (A) of Example 11-2.

11-5 Sketch the function $P_1(\Psi_r, T_r)$ over the interval from $\Psi_r = 0$ to $\Psi_r = 1/(1 - K_{rh})$. Produce the traces for $T_r < T_{BP}$, $T_{DP} < T_r < T_h$, and $T_r > T_h$, where T_{BP} is temperature of the distillate at the column pressure.

11-6 Develop the functions $p_1(\Psi_s, T_s)$ and $p_2(\Psi_s, T_s)$ which are given by Eqs. (11-27) and (11-28), and produce the trace of $p_1(\Psi_s, T_s)$ over the interval $0 < \Psi_s < 1/(K_{sl} - 1)$, where T_s lies between T_l and T_{BP} (the bubble-point temperature of the bottoms at the column pressures).

11-7 In the original treatment of the separated light components in a conventional column at minimum reflux by Holland,[4] the material balances were written around the top of the column and each plate down to and including plate $s - 1$. (a) Formulate these balances and (b) show that the set of equations found in part (a) is equivalent to the set given by Eq. (11-12).

11-8 Originally,[4] the material balances for the separated heavies in a column at minimum reflux were written around the bottom and each plate $(s, s - 1, s - 2, ..., f + 1, f, f - 1, ..., r + 2, r + 1)$. Repeat parts (a) and (b) of Prob. 11-7, and show that the equations found in part (b) are equivalent to those given by Eq. (11-4).

11-9 Develop the component-material balances and corresponding matrix equations for the distributed components, the separated lights, and the separated heavies for the case where a sidestream W_1 is withdrawn a finite number of plates below the top of the column.

11-10 Repeat Prob. 11-9 for the case where W_1 is withdrawn a finite number of plates above the feed plate.

11-11 (a) The following problem is based on the paper by McDonough et al.[8] For the case where the keys are the lightest and heaviest components of a mixture in a conventional distillation column at minimum reflux the pinches occur at and adjacent to the feed plate. For the case where the total flow rates are taken to be constant throughout each section of the column, the vapor and liquid flow rates in the pinches are related through the use of the quantity q which is defined as follows

$$L_s = L_r + qF \qquad V_r = V_s + (1 - q)F$$

where r and s refer to the rectifying and stripping pinches, respectively. For bubble-point liquid feeds and subcooled feeds, the composition y_{fi} of the vapor leaving plate f is equal to the composition of the vapor entering plate $f - 1$. For dew-point and superheated feeds, the composition of the liquid leaving plate $f - 1$ is the same as that of the liquid entering plate f. For a partially vaporized feed, use model 2 (Fig. 2-2). Construct the diagrams describing the assumed feed-plate behavior.

(b) By use of these relationships for the feed plate and the necessary equilibrium and material-balance relationships show that the q-line

$$X_i = (1 - q)y_{ri} + qx_{ri}$$

holds for bubble-point liquid, drew-point vapor, and subcooled and superheated feeds.

(c) Show that the q-line may be rearranged to the following form

$$p(T) = \sum_{i=1}^{c} \frac{X_i}{q + (1 - q)K_{ri}} - 1$$

11-12 Construct proofs for the relationships stated in the following table:

Thermal condition of the feed	Relationships (where the keys are the and heaviest components of the mixture)
Boiling-point liquid	$y_{ri} = y_{si}$, $x_{ri} = x_{si}$, $X_i = x_{ri}$, and $T_F = T_r = T_s$, where T_F = flash or feed temperature
Subcooled liquid	$y_{ri} = y_{si}$, $x_{ri} = x_{si}$, $T_r = T_s$, and T_r is that $T > 0$ which makes $p(T) = 0$ (see part (c) of Prob. 11-11)
Partially vaporized	$y_{ri} = y_{si} = y_{Fi}$, $x_{ri} = x_{si} = x_{Fi}$, and $T_r = T_s = T_F$
Dew-point vapor	$x_{ri} = x_{si}$, $y_{ri} = y_{si}$, $T_r = T_s$, $X_i = y_{ri}$, and $T_F = T_r = T_s$
Superheated vapor	$x_{ri} = x_{si}$, $y_{ri} = y_{si}$, $T_r = T_s$, and T_r is that $T > 0$ which makes $p(T) = 0$ (see part (c) of Prob. 11-11)

Except for a partially vaporized feed, a direct method of proof may be used which makes use of the relationships of part (a) of Prob. 11-11 and certain material balance and equilibrium relationships. A proof for the relationships given for a partially vaporized feed may be obtained by use of the indirect method. An outline of this proof follows. In order to prove that $T_F = T_r = T_s$, make the supposition that $T_F < T_r$, and then obtain a contradiction. Next suppose T_F to be greater than T_r, and obtain a contradiction. This leads to the conclusion that $T_F = T_r$. Similarly, show $T_F = T_s$. After these two proofs have been obtained, the equality of the compositions readily follows.

11-13 When the keys are the lightest and heaviest components of the mixture and the total-flow rates are constant within each section of the column, a direct solution may be obtained by use of the relationships given in the table in Prob. 11-12(c).

(a) For the following example, show that the specifications permit the keys to be the lightest and heaviest components, that is, show that $A_{ri} < 1$, $A_{si} < 1$ for all i.

Component	K_i at $T_r = T_s$	X_i	Specifications
1	1.5	1/3	Thermal condition of feed = boiling-point liquid. $D = 50$,
2	1.0	1/3	$V_1 = V_r = 90$ (thus $L_s = 140$, $V_s = 90$). The keys
3	0.5	1/3	are components 1 and 3.

(b) Find the product distribution.

11-14 Consider again the case where the keys are the lightest and heaviest of the mixture. If a superheated feed is used instead of a boiling-point liquid feed, it is known that $T_r = T_s$. In this case the temperature T_r is unknown, but it may be determined by trial by finding the T that makes $p(T) = 0$, where $p(T)$ is the function given in Prob. 11-11(c). Find the K_b at $T_r = T_s$ and the corresponding product distribution for the following example.

Component	α_i	X_i	Specifications
1	3.0	1/3	Superheated feed, $q = -0.05$. $D = 50$,
2	2.0	1/3	$V_1 = V_r = 120$, $L_1 = L_r = 70$, $L_s = 65$.
3	1.0	1/3	The keys are components 1 and 3.

For mixtures in which the relative volabilities are constant, it is convenient to replace K_i in the expression for $p(T)$ by $\alpha_i K_b$ and find the value of K_b that makes $p(K_b) = 0$, where

$$p(K_b) = \sum_{i=1}^{c} \frac{X_i}{q + (1 - q)\alpha_i K_b} - 1$$

Also find the composition of the liquid in the rectifying pinch.

REFERENCES

1. A. Acrivos and N. R. Amundson: "On the Steady State Fractionation of Multicomponent and Complex Mixtures in an Ideal Cascade: Part 2—The Calculational of the Minimum Reflux Ratio," *Chem. Eng. Sci.*, **4**(2): 68 (1955).
2. J. B. Bachelor: "How to Figure Minimum Reflux," *Pet. Refiner*, **36**(6): 161 (1957).
3. G. G. Brown and D. E. Holcomb: "Vapor-Liquid Equilibria in Hydrocarbon Systems," *Pet. Eng.*, **11**: 23 (August 1940).

4. Charles D. Holland: *Multicomponent Distillation*, Prentice-Hall, Englewood Cliffs, N.J., 1963.
5. W. K. Lewis and G. L. Matheson: "Studies in Distillation-Design of Rectifying Columns for Natural and Refinery Gasoline," *Ind. Eng. Chem.*, **24**:494 (1932).
6. J. A. McDonough and C. D. Holland: "Figure Separations This New Way: Part 9—How to Figure Minimum Reflux," *Pet. Refiner*, **41**(3):153 (1962).
7. ——— and ———: "Figure Minimum Separations This New Way: Part 10—Minimum Reflux for Complex Columns," *Pet. Refiner*, **41**(4):135 (1962).
8. ——— ——— and H. L. Bauni: "Determination of the Conditions of Minimum Reflux When the Keys are the Most and Least Volatile Components," *Chem. Eng. Sci.* **16**:143 (1961).
9. P. G. Murdoch and C. D. Holland: "Multicomponent Distillation: III—Equations in Product Form and Simplified Application," *Chem. Eng. Prog.*, **48**(5):254 (1952).
10. ——— and ———: "Multicomponent Distillation: IV—Determination of Minimum Reflux," *Chem. Eng. Prog.*, **48**(6):287 (1952).
11. C. S. Robinson and E. R. Gilliland, *Elements of Fractional Distillation*, p. 24, McGraw-Hill Book Company, New York, 1950.
12. R. N. Shiras, D. H. Hanson, and G. H. Gibson, "Calculation of Minimum Reflux in Distillation Columns," *Ind. Eng. Chem.*, **42**:871 (1950).
13. M. Tavana and D. N. Hanson, "The Exact Calculation of Minimum Flows in Distillation Columns," *Ind. Eng. Chem. Process Des. Dev.*, **18**(1):154 (1979).
14. E. W. Thiele and R. L. Geddes, "Computation of Distillation Apparatus for Hydrocarbon Mixtures," *Ind. Eng. Chem.*, **25**:289 (1933).
15. A. J. V. Underwood: "Fractional Distillation of Multicomponent Mixtures," *Chem. Eng. Prog.*, **44**:603 (1948).

TWELVE

DESIGN OF SIEVE AND VALVE TRAYS

Tray design encompasses the determination of the column diameter and the tray spacing as well as a number of mechanical considerations. The scope of the material in this chapter is limited primarily to the fundamentals involved in the design of single-pass sieve trays. The fundamentals involved in the design of valve trays are essentially the same as those involved in sieve trays. No attempt is made to treat bubble-cap trays, since valve and sieve trays have been used extensively in new installations since the early 1950s. Up until that time, bubble-cap trays were used almost exclusively. The design of bubble-cap trays has been treated by a number of authors; see for example Van Winkle.[17]

Pictures of a typical sieve tray and valve tray are shown in Figs. 12-1 and 12-2, respectively. Sieve trays consist of metal plates with small circular perforations. The valve of a valve tray consists of a self-regulating variable orifice (Fig. 12-2) which adjusts its opening in proportion to the total flow rate of the vapor. Most of the equations for sieve trays are also applicable for valve trays. A treatment of sieve trays is presented in Sec. 12-1 and the modifications of these equations as well as additional equations needed to describe valve trays are presented in Sec. 12-2.

Figure 12-1 A typical single-pass sieve tray with cross flow [*by courtesy Glitsch, Inc., Dallas, Texas*].

Figure 12-2 A typical single-pass valve tray (Koch Flexitray, type T caps) with cross flow [*by courtesy Koch Engineering Co., Inc., Wichita, Kansas*].

12-1 SINGLE-PASS SIEVE TRAYS WITH CROSSFLOW

The single-pass sieve tray with crossflow appears to be the most widely used type of tray today. In a tray of this type, the vapor passes upward through the perforations (or valves) and then through the liquid on the tray. The liquid flows down through the column and enters each plate by flowing under the downcomer weir. Then the liquid flows across the plate and over the outlet weir and into the downcomer to the plate below as shown in Fig. 12-3. In order for the liquid to flow across each plate in succession as it passes down through the column, a liquid head in each downcomer is required as depicted in Fig. 12-3.

As the vapor passes through the liquid, a "froth" or "foam" is formed as depicted in Fig. 12-3. At the outlet weir and in the downcomer, the vapor disengages itself by the formation of vapor bubbles. To overcome the frictional forces encountered by the vapor as it passes through each stage in succession, a pressure driving force is required. (Note in Fig. 12-3, $P_2 > P_1$.)

Other types of plates consist of those in which the liquid makes more than one pass and those in which counterflow of the vapor and liquid through the same perforations is employed.

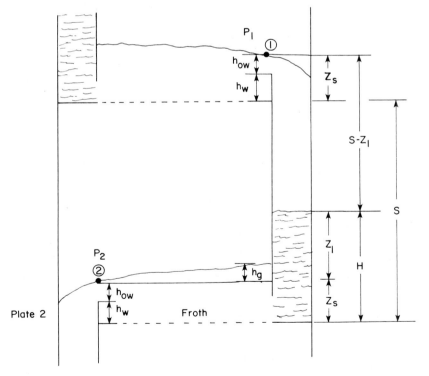

Figure 12-3 A graphical display of the meaning of the symbols used in the fluid dynamic analysis of a single-pass sieve tray with crossflow.

Fluid Dynamic Analysis of a Sieve Tray

In order to avoid flooding, the designer must provide for a sufficient tray spacing S; see Fig. 12-3. To set the tray spacing, the height of liquid H in the downcomer must be determined through the use of a fluid dynamic analysis. First, Bernoulli's theorem is applied to the liquid as it flows from point (1) of plate 1 to point (2) of plate 2 (see Fig. 12-3). The datum for measuring all heads is taken to be the height of liquid over the weir on plate 2. Thus

$$\frac{P_1}{\rho_L} + \frac{g}{g_c} Z_1 + \frac{g}{g_c} (S - Z_1) \frac{\rho_V}{\rho_L} = \frac{P_2}{\rho_L} + \sum_i F_i \tag{12-1}$$

where $\sum F_i$ = frictional losses
g = acceleration of gravity
g_c = Newton's law conversion factor
P = pressure
S = tray spacing
Z_1 = distance shown in Fig. 12-3
ρ_L = mass density of the vapor-free liquid
ρ_V = mass density of the vapor

In Eq. (12-1) the change in kinetic energy from point (1) to point (2) was taken to be negligible. The frictional losses, $\sum_i F_i$, consist of the head lost by the liquid in flowing down the downcomer which is taken to be negligible, the head lost by flowing under the downcomer weir, and the head lost in flowing across the plate. Thus

$$\sum_i F_i = \frac{g}{g_c} h_{dc} + \frac{g}{g_c} h_g \tag{12-2}$$

where h_{dc} = downcomer head loss in equivalent (mm or in) of vapor-free liquid
h_g = hydraulic gradient, the head loss by the liquid in flowing across the plate in equivalent (mm or in) of vapor-free liquid

Equation (12-1) may be solved for Z_1 to give

$$Z_1 = \frac{[(P_2 - P_1)/\rho_L]g_c/g + h_{dc} + h_g - S\rho_V/\rho_L}{(1 - \rho_V/\rho_L)} \tag{12-3}$$

The height of liquid H in the downcomer is found by adding Z_s (where $Z_s = h_w + h_{ow}$ as shown in Fig. 12-3 to both sides of Eq. (12-3) and rearranging to obtain

$$H = \frac{[(P_2 - P_1)/\rho_L]g_c/g + h_{dc} + h_g + Z_s - (S + Z_s)\rho_V/\rho_L}{(1 - \rho_V/\rho_L)} \tag{12-4}$$

The total pressure drop $P_2 - P_1$ consists of the dry pressure drop h_0 of the vapor as it passes through the perforations (or orifices) and the drop h_L the vapor experiences in passing through the liquid and froth on the plate. Thus

$$\frac{P_2 - P_1}{\rho_L} = \frac{g}{g_c} h_0 + \frac{g}{g_c} h_L \tag{12-5}$$

Elimination of $(P_2 - P_1)/\rho_L$ from Eqs. (12-4) and (12-5) gives

$$H = \frac{h_0 + h_L + h_{dc} + h_g + Z_s - (S + Z_s)\rho_V/\rho_L}{(1 - \rho_V/\rho_L)} \tag{12-6}$$

When the vapor density ρ_V is negligible relative to the liquid density, Eq. (12-6) reduces to

$$H = h_0 + h_{dc} + h_L + h_w + h_{ow} + h_g \tag{12-7}$$

where Z_s has been replaced by its equivalent $(h_w + h_{ow})$. The result given by Eq. (12-7) is displayed graphically in Fig. 12-4. The methods which have been proposed for computing head losses h_0, h_{dc}, h_L, and h_g follow.

CALCULATION OF h_{ow}, HEIGHT OVER THE WEIR

The equivalent height of vapor-free liquid over the weir may be calculated for circular columns by use of a modified version of the Francis weir formula which was proposed by Bolles.[1] For a straight segmental weir

$$h_{ow} = 0.48 F_w \left(\frac{Q}{l_w}\right)^{2/3} \tag{12-8}$$

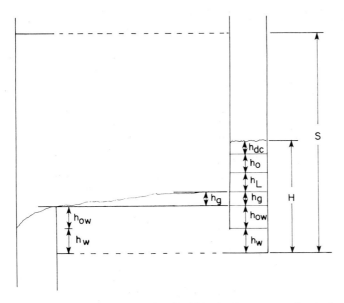

Figure 12-4 Height of vapor-free liquid in the downcomer of a single-pass sieve tray with crossflow.

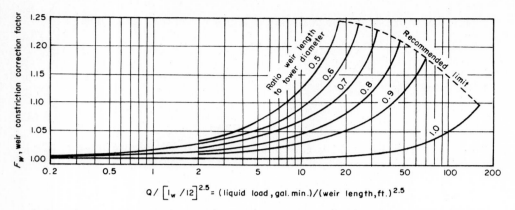

$$Q/\left[l_w/12\right]^{2.5} = \text{(liquid load, gal. min.)/(weir length, ft.)}^{2.5}$$

Figure 12-5 The correction factor for effective weir length. [W. L. Bolles, *Pet. Refiner*, **25**:613 (1946), *by courtesy Hydrocarbon Process*.]

where F_w = weir constriction correction factor
   h_{ow} = equivalent height of clear liquid, in
   l_w = length of weir, in
   Q = liquid flow rate, gal/min

A graph for evaluation of the correction factor F_w is presented in Fig. 12-5. The factor F_w is used to correct for the distorted flow pattern of the liquid as it approaches the weir.

When the crest over a straight segmented weir is less than a quarter of an inch, Fair et al.[2] recommend the use of a serrated weir, and for weirs of this type

$$h_{ow} = 0.7F_w\left(\frac{Q}{\tan \theta/2}\right)^{0.4} \tag{12-9}$$

where θ is the angle of serration and the other symbols are defined beneath Eq. (12-8).

CALCULATION OF h_0, THE HEAD EQUIVALENT TO THE DRY HOLE PRESSURE DROP

The dry hole pressure drop for a sieve tray may be calculated by use of the following equation for thick plate orifices

$$h_0 = 0.186\left(\frac{u_0}{C_0}\right)^2 \frac{\rho_V}{\rho_L} \tag{12-10}$$

where h_0 = dry hole pressure drop of vapor across the perforations in inches of equivalent vapor-free liquid
   u_0 = linear velocity of the vapor through the perforation in feet per second

Values of the discharge coefficient C_0 are given by the chart presented in Fig. 12-6 which was prepared by Leibson et al.[12]

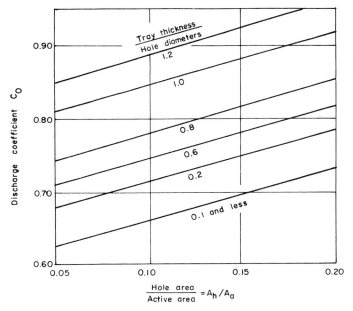

Figure 12-6 Discharge coefficients for the flow of vapor through sieve trays. [I. Leibson, R. W. Kelley, and L. A. Bullington, *Pet. Refiner*, **36**(2): 127 (1957), *by courtesy Hydrocarbon Process.*]

CALCULATION OF h_{dc}, THE HEIGHT EQUIVALENT TO THE PRESSURE DROP RESULTING FROM THE FLOW OF THE LIQUID UNDER THE DOWNCOMER

For a design in which no inlet weir is provided, the pressure drop resulting from the flow of the liquid through the clearance between the downcomer and the plate may be calculated by use of the conventional formula for submerged weirs

$$h_{dc} = 0.057 \left(\frac{Q}{A_{dc}} \right)^2 \tag{12-11}$$

where h_{dc} = head loss in inches of vapor-free liquid
Q = flow rate of the liquid under the downcomer weir in gallons per minute
A_{dc} = clearance area between the downcomer and the floor of the tray in square inches

If the tray is equipped with an inlet weir, Leibson et al.[12] recommend that Eq. (12-11) be modified as follows

$$h_{dc} = 0.068 \left(\frac{Q}{A_{dc}} \right)^2 \tag{12-12}$$

The larger constant in Eq. (12-12), compared to the constant in Eq. (12-11), takes into account a loss of velocity head due to the partial reversal of the liquid flow which occurs when an inlet weir is provided.

CALCULATION OF h_L, THE HEAD LOSS RESULTING FROM THE FLOW OF THE VAPOR THROUGH THE AERATED LIQUID

The pressure drop through the aerated liquid, h_L, has been correlated as a function of the calculated height $(h_w + h_{ow} + \frac{1}{2}h_g)$ of vapor-free liquid over the perforations by Fair[3] as follows

$$h_L = \beta(h_w + h_{ow} + \tfrac{1}{2}h_g) \tag{12-13}$$

where h_L = head loss in inches of vapor-free liquid
β = aeration factor, dimensionless

A graph for estimating β is given in Fig. 12-7. Also given in Fig. 12-7 is a curve for estimating the relative froth density ϕ which is defined as follows

$$\phi = \frac{h_L}{h_f} \tag{12-14}$$

h_f = actual height of the froth, in

The following theoretical relationship between ϕ and β was developed by Hutchinson et al.[9]

$$\beta = \frac{\phi + 1}{2} \tag{12-15}$$

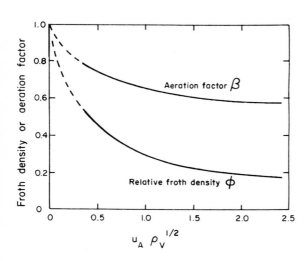

Figure 12-7 Aeration factor and froth density for bubble-cap, sieve, and valve plates, u_0 = linear vapor velocity through the active area, ft/s; ρ_V = vapor density, lb/ft³. [B. D. Smith, *Design of Equilibrium Stage Processes*, McGraw-Hill Book Company, New York, 1963, by courtesy McGraw-Hill Book Company.]

CALCULATION OF h_g, THE HYDRAULIC GRADIENT

Hughmark and O'Connell[10] presented the following correlation for computing the hydraulic gradient for a sieve plate

$$h_g = \frac{fu_f^2 l_f}{12g_c r_h} \tag{12-16}$$

where h_g = hydraulic gradient in inches of vapor-free liquid
  f = friction factor; see Fig. 12-8
  u_f = velocity of the aerated mass, ft/s
  g_c = Newton's law conversion factor, 32.17
  l_f = length of flow path across plate, ft
  r_h = hydraulic radius of the aerated mass, ft; defined by Eq. (12-18)

A correlation of the friction factor f as a function of Reynolds number is shown in Fig. 12-8. The Reynolds number used in this correlation is defined as follows

$$N_{\text{Re}} = \frac{r_h u_f \rho_L}{\mu_L} \tag{12-17}$$

where ρ_L = mass density of the vapor-free liquid, lb/ft^3
  μ_L = viscosity of the vapor-free liquid, lb/(ft s)

The hydraulic radius r_h of the aerated mass is defined as follows

$$r_h = \frac{\text{cross section}}{\text{wetted perimeter}} = \frac{h_f D_f}{2h_f + 12D_f} \tag{12-18}$$

where D_f = arithmetic average of the tower diameter and the weir length, ft
  h_f = froth height in inches; estimated by use of Eq. (12-14) and Fig. 12-7.

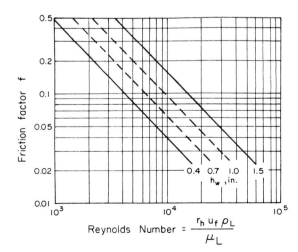

Figure 12-8 Friction factor used in the calculation of the hydraulic gradient, h_g, for sieve trays with crossflow. [B. D. Smith, *Design of Equilibrium Stage Processes*, McGraw-Hill Book Company, New York, 1963, *by courtesy McGraw-Hill Book Company.*]

The velocity of the aerated mass u_f in feet per second is taken to be the same as that of the vapor-free liquid, and it is computed as follows

$$u_f = \frac{12q}{h_L D_f} \tag{12-19}$$

where q is the liquid flow rate in cubic feet per second.

Column Capacity

The capacity of a column for handling maximum vapor and liquid flows is of importance because these flow rates determine the column diameter and the tray spacing. At a given liquid rate, the vapor rate may be increased to the point where an excessive amount of liquid is entrained and finally flooding occurs. This type of flooding is commonly called *entrainment flooding*. Observations show that the downcomer may be only partially filled with liquid when such flooding begins. As flooding develops further, the downcomers become filled with liquid and all of the liquid fed to the column is carried out of the column with the overhead vapor. This type of flooding is further characterized by a large increase in the pressure drop across the column as the column becomes flooded.

Similarly, with the gas rate held fixed, the liquid rate may be increased until flooding occurs. In this type of flooding, called *downflow flooding*, the liquid in the downcomer of each plate backs up to the plate above and a sharp increase in pressure drop across the column is observed.

On the other hand as the flow rates are decreased, a point is reached at which the contacting device becomes ineffective in providing good contact between the two phases. Each type of plate is characterized by a different set of flow rates below which it will not operate satisfactorily. In the case of crossflow-sieve plates, the reduction of the gas flow rate beyond a certain value causes excessive weeping and dumping of liquid which reduces the dispersion of the vapor throughout the liquid phase. With the reduced dispersion comes a reduction in the interfacial area which in turn reduces the rate of mass transfer between phases. Valve trays can be operated at relatively low gas rates because the valve opening closes as the liquid rate decreases. Bubble-cap plates can operate at very low gas flow rates because of their seal arrangement. A diagram depicting the safe operating zone is shown in Fig. 12-9.

TRAY SPACING AND DOWNFLOW FLOODING

In order to prevent downflow flooding, the trays must be adequately spaced and the downcomers must be appropriately sized. It is customary in many designs to take the tray spacing S to be equal to twice the liquid holdup H in the downcomer $(S = 2H)$. However, this choice of S may be more conservative than necessary or it may be inadequate. If it is known that the flow over the outlet weir is relatively free of vapor, then the tray spacing may be taken approximately equal to H. On the other hand, if it is known that the downcomer is filled with froth having a much lower density than the liquid, a tray spacing equal to as

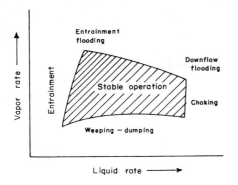

Figure 12-9 Operational characteristics of a column. [From *Chemical Engineers' Handbook*, R. H. Perry and C. H. Chilton (eds.), 5th ed., sec. 18, by J. R. Fair, D. E. Steinmeyer, W. R. Penney, and J. A. Brink, 1973; *by courtesy McGraw-Hill Book Company*.]

much as $5H$ may be required. If the tray spacing is too small, the downcomer will become filled with liquid and froth and downflow flooding will result.

ENTRAINMENT FLOODING

Determination of the column diameter through the use of the velocity at which a droplet will become suspended in a vapor stream was first suggested by Souders and Brown.[16] The expression for this velocity is easily developed by considering the case of a freely falling spherical droplet of mass m through a vapor at rest. Let the projected area of the droplet be denoted by A_d and its diameter by D_d. Let the drag force on the droplet be denoted by F_D, the buoyancy force by F_b, and the gravitational force by F_g. If the downward direction of motion is taken as the positive direction, then the application of Newton's second law of motion gives

$$F_g - F_b - F_D = \frac{1}{g_c}\frac{d(mu)}{dt} = \frac{1}{g_c}m\frac{du}{dt} \tag{12-20}$$

where u is the velocity of the droplet in the positive direction and the conversion factor g_c has the value of 32.17 when m is in pounds force. The forces are given by

$$F_b = \left(\frac{\pi D_d^3}{6}\right)\rho_V\frac{g}{g_c} \tag{12-21}$$

$$F_g = \left(\frac{\pi D_d^3}{6}\right)\rho_L\frac{g}{g_c} \tag{12-22}$$

$$F_D = C_D A_d \rho_V \frac{u^2}{2g_c} = C_D\left(\frac{\pi D_d^2}{4}\right)\rho_V\frac{u^2}{2g_c} \tag{12-23}$$

where C_D is the customary drag coefficient, ρ_V is the mass density of the vapor, ρ_L is the mass density of the liquid, and g is the acceleration of gravity.

When the droplet reaches it terminal velocity, $du/dt = 0$, and Eqs. (12-20) through (12-23) may be solved for the terminal velocity u to give

$$u = \left(\frac{4gD_d}{3C_D}\right)^{1/2}\left(\frac{\rho_L - \rho_V}{\rho_V}\right)^{1/2} \tag{12-24}$$

Since the first term on the right-hand side is very nearly constant for any one droplet, Eq. (12-24) is commonly stated in the following form

$$u = C_{SB}\left(\frac{\rho_L - \rho_V}{\rho_V}\right)^{1/2} \tag{12-25}$$

where

$$C_{SB} = \left(\frac{4gD_d}{3C_D}\right)^{1/2}$$

Let D_p be regarded as the average droplet diameter. Then all droplets having a diameter less than D_p become entrained in the vapor and are carried to the plate above. This observation suggests the use of Eq. (12-25) in the correlation of the vapor velocity at which entrainment flooding occurs. The original correlation of Souders and Brown[16] consisted of a plot of C_{SB} against tray spacing with surface tension as a parameter. Fair and Mathews[5] presented a correlation for bubble-cap trays in 1958 and sieve trays[4] in 1961 in which C_{SB} is plotted against the Sherwood function, $(L/G)(\rho_V/\rho_L)^{1/2}$, where L and G are the mass rates of flow of the liquid and vapor, respectively, in pounds per hour. In Fair's correlations, tray spacing is used as a parameter and a surface tension correction is applied. For crossflow sieve trays, Fair's correction for estimating the entrainment flood point is shown in Fig. 12-10. The factor $(20/\sigma)^{0.2}$ which appears in

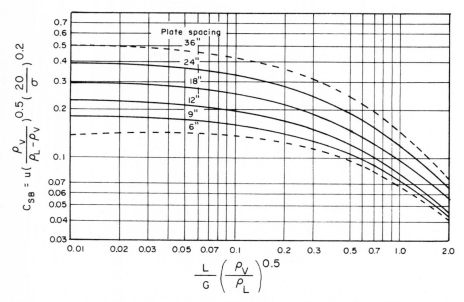

Figure 12-10 Values of C_{SB} at flooding conditions for sieve and bubble-cap plates. $L/G =$ ratio of mass flow rate of liquid to vapor. u is in feet per second, and σ is in dynes per centimeter. [J. R. Fair, *Pet. Chem. Eng.* **33**(10):45 (1961), *by courtesy Petroleum Engineer.*]

the ordinate function corrects for the deviation of the system surface tension from the base value of 20 (dynes per cm) to the actual surface tension σ of the liquid. In Fig. 12-10, the linear velocity u is based on the active area available for deentrainment (usually the tower cross-sectional area minus one downcomer) in feet per second.

Fair[4] restricted the correlation presented in Fig. 12-10 to systems which satisfy the following conditions

1. Weir height is less than 15 percent of tray spacing
2. System is low to nonfoaming
3. Hole diameter is one-quarter inch in diameter or less (sieve trays)
4. Holes are evenly distributed and occupy at least 10 percent of the active area, the area between the downcomer and the outlet weir.

If the hole area A_h is less than 10 percent of the active area A_a, then Fair[4] recommends that the chart values be corrected as follows

A_h/A_a	$u/(u)_{\text{chart}}$
0.1	1.0
0.08	0.9
0.06	0.8

where A_a is the area between the downcomer weir and the outlet weir.

A correlation similar to the one proposed by Fair[4] is presented in Fig. 12-11. This correlation was proposed by Smith et al.[15] and it is applicable to columns having either bubble caps, sieve, or valve trays. In this correlation, the parameter is settling height rather than tray spacing used by Fair.[4] Settling height is defined as the tray spacing minus the clear (vapor-free) liquid depth. For this correlation, the clear liquid depth is defined as $(h_w + h_{ow})$, the height of the weir plus the height at which clear liquid overflows the weir. It is further specified that h_{ow} is the value found by use of the Francis weir formula [see Eqs. (12-8) and (12-9)]. In the same year that Smith's correlation appeared, Gerster[6] proposed a similar correlation in which the parameter $(h_w + h_{ow})$ was used.

The correlation presented in Fig. 12-11 is based on data collected by Smith et al.[15] on a number of towers which were operating under conditions near their physical limits and occasionally overflowing. Smith et al. state that the average deviation in the flooding velocity data from manufacturers of valve trays was of six percent. Previously published data used by Fair[4] to formulate the correlation of Fig. 12-10 were represented less satisfactorily by Fig. 12-11 with an average deviation of 25 percent, consistently low. Smith et al. had no explanation to offer except to point out that some of the data used by Fair were taken on towers of experimental design rather than from towers of commercial design which were used in his study.

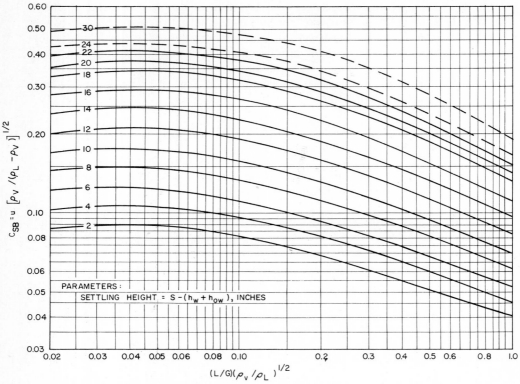

Figure 12-11 Flooding limits for sieve, valve, and bubble-cap plates. L/G = ratio of mass flow rate of liquid to vapor. u is in feet per second, and S is the tray spacing in inches. [Smith et al., *Petroleum Refiner* **40**(5): 183 (1963), *by courtesy Hydrocarbon Process.*]

A FURTHER INSIGHT TO ENTRAINMENT AND DROPLET RUPTURE

A further insight into droplet entrainment and rupture is provided by the analysis of Manning[13] who developed a relationship for the absolute maximum velocity at which a droplet of a given diameter will remain suspended without rupturing into smaller droplets.

For a given vapor velocity, the diameter which a liquid droplet must have in order to become suspended and have a zero velocity with respect to the floor of the tray is found by rearrangement of Eq. (12-24)

$$D_d = \frac{3u^2 C_D}{4g}\left(\frac{\rho_V}{\rho_L - \rho_V}\right) \qquad (12\text{-}26)$$

Hinze[8] found that for liquid particles falling through vapors, $C_D \cong 0.7$.

Hinze[8] and others have observed that as the vapor flowing past a droplet is increased, a critical value of the Weber number is reached at which rupture occurs. The Weber number is defined as follows

$$N_{We} = \frac{D_d \rho_V u^2}{\sigma} \tag{12-27}$$

where σ is the surface tension expressed in consistent units. Since the difference in pressure P_i on the inside of a liquid droplet and the pressure P on the outside of a droplet are related to the surface tension by the well-known relationship

$$P_i - P = \frac{4\sigma}{D_d} \tag{12-28}$$

it is possible to restate Eq. (12-27) in the following form

$$N_{We} = 8 \left[\frac{\rho_V / \rho_L (u^2 / 2g_c)}{[(P_i - P)/\rho_L g_c]} \right] \tag{12-29}$$

Thus, the Weber number is seen to be the ratio of the impact kinetic energy of the vapor stream to the surface tension energy tending to hold the droplet together. As the impact kinetic energy is increased until it exceeds the surface tension energy, the droplet will rupture. Although Eq. (12-29) suggests that the critical value of the Weber number should be 8, Hinze[8] points out that the critical value of the Weber number varies somewhat with liquid viscosity. For the set of substances water, carbon tetrachloride, and glycerin plus 20 percent water, Hinze[8] concluded on the basis of some rough experimental data that rupture occurs at an average Weber number of approximately 20.

Since a droplet ruptures at values of the Weber number which exceed the critical value, Eq. (12-27) may be solved for the maximum value of the droplet diameter to give

$$D_{d,\,max} = \frac{\sigma (N_{We})_{crit}}{\rho_V u^2} \tag{12-30}$$

where $D_{d,\,max}$ = maximum diameter of the droplet
$\quad\quad (N_{We})_{crit}$ = critical value of the Weber number

Equations (12-26) and (12-30) define the maximum diameter which a droplet may have and remain suspended in a vapor stream without rupture. The maximum diameter of a droplet as well as the maximum vapor velocity which can be tolerated without rupture occurs at the intersection of Eqs. (12-26) and (12-30); see Fig. 12-12.

When D_d given by Eq. (12-26) is equal to $D_{d,\,max}$ given by Eq. (12-30), the vapor velocity is absolutely the highest which can be tolerated. For instance, if the vapor velocity is increased, the droplets will have a positive velocity upward or the droplets that were of the maximum diameter will now rupture. Thus, the intersection of the two curves in Fig. 12-12 represents the absolute maximum

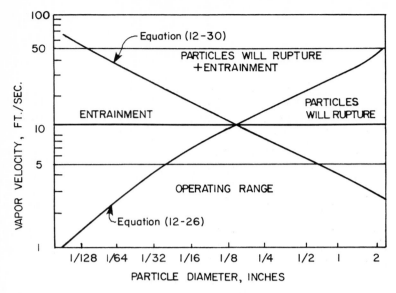

Figure 12-12 Four areas of tray hydraulics are described by Eqs. (12-26) and (12-30). The intersection of the two curves represents the highest possible vapor velocity that can be tolerated. [E. Manning, *Ind. Eng. Chem.*, **56**(4): 15 (1964), *by courtesy American Chemical Society.*]

velocity that can exist. The intersection is found by solving Eqs. (12-26) and (12-30) for the velocity subject to the condition that $D_d = D_{d,\,max}$.

$$u_{max} = \sqrt[4]{\frac{4g\sigma(N_{We})_{crit}(\rho_L - \rho_V)}{3\rho_V^2 C_D}} \qquad (12\text{-}31)$$

If all droplets were of the same size (the maximum size that can exist), the vapor velocity at which the droplets would become suspended but would not rupture is given by Eq. (12-31). Above any actual tray, a range of droplet sizes exists, and the maximum vapor velocity which can actually be reached is well below the values predicted by Eq. (12-31). Figure 12-13, taken from Manning,[13] shows a comparison of the theoretical maximum vapor velocity given by Eq. (12-31) and the actual maximum vapor velocity for several high-capacity trays.

MINIMUM VAPOR VELOCITY AND WEEPING

As the vapor rate is reduced below the weep point, excessive liquid weeping through the perforations occurs which reduces the plate efficiency significantly. Although some weeping of the liquid through the perforations occurs on all sieve trays at normal operating conditions, it does not affect the plate efficiency unless the vapor rate is reduced until the weep point is reached. This operating condition may be predicted through the use of a correlation proposed by Fair.[3] The correlation is based on the observation that liquid will not drain through the perforations if the liquid head above the perforation is less than the sum of h_0

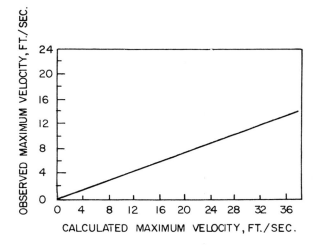

Figure 12-13 Comparison of theoretical maximum vapor velocity with maximum for conventional trays. [E. Manning, *Ind. Eng. Chem.*, **56**(4):15 (1964), *by courtesy American Chemical Society.*]

(the pressure drop across the perforation) and h_σ (the vapor-free liquid head required to overcome the surface tension of the liquid over the perforation). In the correlation shown in Fig. 12-14, the liquid head over the perforation is taken to be the head at the outlet weir, $(h_w + h_{ow})$. If the design point lies above the appropriate line in Fig. 12-14, then weeping will not be of a sufficient magnitude to have an appreciable effect on the plate efficiency.

The pressure drop of the vapor across the orifice expressed in inches of vapor-free liquid is computed by use of Eq. (12-10). The head required to overcome the surface tension force at the perforation may be estimated by use of an

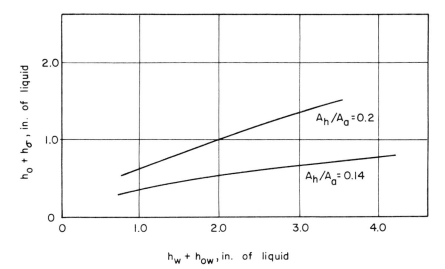

Figure 12-14 Weeping correlation for sieve plates. [From B. D. Smith, *Design of Equilibrium Stage Processes*, McGraw-Hill Book Company, New York, 1963, *by courtesy McGraw-Hill Book Company.*]

expression proposed by Fair et al.,[2] which was apparently obtained by restating Eq. (12-28) in inches of vapor-free liquid

$$h_\sigma = \frac{0.040\sigma}{\rho_L D_0} \qquad (12\text{-}32)$$

where D_0 = diameter of perforation, in
σ = liquid surface tension, dyne/cm

[Actually, on the basis of conversion of units alone, a coefficient of 0.053 rather than 0.04 is obtained for Eq. (12-32).]

Typical Values of Selected Variables

As a guide in the design of sieve trays, the values of selected variables encountered are presented. These values were taken from Leibson et al.[12]

The perforations in sieve trays are seldom less than $\frac{3}{16}$ of an inch in diameter and they are located on equilateral triangles such that the pitch to diameter ratio is greater than 2 but less than 5 with an optimum of about 3.8. The trays are generally designed with tray thickness to hole diameter of 0.1 to 0.7.

Because of its large contribution to the total pressure drop across a tray, the liquid seal $(h_w + h_{ow}$, at the outlet weir) should not normally exceed 4 inches. The outlet weir height is normally greater than $1\frac{1}{2}$ inches but less than 3 inches.

The minimum clearance between the downcomer outlet and the deck of the plate is $1\frac{1}{2}$ inches. If the downcomer to tray clearance is greater than the liquid seal height on the tray at the minimum liquid rate, an inlet weir of suitable height should be provided to ensure that the downcomer is actually sealed in order to avoid vapor blowback through the downcomer.

12-2 VALVE TRAYS

Valve trays have essentially all of the advantages of both bubble-cap trays and sieve trays but hardly any of the disadvantages. They have about the same capacity, efficiency, and low pressure drop offered by sieve trays. Like bubble-cap trays, valve trays can be operated over a wider range of operating conditions (including very low flow rates) than sieve trays. The relatively simple construction of valve trays leads to a cost which is not a great deal higher than the comparable sieve tray. Some specific claims of Glitsch[7] are as follows:

1. Maximum efficiency at low loads and high efficiency at conditions 5 to 10 percent below flooding results in usable capacity.
2. A combination of lower pressure drop per stage and high efficiency for vacuum systems results in low tower pressure drops.
3. The mechanical design of the tabs are such as to virtually eliminate sticking problems. At zero to relatively low vapor rates, the V-type unit is sealed on

three tabs which hold the disc above the deck by a distance of approximately 0.1 inches. The maximum clearance is 0.32 inches.

Some specific claims of Koch Engineering[11] are as follows:

1. Flexitrays can handle loads up to 10 percent higher than sieve trays and 15 to 20 percent higher than bubble-cap trays.
2. The efficiencies are essentially constant over a wide range of operating conditions.
3. Low pressure drop and negligible hydraulic gradient are characteristic of Flexitrays.
4. Stops on the caps prevent caps spinning in tray holes.

At the present time, valve trays constitute about 70 percent of the market, sieve trays 25 percent, and bubble-cap trays 5 percent. The standard valve tray costs about 20 percent more than the standard sieve tray with $\frac{3}{8}$- to $\frac{1}{2}$-inch or larger holes. The cost of a bubble-cap tray is approximately 300 percent more than the corresponding valve tray.

Although the design equations presented by the manufacturers differ in form from those presented for sieve trays, they are based on the same fundamentals. Except for the pressure drop through the perforations, the equations presented in Sec. 12-1 for sieve trays should yield approximately the same total head H as those of the manufacturer. The mechanism for the flow of vapor through the perforations differs from the mechanism for the flow of vapor through the valves. For the *V*-1 and *V*-4 units, Glitsch[7] has proposed the following equation for computing the pressure drop of the vapor as it passes through the opening of the valves of a dry plate

Units part open:

$$h_0 = 1.35\frac{t_m \rho_m}{\rho_L} + K_1 u_0^2 \frac{\rho_V}{\rho_L} \qquad (12\text{-}33)$$

Units full open:

$$h_0 = K_2 u_0^2 \frac{\rho_V}{\rho_L} \qquad (12\text{-}34)$$

where h_0 = pressure loss in inches of vapor-free liquid
t_m = valve thickness, in
ρ_m = density of valve metal, lb/ft^3
K_1, K_2 = pressure drop coefficients
u_0 = velocity of the vapor through the opening of the ballast unit, ft/s

Values of K_1 and K_2 are given below with the thickness corresponding to several densities of commonly used metals.

Pressure drop coefficients

Type unit	K_1	K_2 for deck thickness of			
		0.074 in	0.104 in	0.134 in	0.25 in
V-1	0.2	1.05	0.92	0.82	0.58
V-4	0.1	0.50	0.50	0.50	

Thickness		Valve material			
Gage	t_m, inches	Metal	Density lb/ft^3	Metal	Density lb/ft^3
20	0.037	C.S.	480	Hastelloy	560
18	0.050	S.S.	510	Aluminum	168
16	0.060	Nickel	553	Copper	560
14	0.074	Monel	550	Lead	708
12	0.104	Titanium	283		
10	0.134				

The hole or open area A_h (in square feet) used to compute the velocity through the opening, u_0, which appears in Eq. (12-33) is computed as follows

$$A_h = \frac{\text{No. ballast units}}{78.5} \qquad (12\text{-}35)$$

Example 12-1 This example is based on the results of a design calculation given by Leibson et al.[12] For the following single-pass sieve tray with crossflow:
(a) Compute the total holdup H of vapor-free liquid in the downcomer, and compare $2H$ with the specified tray spacing S. Estimate the height of froth in the downcomer on the basis of the assumption that the froth density in the downcomer is the same as that over the active area.
(b) Compute the flooding velocity of the vapor.
(c) Compute the minimum vapor rate.

Given:

Properties (at the operating conditions 270°F and 32.7 lb/in^2 abs)

$\rho_L = 40.8$ lb/ft^3, $\mu_L = 0.24$ centipoise, $\sigma = 12.5$ dyne/cm, gravity $= 56.3°$ API, and $\rho_V = 0.40$ lb/ft^3

Tray Geometry

Diameter = 15.5 ft
Hole diameter = $\frac{3}{16}$ in = 0.1875 in
Outlet weir length = 124 in = 10.33 ft
Outlet weir height = 2.5 in
Length of liquid flow path = 13.67 ft
Clearance between the downcomer outlet and the deck = 2.25 in

Tray thickness = 0.1345 in (10 gauge)
Total area = 188.6 ft^2
Active area = 116 ft^2
Perforation (hole) area = 7.8 ft^2
Area per downcomer = 14 ft^2

Flow Rates

Maximum flow rate of liquid = 1.5 ft^3/s = 673.25 gal/min
Minimum flow rate of liquid = 0.526 ft^3/s = 236 gal/min
Maximum flow rate of the vapor = 250 ft^3/s
Minimum flow rate of the vapor = 125 ft^3/s

(a) Calculation of the Total Holdup H of Vapor-Free Liquid in the Downcomer

In order to determine H, it is necessary to compute the values of h_o, h_{ow}, h_{dc}, h_L and h_g as described in the text.

1. *Calculation of h_o, the Pressure Drop through the Perforations at the Maximum Vapor Rate*

In order to find C_o by use of Fig. 12-6, the following quantities must be evaluated

$$\frac{A_h}{A_a} = \frac{\text{hole area}}{\text{active area}} = \frac{7.8}{116} = 0.067$$

$$\frac{t}{D_o} = \frac{\text{tray thickness}}{\text{hole diameter}} = \frac{0.1345}{0.1875} = 0.717$$

From Fig. 12-6, $C_o = 0.73$
The linear velocity u_o through the perforations is given by

$$u_o = \frac{250 \text{ ft}^3/\text{s}}{7.8 \text{ ft}^2} = 32.05 \text{ ft/s}$$

Then by Eq. (12-10),

$$h_o = 0.186\left(\frac{u_o}{C_o}\right)^2 \frac{\rho_V}{\rho_L} = (0.186)\left(\frac{32.05}{0.73}\right)^2 \left(\frac{0.40}{40.8}\right) = 3.5 \text{ in}$$

2. *Repeat of Item 1 for the Minimum Vapor Rate*

In this case $C_o = 0.73$, and

$$u_o = \frac{127 \text{ ft}^3/\text{s}}{7.8 \text{ ft}^2} = 16.03 \text{ ft/s}$$

Then by Eq. (12-10)

$$h_o = 0.186\left(\frac{u_o}{C_o}\right)^2 \frac{\rho_V}{\rho_L} = (0.186)\left(\frac{16.03}{0.73}\right)^2 \left(\frac{0.40}{40.8}\right) = 0.88 \text{ in}$$

3. Calculation of the Height Over the Weir h_{ow} at the Maximum Liquid Rate

In order to determine the correction factor F_w of Eq. (12-8) by use of Fig. 12-5, the value of the following quantity is needed

$$\frac{Q}{(l_w/12)^{2.5}} = \frac{673.25 \text{ gal/min}}{(124/12)^{2.5}} = 1.96$$

and

$$\frac{\text{Weir length (ft)}}{\text{Tower diameter (ft)}} = \frac{(124/12) \text{ ft}}{15.5 \text{ ft}} = 0.667$$

From Fig. 12-5, $F_w = 1.025$. Then by use of Eq. (12-8), one obtains

$$h_{ow} = 0.48 \, F_w \left(\frac{Q}{l_w}\right)^{2/3} = (0.48)(1.025)\left(\frac{673.25 \text{ gal/min}}{124 \text{ in}}\right)^{2/3} = 1.51 \text{ in}$$

4. Repeat of Item 3 for the Minimum Liquid Rate

In this case

$$\frac{Q}{(l_w/12)^{2.5}} = \frac{236 \text{ gal/min}}{(124/12)^{2.5}} = 0.688$$

and since

$$\frac{\text{Weir length}}{\text{Tower diameter}} = 0.667$$

and Fig. 12-5 gives

$$F_w = 1.02$$

it follows that

$$h_{ow} = (0.48)F_w \left(\frac{Q}{l_w}\right)^{2/3} = (0.48)(1.02)\left(\frac{236}{124}\right)^{2/3} = 0.75 \text{ in}$$

5. Calculation of h_{dc}, the Pressure Drop of the Liquid Resulting from its Flow Under the Downcomer at the Maximum Liquid Rate

The area between the downcomer outlet and the deck of the tray is found as follows

$$A_{dc} = (2.25 \text{ in})(124 \text{ in}) = 279 \text{ in}^2$$

Then, by use of Eq. (12-11)

$$h_{dc} = 0.057\left(\frac{Q}{A_{dc}}\right)^2 = (0.057)\left(\frac{673.25}{279}\right)^2 = 0.332 \text{ in}$$

6. Repeat of Item 5 for the Minimum Liquid Rate

$$h_{dc} = 0.057\left(\frac{Q}{A_{dc}}\right)^2 = (0.057)\left(\frac{236}{279}\right)^2 = 0.041 \text{ in}$$

7. *Calculation of* h_L, *the Head Loss from the Flow of Vapor Through the Aerated Liquid*

For the first estimate of h_L, neglect the hydraulic gradient h_g. In order to evaluate β of Eq. (12-13), the abscissa of Fig. 12-7 is evaluated. The linear velocity through the active area is given by

$$u_0 = \frac{250 \text{ ft}^3/\text{s}}{116 \text{ ft}^2} = 2.16 \text{ ft/s}$$

Then

$$u_0 \rho_V^{1/2} = (2.16)(0.40)^{1/2} = 1.37$$

From Fig. 12-7

$$\beta = 0.61$$

Thus, by use of Eq. (12-13) with the assumption $h_g \cong 0$, one obtains

$$h_L = \beta(h_w + h_{ow}) = (0.61)(2.5 + 1.51) = 2.45 \text{ in}$$

8. *Calculation of the Froth Height,* h_f

For the value of the abscissa computed in item 7, the value $\phi = 0.22$ is read from Fig. 12-7. Then by Eq. (12-14)

$$h_f = h_L/\phi = 2.446 \text{ in}/0.22 = 11.12 \text{ in}$$

9. *Calculation of the Hydraulic Gradient,* h_g

In order to evaluate the hydraulic gradient h_g by use of Eq. (12-16), it is necessary to determine the friction factor f. The latter is determined by use of Eqs. (12-17) through (12-19) and Fig. 12-8 as follows

$$D_f = \frac{\text{diameter of the column} + \text{length of the weir}}{2}$$

$$= \frac{15.5 + 10.33}{2} = 12.9 \text{ ft}$$

$$r_h = \frac{h_f D_f}{2h_f + 12D_f} = \frac{(11.12)(12.9)}{(2)(11.12) + (12)(12.9)} = 0.81 \text{ ft}$$

$$u_f = \frac{12q}{h_L D_f} = \frac{(12)(1.5)}{(2.45)(12.9)} = 0.57 \text{ ft/s}$$

Since the viscosity of the liquid was not given by Leibson et al.,[12] it was estimated to be 0.24 centipoise. Thus

$$\mu_L = 0.24 \text{ centipoise} = 1.6 \times 10^{-4} \text{ lb/(ft)(s)}$$

and

$$N_{\text{Re}} = \frac{r_h u_f \rho_L}{\mu_L} = \frac{(0.81)(0.57)(40.8)}{1.61 \times 10^{-4}} = 11.7 \times 10^4$$

By extrapolation of Fig. 12-8 to $h_w = 2.5$, one obtains

$$f = 0.035$$

Since it is given that the length of the liquid flow path $l_f = 13.67$ ft, the hydraulic gradient may now be evaluated by use of Eq. (12-16) to give

$$h_g = \frac{f u_f^2 l_f}{12 g_c r_h} = \frac{(0.035)(0.57)^2(13.67)}{(12)(32.17)(0.81)} = 0.0005 \text{ in} \cong 0$$

Since $h_g \cong 0$, the assumption made in item 7 in the calculation of h_L is valid.

10. *Calculation of H, the Height of Vapor-Free Liquid in the Downcomer at the Maximum Liquid and Vapor Rates*

By Eq. (12-7)

$$H = h_o + h_{dc} + h_L + h_{ow} + h_w + h_g$$

$$= 3.5 + 0.332 + 2.45 + 2.5 + 1.51 = 10.29 \text{ in}$$

and

$$2H = 20.58 \text{ in}$$

Thus, the specified tray spacing of $S = 24$ inches is adequate.

The maximum holdup of liquid and froth in the downcomer may be estimated by use of the relative froth density ϕ for the active area of the plate

$$\text{Maximum height} = \frac{10.29}{0.22} = 46.8 \text{ in}$$

If the relative froth density were this small ($\phi = 0.22$) for the downcomer, the column would flood.

(b) Calculation of the Flooding Velocity of the Vapor

Figure 12-10 is used in the calculation of this velocity. To use this figure, the value of the abscissa is needed.

$$\frac{L}{G}\left(\frac{\rho_V}{\rho_L}\right)^{0.5} = \frac{(1.5 \text{ ft}^3/\text{s})(40.8 \text{ lb/ft}^3)}{(250 \text{ ft}^3/\text{s})(0.4 \text{ lb/ft}^3)}\left(\frac{0.4}{40.8}\right)^{0.5} = 0.061$$

For this value of the abscissa and a tray spacing of 24 inches, Fig. 12-10 gives

$$C_{SB} = 0.38$$

Thus, the flooding velocity is given by

$$u = C_{SB}\left(\frac{\rho_L - \rho_V}{\rho_V}\right)^{0.5}\left(\frac{\sigma}{20}\right)^{0.2} = (0.38)\left(\frac{40.8 - 0.4}{0.4}\right)^{0.5}\left(\frac{12.5}{20}\right)^{0.2}$$

and the chart or graphical value of u is

$$u_{chart} = 3.48 \text{ ft/s}$$

Since

$$A_h/A_a = \frac{7.8}{116} = 0.067$$

Fair[4] recommends that the flooding value of u be obtained by multiplying the chart value by 0.835 [see the tabular information given below Eq. (12-25)]. Thus

$$u_{flooding} = (3.48 \text{ ft/s})(0.835) = 2.9 \text{ ft/s}$$

If a design value of u equal to 80 percent of the flooding velocity is used, then

$$u_{design} = (0.8)(2.9) = 2.33 \text{ ft/s}$$

The specified value for the maximum vapor velocity is

$$u = \frac{250 \text{ ft}^3/\text{s}}{(188.6 - 14) \text{ ft}^2} = 1.43 \text{ ft/s}$$

Thus, the specified value for the maximum velocity of 1.43 ft/s is well below the value of 2.33 ft/s computed for design.

(c) Calculation of the Minimum Vapor Rate

The minimum vapor rate may be estimated by use of Fig. 12-14. To use this graph, the following quantity must be evaluated

$$h_\sigma = \frac{0.040\sigma}{\rho_L D_h} = \frac{(0.040)(12.5)}{(40.8)(0.1875)} = 0.0654 \text{ in}$$

Thus,

$$h_0 + h_\sigma = 3.5 + 0.0654 = 3.57 \text{ in}$$

The point $[(h_w + h_{ow} = 4.02), (h_0 + h_\sigma = 3.57)]$ lies well above the curve for $A_h/A_a = 0.067$. Thus, weeping will not be deleterious to plate efficiency.[3]

NOTATION

A_a the active area of a sieve tray, the area between the downcomer weir and the outlet weir (the section of the plate which contains the perforations)

A_{dc} clearance area between the downcomer and the deck of the tray, in^2

A_h hole area of the perforations of a sieve plate or the openings of a valve plate in the units given with the respective equations in which A_h appears

C_D	drag coefficient; defined by Eq. (12-23)
C_0	discharge coefficient for the flow of vapor through the perforations of a sieve tray; see Eq. (12-10)
D_d	diameter of droplet
D_f	arithmetic average of the tower diameter and the weir length, ft
D_0	diameter of the perforation of a sieve tray, in
f	friction factor; appears in Eq. (12-16)
F_w	weir constriction correction factor; appears in Eqs. (12-8) and (12-9)
$\sum_i F_i$	frictional losses accompanying the flow of the liquid from point (1) to point (2) of Fig. 12-3
g_c	Newton's law conversion factor
g	acceleration of gravity
h_{dc}	head loss by the liquid in flowing under the downcomer, inches of vapor-free (or clear) liquid
h_f	froth height, total height of froth plus liquid in the downcomer, in
h_g	height of the hydraulic gradient, inches of vapor-free liquid
h_L	head loss of the vapor as it flows through the aerated liquid above the perforations of a sieve tray, inches of vapor-free liquid
h_0	dry hole pressure drop through the perforations of a sieve tray or the ballast units of a valve tray, inches of vapor-free liquid
h_{ow}	height of liquid over the weir, inches of vapor-free liquid
h_w	height of the weir, in
K_1, K_2	coefficients in the pressure drop expression for V-1 and V-4 ballast units; see Eqs. (12-33) and (12-34)
l_f	length of flow path across the plate, ft
l_w	length of weir, in
N_{Re}	Reynolds number; defined by Eq. (12-17)
N_{We}	Weber number; defined by Eq. (12-27)
P	pressure in consistent units in the equation in which it appears
Q	volumetric flow rate of the vapor-free liquid, gal/min
r_h	hydraulic radius of the aerated mass in inches; see Eq. (12-18)
u_f	linear velocity of the aerated mass in feet per second; see Eq. (12-19)
u_0	line or velocity of the vapor through the perforations of a sieve tray or the opening of a valve tray, ft/s
t_m	valve thickness, in
Z	height of liquid; see Fig. 12-3

Greek symbols

β	aeration factor; defined by Eq. (12-13); dimensionless
ρ_V	mass density of the vapor, lb/ft^3
ρ_L	mass density of the liquid, lb/ft^3
ϕ	relative froth density; defined by Eq. (12-14)
ρ_m	density of metal, lb/ft^3

PROBLEMS

12-1 The correlation presented in Fig. 12-6 was developed by Leibson et al.[12] on the basis of a formula for C_0 obtained by adding the contraction loss from point i to point o of Fig. P12-1 to the expansion loss from point o to point d. Equations for the contraction loss $\Delta P_c/\rho_L$ and the expansion loss $\Delta P_e/\rho_L$ are available in standard texts and handbooks on fluid flow. The expressions used are as follows

$$\frac{\Delta P_c}{\rho_L} = K \frac{u_0^2}{2g_c}$$

where K = contraction coefficient and is dependent on the area ratio, (A_0/A_i) for Reynold's numbers based on A_0 and greater than 10,000

ρ_L = fluid density of the liquid

u_0 = linear velocity at point o of Fig. P12-1

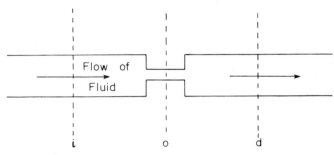

Figure P12-1

If the total drop across the perforation is denoted by $\Delta P_0/\rho_L$ (where $\Delta P_0 = \Delta P_e + \Delta P_c$), obtain a formula for C_0 of the orifice equation

$$\frac{\Delta P_0}{\rho_L} = \frac{1}{C_0^2} \frac{u_0^2}{2g_c}$$

in terms of the contraction coefficient K and the area A_o and A_i.

Answer:
$$C_0 = \frac{1}{[K + (1 - A_o/A_i)^2]^{1/2}}$$

12-2 Develop the expression given by Eq. (12-31) for Manning's correlation.[13]

12-3 On the basis of a vapor velocity of 2.16 ft/s (250 ft³/s/116 ft² = 2.16 ft/s) and a critical Weber number of 20 for the system given in Example 12-1, compute the maximum droplet diameter by use of Eq. (12-30).

12-4 For the system given in Example 12-1, calculate the absolute maximum vapor velocity as given by Eq. (12-31). Take $C_D = 0.7$, and $(N_{We})_{crit} = 20$.

REFERENCES

1. W. L. Bolles: "Rapid Graphical Method of Estimating Tower Diameter and Tray Spacing of Bubble-Plate Fractionators," *Pet. Refiner,* **25**(12): 103 (1946).
2. *Chemical Engineers Handbook,* Robert H. Perry and Cecil H. Chilton (eds.), 5th ed., sec. 18, by J. R. Fair, D. E. Steinmeyer, W. R. Penney, and J. A. Brink, 1973.

3. J. R. Fair: Chap. 15 of *Design of Equilibrium Stage Processes*, by B. D. Smith, McGraw-Hill Book Company, New York, 1963.
4. ————: "How to Predict Sieve Tray Entrainment and Flooding," *Pet. Chem. Eng.* **33**(10):45 (1961).
5. ———— and R. L. Mathews: "Better Estimate of Entrainment from Bubble-Cap Trays," *Pet. Refiner*, **37**(4):153 (1958).
6. J. A. Gerster: *Recent Advances in Distillation*, Davis-Swindin Memorial Lecture, University of Loughborough (1964).
7. F. W. Glitsch and Sons: "Ballast Tray Design Manual," Bull. 4900 (Rev.), Dallas, Texas.
8. J. O. Hinze: "Critical Speeds and Sizes of Liquid Globules," *Appl. Sci. Res.* **A1**:273 (1949).
9. M. H. Hutchinson, A. G. Buron, and B. P. Miller: "Aerated Flow Principles Applied to Sieve Plates." Paper presented at Los Angeles AIChE Meeting, May 1949.
10. G. A. Hughmark and H. E. O'Connell: "Design of Perforated Plate Fractionating Towers," *Chem. Eng. Prog.* **53**:127 (1927).
11. Koch Engineering Co. Inc., Bulletin KT-5, 4111 East 37th St. North, Wichita, Kansas 67208.
12. I. Leibson, R. E. Kelley, and L. A. Bullington, "How to Design Perforated Trays," *Pet. Refiner*, **36**(2):127 (1957).
13. E. Manning: "High Capacity Distillation Trays," *Ind. Eng. Chem.*, **56**(4):14 (1964).
14. T. K. Sherwood, G. H. Shipley, and F. A. L. Holloway, "Flooding Velocities in Packed Columns," *Ind. Eng. Chem.*, **30**(7):765 (1938).
15. R. B. Smith, T. Dresser, and S. Ohlswager, "Tower Capacity Rating Ignores Trays," *Hydrocarbon Process. and Pet. Refiner*, **40**(5):183 (1963).
16. Mott Souders, Jr., and G. G. Brown, "Design of Fractionating Columns," *Ind. Eng. Chem.* **26**(1):98 (1934).
17. Van Winkle, Matthew, *Distillation*, McGraw-Hill Book Company, New York, 1963.
18. S. Zanelli and R. Del Bianco: "Perforated Plate Weeping," *Chem. Eng. J.*, **6**:181 (1973).

THIRTEEN

APPLICATION OF THE FUNDAMENTALS OF MASS TRANSFER TO PLATE EFFICIENCIES AND PACKED COLUMNS

The fundamental relationships required to describe the mass transfer which occurs on the plate of a distillation column are the same relationships required to describe the mass transfer which occurs in a packed distillation column. For many years, packed columns were used primarily for corrosive applications where ceramic packing was advantageous and for small-diameter columns where it was inconvenient to install plates. Because of the favorable economics of installation and low pressure drop, packed columns should be considered as an alternative to a column with plates.

In Sec. 13-1 the fundamental concepts of mass transfer are presented. In Sec. 13-2 these fundamentals are used in the development of predictive models for plate efficiencies. In the final section, 13-3, a treatment of packed columns is presented.

13-1 FUNDAMENTAL RELATIONSHIPS FOR MASS TRANSFER BETWEEN VAPOR AND LIQUID PHASES

Four of the simplest and best known of the theories of mass transfer from flowing streams are (1) the *stagnant-film model*, (2) the *penetration model*, (3) the *surface-renewal model*, and (4) the *turbulent boundary-layer model*.

The *stagnant-film model* was proposed by Nernst[28] in 1904. The model may be described by visualizing the passage of a gas stream containing components A

and B past a liquid surface. Suppose that A is soluble and B is not soluble in the liquid. In this model, it is supposed that the gas next to the surface is stagnant and that the transfer of A through this thin stagnant layer or film of gas occurs by molecular diffusion. From the theory of molecular diffusion, the following expression is obtained for N_A, the rate of diffusion of A per unit area of liquid surface

$$N_A = \frac{D_{AB}P(C_{A1} - C_{A2})}{RTyC_{BM}} \tag{13-1}$$

where D_{AB} = molecular diffusivity of A with respect to B in a mixture of A and B
P = pressure
C_{A1} = bulk concentration of A in the gas stream
C_{A2} = concentration of A at the surface of the liquid
C_{BM} = logarithmic mean of concentrations of B at the boundaries of the film
y = thickness of the stagnant film
T = temperature, and R is the gas constant

The model has two major faults. It provides no method for the calculation of the film thickness y and it predicts that the rate of diffusion should be proportional to the first power of the diffusivity. The experimental results of Gilliland and Sherwood[17] and of Vivian and King[36] suggest that the diffusivity should be raised to the powers of 0.5 and 0.56, respectively.

Since y is unknown, it is customary to state Eq. (13-1) in terms of mass transfer coefficients as follows

$$N_A = k'_L(C_{A1} - C_{A2}) = k'_G(p_{A1} - p_{A2}) \tag{13-2}$$

The *penetration theory* was proposed by Higbie[19] who observed that the mass transfer of a solute A from a liquid to a gas in industrial processes could be regarded as a succession of intermittent processes wherein each bubble of vapor is exposed to a succession of liquid surfaces. Each liquid surface is visualized as forming at the top of the bubble and remaining in contact with it for amount of time t, equal to that required for the bubble to move through a distance equal to one bubble diameter. The mass transfer process over the time period t was described by the partial differential equation known as Fick's second law. The solution obtained for a suitable set of boundary conditions led to the following expression for N_A, the rate of transfer of A from the liquid to the vapor per unit of interfacial area

$$N_A = 2(C_{A1} - C_{A0})\sqrt{\frac{D}{\pi t}} \tag{13-3}$$

where C_{A0} = concentration of A in the liquid
C_{A1} = concentration of A at the phase boundary between the gas bubble and the liquid
t = average contact time between a gas bubble and a liquid surface

Since the model provides no means for computing the contact time t, it is not suitable for predicting mass transfer coefficients. It does agree well with experimental evidence[17, 34] which suggests that N_A should be proportional to the diffusivity raised to approximately the 0.5 power.

Danckwerts[8] reasoned that some surfaces may remain in contact longer than other surfaces and generalized the penetration theory by introducing a probability function which expresses the fact that there exists a distribution of contact times lying between zero and infinity. In the resulting expression obtained for N_A, the term $2\sqrt{D/\pi t}$ is replaced by $\sqrt{Ds}$, where s is the fractional renewal rate. Since the unknown t is replaced by the unknown s, the model is not suitable for the prediction of mass transfer coefficients.

The *turbulent boundary layer model* accounts for the transfer of a solute molecule A from a turbulent stream to a fixed surface. Eddy diffusion is rapid in the turbulent stream and molecular diffusion is relatively insignificant. It is supposed that the turbulence is damped out in the immediate vicinity of the surface. In the intermediate neighborhood between the turbulent stream and the fixed surface, it is supposed that transport is by both molecular and eddy diffusion which take place in parallel. The total rate of transfer (moles of A transferred per unit time per unit area) is given by an extended form of Fick's law

$$N_A = -(D + E_D)\frac{dC_A}{dy} \tag{13-4}$$

where E_D is the eddy diffusivity. Although the eddy diffusivity is not generally available from experimental observations, it may be estimated through its relationship to eddy viscosity, which can be derived from velocity profiles.[29] Further treatment of the general topic of mass transfer is outside the scope of this book. An excellent treatment of this topic by Sherwood et al.[29] has been published recently.

In the remainder of this chapter, the two-film theory with equations of the form of Eq. (13-2) is applied to packed columns and to plate efficiencies. The essence of the two-film theory is the additivity of the vapor and liquid film resistances which was first proposed by Lewis and Whitman.[26, 37]

Applications of the Two-Film Theory to Mass Transfer between Vapor and Liquid Phases

For definiteness in the following development consider a packed column such as the one shown in Fig. 13-1, and suppose that it is operating at steady state. The rate of flow of the liquid absorbent stream is denoted by L_0, the rich absorbent stream by L_N, the rich gas stream by V_{N+1}, and the lean gas stream by V_1, all in moles per unit time.

In the following developments, it is supposed that the mixing is perfect in the radial direction, but that no mixing occurs in the vertical direction. At any particular Z of this column, the transfer process whereby component i is transferred from the liquid to the vapor phase may be represented graphically as shown in Fig. 13-2.

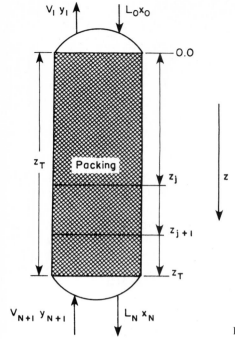

Figure 13-1 Sketch of a packed absorption column.

As implied by Fig. 13-2, the rate of mass transfer of a component from the liquid to the vapor phase consists of a sequence of rate processes which occur in series. First component i is transferred from the liquid phase to the interface at the liquid side. The fugacity potential or driving force for this transfer process is taken to be $\hat{f}_i^L - \hat{f}_i^{\mathscr{L}}$, where the fugacity $\hat{f}_i^L$ is evaluated at the bulk conditions of the liquid phase and $\hat{f}_i^{\mathscr{L}}$ is evaluated at the conditions at the liquid side of the interface. The rate of mass transfer per unit time per unit height of packing R_i^L for this step of the process is expressed as follows

$$R_i^L = k_L a_i S(\hat{f}_i^L - \hat{f}_i^{\mathscr{L}}) \tag{13-5}$$

where S = cross-sectional area of the column
a = interfacial area per unit volume of empty column
k_L = rate constant with the units of moles of component i transferred per unit time per atmosphere per unit area

Since k_L and a are commonly determined as a product rather than individually, the product is customarily regarded as a single constant with the subscript i on the symbol a denoting the value of the product for component i.

The fugacity driving force for transfer across the interface is $\hat{f}_i^{\mathscr{L}} - \hat{f}_i^{\mathscr{V}}$, and the rate of transfer is given by

$$R_i^I = k_I a_i S(\hat{f}_i^{\mathscr{L}} - \hat{f}_i^{\mathscr{V}}) \tag{13-6}$$

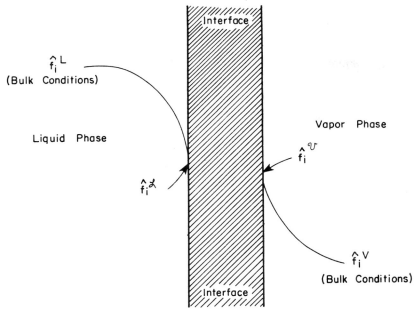

Figure 13-2 Sketch of the fugacity profile involved in the transfer of a component from the liquid phase to the vapor phase.

Similarly, for the transfer of component i from the vapor side of the interface to the vapor phase, the driving force is $\hat{f}_i^{\gamma} - \hat{f}_i^{V}$, and the rate is given by

$$R_i^G = k_G a_i S(\hat{f}_i^{\gamma} - \hat{f}_i^{V}) \tag{13-7}$$

Since the transfer processes occur in series and are at steady state, it follows that

$$R_i^L = R_i^I = R_i^G = R_i \tag{13-8}$$

and that

$$R_i = k_L a_i S(\hat{f}_i^{L} - \hat{f}_i^{\mathscr{L}}) = k_I a_i S(\hat{f}_i^{\mathscr{L}} - \hat{f}_i^{\gamma}) = k_G a_i S(\hat{f}_i^{\gamma} - \hat{f}_i^{V}) \tag{13-9}$$

ADDITIVITY OF THE RESISTANCES

The developments which follow are based on the additivity of the resistances (the reciprocal of the mass transfer coefficients) which was first proposed by Lewis and Whitman.[26, 37] When the rate expressions for the three processes described in Fig. 13-2 are added, one obtains the following result upon rearrangement

$$R_i = \left[\frac{1}{k_L a_i} + \frac{1}{k_I a_i} + \frac{1}{k_G a_i} \right] = S[(\hat{f}_i^{L} - \hat{f}_i^{\mathscr{L}}) + (\hat{f}_i^{\mathscr{L}} - \hat{f}_i^{\gamma}) + (\hat{f}_i^{\gamma} - \hat{f}_i^{V})] \tag{13-10}$$

This expression may be restated in the following form

$$R_i = K_G a_i S(\hat{f}_i^{L} - \hat{f}_i^{V}) \tag{13-11}$$

where the overall mass transfer coefficient $K_G a_i$ is defined as follows

$$\frac{1}{K_G a_i} = \frac{1}{k_G a_i} + \frac{1}{k_I a_i} + \frac{1}{k_L a_i} \tag{13-12}$$

When the rate constant k_I for transfer across the interface is exceedingly large or the resistance to transfer across the interface is negligible, then

$$\lim_{k_I \to \infty} K_G a_i = \frac{1}{1/k_L a_i + 1/k_G a_i} \tag{13-13}$$

This limiting case is sometimes referred to as the *two-film theory* for mass transfer. The resistances to mass transfer in the vapor and liquid phases are concentrated in the vapor and liquid films. Experimental evidence presented by Tung and Drickamer[34] and Emmett and Pigford[13] suggest that k_I is very large except at very high rates of mass transfer. Thus, for all practical purposes,

$$\hat{f}_i^Y \cong \hat{f}_i^{\mathscr{L}} \tag{13-14}$$

Setting $\hat{f}_i^Y \cong \hat{f}_i^{\mathscr{L}}$ should not be taken to mean that the rate of mass transfer across the interface is equal to zero. On the contrary, the rate of mass transfer across the interface is equal to the rate of mass transfer across the vapor and liquid phases. That is, in the limit as k_I approaches infinity, the difference $\hat{f}_i^Y - \hat{f}_i^{\mathscr{L}}$ approaches zero and the corresponding indeterminate $k_I a_i (\hat{f}_i^Y - \hat{f}_i^{\mathscr{L}})$ approaches the finite number R_i [see Eq. (13-9)].

Instead of using the difference in fugacities as the driving forces in the rate expressions, concentrations and partial pressures are commonly used. Again, dynamic equilibrium is assumed to exist at the interface and the rate of transfer across the interface is omitted in this and subsequent developments. The rate of mass transfer of component i from the liquid to the vapor phase is given by

$$R_i = k'_L a_i S(C_i^L - C_i^{\mathscr{L}}) = k'_G a_i S(p_i^Y - p_i^Y) \tag{13-15}$$

where C_i has the units of moles per unit volume.

RELATIONSHIPS BETWEEN THE MASS TRANSFER COEFFICIENTS

If the following reasonable unrestrictive assumptions are made, then fairly simple relationships exist between $k_L a_i$, $k_G a_i$, $K_G a_i$ and $k'_L a_i$, $k'_G a_i$, $K_G a_i$. Such relationships are needed because most of the experimental data have been collected by measuring concentrations and partial pressures. At any Z of the column, the following relationships among the variables are assumed

1. $\qquad P^V = P^Y = P^{\mathscr{L}} = P^L = P$

2. $\qquad T^V = T^Y = T^{\mathscr{L}} = T^L = T$

$$\tag{13-16}$$

3. $\qquad \dfrac{\gamma_i^Y}{\gamma_i^V} = 1 \qquad \dfrac{\gamma_i^{\mathscr{L}}}{\gamma_i^L} = 1$

4. $\qquad \rho^L = \rho^{\mathscr{L}}$

where ρ^L is equal to the molar density of the liquid phase. Since the fugacities of the pure components depend on temperature and pressure alone, it follows from conditions 1 and 2 that $f_i^V = f_i^V$ and $f_i^{\mathscr{L}} = f_i^L$ at any Z. Thus, on the basis of the assumptions given by Eq. (13-16), it is possible to restate Eq. (13-15) as follows

$$R_i = \frac{k_L' a_i \rho^L S}{\gamma_i^L f_i^L}(\hat{f}_i^L - \hat{f}_i^{\mathscr{L}}) = \frac{k_G' a_i PS}{\gamma_i^V f_i^V}(\hat{f}_i^V - \hat{f}_i^V) \tag{13-17}$$

Comparison of Eqs. (13-9) and (13-17) gives the relationships shown in item I of Table 13-1 for $k_L a_i$, $k_L' a_i$, and $k_G' a_i$.

When the rates of mass transfer across the liquid and vapor films as given by Eq. (13-17) are divided by $k_L' a_i \rho^L S/\gamma_i^L f_i^L$ and $k_G' a_i PS/\gamma_i^V f_i^V$, respectively, and added, one obtains the following result upon rearrangement

$$R_i = K_G' a_i S(p_i^\dagger - p_i^V) \tag{13-18}$$

where $K_G' a_i$ is defined in item I of Table 13-1 and

$$p_i^\dagger = \frac{\gamma_i^L P K_i x_i}{\gamma_i^V} \qquad K_i = f_i^L/f_i^V$$

Similarly, R_i may be stated in terms of concentration units as follows

$$R_i = K_L' a_i S(C_i^L - C_i^\dagger) \tag{13-19}$$

where $K_L' a_i$ is defined in item I of Table 13-1 and

$$C_i^\dagger = \frac{\gamma_i^V \rho^L y_i}{\gamma_i^L K_i}$$

[Note that $\sum_{i=1}^c p_i^\dagger$ is not necessarily equal to P and that $\sum_{i=1}^c C_i^\dagger$ is not necessarily equal to ρ^L because the liquid and vapor were not assumed to be at the bubble-point and dew-point temperatures, respectively.] By comparison of the above equations, other relationships are readily obtained.

Component-Material Balances

Consider again the packed column which is graphically represented in Fig. 13-1. Again it is supposed that the mixing is perfect in the radial direction and that no mixing occurs in the vertical direction. In Fig. 13-1, the distance Z is measured from the top of the column down. The liquid flows in the positive direction of Z, and the vapor flows in the counter direction (or in the negative direction of Z).

A material balance on component i in the vapor phase over the element of volume $S(Z_{j+1} - Z_j)$ is given by

$$Vy_i\bigg|_{Z_{j+1}} - Vy_i\bigg|_{Z_j} + \int_{Z_j}^{Z_{j+1}} R_i \, dZ = 0 \tag{13-20}$$

Application of the *mean value theorem of differential calculus* (see Theorem A-2 of Appendix A) to the first two terms of Eq. (13-20) and the *mean value theorem of*

Table 13-1 Summary of the mass transfer relationships

I. Relationships based on the assumptions given by Eq. (13-16)

1. Mass Transfer Coefficients

$$k_L a_i = \left(\frac{\rho^L}{\gamma_i^L f_i^L}\right) k_L' a_i' \qquad k_G a_i = \left(\frac{P}{\gamma_i^V f_i^V}\right) k_G' a_i \qquad \bar{K}_G a_i = \left(\frac{P}{\gamma_i^V f_i^V}\right) K_G' a_i$$

$$\frac{1}{K_G a_i} = \frac{1}{k_G a_i} + \frac{1}{k_L a_i} \qquad \frac{1}{K_G' a_i} = \frac{1}{k_G' a_i} + \left(\frac{\gamma_i^L K_i P}{\gamma_i^V \rho^L}\right)\frac{1}{k_L' a_i}$$

$$\frac{1}{K_L' a_i} = \frac{1}{k_L' a_i} + \left(\frac{\gamma_i^V \rho^L}{\gamma_i^L P K_i}\right)\frac{1}{k_G' a_i}$$

2. Number of Transfer Units

$$n_{Li} = k_L a_i \gamma_i^L f_i^L \left(\frac{S Z_T}{L}\right) = -\int_①^② \frac{dx_i}{x_i - x_i^{\mathscr{L}}} \qquad \text{where} \quad \begin{array}{l} ① = \text{value of } x_i \text{ at } Z = 0; \\ ② = \text{value of } x_i \text{ at } Z = Z_T. \end{array}$$

$$n_{Gi} = k_G a_i \gamma_i^V f_i^V \left(\frac{S Z_T}{V}\right) = -\int_①^② \frac{dy_i}{y_i^V - y_i} \qquad \text{where} \quad \begin{array}{l} ① = \text{value of } y_i \text{ at } Z = 0; \\ ② = \text{value of } y_i \text{ at } Z = Z_T. \end{array}$$

$$n_{OGi} = K_G a_i \gamma_i^V f_i^V \left(\frac{S Z_T}{V}\right) = -\int_①^② \frac{dy_i}{Y_i - y_i} \qquad \text{where } Y_i = \frac{\gamma_i^L K_i x_i}{\gamma_i^V}$$

$$n_{Li}' = k_L' a_i \rho^L \left(\frac{S Z_T}{L}\right) = -\int_①^② \frac{dx_i}{x_i - x_i^{\mathscr{L}}}$$

$$n_{Gi}' = k_G' a_i P \left(\frac{S Z_T}{V}\right) = -\int_①^② \frac{dy_i}{y_i^V - y_i}$$

$$n_{OGi}' = K_G' a_i P \left(\frac{S Z_T}{V}\right) = -\int_①^② \frac{dy_i}{Y_i - y_i}$$

$$n_{OLi}' = K_L' a_i \rho^L \left(\frac{S Z_T}{L}\right) = -\int_①^② \frac{dx_i}{x_i - X_i} \qquad \text{where } X_i = \frac{\gamma_i^V y_i}{\gamma_i^L K_i}$$

II. Relationships based on the use of the tangent line $y_i^* = m_i x_i + b_i$ and associated assumptions

1. Mass Transfer Coefficients

$$\frac{1}{K_{GM}' a_i} = \frac{1}{k_G' a_i} + \left(\frac{m_i P}{\rho^L}\right)\frac{1}{k_L' a_i}$$

2. Number of Transfer Units

$$n_{Gi}' = \text{same as shown in Item I}$$

$$n_{Li}' = \text{same as shown in Item I}$$

$$n_{OGMi}' = K_{GM}' a_i P \left(\frac{S Z_T}{L}\right) = -\int_①^② \frac{dy_i}{y_i^* - y_i} \qquad \text{where } y_i^* \text{ is a hypothetical mole fraction defined by Eq. (13-27).}$$

$$n_{OLMi}' = K_{LM}' a_i \rho^L \left(\frac{S Z_T}{L}\right) = -\int_①^② \frac{dx_i}{x_i - x_i^*} \qquad \text{where } x_i^* \text{ is a hypothetical mole fraction defined below Eq. (13-30).}$$

Table 13-1 (*continued*)

III. Relationships based on Henry's law, $p_i = H_i C_i$ and associated assumptions

1. *Mass Transfer Coefficients*

$$\frac{1}{K'_{GH} a_i} = \frac{1}{k'_G a_i} + \frac{H_i}{k_L a_i}$$

$$\frac{1}{K'_{LH} a_i} = \frac{1}{k'_L a_i} + \frac{1}{H_i k'_G a_i}$$

2. *Number of Transfer Units*

$$n'_{Li} = \text{same as shown in Item I}$$

$$n'_{Gi} = \text{same as shown in Item I}$$

$$n'_{OG\,Mi} = K'_{GH} a_i P\left(\frac{SZ_T}{V}\right) = -\int_{①}^{②} \frac{dy_i}{y_i^* - y_i} \qquad \begin{array}{l} \text{where } y_i^* \text{ is defined below} \\ \text{Eq. (13-32)} \end{array}$$

$$n'_{OG\,Hi} = K'_{LH} a_i \rho^L\left(\frac{SZ_T}{L}\right) = -\int_{①}^{②} \frac{dx_i}{x_i - x_i^*} \qquad \begin{array}{l} \text{where } x_i^* \text{ is defined below} \\ \text{Eq. (13-32)} \end{array}$$

---

integral calculus (see Theorem A-3 of Appendix A) to the integral appearing in Eq. (13-20) yields

$$\Delta Z \left.\frac{d(Vy_i)}{dZ}\right|_{Z_j + \alpha\,\Delta Z} + \Delta Z R_i \Big|_{Z_j + \beta\,\Delta Z} = 0 \qquad (13\text{-}21)$$

where $0 < \alpha < 1$ and $0 \le \beta \le 1$. Division of each term of Eq. (13-21) by ΔZ followed by the limiting process whereby ΔZ is allowed to go to zero yields

$$\left.\frac{d(Vy_i)}{dZ}\right|_{Z_j} + R_i \Big|_{Z_j} = 0 \qquad (13\text{-}22)$$

Since Z_j and Z_{j+1} were arbitrarily selected $(0 < Z_j < Z_{j+1} < Z_T)$, Eq. (13-22) holds for all Z $(0 < Z < Z_T)$ which is implied by restating the equation in the following form

$$\frac{d(Vy_i)}{dZ} + R_i = 0 \qquad (0 < Z < Z_T) \qquad (13\text{-}23)$$

Similarly, the limit of the material balance on component i over the interval from Z_j to Z_{j+1} as the increment over which the balance is made is allowed to go to zero, gives the following differential equation for component i in the liquid phase

$$\frac{d(Lx_i)}{dZ} + R_i = 0 \qquad (0 < Z < Z_T) \qquad (13\text{-}24)$$

Number of Transfer Units

Expressions for the number of transfer units are obtained by the restatement of Eqs. (13-23) and (13-24) in an integrated form for each of the equivalent expressions for R_i. For example, consider the case where the expression given by Eq. (13-11) is used for R_i. This expression may be stated in terms of mole fractions as follows

$$R_i = K_G a_i S(\hat{f}_i^L - \hat{f}_i^V) = K_G a_i S(\gamma_i^L f_i^L x_i - \gamma_i^V f_i^V y_i)$$
$$= K_G a_i S \gamma_i^V f_i^V (Y_i - y_i) \qquad (13\text{-}25)$$

where

$$Y_i = \left(\frac{\gamma_i^L K_i}{\gamma_i^V}\right) x_i \qquad K_i = f_i^L / f_i^V$$

Elimination of R_i from Eqs. (13-23) and (13-25) followed by the separation of variables and then integration from $Z = 0$ to $Z = Z_T$ yields the following result upon rearrangement

$$K_G a_i \gamma_i^V f_i^V \left(\frac{S Z_T}{V}\right) = -\int_{\textcircled{1}}^{\textcircled{2}} \frac{dy_i}{Y_i - y_i} \qquad (13\text{-}26)$$

where $\textcircled{1}$ = value of y_i at $Z = 0$
$\qquad \textcircled{2}$ = value of y_i at $Z = Z_T$

Also, the variation of the quantity on the left-hand side of Eq. (13-26) with Z was neglected in the integration process. Since the right-hand side of Eq. (13-26) involves a ratio of mole fractions, it is dimensionless. Consequently, the left-hand side is dimensionless. These dimensionless quantities are given the name of the *number of transfer units* and denoted by the symbol n_{OGi} (see Table 13-1). The subscript *OGi* denotes the fact that the transfer unit for component i is based on the overall mass transfer coefficient K_G for the gas phase. Other expressions given in Table 13-1 for the number of transfer units are obtained by use of other expressions for R_i in Eqs. (13-23) and (13-24).

Use of Tangent Line $(y_i^* = m_i x_i + b_i)$ Relationships in the Definitions of $K_G' a_i$ and $K_L' a_i$

In this method of relating the mole fractions in the vapor and liquid phases, a linear relationship is assumed between the mole fraction of component i in a given phase and the mole fraction that it would have in the other phase if the two phases were in equilibrium. For the case of a binary mixture, the linear relationship is represented by the tangent line to the equilibrium curve

$$y_i^* = m_i x_i + b_i \qquad (13\text{-}27)$$

where the quantity y_i^* is defined as the mole fraction that component i would have in the vapor phase if the vapor were in equilibrium with a liquid having the mole fraction x_i. From this definition of y_i^*, it follows that

$$y_i^* = K_i x_i$$

and

$$\sum_{i=1}^{c} y_i^* = 1 = \sum_{i=1}^{c} K_i x_i$$

Thus, y_i^* may be evaluated by making a bubble-point calculation on a liquid having the mole fractions $\{x_i\}$.

Let it be assumed that the values of m_i and b_i apply with good accuracy for both x_i and $x_i^{\mathscr{L}}$, which amounts to assuming that the equilibrium curve is straight over this range of values of x_i and that $\rho^L = \rho^{\mathscr{L}}$. Then Eq. (13-15) may be restated in the following form

$$R_i = \frac{k_L' a_i \rho^L S}{m_i} [(m_i x_i + b_i) - (m_i x_i^{\mathscr{L}} + b_i)]$$

$$= \frac{k_L' a_i \rho^L S}{m_i} (y_i^* - y_i^{\gamma}) = k_G' a_i PS(y_i^{\gamma} - y_i) \tag{13-28}$$

These two expressions for R_i may be used to develop an expression for the overall mass transfer coefficient in a manner analogous to that demonstrated in the development of Eq. (13-11). In this case

$$R_i = K_{GM}' a_i S(p_i^* - p_i^{\gamma}) = K_{GM}' a_i PS(y_i^* - y_i) \tag{13-29}$$

where $K_{GM}' a_i$ is defined in item II of Table 13-1. In a similar manner, Eq. (13-28) may be used to obtain

$$R_i = K_{LM}' a_i S(C_i^L - C_i^*) = K_{LM}' a_i \rho^L (x_i - x_i^*) \tag{13-30}$$

where $K_{LM}' a_i$ is defined in Prob. 13-5 and x_i^* is the mole fraction that component i would have if the liquid phase were in equilibrium with a vapor phase having the mole fraction y_i. From this definition of x_i^*, it follows that

$$x_i^* = y_i / K_i$$

and

$$\sum_{i=1}^{c} x_i^* = 1 = \sum_{i=1}^{c} y_i / K_i$$

Thus x_i^* may be evaluated by making a dew-point calculation on a vapor having the mole fraction y_i.

Use of Henry's Law, $p_i = H_i C_i$, in the Definition of $K_{HG}' a_i$ and $K_{HL}' a_i$

If it is assumed that the same value of H_i may be used for both C_i^L and $C_i^{\mathscr{L}}$, then the following expressions may be obtained for the rate equations involving the overall mass transfer coefficients $K'_{HG} a_i$ and $K'_{HL} a_i$, namely,

$$R_i = K'_{GH} a_i S(p_i^* - p_i^V) = K'_{GH} a_i PS(y_i^* - y_i) \tag{13-31}$$

$$R_i = K'_{LH} a_i S(C_i^L - C_i^*) = K'_{LH} a_i \rho^L S(x_i - x_i^*) \tag{13-32}$$

where $K'_{GH} a_i$ and $K'_{LH} a_i$ are defined in item III of Table 13-1, and

$$y_i^* = \frac{H_i C_i^L}{P} \qquad x_i^* = \frac{p_i^V}{H_i p_L}$$

Height of a Transfer Unit

The *height of a transfer unit*, denoted by the symbol HTU, was introduced by Chilton and Colburn.[7] For the gas phase, this quantity is defined as follows

$$(HTU)_{Gi} = \frac{Z_T}{n_{Gi}} \tag{13-33}$$

For each of n_i listed in Table 13-1, a corresponding $(HTU)_i$ may be defined.

Other relationships are readily deduced from those presented in Table 13-1. For example, by use of the relationships given in item I of Table 13-1, it can be shown that

$$\frac{1}{n_{OGi}} = \frac{1}{n_{Gi}} + \left(\frac{\gamma_i^L K_i V}{\gamma_i^V L}\right)\frac{1}{n_{Li}} \qquad \frac{1}{n'_{Gi}} = \frac{1}{n'_{Gi}} + \left(\frac{\gamma_i^L K_i V}{\gamma_i^V L}\right)\frac{1}{n'_{Li}} = \frac{1}{n'_{OGi}} \tag{13-34}$$

From this relationship it is evident that $n_{Li} = n'_{Li}$, $n_{Gi} = n'_{Gi}$, and $n_{OGi} = n'_{OGi}$.

The relationships in item II of Table 13-1 for the tangent line may be used to show that

$$\frac{1}{n'_{OGMi}} = \frac{1}{n'_{Gi}} + \left(\frac{m_i V}{L}\right)\frac{1}{n'_{Li}} \tag{13-35}$$

The additional subscript M has been added to emphasize the fact that the number of overall transfer units computed by Eq. (13-35) may differ slightly from the number computed by use of Eq. (13-34). The difference results from the fact that over any one stage of a distillation column, the slope of the equilibrium line generally changes, which gives rise to more than one possible choice for m_i as demonstrated in the last section of Sec. 13-2.

13-2 PLATE EFFICIENCIES

This presentation is limited to two models, the *Murphree plate efficiency* and the *modified Murphree plate efficiency*, and their applications to columns in the service of separating both binary and multicomponent mixtures. For convenience of application, these efficiencies are restated in terms of the *vaporization plate efficiency*. Definitions of the vaporization point and plate efficiency follow immediately and the Murphree plate efficiencies are defined in a subsequent section as they arise in the development of the *perfectly mixed liquid phase model*.

Definitions of the Vaporization Efficiencies

The utility of the vaporization efficiency rests on the fact that the equations for columns with perfect plates are readily transformed to those for columns with actual plates by the introduction of a single multiplier of K_i. Unlike the Murphree plate efficiency, the vaporization plate efficiency does not arise from a particular model for the behavior of a plate. Instead it is a function which may be evaluated for any model which is assumed to describe the behavior of a plate. In this respect, the definition of the vaporization plate efficiency is analogous to the thermodynamic activity coefficient.

DEFINITION OF THE VAPORIZATION POINT EFFICIENCY

The definition of the vaporization point efficiency is axiomatic. For let a and b be any two nonzero numbers which are finite and positive. Then if a is unequal to b, there exists a positive number c such that $a = cb$. Similarly, if the fugacity $\hat{f}_i^V$ of component i in the vapor phase at any point above the liquid on a plate is unequal to its fugacity $\hat{f}_i^L$ in the liquid, then there exists a multiplier E_i, called the *vaporization point efficiency*, such that

$$\hat{f}_i^V = E_i \hat{f}_i^L \tag{13-36}$$

where $\hat{f}_i$ = fugacity of component i at a given point in the vapor phase
$\hat{f}_i^L$ = fugacity of component i at a point in the liquid phase which is just
beneath the point in the vapor phase

From the definitions of $\hat{f}_i^V$ and $\hat{f}_i^L$, it follows that

$$y_i = E_i \frac{\gamma_i^V}{\gamma_i^L} K_i x_i \tag{13-37}$$

where γ_i^V and γ_i^L are the vapor and liquid activity coefficients and K_i is the ideal solution K value ($K_i = f_i^V/f_i^L$, where f_i^V and f_i^L are the fugacities of pure component i evaluated at the temperature and pressure of the mixture).

DEFINITION OF THE VAPORIZATION PLATE EFFICIENCY

The *vaporization plate efficiency* is defined in a manner analogous to the vaporization point efficiency. Let $\hat{f}_{ji}^V$ be the fugacity of component i in the perfectly mixed vapor phase which leaves any plate j of a distillation column and $\hat{f}_{ji}^L$ be the fugacity of component i in the liquid phase leaving plate j. If $\hat{f}_{ji}^V$ is unequal to $\hat{f}_{ji}^L$ (where both $\hat{f}_{ji}^V$ and $\hat{f}_{ji}^L$ are nonzero, finite and positive), then there exists a positive number E_{ji} such that

$$\hat{f}_{ji}^V = E_{ji} \hat{f}_{ji}^L \tag{13-38}$$

Introduction of the thermodynamic definitions of $\hat{f}_{ji}^V$ and $\hat{f}_{ji}^L$ followed by rearrangement yields

$$y_{ji} = E_{ji} \frac{\gamma_{ji}^L}{\gamma_{ji}^V} K_{ji} x_{ji} \tag{13-39}$$

The vaporization efficiency was perhaps first employed by either McAdams[6] or Carey.[5] In 1949, Edmister[12] proposed a similar efficiency and called it the *absorption efficiency*. Vaporization efficiencies have been applied to batch-steam distillation[21] and to conventional columns with plates.[22, 33]

Temperature Conventions

In the Murphree and modified Murphree plate efficiencies, it is supposed that the vapor and liquid leaving each plate are in thermal equilibrium: $T_j^V = T_j^L = T_j$. Since the sum of the y_{ji}'s over all components is equal to unity, it is evident from Eq. (13-39) that

$$1 = \sum_{i=1}^{c} E_{ji} \frac{\gamma_{ji}^L}{\gamma_{ji}^V} K_{ji} x_{ji} \tag{13-40}$$

In many applications, the vapor forms an ideal solution ($\gamma_{ji}^V = 1$ for all i), and Eq. (13-40) reduces to

$$1 = \sum_{i=1}^{c} E_{ji} \gamma_{ji}^L K_{ji} x_{ji} \tag{13-41}$$

The temperature of plate j is that value of T_j which satisfies Eq. (13-41) [or (13-40)].

The determination of the temperature by the Murphree convention is based on the assumption that the liquid is at its bubble-point temperature at every point on the plate. Then for the case where the liquid phase is perfectly mixed but the vapor phase is not, the actual y above the vapor may differ from y_i^*, the mole fraction component that i would have in the vapor if the vapor were in equilibrium with the liquid. The value of y_{ji}^* is given by the equilibrium relationship

$$y_{ji}^* = \frac{\gamma_{ji}^L}{\gamma_{ji}^V} K_{Mji} x_{ji} \tag{13-42}$$

where K_{Mji} is evaluated at the temperature the liquid would have if it were at its bubble-point temperature. That is, for the case where the vapor phase forms an ideal solution, the temperature of plate j by the Murphree temperature convention is that T_j which satisfies the following expression

$$1 = \sum_{i=1}^{c} \gamma_{ji}^L K_{Mji} x_{ji} \tag{13-43}$$

Modified Murphree Plate and Point Efficiencies for the Perfectly Mixed Liquid Phase Model

First the equations for the modified Murphree plate efficiency are developed and then the corresponding equations for the Murphree plate efficiency are developed. Expressions for the vaporization efficiencies for each of these efficiencies are then presented. The section is concluded by the presentation of numerical examples for binary mixtures which demonstrate that essentially the same compositions of the vapor leaving a plate are predicted by vaporization efficiencies corresponding to the modified Murphree efficiency model as are predicted by the Murphree plate efficiencies. This result suggests that the correlations for the film coefficients developed by others[4] for use with Murphree efficiencies may be used in the prediction of vaporization efficiencies for multi-component systems.

THE MODIFIED MURPHREE PLATE EFFICIENCY AND THE CORRESPONDING VAPORIZATION EFFICIENCY

The modified Murphree plate efficiency model consists of the following set of assumptions.

1. The liquid phase is perfectly mixed.
2. No mixing of the vapor occurs as it passes up through the liquid phase, but upon leaving the surface of the liquid, the vapor becomes perfectly mixed before entering the next plate.
3. The variation of the total-flow rate across any one stage may be neglected in the efficiency model for that stage.
4. The vapor and liquid phases are in thermal equilibrium at the temperature T_j of the liquid leaving plate j which satisfies Eq. (13-42).
5. The overall mass transfer coefficient and the interfacial area are finite and positive constants for each plate.

In the following developments, the number of mass transfer units is taken to be an independent variable and the compositions dependent variables. The equations for mass transfer on the plate of a distillation column are of the same form as those for a packed column. Thus, on the basis of the above assumptions, the component-material balance at any distance Z from the surface of the liquid on plate j may be represented by the following differential equation

$$V_j \frac{dy_i}{dZ} + (K_G a f^V)_{ji} S(K_{ji} x_{ji} - y_i) = 0 \qquad (13\text{-}44)$$

which is obtained by elimination of R_i from Eqs. (13-23) and (13-25). Also the activity coefficients appearing in Eq. (13-25) were omitted in the interest of simplicity in the above and in the following developments. (The effect of activity coefficients may be included by replacing K_{ji} by $\gamma_{ji}^L K_{ji} / \gamma_{ji}^V$ wherever K_{ji} appears.)

Suppose that $K_{ji}x_{ji} > y_i$ over the interval of integration from $Z = 0$ (the surface of the liquid on plate j) to $Z = Z_T$ (the floor of the plate on plate j). Separation of the variables in Eq. (13-44) followed by integration yields

$$\int_{y_{ji}}^{y_{j+1,\,i}} \frac{dy_i}{K_{ji}x_{ji} - y_i} = -\int_0^{Z_T} \frac{(K_G af^V)_{ji}}{V_j} S\,dZ \tag{13-45}$$

Since the liquid is assumed to be perfectly mixed, $K_{ji}x_{ji}$ is constant over the interval from $Z = 0$ to $Z = Z_T$. Thus, Eq. (13-45) may be integrated and rearranged to give

$$\frac{y_{ji} - y_{j+1,\,i}}{K_{ji}x_{ji} - y_{j+1,\,i}} = 1 - \exp\left(-n_{OGji}\right) \tag{13-46}$$

where

$$n_{OGji} = (K_G af)_{ji}\left(\frac{SZ_T}{V_j}\right)$$

Also, it follows from Eq. (13-34) that the number of overall transfer units may be expressed in terms of the number of gas-phase and liquid-phase units as follows

$$\frac{1}{n_{OGji}} = \frac{1}{n_{Gji}} + \left(\frac{K_{ji}V_j}{L_j}\right)\frac{1}{n_{Lji}} \tag{13-47}$$

Let the modified Murphree plate efficiency be defined by

$$E_{ji}^M = \frac{y_{ji} - y_{j+1,\,i}}{K_{ji}x_{ji} - y_{j+1,\,i}} \tag{13-48}$$

Then, it follows from Eqs. (13-46) and (13-48) that one should be able to predict the modified Murphree plate efficiency on the basis of the total number of overall transfer units, namely,

$$E_{ji}^M = 1 - \exp\left(-n_{OGji}\right) \tag{13-49}$$

When both the numerator and denominator of the expression for E_{ji}^M given by Eq. (13-48) are divided by $K_{ji}x_{ji}$ and the expression so obtained is solved for E_{ji}, the following formula is obtained

$$E_{ji} = \frac{y_{j+1,\,i}}{K_{ji}x_{ji}} + E_{ji}^M\left[1 - \frac{y_{j+1,\,i}}{K_{ji}x_{ji}}\right] \tag{13-50}$$

Replacement of E_{ji}^M in Eq. (13-50) by its equivalent as given by Eq. (13-49) yields the predictive formula for E_{ji}, namely,

$$E_{ji} = 1 - e^{-n_{OGji}}\left[1 - \frac{y_{j+1,\,i}}{K_{ji}x_{ji}}\right] \tag{13-51}$$

THE MODIFIED MURPHREE POINT EFFICIENCY AND
THE CORRESPONDING VAPORIZATION POINT EFFICIENCY

Instead of making a material balance over the entire plate, a balance may be made over a section of the plate which is ΔZ high and ΔW wide. The variable W is measured from the outlet weir at which $W = W_T$. If it is supposed that the vapor is distributed uniformly over the plate, then the flow rate per unit length is given by V/W_T. Thus, the material balance over the element from W_j to $W_j + \Delta W$ is given by

$$\int_{W_j}^{W_j+\Delta W} \left(\frac{V y_i}{W_T} \right) dW \bigg|_{z_{j+1}} - \int_{W_j}^{W_j+\Delta W} \left(\frac{V y_i}{W_T} \right) dW \bigg|_{z_j} + \int_{W_j}^{W_j+\Delta W} \int_{z_j}^{z_j+\Delta Z} \left(\frac{R_i}{W_T} \right) dZ = 0$$

(13-52)

Application of the mean value theorems of integral and differential calculus followed by the limiting process wherein ΔZ and ΔW are allowed to go to zero yields

$$\frac{V dy_i}{dZ} + R_i = 0 \qquad (0 < Z < Z_T, 0 < W < W_T)$$

(13-53)

Separation of variables followed by integration from $Z = 0$ to $Z = Z_T$ at a given W yields

$$\int_{y_i}^{y_{j+1, i}} \frac{dy_i}{K_i x_i - y_i} = -\int_{Z=0}^{Z=Z_T} \frac{(K_G a f^V)_i}{V} S \, dZ$$

(13-54)

Since $K_i x_i$ does not change with Z at a given W, this expression may be integrated and rearranged to give the definition of the modified Murphree point efficiency.

$$E_{Pi}^M = \frac{y_i - y_{j+1, i}}{K_i x_i - y_{j+1, i}} = 1 - e^{-n_i}$$

(13-55)

where n_i is the point value of the number of overall transfer units. At any W

$$n_i(W) = \frac{(K_G a f^V)_i S Z_T}{V} \qquad \frac{1}{n_i} = \frac{1}{n_{Gi}} + \left(\frac{K_i V}{L} \right) \frac{1}{n_{Li}}$$

(13-56)

By use of the definition of the point vaporization efficiency ($y_i = E_i K_i x_i$) and Eq. (13-55), the following predictive expression is obtained for the vaporization point efficiency for plate j

$$E_i = 1 - e^{-n_i} \left(1 - \frac{y_{j+1, i}}{K_i x_i} \right)$$

(13-57)

Murphree Plate Efficiency for the Perfectly Mixed Liquid Phase Model

The assumptions for the Murphree plate efficiency model are the same as those for the modified Murphree plate efficiency model except for the additional assumption that the liquid leaving each stage is at its bubble-point temperature. The beginning equation for the development of the expression for the Murphree plate efficiency is obtained by replacing R_i in Eq. (13-23) by the expression given by Eq. (13-29) to give

$$V_j \frac{dy_i}{dZ} + (K'_G a)_{ji} PS(y_i^* - y_i) = 0 \qquad (13\text{-}58)$$

where the variation of V over plate j has been neglected in accordance with the assumptions. Integration followed by rearrangement in a manner analogous to that demonstrated for the modified Murphree efficiency model yields the following result

$$E_{Mji} = \frac{y_{ji} - y_{j+1,i}}{y_{ji}^* - y_{j+1,i}} = 1 - \exp\left(-n'_{OGMji}\right) \qquad (13\text{-}59)$$

where

$$n'_{OGMji} = (K'_G a)_{ji} P\left(\frac{SZ_T}{V_j}\right)$$

Also

$$\frac{1}{n'_{OGMji}} = \frac{1}{n'_{Gji}} + \left(\frac{m_{ji} V_j}{L_j}\right)\frac{1}{n'_{Lji}} \qquad (13\text{-}60)$$

and y_{ji}^* is given by either

$$y_{ji}^* = K_{Mji} x_{ji} \qquad (13\text{-}61)$$

or

$$y_{ji}^* = m_{ji} x_{ji} + b_i \qquad (13\text{-}62)$$

As shown in Problem 13-9, E_{ji} may be expressed in terms of E_{Mji}.

THE MURPHREE POINT EFFICIENCY

In this case an assumption in addition to those stated for the modified Murphree plate efficiency model is made. This assumption is that the liquid is perfectly mixed vertically but not horizontally, and that the liquid is at its bubble-point temperature at each point on the plate. Thus, along any one vertical line, the bubble-point temperature is the same, but it varies from point to point in the horizontal direction W.

By following the same general approach as that demonstrated for the modified Murphree point efficiency, one obtains the following expression for the Murphree point efficiency

$$E_{MPi} = \frac{y_i - y_{j+1,i}}{y_i^* - y_{j+1,i}} = 1 - \exp\left(-n'_{Mi}\right) \qquad (13\text{-}63)$$

where

$$n'_{Mi} = (K'_G a)_i P\left(\frac{S Z_T}{V_j}\right)$$

Also,

$$\frac{1}{n'_{Mi}} = \frac{1}{n'_{Gi}} + \left(\frac{m_i V_j}{L_j}\right)\frac{1}{n'_{Li}} \tag{13-64}$$

and is given by

$$y_i^* = m_i x_i + b_i \tag{13-65}$$

or

$$y_i^* = K_{Mi} x_i \tag{13-66}$$

Comparison of the Murphree Plate Efficiencies and the Vaporization Efficiencies Corresponding to the Modified Murphree Efficiency Model

Comparison of the Murphree efficiency [Eq. (13-59)] with the modified Murphree efficiency [(13-48) and (13-49)] shows that the Murphree efficiency depends on n'_{OGMi} [Eq. (13-59)] while the modified Murphree efficiency depends on n'_{OGi} [Eq. (13-49)]. The number of overall transfer units n'_{OGi} and n'_{OGMi} for the two efficiencies depend on the same set of transfer units (n'_{Gi} and n'_{Li}) and are the same except that the quantity $\gamma_i^L K_i / \gamma_i^V$ appears in n'_{OGi} in the same position that m_i appears in n'_{OGMi}.

In both models, a state of dynamic equilibrium is assumed to exist at the interface ($\hat{f}_i^V = \hat{f}_i^{\mathscr{L}}$) and they have this one point ($x_{ji}^{\mathscr{L}}$, y_{ji}^V) at the interface in common. In the model for the vaporization efficiency, the T_j found by Eq. (13-40) is the same throughout the vapor and liquid phases on plate j. Since equilibrium is assumed to exist at the interface, it follows that the vaporization efficiency must be equal to unity at the interface and, consequently, the K_{ji}'s found by use of Eq. (13-40) must satisfy the equilibrium relationship $y_{ji}^V = \gamma_{ji}^{\mathscr{L}} K_{ji} x_{ji}^{\mathscr{L}}$ (where it is supposed that the vapor phase forms an ideal solution). Since the liquid is assumed to be at its bubble-point temperature throughout the liquid phase in the Murphree model for the plate efficiency, it follows that $y_{ji}^V = \gamma_{ji}^{\mathscr{L}} K_{Mji} x_{ji}^{\mathscr{L}} = \gamma_{ji}^{\mathscr{L}} K_{ji} x_{ji}^{\mathscr{L}}$ or K_{Mji} is equal to K_{ji}. That is, the two models have the same interface temperatures and, furthermore, the temperature computed by use of vaporization efficiencies [Eq. (13-40)] is seen to be the bubble-point temperature of the interface [see Fig. 13-3].

The vaporization plate efficiency (based on the modified Murphree efficiency model) and the Murphree plate efficiency have two points [(x_{j1}, $y_{j+1, 1}$) and ($x_{j1}^{\mathscr{L}}$, y_{j1}^V)] in common as may be seen from Fig. 13-3. The expression for the line connecting these two points is

$$\frac{L n'_L}{V n'_G} = \frac{y_i^V - y_i}{x_i - x_i^{\mathscr{L}}} \tag{13-67}$$

After the interface point has been reached, the paths for the two models differ. In the case of the vaporization plate efficiency (based on the modified Murphree efficiency model), a projection is made from the point $(x_{ji}^{\mathscr{L}}, y_{ji}^{\gamma})$ to the point (x_{j1}, y_{j1}) with the slope K_{j1} that satisfies Eq. (13-40). In the case of the Murphree plate efficiency, a projection is made from the point $(x_{ji}^{\mathscr{L}}, y_{ji}^{\gamma})$ to the point (x_{j1}, y_{j1}^{*}) along the tangent line, $y_1^{*} = m_1 x_1 + b_1$. Since the equilibrium curve is seldom straight, an appropriate value of m_1 lying between the values at $(x_{ji}^{\mathscr{L}}, y_{ji}^{\gamma})$ and (x_{j1}, y_{j1}^{*}) is selected. The values of y_{j1} predicted by use of vaporization efficiencies are calculated by use of Eqs. (13-40), (13-47), and (13-51).

Although the Murphree model contains an additional assumption (that the liquid leaving plate j is at its bubble-point temperature) over the modified Murphree model, the corresponding values of y_{j1} predicted by both models on the basis of the same sets coefficients $\{n_{Li}', n_{Gi}'\}$ and points $\{x_{j1}, y_{j+1, 1}\}$ appear to be in almost perfect agreement for the two examples presented (see Tables 13-2 and 13-3). These examples were taken from Ref. 24. The number of transfer units for each film in these examples was taken to be independent of component identity just as they are for the existing correlations for binary mixtures which are given below. In Example 13-1 (the benzene-toluene system), the vapor and liquid phases closely approximate ideal solutions, but the liquid phase of the ethanol-water system in Example 13-2 is highly nonideal.

The following calculational procedure was used to compute the vaporization plate efficiencies (corresponding to the modified Murphree plate efficiency model):

1. On the basis of the given set $\{n_G', n_L'\}$ and a selected value of x_{j1}, compute $y_{j+1, 1}$ by use of the operating line.
2. Find the temperature T_j which satisfies Eq. (13-40). For each assumed value of T_j, each vaporization efficiency E_{ji} must be evaluated by use of Eqs. (13-34) and (13-51). (Note in Example 13-2 in which the liquid phase is highly nonideal each K_{ji} should be preceded by a γ_{ji}^{L}.)

The values of y_{j1} found by use of the Murphree plate efficiency model were calculated as follows:

1. On the basis of the given set $\{n_G', n_L'\}$ and a selected value of x_{j1}, compute $y_{j+1, 1}$ by use of the operating line.
2. Find the temperature T_{Mj} which satisfies Eq. (13-43).
3. Evaluate m_1 by use of Eq. (13-68) [or (13-69)] given below.
4. Compute n_{OGMi}' by use of Eq. (13-35) and y_{j1} by use of Eq. (13-59).

For definiteness, the value of the slope m_1 corresponding to the outlet composition $\{x_{ji}\}$ was used. The values of m_1 for Example 13-1 were computed by use of the following formula for m_1 as suggested in Ref. 4.

$$m_1 = \frac{dy_1^{*}}{dx_1} = \frac{d\left[\alpha_1 x_1 \left(\sum_{i=1}^{2} \alpha_i x_i\right)^{-1}\right]}{dx_1} = \frac{\alpha_1}{[1 + x_1(\alpha_1 - 1)]^2} \tag{13-68}$$

For Example 13-2 whose liquid phase is highly nonideal, the following formula for m_1 was used

$$m_1 = \frac{d\left[\gamma_1 \alpha_1 x_1 \left(\sum_{i=1}^{2} \gamma_i \alpha_i x_i\right)^{-1}\right]}{dx_1} = \frac{\gamma_1 \alpha_1 + \alpha_1 x_1 (d\gamma_1/dx_1)}{\sum_{i=1}^{2} \gamma_i \alpha_i x_i}$$

$$- \frac{\gamma_1 \alpha_1 x_1 [\gamma_1 \alpha_1 + \alpha_1 x_1 (d\gamma_1/dx_1) - \gamma_2 + (1 - x_2)(d\gamma_2/dx_1)]}{\left(\sum_{i=1}^{2} \gamma_i \alpha_i x_i\right)^2} \quad (13\text{-}69)$$

where component 2 (the least volatile) was taken to be the base component in Eqs. (13-68) and (13-69). (For the base component, $\alpha_2 = K_2/K_2 = 1$.) Examination of Fig. 13-3 shows that the value of m_1 at x_{j1} may differ appreciably from its value at $x_{j1}^{\mathscr{L}}$, particularly at large values of n_G' relative to n_L' (and/or large values of V relative to L). The shape of the equilibrium curve determines the precise variation of m_1 with the mole fraction of component 1 in the liquid. For those cases in which $n_L' < n_G'$, values of m_1 of good accuracy may be obtained for Example 13-2 by locating the point $(x_{j1}^{\mathscr{L}}, y_{j1}^{\gamma})$ graphically by use of the slope $(-Ln_L'/Vn_G')$ and then taking m_1 to be equal to the slope of the straight line that passes through the points $(x_{j1}^{\mathscr{L}}, y_{j1}^{\gamma})$ and (x_{j1}, y_{j1}^{*}) as may be visualized by use of Fig. 13-3. The values of y_{j1} determined by use of these values of m_1 are shown in parentheses in Table 13-2 for the selected values of x_{j1}. The values of y_{j1} enclosed in parentheses are seen to give an improvement in the agreement of the values of y_{j1} predicted by use of the Murphree and the vaporization efficiencies.

If n_G' (and/or V) is small relative to n_L' (and/or L), it may be seen from Fig. 13-3 that the value of m_1 at $x_{j1}^{\mathscr{L}}$ approaches its value at x_{j1}, and the value of m_1 given by Eq. (13-69) approaches the correct value of m_1. If n_G' is small relative to n_L', it is seen from Table 13-3 that y_{j1} computed by the Murphree plate efficiency approaches the value of y_{j1} computed by use of the vaporization plate efficiency.

Since the assumptions given by Eq. (13-16) become approximations for highly nonideal solutions such as ethanol and water, n_G and n_L become only approximately equal to n_G' and n_L', respectively. Thus, some disagreement in the y_{j1}'s predicted by use of the two types of efficiencies might be expected. The error in the y_{j1}'s resulting from setting $n_G' = n_G$ and $n_L' = n_L$ appears to be small relative to the precise method used to compute the m_1's.

Examples 13-1 and 13-2 demonstrate that for the same set of film coefficients (n_L', n_G') and compositions $\{x_{j1}, y_{j+1, 1}\}$ approximately the same values of y_{j1} are obtained by use of vaporization plate efficiencies (based on the modified Murphree model) as are obtained by use of Murphree plate efficiencies for binary mixtures. This result suggests that the existing correlations for n_L' and n_G' given in the *Bubble-Tray Design Manual*[4] may be used to estimate the E_{ji}'s for the modified Murphree model for multicomponent mixtures. This statement arises from the fact that the formulas for the vaporization plate efficiencies which were used in the computations for binary mixtures are precisely the same ones that

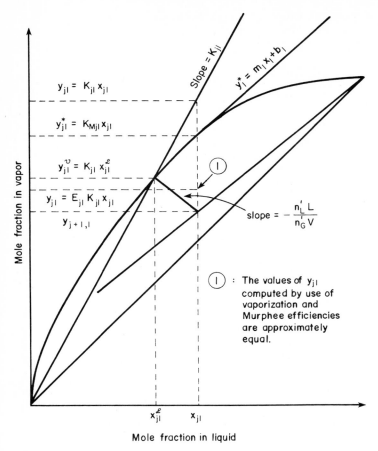

Figure 13-3 Comparison of Murphree plate efficiencies with vaporization plate efficiencies. [C. D. Holland and K. S. McMahon, *Chem. Eng. Sci.*, **25**:431 (1970), *by courtesy Chemical Engineering Science.*]

are also applicable for multicomponent mixtures. The use of vaporization plate efficiencies (based on the modified Murphree model) rather than Murphree plate efficiencies avoids the uncertainty encountered in the determination of m_1 when one attempts to apply the Murphree efficiencies to multicomponent mixtures. The use of vaporization plate efficiencies (based on the modified Murphree model) also avoids the uncertainty encountered in the determination of m_1 when the Murphree plate efficiencies are applied to multicomponent mixtures. The postulate in the Murphree efficiency model that the liquid leaving a plate is at its bubble-point temperature uses up one degree of freedom and consequently only $(c - 1)$ Murphree plate efficiencies may be fixed independently for each plate. The difficulty in selecting the independent and dependent Murphree efficiencies is avoided by the use of the vaporization efficiencies (based on the modified Murphree model).

Table 13-2 Statement and solution of Example 13-1[24]

I. STATEMENT OF EXAMPLE 13-1

A distillation column operating at an absolute pressure of 760 mmHg is to be used to separate a benzene-toluene mixture. The column has a total condenser, and in the rectifying section, $L = L_j = 100$ mol/h. The distillate rate $D = 50$ mol/h and the mole fraction of benzene (component no. 1) in the distillate is 0.95. Find the y_{j1} by use of the vaporization plate efficiencies (based on the modified Murphree model) and by use of Murphree plate efficiencies for the set $\{n'_L, n'_G\}$ of transfer units enumerated below at $x_j = 0.4, 0.6,$ and 0.9. (The operating line for the rectifying section intersects the equilibrium curve at an x_{j1} slightly greater than 0.3 but less than 0.4.) Use the following expressions for vapor pressures of benzene and toluene which are based on data taken from p. 578 of *Chemical Engineers Handbook* (3d ed.), J. H. Perry (ed.)

$$P_1(\text{benzene}) = 36.3 \times 10^6 \exp\left(-\frac{6852}{T}\right) \qquad (P \text{ in mmHg and } T \text{ in } °R)$$

$$P_2(\text{toluene}) = 48.151 \times 10^6 \exp\left(-\frac{7636.36}{T}\right) \qquad (P \text{ in mmHg and } T \text{ in } °R)$$

Compute m by use Eq. (13-68).

II. SOLUTION OF EXAMPLE 13-1

Specifications			Vaporization plate efficiency†				Murphree (plate)		
n'_G	n'_L	x_{ji}	$T_j (°F)$	$E_{j,1}$	$E^M_{j,1}$	$y_{j,1}$	$T_{Mj} (°F)$	$E_{Mj,1}$	$y_{j,1}$
1	1	0.4	204.4310	0.9433	0.2569	0.5957	204.3829	0.3329	0.5960
1	1	0.6	194.9532	0.9126	0.2804	0.7445	192.9477	0.3879	0.7451
1	1	0.9	181.1069	0.9585	0.3166	0.9355	180.0270	0.4521	0.9356
0.2	10	0.4	203.4859	0.9483	0.1739	0.5901	203.3829	0.1766	0.5900
0.2	10	0.6	193.0473	0.9222	0.1750	0.7297	192.9477	0.1779	0.7297
0.2	10	0.9	180.1058	0.9630	0.1762	0.9243	180.0270	0.1791	0.9242
10	0.2	0.4	205.3324	0.9180	0.0793	0.5879	203.3829	0.1256	0.5881
10	0.2	0.6	197.2668	0.8604	0.0894	0.7283	192.9477	0.1724	0.7293
10	0.2	0.9	183.0296	0.9197	0.1110	0.9268	180.0270	0.2541	0.9273

† Based on the modified Murphree model.

Table 13-3 Statement and solution of Example 13-2[24]

I. STATEMENT OF EXAMPLE 13-2

A column operating at total reflux at an absolute pressure of 760 mmHg is to be used to separate an ethanol-water mixture. Find the set of y_{j1}'s corresponding to the sets $\{n'_L, n'_G\}$ listed below by use of the vaporization plate efficiencies at the following values $x_{j1} = 0.2, 0.3, 0.4,$ and 0.9. Curve fits for the activity coefficients which are based on the experimental results of J. S. Carey (Sc.D. Thesis in Chemical Engineering, M.I.T., 1929).

$$\frac{\ln \gamma_1}{x_2^2} = 0.2924 + 1.1939x_2$$

$$\frac{\ln \gamma_2}{x_2^2} = 0.2925 + 1.1939(0.5 + x_2)$$

Use the following curve fits for the vapor pressure of the pure components:

1. $P_1(\text{ethanol}) = 745 \times 10^6 \exp(-8728.11/T)$ (P in mmHg, T in °R)

2. $P_2(\text{water}) = 530.5 \times 10^6 \exp(-9038.0/T)$ (P in mmHg, T in °R)

These curve fits are based on data presented in the International Critical Tables vol. III, pp. 212–217, McGraw-Hill Book Company, New York, 1928.

Compute m_1 by use of Eq. (13-69) and also take $L_j/V_j = 1$ in the calculation of n_{OGi} and n'_{OGMi}.

II. SOLUTION OF EXAMPLE 13-2

Specifications			Vaporization plate efficiency†				Murphree plate efficiency		
n'_G	n'_L	x_{ji}	T_j (°F)	$E_{j,1}$	E_{j1}^M	$y_{j,1}$	T_{Mj} (°F)	$E_{Mj,1}$	$y_{j,1}$
1	1	0.2	192.7814	0.4400	0.2043	0.2971	182.2920	0.0960	0.2329 (0.3087)
1	1	0.3	185.8769	0.5846	0.2622	0.4014	179.3242	0.1778	0.3529 (0.3978)
1	1	0.4	181.0562	0.7129	0.3086	0.4876	177.6956	0.2791	0.4660
1	1	0.9	172.8911	0.9990	0.3932	0.9006	173.2978	0.4183	0.9040
0.2	10.0	0.2	182.6709	0.4752	0.1727	0.2599	182.2920	0.1561	0.2535
0.2	10.0	0.3	179.2672	0.5896	0.1750	0.3520	179.3242	0.1687	0.3502
0.2	10.0	0.4	177.2978	0.6983	0.1762	0.4407	177.6956	0.1748	0.4414
0.2	10.0	0.9	172.8906	0.9987	0.1780	0.9003	173.2978	0.1785	0.9017
10.0	0.2	0.2	199.2087	0.2973	0.0504	0.2287	182.2920	0.0221	0.2076 (0.2337)
10.0	0.2	0.3	192.0522	0.4296	0.0735	0.3353	179.3242	0.3473	0.3141 (0.3471)
10.0	0.2	0.4	185.5680	0.5784	0.0999	0.4352	177.6956	0.0918	0.4217
10.0	0.2	0.9	172.8916	0.9986	0.1778	0.9003	173.2978	0.2062	0.9020

---

† Based on the modified Murphree plate efficiency model.

Summary of Existing Correlations for the Prediction of the Number of Transfer Units

Correlations proposed in the *Bubble-Tray Design Manual*[4] may be used for the prediction of n_G and n_L for binary mixtures. These equations may be used to estimate the number of transfer units for multicomponent mixtures as shown by Graham et al.[18] For the prediction of n_G, Eq. (2) on p. 27 of the *Manual*[4] may be used; for the prediction of n_L, the correlation given by Eq. (9) on p. 34 is available; this correlation contains the liquid contact time as a parameter, which may be computed by use of Eq. (9) on p. 35; the liquid holdup may be estimated by use of Eq. (10) on p. 35. For sieve trays, which are not treated in the Manual,[4] a correlation recommended by Gerster et al.[15] has been presented by Smith.[31]

> **Example 13-3** To demonstrate the use of the predictive method, Example 13-3, which is taken from Ref. 24, is presented. A statement of this example appears in Table 13-4. Physical properties which are listed in the table were estimated as follows. The viscosity of each pure component i in the vapor phase was predicted by use of the correlation proposed by Bird et al.[1, 2, 20] The pure component viscosities so obtained were used to predict the viscosities of the vapor mixtures by use of the semi-empirical formula proposed by Wilke.[40]
>
> Vapor phase diffusivities for each binary pair were predicted by use of the modified form of the Hirschfelder-Bird-Spotz equation[20] which was proposed by Fuller et al.[14] The diffusivities so obtained were used to predict the diffusivity of each component i in the vapor mixtures by use of the equation presented by Wilke.[38]
>
> The viscosities and diffusivities for components in multicomponent liquid mixtures were predicted by an extension of existing formulas for the prediction of these properties for binary and ternary mixtures.
>
> Viscosities for each of the pure components in the liquid phase were obtained from the literature. Then the viscosities of liquid mixtures were predicted by use of the generalized form of the Arrhenius equation as described by Graham et al.[18]
>
> Diffusivities for each component i in the mixtures were estimated by the correlations proposed by Tang,[32] Wilke and Chang,[39] Wilke,[40] and Fuller et al.[14] as described by Graham et al.[18]
>
> In Table 13-5 the product distributions obtained experimentally are compared with those found by use of perfect plates and by use of the perfectly mixed liquid phase model [Eq. (13-51)]. The experimental results show that the plate behavior of the actual column was better than perfect plates. The observed behavior can be explained on the basis of the existence of a concentration gradient in the direction of flow of the liquid across each plate which was neglected in the perfect plate and perfectly mixed liquid phase models.

To demonstrate the behavior of the vaporization efficiencies throughout the column, those efficiencies obtained for *o*-xylene and the C_9 + aromatics are listed in Table 13-6.

Table 13-4 Statement of Example 13-3[18]

1. *Operating conditions and geometry*

$L_1/D = 13.35$
$f = 16$ (feed plate)
$N = 42$
Total condenser at 19.7 lb/in² abs
i.d. of column = 5.5 ft.
Plates = Koch Flexitray-AF, single-pass flow

$h_w = 3$ in
$Z_1 = 4$ ft and 1 in (distance between weirs)
ΔP (across condenser) = 2 lb/in²
ΔP (per plate) = 0.225 lb/in²
Pressure in reboiler = 26.7 lb/in² abs
Thermal condition of feed: liquid at 148.9°F and 54.7 lb/in² abs

Component	Component no.	Feed (lb/h)	Feed (mol/h)	Distillate (lb/h)	Distillate (mol/h)	Bottoms (lb/h)	Bottoms (mol/h)
o-xylene	1	4360.5	41.07479	4280.5	40.32121	....	0.75358
m-xylene	2	170†	0.80068	85.0	0.80068		
p-xylene	3		0.80068	85.0	0.80068		
Ethylbenzene	4	8.5	0.08007	8.5	0.08007		
C_9 + aromatics‡	5	3935.5	32.7399	15.5	0.12896	3920.0	32.61503
Nonaromatics‡	6	25.5	0.22723	25.5	0.22723		

2. *Sources of the physical properties data*

1. Vapor and liquid enthalpies except for the liquid enthalpy of ethylcyclohexane, were taken from API Research Project 44. (Selected Values of Properties of Hydrocarbon and Related Compounds. American Research Project 44. Thermodynamics Research Center. Texas A&M University.) The liquid enthalpy of ethylcyclohexane was estimated by use of Theisen's correlation.§

2. The vapor pressures were computed by use of the Antoine equation, and the constants for this equation were taken from *Lange's Handbook of Chemistry* by N. A. Lange (Handbook Publishers, 10th Edition, McGraw-Hill, New York, 1967).

3. Liquid viscosities for the pure components were taken from API Research Project 44.

4. Critical properties for all components except ethylcyclohexane were taken from API Research Project 44. The critical properties for ethylcyclohexane were estimated by use of Lydersen's method.§

5. Atomic diffusion volumes were taken from Ref. 14.

6. Molar volumes were taken from Ref. 39.

---

† The experimental value for both *m*-xylene and *p*-xylene was 170 lb/h and an equal distribution (85 lb/h) was assumed for computational purposes.

‡C_9 + aromatics were taken to be cumene, and the nonaromatics were taken to be ethylcyclohexane. (These choices were based on experimental evidence which was available.)

§ *The Properties of Gases and Liquids, Their Estimation and Correlation*, 2d ed. by R. C. Reid and T. K. Sherwood. McGraw-Hill Book Company, New York, 1966.

Table 13-5 Comparison of the experimental product distributions with those predicted by use of perfect plates and by the modified Murphree plate efficiency model

	X_{Di}		
Component	Experimental	Perfect plates	Modified Murphree model
o-xylene	0.95190	0.92217	0.88555
m-xylene	0.037781	0.01889	0.01886
p-xylene		0.01889	0.01884
Ethylbenzene	0.00189	0.00189	0.00189
C_9 + aromatics	0.00304	0.03280	0.06949
Nonaromatics	0.00536	0.00536	0.00536

	x_{Bi}		
Component	Experimental	Perfect plates	Modified Murphree model
o-xylene	0.02258	0.06032	0.10681
m-xylene	0.0	0.00001	0.00005
p-xylene	0.0	0.00002	0.00008
Ethylbenzene	0.0	0.0	0.0
C_9 + aromatics	0.97742	0.93965	0.89306
Nonaromatics	0.0	0.0	0.0

Table 13-6 Vaporiation efficiencies for the modified Murphree efficiency model

Plate	Component o-xylene	C_9 + aromatics	Plate	Component o-xylene	C_9 + aromatics
1*	1.000	1.00	21	0.988	1.02
2	0.998	1.06	23	0.985	1.02
3	0.997	1.06	25	0.983	1.02
5	0.995	1.05	27	0.980	1.02
7	0.994	1.05	29	0.978	1.02
9	0.992	1.04	31	0.975	1.01
11	0.991	1.04	33	0.973	1.01
13	0.989	1.04	35	0.970	1.01
15	0.988	1.03	37	0.968	1.01
16	0.993	1.02	39	0.965	1.01
17	0.992	1.02	41	0.963	1.01
19	0.990	1.02	42	1.000	1.00

13-3 PACKED DISTILLATION COLUMNS

The packed column is now a widely used device for mass transfer in distillation, absorption, and stripping. Typical of the packings which are available today are those shown in Figs. 13-4 and 13-5. Recently, Bolles and Fair[3] presented the results of an excellent evaluation of the design equations for packed columns. The most reliable equations or correlations for making the following types of determinations

1. Capacity, or flood point
2. Pressure drop
3. Mass transfer

were determined. The equations and correlations for making these determinations are presented below.

In the study conducted by Bolles and Fair, 545 observations from 13 original sources were used. Column diameters ranged from 0.82 to 4 feet. Packings used consisted of ceramic Raschig rings, metal Raschig rings, ceramic Berl saddles, and metal Pall rings. Packing sizes ranged from 0.6 to 3 inches. Data were obtained from distillation columns, absorbers, and strippers in which the operating pressures ranged from 0.97 to 315 lb/in² abs.

Flood Velocity Model

The flood velocity is defined as that vapor velocity above which liquid accumulates uncontrollably in the packed bed and continued operation becomes impossible. The variables which determine the flood velocity are the packing geometry, system properties, and liquid viscosity. The first published model was that of Walker et al.[35] in 1937. In 1938, Sherwood et al.[30] made a minor adjustment to the correlations. Lobo et al.[27] made a further improvement in 1945. Further modification to the correlation was made by Eckert in 1963[10] and in 1970.[11] The final Eckert correlation for flood velocity

$$\left(\frac{G_{Vsf}^2}{g}\right)(F_p)\left(\frac{1}{\rho_V \rho_L}\right)\left(\frac{\rho_L}{\rho_W}\right)^{-1}\left(\frac{\mu_L}{\mu_W}\right)^{0.2} = f\left[\frac{w_L}{w_V}\sqrt{\frac{\rho_V}{\rho_L}}\right] \tag{13-70}$$

is shown in Fig. 13-6 and the symbols are defined in the caption of this figure. The flood-mass velocity of the vapor [lb/(h)(ft²)] based on the superficial area is denoted by G_{Vsf}. The packing factors F_p, given by Eckert[11] are presented in Table 13-7. This correlation was found by Bolles and Fair[3] to be more reliable than those proposed by Walker et al.,[35] Sherwood et al.,[30] and Lobo et al.[27] Bolles and Fair report that a safety factor of 1.32 should be used for this model for 95 percent confidence.

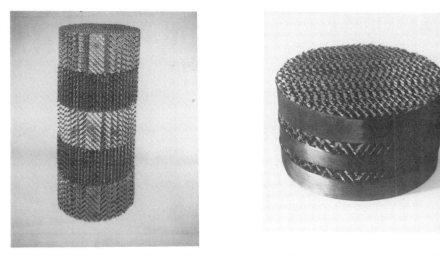

Figure 13-4 Two types of packing used for packed columns [*by courtesy Koch Engineering Company, Inc.*]

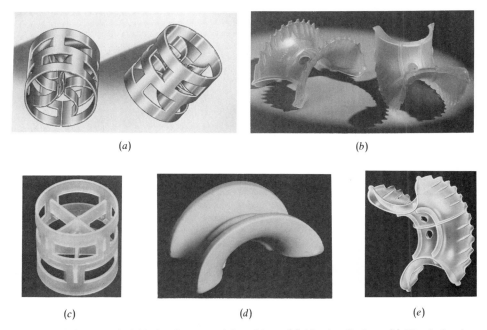

(a)

(b)

(c)

(d)

(e)

Figure 13-5 Some typical kinds of commercial packings: (a) Metal pall rings, (b) Plastic Intalox Saddles, (c) Raschig ring, (d) Ceramic Intalox Saddle, (e) Plastic Intalox Saddle. [*By courtesy of Norton Company, Chemical Process Products Division.*]

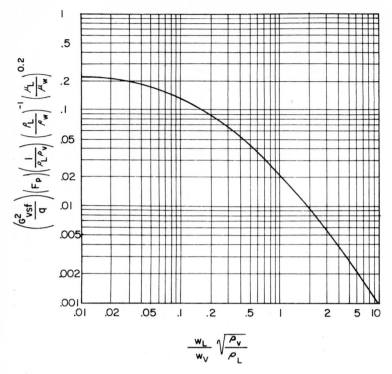

Figure 13-6 The Eckert flood velocity correlation. (F_p = packing factor, ft^{-1}; see Table 13-7. g = acceleration of gravity, ft/s^2. G_{vsf} = mass velocity of the vapor, superficial, flood, lb/s-ft^2. w_L = liquid mass flow rate, lb/s. w_V = vapor mass velocity, lb/s. ρ_V = density of vapor, lb/ft^3. ρ_L = density of liquid, lb/ft^3. μ_L = viscosity of vapor, lb/ft-s. μ_W = viscosity of water, lb/ft-s.) [J. S. Eckert, *Chem. Eng. Prog.*, **63**(3):39 (1949), *by courtesy American Institute of Chemical Engineers.*]

Table 13-7 The Eckert packing factors F_p in ft^{-1} for dumped packing[11]

	F_p in ft^{-1}			
Packing size inches	Ceramic Raschig rings	Metal Raschig rings	Ceramic Berl saddles	Metal pall rings
$\frac{1}{2}$	580 $(\frac{3}{32})$	300 $(\frac{1}{32})$	240	85
$\frac{5}{8}$	380 $(\frac{3}{32})$	170 $(\frac{1}{32})$	200	70
$\frac{3}{4}$	255 $(\frac{3}{32})$	155 $(\frac{1}{32})$	170	60
1	155 $(\frac{1}{8})$	137 $(\frac{1}{16})$	110	48
$1\frac{1}{4}$	125 $(\frac{3}{16})$	110 $(\frac{1}{16})$	85	39
$1\frac{1}{2}$	93 $(\frac{3}{16})$	83 $(\frac{1}{16})$	65	28
2	65 $(\frac{1}{4})$	57 $(\frac{1}{16})$	45	20
3	36 $(\frac{5}{16})$	32 $(\frac{1}{16})$	25	16

† Wall thickness, inches.

Pressure Drop Correlations

The drop in pressure with the depth of packing is a key design variable, particularily for columns in vacuum service. The most reliable correlation is the generalized pressure-drop correlation of Leva[25] as refined by Eckert,[11] which is referred to as the Eckert model. This correlation, presented in Fig. 13-7 is of the form

$$\left(\frac{G_{Vs}}{g}\right)^2 (F_p)\left(\frac{1}{\rho_V \rho_L}\right)\left(\frac{\rho_L}{\rho_W}\right)^{-1}\left(\frac{\mu_L}{\mu_W}\right)^{0.2} = f\left(\frac{w_L}{w_V}\sqrt{\frac{\rho_V}{\rho_L}}, P\right) \tag{13-71}$$

where the packing factors F_p are the same as those used in the flooding correlation (see Table 13-7). The mass velocity of the vapor based on the superficial area is denoted by G_{Vs} [lb/(s)(ft^2)]. The pressure P appears implicitly as a parameter of the graphical representation of Eq. (13-71) in Fig. 13-7, and the remaining symbols are defined in the caption of this figure. Bolles and

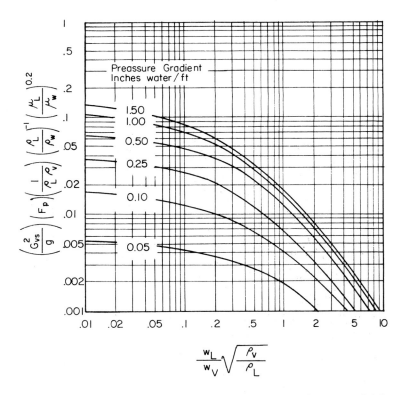

Figure 13-7 The Eckert correlation of the pressure gradient. (G_{vs} = superficial mass velocity of the vapor, lb/s-ft^2. Remaining symbols are defined in Fig. 13-6.) [J. S. Eckert, *Chem. Eng. Prog.*, **63**(3): 39 (1970), *by courtesy American Institute of Chemical Engineers*.]

Fair[3] report that a safety factor of 2.2 is required for this model for 95 percent confidence.

Mass Transfer Models

From the definition of the height of a transfer unit HTU, given by Eq. (13-33), it is evident that the multiplication of Eq. (13-35) by the total depth of packing yields the following relationship used by Bolles and Fair in their analysis

$$H_{OG} = H_G + \left(\frac{mV}{L}\right)H_L \tag{13-72}$$

where

$$H_{OG} = Z_T/n'_{OGi} \qquad H_G = Z_T/n'_G \qquad H_L = Z_T/n'_L$$

Bolles and Fair[3] proposed the following correlations for H_L and H_G, which were superior to all other correlations tested

$$H_L = \phi C_{fL}(Z_T/10)^{0.15}\sqrt{Sc_L} \tag{13-73}$$

$$H_G = \psi(d')^m(Z_T/10)^{1/3}(3600G_L f_\mu f_\rho f_\sigma)^{-n}\sqrt{Sc_V} \tag{13-74}$$

where ϕ = packing parameter; see Fig. 13-8
C_{fL} = a factor, which is given by Fig. 13-9
Z_T = total height of packing, ft
$Sc_L = \mu_L/\rho_L D_L$, Schmidt number for the liquid
$Sc_V = \mu_V/\rho_V D_V$, Schmidt number for the vapor
ψ = packing parameter; see Fig. 13-10
d' = min $(d, 2)$, the minimum of d or 2
d = diameter of the column, ft
m = 1.24 for rings and 1.11 for saddles
n = 0.6 for rings and 0.5 for saddles
G_L = mass velocity of the liquid, lb/s-ft²
$f_\mu = (\mu_L/\mu_W)^{0.16}$; μ_W = 1.0 centipoise
$f_\rho = (\rho_L/\rho_W)^{-1.25}$; ρ_W = 62.4 lb/ft³
$f_\sigma = (\sigma_L/\sigma_W)^{-0.8}$; σ_W = 72.8 dyne/cm
σ_L = surface tension, dyne/cm

In the development of the correlation, the values of the physical properties μ_L, ρ_L, σ_L were evaluated at the operating conditions while the corresponding values of μ_W, ρ_W, and σ_W were held fixed at the values shown above. The improved mass transfer correlation of Bolles and Fair[3] requires a safety factor of 1.7 for 95 percent confidence.

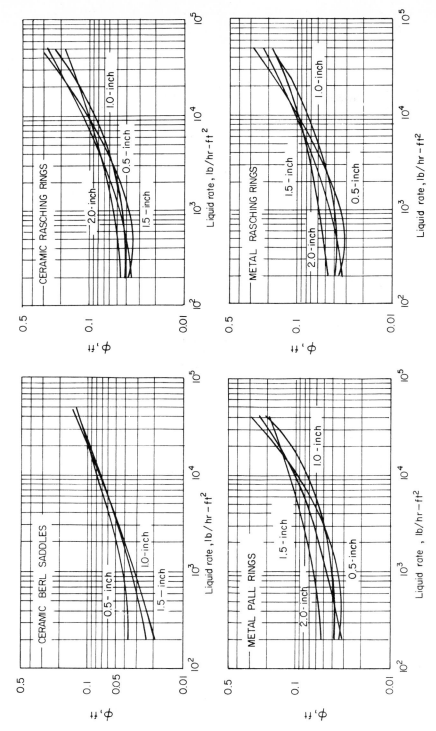

Figure 13-8 Correlation of the packing parameter ϕ for the liquid phase. (G_L = mass velocity of the liquid, lb/h-ft². [W. L. Bolles and J. R. Fair, Third International Distillation Symposium, London, April 1979, No. 213 of E.F.C. Sec. 3.3, p. 35, by courtesy Institute of Chemical Engineers (London).]

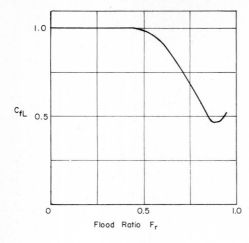

Figure 13-9 Correlation of the load factor C_{fL}. ($F_r = u_{vs}/u_{vsf}$ at a fixed ratio of liquid to vapor mass flow rates, u_{vs} = velocity of vapor based on the superficial area, ft/s, u_{vsf} = velocity of the vapor based on the superficial area at the flood point, ft/s.) [W. L. Bolles and J. R. Fair, Third International Distillation Symposium, London, April 1979, No. 213 of E.F.C. Sec. 3.3, p. 35, *by courtesy Institute of Chemical Engineers* (*London*).]

Example 13-4 This example is based on an example presented by Bolles and Fair.[3] Determine the flood velocity, the column diameter, the pressure drop per foot of packing, H_L, H_G, and H_{OG} for the separation of a benzene-toluene mixture in a packed column. The column conditions to be used for design purposes are as follows:

Composition	99 mole % benzene 1 mole % toluene
Pressure	14.7 lb/in² abs
Temperature	176°F
Liquid load	200,000 lb/h = 55.5 lb/s
Vapor load	240,000 lb/h = 66.6 lb/s

Use the following physical properties:

Density of liquid	50 lb/ft³
Density of vapor	0.168 lb/ft³
Viscosity of liquid	0.31 centipoise
Viscosity of vapor	0.00906 centipoise
Surface tension	21 dynes/cm
Relative volatility	2.5
Diffusion coefficient of liquid	4.26×10^{-5} cm²/s = 4.59×10^{-8} ft²/s
Diffusion coefficient of vapor	4.27×10^{-2} cm²/s = 4.6×10^{-5} ft²/s

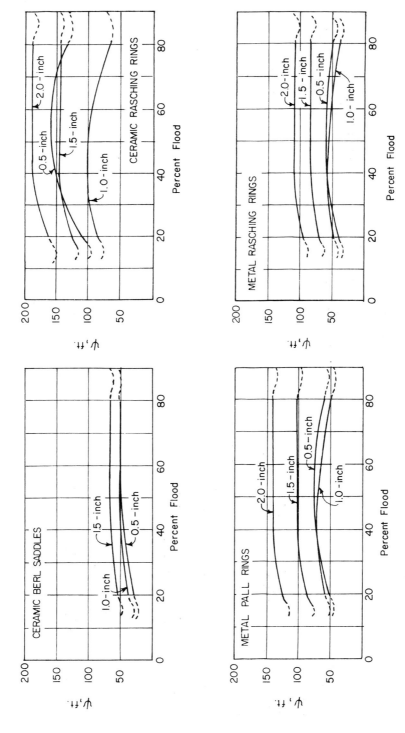

Figure 13-10 Correlation of the packing parameter ψ for vapor phase mass transfer. [W. L. Bolles and J. R. Fair, Third International Distillation Symposium, London, April 1979, No. 213 of E.F.C., Sec. 3.3, p. 35, *by courtesy Institute of Chemical Engineers (London)*.]

Metal, 2-inch Pall rings are to be used as the packing. The free area is 96 percent of the total area. The total packed height will be subdivided into individual reirrigated sections of not more than 30 feet of bed depth each. The column diameter is to be based on 80 percent of the vapor flood velocity.

1. Calculation of the flood velocity by use of the Eckert correlation The abscissa of Fig. 13-6 has the value

$$\frac{w_L}{w_V}\sqrt{\frac{\rho_V}{\rho_L}} = \left(\frac{200{,}000}{240{,}000}\right)\sqrt{\frac{0.168}{50}} = 0.0483$$

For this value of the abscissa, Fig. 13-6 has an ordinate of 0.18. Thus, Eq. (13-71) becomes

$$\left(\frac{G^2_{Vsf}}{g}\right)(F_p)\left(\frac{1}{\rho_L\rho_V}\right)\left(\frac{\rho_L}{\rho_W}\right)^{-1}\left(\frac{\mu_L}{\mu_W}\right)^{0.2} = 0.18$$

The value of $F_p = 20$ is obtained from Table 13-7, and from the data given, it follows that

$$\left(\frac{G^2_{Vsf}}{32.17}\right)(20)\left(\frac{1}{50 \times 0.168}\right)\left(\frac{50}{62.4}\right)^{-1}\left(\frac{0.31}{1.005}\right)^{0.2} = 0.18$$

which gives

$$G_{Vsf} = 1.57 \text{ lb/s-ft}^2$$

$$\text{Design velocity} = \left(\frac{1.57 \text{ lb}}{\text{s-ft}^2 \text{ free surface}}\right)\left(\frac{0.96 \text{ ft}^2 \text{ free surface}}{\text{ft}^2 \text{ cross-sectional area}}\right)$$

$$\times \left(\frac{0.8}{1.32 \text{ safety factor}}\right) = 0.9135 \text{ lb/s-ft}^2 = G_V$$

Then

$$d = \sqrt{\frac{(66.6 \text{ lb/s})4}{(0.9135 \text{ lb/s-ft}^2)\pi}} = 9.64 \text{ ft}$$

2. Calculation of the pressure drop From item I, the abscissa of Fig. 13-7 has the value of 0.0483. To find the pressure drop per foot of packing (the parameter of Fig. 13-7), the ordinate must be evaluated as follows

$$\left(\frac{G^2_{Vs}}{g}\right)(F_p)\left(\frac{1}{\rho_L\rho_V}\right)\left(\frac{\rho_L}{\rho_W}\right)^{-1}\left(\frac{\mu_L}{\mu_W}\right)^{0.2} = \left[\frac{(0.9135/0.96)^2}{32.17}\right](20)\left(\frac{1}{50 \times 0.168}\right)$$

$$\left(\frac{50}{62.4}\right)^{-1}\left(\frac{0.31}{1.005}\right)^{0.2} = 0.0661$$

By Fig. 13-7, the pressure gradient is $= 0.6$ in $H_2O/(\text{ft packing}) = 1.12$ mmHg/(ft packing). Correction of this result by the safety factor of 2.2 gives

$$\text{Pressure gradient} = 1.12 \times 2.2 = 2.46 \text{ mmHg/ft}$$

3. Calculation of H_L In order to evaluate H_L by use of Eq. (13-73), values for ϕ, C_{fL}, and S_{cL} are needed.

Since

$$\text{Liquid-mass velocity} = (200{,}000) \left[\frac{4}{\pi(9.64)^2} \right] = 2.74 \times 10^3 \text{ lb/(h-ft}^2)$$

Fig. 13-8 gives

$$\phi = 0.08 \text{ ft}$$

To find C_{fL}, the value

$$F_r = u_{Vs}/u_{Vsf}$$

is needed

$$u_{Vsf} = G_{Vsf}/\rho_V = \left(\frac{1.57 \text{ lb}}{\text{s-ft}^2} \right) \left(\frac{\text{ft}^3}{0.168 \text{ lb}} \right) = 9.35 \text{ ft/s}$$

The superficial vapor velocity u_{Vs} is found as follows

$$u_{Vs} = \left(\frac{66.6 \text{ lb}}{\text{s}} \right) \left[\frac{4}{\pi(9.64)^2} \right] \left(\frac{1}{0.96} \right) \left(\frac{\text{ft}^3}{0.168 \text{ lb}} \right) = 5.66 \text{ ft/s}$$

Then

$$F_r = \frac{5.66}{9.35} = 0.605$$

For this value of F_r, Fig. 13-9 gives

$$C_{fL} = 0.94$$

The Schmidt number for the liquid is computed as follows

$$Sc_L = \frac{\mu_L}{\rho_L D_L} = \frac{(0.31)(0.672 \times 10^{-3})}{(50)(4.59 \times 10^{-8})} = 90.8$$

Equation (13-73) gives

$$H_L = \phi C_{fL} \left(\frac{Z_T}{10} \right)^{0.15} \sqrt{Sc_L} = (0.08)(0.94)(3)^{0.15}(90.8)^{0.5} = 0.845 \text{ ft}$$

4. Calculation of H_G To compute H_G by use of Eq. (13-74), values for the quantities ψ, f_μ, f_p, f_σ, Sc_V, and G_L must be determined.

At 80 percent of the vapor flood velocity, Fig. 13-10 gives

$$\psi = 140 \text{ ft}$$

and

$$f_\mu = \left(\frac{\mu_L}{\mu_W}\right)^{0.16} = \left(\frac{0.31}{1.0}\right)^{0.16} = 0.829$$

$$f_\rho = \left(\frac{\rho_L}{\rho_W}\right)^{-1.25} = \left(\frac{50}{62.4}\right)^{-1.25} = 1.32$$

$$f_\sigma = \left(\frac{\sigma_L}{\sigma_W}\right)^{-0.8} = \left(\frac{21}{72.8}\right)^{-0.8} = 2.70$$

$$\text{Sc}_V = \frac{\mu_V}{\rho_V D_V} = \frac{(0.00906 \times 0.672 \times 10^{-3})}{(0.168)(4.6 \times 10^{-5})} = 0.788$$

$$G_L = (2.71 \times 10^3 \text{ lb/h-ft}^2)\left(\frac{1 \text{ h}}{3600 \text{ s}}\right) = 0.753 \text{ lb/s-ft}^2$$

Equation (13-74) gives

$$H_G = \frac{\psi(d')^{1.24}(Z_T/10)^{1/3}(\text{Sc}_V)^{1/2}}{(3600 G_L f_\mu f_\rho f_\sigma)^{0.6}}$$

$$= \frac{(140)(2)^{1.24}(3)^{1/3}(0.788)^{1/2}}{(3600 \times 0.753 \times 0.829 \times 1.32 \times 2.70)^{0.6}} = 1.93 \text{ ft}$$

5. Calculation of the overall height of a transfer unit, H_{OG} To compute H_{OG} by use of Eq. (13-72), the value of λ

$$\lambda = \frac{mV}{L}$$

is needed. Let 1 refer to benzene and 2 to toluene. Then

$$m = \frac{dy_1}{dx_1} = \frac{d}{dx_1}\left[\frac{\alpha_1 x_1}{\alpha_1 x_1 + (1 - x_1)}\right] = \frac{\alpha_1}{[1 + x_1(\alpha_1 - 1)]^2}$$

For $y_1 = 0.99$, the corresponding value of x_1 is

$$x_1 = \frac{y_1/\alpha_1}{y_1/\alpha_1 + y_2/\alpha_2} = \frac{0.99/2.5}{0.99/2.5 + 0.01/1.0} = 0.975$$

Then

$$m = \frac{2.5}{[1 + 0.975 \times 1.5]^2} = 0.412$$

Next

$$\text{mol wt of liquid} = 0.975 \times 78.11 + 0.025 \times 92.13 = 78.46$$
$$\text{mol wt of vapor} = 0.99 \times 78.11 + 0.01 \times 92.13 \quad = 78.25$$

and

$$L = (200{,}000)\left(\frac{1}{78.46}\right) = 2{,}549 \text{ lb-mol/h}$$

$$V = (240{,}000)\left(\frac{1}{78.25}\right) = 3{,}067 \text{ lb-mol/h}$$

Then

$$\lambda = (0.412)\left(\frac{3{,}067}{2{,}549}\right) = 0.496$$

and thus

$$H_{OG} = H_G + \lambda H_L = 1.93 + (0.496)(0.845) = 2.35 \text{ ft}$$

Design Procedures for the Absorption of Dilute Gases

Many situations arise where it is necessary to remove diluents such as NH_3 and CO_2 from a carrier gas such as air with a suitable solvent, say water.

Most absorption separations of this type are carried out at pressures near atmospheric where the deviation of the vapor phase from perfect gas behavior is negligible. For such systems, Eqs. (13-23) and (13-29) may be combined to give

$$\frac{d(Vy)}{dZ} + K_G aSP(y^* - y) = 0 \tag{13-75}$$

where the subscript i has been dropped with the understanding that y refers to the solute in the gas stream, and y^* is the mole fraction which the vapor would have if it were in equilibrium with the liquid phase; see Eq. (13-27).

If it is assumed that only one component, the solute, is transferred from the gas phase to the liquid phase, then the derivative in Eq. (13-75) may be stated in terms of dy/dZ. The flow rate V of carrier gas plus solute is related to the total flow rate V' of the carrier gas as follows

$$V = \frac{V'}{1 - y} \tag{13-76}$$

Thus

$$\frac{d(Vy)}{dZ} = V' \frac{d[y/(1 - y)]}{dZ} = \frac{V'}{(1 - y)^2} \frac{dy}{dZ} = \left(\frac{V}{1 - y}\right)\frac{dy}{dZ} \tag{13-77}$$

Elimination of $d(Vy)/dZ$ from Eqs. (13-75) and (13-77) followed by the separation of variables and integration from the top of the column packing at $Z = 0$ to the bottom of the column packing at $Z = Z_T$ (see Fig. 13-1) yields

$$\int_0^{Z_T} \left(\frac{K'_G aPS}{V}\right) dZ = \int_{y_1}^{y_{N+1}} \frac{dy}{(1-y)(y-y^*)} \tag{13-78}$$

where y_1 is equal to the mole fraction of the solute in the gas stream leaving the top of the column while y_{N+1} is equal to the mole fraction of the solute in the gas stream entering the bottom of the column. By use of the mean value of the integrand on the left-hand side of Eq. (13-78) and by replacement of y^* by its equivalent as given by Eq. (13-27), the following result is obtained

$$n'_{OG} = K'_G aP\left(\frac{SZ_T}{V}\right) = \int_{y_1}^{y_{N+1}} \frac{dy}{(1-y)[y-(mx+b)]} \tag{13-79}$$

When m varies significantly over the range of mole fractions x which exist in the column, then the integral in Eq. (13-79) may be evaluated either graphically or numerically. The value of x for any y in the interval y_1 to y_{N+1} is given by the component-material balance

$$\frac{x}{1-x} = \frac{V'}{L'}\left[\frac{y}{1-y} - \frac{y_1}{1-y_1}\right] + \frac{x_0}{1-x_0} \tag{13-80}$$

where L' is the flow rate of the pure solvent.

When both the vapor and liquid phases are very dilute in the solute, the equilibrium sets $\{y^*, x\}$ are located very near the origin, and consequently the tangent line intersects the y axis very near the origin. Thus, Eq. (13-27) reduces to $y^* \cong mx$. Also when the vapor and liquid phases are dilute in the solute, $(1-y) \cong 1$ and $(1-x) \cong 1$, respectively. Under the conditions enumerated plus the assumption that m is constant over the small range of mole fractions involved, Eq. (13-78) may be integrated and rearranged to give

$$n'_{OG} = K'_G aP\left(\frac{SZ_T}{V}\right) = \left(\frac{1}{1-mV/L}\right) \ln \frac{y_{N+1}-mx_N}{y_1-mx_0} \tag{13-81}$$

To obtain the final form of Eq. (13-81), the component-material balance enclosing the entire column, $x_N = (V/L)y_{N+1} + x_0 - (V/L)y_1$, was used.

Example 13-5 An air stream containing two percent NH_3 is to be scrubbed with water in a column which is to be operated isothermally at 85°F and 147 lb/in^2 abs. The water and gas streams are to be passed through the column countercurrently, each at the rate of 240 lb/(h-ft^2). The column is to be packed with 1-inch Raschig rings. On the basis of the following data and correlations deduced from the experimental results by Dwyer and Dodge,[9] compute the height of packing required to remove 99 percent of the NH_3 in the inlet gas stream.

$$y_{NH_3} = 1.26 x_{NH_3}$$

$$\frac{1}{K'_G a} = \frac{1}{0.36 G_V^{0.77} G_L^{0.20}} + \frac{m}{0.356 G_L^{0.78}}$$

where the mass velocities of the vapor and liquid have the units of lb/(h)(ft²), and $K'_G a$ has the units of (lb-mol)/(h)(ft²)(atm).

SOLUTION:

1. *Calculation of y_1 and x_N*

$$\frac{\text{lb-mol of NH}_3 \text{ absorbed}}{(\text{h})(\text{ft}^2)} = (0.99)(V y_{N+1})$$

$$\frac{\text{lb-mol of NH}_3 \text{ in outlet gas}}{(\text{h})(\text{ft}^2)} = V y_{N+1} - 0.99 V y_{N+1}$$

Then

$$y_1 = \frac{V y_{N+1}(1 - 0.99)}{V} = (0.02)(0.01) = 0.0002$$

$$x_N = \frac{\text{lb-mol of NH}_3 \text{ absorbed}}{\text{lb-mol of liquid}} = \frac{(0.99)(V y_{N+1})}{L}$$

$$= (0.99)(0.02)\left(\frac{240/29}{240/18}\right) = 0.01229$$

2. *Calculation of the height of packing*

$$\frac{L}{S}\left(\frac{\text{lb-mol}}{\text{h-ft}^2}\right) = \left[\frac{240 \text{ lb}}{(\text{h})(\text{ft}^2)}\right]\left(\frac{\text{lb-mol}}{18 \text{ lb}}\right) = 13.33 \text{ lb-mol/(h)(ft}^2)$$

$$\frac{V}{S}\left(\frac{\text{lb-mol}}{\text{h-ft}^2}\right) = \left[\frac{240 \text{ lb}}{(\text{h})(\text{ft}^2)}\right]\left(\frac{\text{lb-mol}}{29 \text{ lb}}\right) = 8.28 \text{ lb-mol/(h)(ft}^2)$$

$$\frac{1}{K'_G a} = \frac{1}{(0.36) G_V^{0.77} G_L^{0.20}} + \frac{m}{0.356 G_L^{0.78}}$$

$$\frac{1}{K'_G a} = \frac{1}{(0.36)(240)^{0.77}(240)^{0.20}} + \frac{1.26}{(0.356)(240)^{0.78}} = 0.0629$$

and

$$K'_G a = 15.9 (\text{lb-mol})/(\text{h-ft}^2\text{-atm})$$

Since $y_{N+1} = 0.02$, $V/L = 8.28/13.33 = 0.621$

$$m = 1.26 \qquad y_1 = 0.0002 \qquad x_N = 0.01229$$

$$V/S = 8.28 \qquad K'_G a = 15.9 \qquad \text{and} \qquad P = 1 \text{ atm}$$

the height of packing may be computed by use of Eq. (13-81) as follows

$$Z_T = \left(\frac{1}{K'_G a}\right)\left(\frac{V/S}{1 - mV/L}\right) \ln \frac{y_{N+1} - mx_N}{y_1 - mx_0}$$

$$Z_T = \frac{8.28}{(15.9)[1 - (1.26)(0.621)]} \ln \frac{0.02 - (1.26)(0.01229)}{(0.0002) - (1.26)(0.0)}$$

$$Z_T = 7.46 \text{ ft}$$

Simultaneous Mass and Heat Transfer Between the Vapor and Liquid Phases where the Phases are Composed of Multicomponent Mixtures

Although reliable correlations of mass transfer coefficients for the components of a multicomponent mixture remain to be developed, a development of the equations required to describe simultaneous mass and heat transfer in a packed column follows.

In the following developments, it is supposed that thermal equilibrium exists at the vapor-liquid interface at steady state. That is, the limiting temperature T^V of the vapor as the interface is approached from the vapor side is equal to the limiting temperature $T^{\mathscr{L}}$ of the liquid as the interface is approached from the liquid side of the interface. In the interest of simplicity, the enthalpies of the mixtures are assumed to obey the ideal solution relationships.

Over an element of packing from Z_j to $Z_j + \Delta Z$, the liquid phase, vapor phase, and interface may be visualized as shown in Fig. 13-11. The energy balance enclosing the vapor phase over the element ΔZ_j and over the vapor side of the vapor film [see enclosure (1) of Fig. 13-11] is given by

$$(VH)\Big|_{Z_j+\Delta Z} - (VH)\Big|_{Z_j} + \int_{Z_j}^{Z_j+\Delta Z} \sum_{i=1}^{c} R_i H_i \, dZ - \int_{Z_j}^{Z_j+\Delta Z} q_G \, dZ = 0 \quad (13\text{-}82)$$

where $H = \sum_{i=1}^{c} H_i y_i$, ideal solution

H_i = enthalpy of pure component i at the temperature T^V and pressure at Z

$q_G = h_G a \, (T^V - T^I)$, Btu/(h)(ft of packing)

a = interfacial area for mass transfer, ft^2 of surface per ft of packing

h_G = heat transfer coefficient, Btu/(h)(°F)

T^V = bulk temperature of the vapor phase, °F

T^I = temperature of interface, °F

Application of the *mean value theorem of differential calculus* to the first two terms and the *mean value theorem of integral calculus* to each of the integrals of Eq. (13-82) followed by the limiting process whereby ΔZ is allowed to go to zero yields the following result upon recognition that Z_j and $Z_j + \Delta Z$ were selected arbitrarily in the domain $(0 < Z < Z_T)$ of interest

$$\frac{d(VH)}{dZ} + \sum_{i=1}^{c} R_i H_i - q_G = 0 \qquad (0 < Z < Z_T) \qquad (13\text{-}83)$$

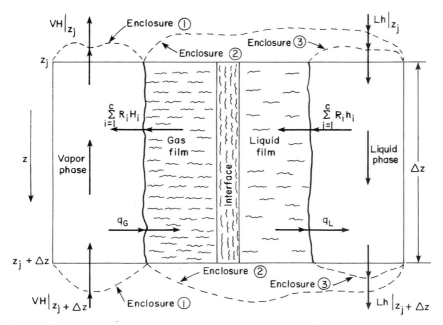

Figure 13-11 Simultaneous mass and heat transfer across an interface. [C. D. Holland, *Fundamentals and Modeling of Separation Processes*, Prentice-Hall, Inc. (1975).]

Similarly, when the energy-balance enclosure for the liquid phase includes the liquid phase, the liquid film, the interface, and the vapor film [see enclosure (2) of Fig. 13-11], one obtains

$$-\frac{d(Lh)}{dZ} - \sum_{i=1}^{c} R_i H_i + q_G = 0 \qquad (0 < Z < Z_T) \qquad (13\text{-}84)$$

Likewise for enclosure (3) of Fig. 13-11, the following result is obtained

$$-\frac{d(Lh)}{dZ} - \sum_{i=1}^{c} R_i h_i + q_L = 0 \qquad (0 < Z < Z_T) \qquad (13\text{-}85)$$

where $q_L = h_L a(T^I - T^L)$, Btu/(h)(ft of packing)

h_L = heat transfer coefficient for the liquid phase, Btu/(h)(°F)

$h = \sum_{i=1}^{c} h_i x_i$

h_i = enthalpy of pure component i in the liquid phase at the temperature T^L and pressure P at Z

Equations (13-83), (13-84), and (13-85) plus the component-material balances given by Eqs. (13-23) and (13-24) constitute the set of equations needed to describe simultaneous mass and heat transfer in a packed column in the service of separating a multicomponent mixture.

The enthalpy balances given by Eqs. (13-83), (13-84), and (13-85) may be restated explicitly in terms of the temperature T^V, T^L, and T^I in the following manner. First observe that

$$\frac{d(VH)}{dZ} = \frac{d\sum_{i=1}^{c} v_i H_i}{dZ} = \sum_{i=1}^{c} v_i \frac{dH_i}{dZ} + \sum_{i=1}^{c} H_i \frac{dv_i}{dZ} \qquad (13\text{-}86)$$

If the effect of pressure and composition on enthalpy is neglected

$$\frac{dH_i}{dZ} = \frac{dH_i}{dT^V}\frac{dT^V}{dZ} = C^V_{pi}\frac{dT^V}{dZ} \qquad (13\text{-}87)$$

Multiplication of each member of Eq. (13-23) by H_i followed by the summation over all components yields

$$\sum_{i=1}^{c} H_i \frac{dv_i}{dZ} = -\sum_{i=1}^{c} R_i H_i \qquad (13\text{-}88)$$

The relationships given by Eqs. (13-86), (13-87), and (13-88) may be used to reduce Eq. (13-83) to the following form

$$\left(\sum_{i=1}^{c} v_i C^V_{pi}\right)\frac{dT^V}{dZ} - q_G = 0 \qquad (13\text{-}89)$$

Similarly, Eq. (13-84) may be reduced to

$$-\left(\sum_{i=1}^{c} l_i C^L_{pi}\right)\frac{dT^L}{dZ} - \sum_{i=1}^{c} R_i(H_i - h_i) + q_G = 0 \qquad (13\text{-}90)$$

and Eq. (13-85) reduces to

$$-\left(\sum_{i=1}^{c} l_i C^L_{pi}\right)\frac{dT^L}{dZ} + q_L = 0 \qquad (13\text{-}91)$$

The above equations have been used as summarized by Holland[22] in the analysis of the experimental results obtained for a packed absorber.

NOTATION

a interfacial area for mass transfer in ft^2 of surface per ft^3 of empty column; interfacial area for heat transfer in ft^2 of surface per ft of packing (although these areas are not necessarily equal, the same symbol is used for each in the interest of simplicity)

b intercept of the tangent line, $y^* = mx + b$

C_A concentration of component A, lb-mol/unit vol

C_{fL} a factor in Eq. (13-73)

C_{pi} molar heat capacity of component i, Btu/(lb-mol)(°F)

d column diameter, ft

D	diffusivity, ft^2/s; D_{AB} = diffusion coefficient of A relative to B
E_D	eddy diffusivity, ft^2/s; see Eq. (13-4)
E_i	vaporization point efficiency, defined by Eq. (13-36)
E_{ji}	vaporization plate efficiency defined by Eq. (13-38)
E_{ji}^M	modified Murphree plate efficiency; defined by Eq. (13-48)
E_{Mji}	Murphree efficiency; defined by Eq. (13-59)
E_{Pi}^M	modified Murphree point efficiency; defined by Eq. (13-55)
E_{MPi}	Murphree point efficiency; defined by Eq. (13-63)
$\hat{f}_i$	fugacity of component i in a mixture; evaluated at the temperature, pressure, and composition of the mixture, atm
f_i	fugacity of a pure component; evaluated at the temperature and pressure of the system, atm
f_μ, f_ρ, f_σ	factors appearing in Eq. (13-74)
F_p	a factor appearing in the Eckert correlation for the flooding velocity [Eq. (13-70)]; see Table 13-7
F_r	flood ratio, at a fixed ratio of liquid to vapor mass flow rates
G	mass velocity, $lb/(s)ft^2$ or $lb(h)(ft^2)$
G_V	mass velocity of the vapor
G_{Vs}	mass velocity of the vapor based on the superfacial area
G_L	mass velocity of the liquid
h_G	heat transfer coefficient for the vapor phase; defined below Eq. (13-82), $Btu/(h)(°F)$
h_L	heat transfer coefficient for the liquid phase; defined below Eq. (13-85), $Btu/(h)(°F)$
H_G	height of a gas phase transfer unit; also denoted by $(HTU)_G$
H_L	height of a liquid phase transfer unit; also denoted by $(HTU)_L$
H_{OG}	height of an overall transfer unit; defined by Eq. (13-72)
$(HTU)_G$	height of a gas phase transfer unit; defined by Eq. (13-33)
k_G, k'_G	mass transfer coefficients for the gas phase; defined by Eqs. (13-7) and (13-15), respectively, $lb\text{-}mol/(h)(ft^2)(atm)$
k_L, k'_L	mass transfer coefficients for the liquid phase; defined by Eqs. (13-5) and (13-15), respectively, k_L has the same units as k_G and k'_L has the units of $lb\text{-}mol/(h)(ft)$
k_I	mass transfer coefficient for the interface; defined by Eq. (13-6), $lb\text{-}mol/(h)(ft^2)(atm)$
K_G	overall mass transfer coefficient; defined by Eq. (13-13), $lb\text{-}mol/(h)(ft^2)(atm)$
K'_G	overall mass transfer coefficient; see Table 13-1
K_L, K'_L	overall mass transfer coefficients; see Table 13-1
K_i	ideal solution K value; defined below Eq. (13-18)
K_{Mji}	ideal solution K value; evaluated at the temperature given by the Murphree temperature convention; see Eq. (13-43)
l_i	flow rate of component i, $lb\text{-}mol/h$
L	flow rate of liquid, $lb\text{-}mol/h$
L'	flow rate of the solute-free liquid, $lb\text{-}mol/h$

m	slope of the tangent line, $y^* = mx + b$
N_i	rate of mass transfer, lb-mol/(h)(ft^2)
n_G, n_G'	number of gas phase transfer units; see Table 13-1
n_L, n_L'	number of liquid phase transfer units; see Table 13-1
n_{OG}	number of overall transfer units; see Table 13-1
n_{OGM}'	number of overall transfer units; defined below Eq. (13-59)
n_i	point value of the number of overall transfer units, defined by Eq. (13-56)
p_A	partial pressure of component A, atm
P	total pressure, atm
q	rate of heat transfer for the gas phase, Btu/(h)(ft of packing)
q_L	rate of heat transfer to the liquid phase; Btu/(h)(ft of packing)
Q	flow rate of clear liquid, gal/min
R_i	rate of mass transfer, lb-mol/(h)(ft of packing)
S	cross-sectional area of column, ft
Sc	Schmidt number
t	average contact time between a gas bubble and a liquid surface; see Eq. (13-3)
u_V	vapor velocity, ft/s
u_{Vs}	vapor velocity based on the superficial area, ft/s
v_i	flow rate of component in the vapor, lb-mol/h
V	total flow rate of vapor, lb-mol/h
V'	flow rate of the solute free vapor, lb-mol/h
w_L	mass flow rate of liquid, lb/h
w_V	mass flow rate of vapor, lb/h
W	length of flow path; measured from the outlet of the downcomer, ft
y_i	mole fraction of component i in the vapor
Z_T	total depth of packing, ft

Greek Symbols

γ_i	activity coefficient; see Chap. 14
λ	mV/L
μ	viscosity
ρ	density
σ	surface tension
ϕ	packing parameter; see Fig. 13-8
ψ	packing parameter, see Fig. 13-10

Subscripts

G	gas (or vapor) phase
s	based on superficial area
L	liquid
V	vapor
f	flood
w	water

Superscripts

I	interface
L	liquid
$\mathscr{L}$	liquid side of interface
V	vapor
$\mathscr{V}$	vapor side of interface

PROBLEMS

13-1 From the appropriate relationships given in the text, obtain the expressions given in item I of Table 13-1.

13-2 Begin with the defining equations for the number of mass transfer units and the assumptions enumerated in the text and obtain the result given by Eq. (13-34), namely,

$$n_{OGi} = n'_{OGi}$$

13-3 Use the relationships in the text to obtain the expression given in item II of Table 13-1 for $K'_G a_i$.

13-4 Obtain the relationship given by Eq. (13-35) for $n'_{OG\,Mi}$.

13-5 (a) By use of Eqs. (13-15) and (13-27), show that

$$\frac{1}{K'_{LM}a_i} = \frac{1}{k'_L a_i} + \left(\frac{\rho_L}{m_i P}\right)\frac{1}{k'_G a_i}$$

(b) Use the result of part (a) to develop the following expression

$$\frac{1}{n'_{OLMi}} = \frac{1}{n'_{Li}} + \left(\frac{L}{m_i V}\right)\frac{1}{n'_{Gi}}$$

where

$$n'_{OLMi} = K'_{LM}a_i \rho^L \left(\frac{SZ_T}{L}\right)$$

13-6 Obtain the expression given by Eq. (13-67) for the line connecting the points $(x_{j1}, y_{j+1, 1})$ and $(x^{\mathscr{L}}_{j1}, y^{\mathscr{V}}_{j1})$ of Fig. 13-3.

13-7 Begin with first principles and obtain the expressions given by Eqs. (13-68) and (13-69) for m_1.

13-8 Carry out the developments given by Eqs. (13-82) through (13-91) for the case where both the vapor and liquid phases form nonideal solutions.

13-9 Begin with Eqs. (13-39), (13-59), and (13-61) and show that

$$E_{ji} = \frac{y_{j+1, i}}{K_{ji} x_{ji}} + E_{Mji}\left[\frac{K_{Mji}}{K_{ji}} - \frac{y_{j+1, i}}{K_{ji} x_{ji}}\right]$$

and that

$$E_{ji} = \frac{K_{Mji}}{K_{ji}} - e^{-n'_{OG\,Mji}}\left[\frac{K_{Mji}}{K_{ji}} - \frac{y_{j+1, i}}{K_{ji} x_{ji}}\right]$$

REFERENCES

1. R. B. Bird, J. O. Hirschfelder, and C. F. Curtiss: "Theoretical Calculation of the Equation of State and Transport Properties of Gases and Liquids," *Trans. ASME*, **76**: 1011 (1954).
2. R. B. Bird, W. E. Stewart, and F. N. Lightfoot: *Transport Phenomena*, John Wiley and Sons, New York, 1960.
3. W. L. Bolles and J. R. Fair: "Performance and Design of Packed Distillation Columns," presented at the Third International Distillation Symposium, London, April, 1979, No. 213 of E.F.C., Sec. 3.3, p. 35, Institute of Chemical Engineers (London).
4. *Bubble-Tray Design Manual*, American Institute of Chemical Engineers, 345 East 47 Street, New York, 1958.
5. J. S. Carey: "Plate Efficiencies in Bubble Cap Rectifying Columns," Sc.D. Thesis, Massachusetts Institute of Technology, 1930.
6. *Chemical Engineers' Handbook*, 3d ed., J. H. Perry (ed.), McGraw-Hill Book Company, p. 587, New York, 1950.
7. T. H. Chilton and A. P. Colburn: "Distillation and Absorption in Packed Columns—A Convenient Design and Correlation Method," *Ind. Eng. Chem.*, **27**: 255 (1935).
8. P. V. Danckwerts: "Significance of Liquid-Film Coefficients in Gas Absorption," *Ind. Eng. Chem.*, **43**(6): 1960.
9. O. E. Dwyer and B. F. Dodge: "Rate of Absorption of Ammonia by Water in a Packed Tower," *Ind. Eng. Chem.*, **33**(4): 485 (1941).
10. J. S. Eckert: "A New Look at Distillation—4 Tower Packings ... Comparative Performance," *Chem. Eng. Prog.* **59**(5): 76 (1963).
11. ———: "Selecting the Proper Distillation Column Packing," *Chem. Eng. Prog.* **63**(3): 39 (1970).
12. W. C. Edmister: "Hydrocarbon Adsorption and Fractionation Process Design Methods—Part 18, Plate Efficiency," *Petroleum Engineer*, **21**: C-45 (1949).
13. R. E. Emmett and R. L. Pigford: "A Study of Gas Absorption in Falling Liquid Films," *Chem. Eng. Prog.* **50**: 87 (1954).
14. E. N. Fuller, D. D. Schettler, and J. C. Giddings: "A New Method for Prediction of Binary Gas—Phase Diffusion Coefficients," *Ind. Eng. Chem.*, **58**: 19 (1966).
15. J. A. Gerster, A. B. Hill, M. N. Hochgraf, and D. N. Robinson, *Tray Efficiencies in Distillation Columns—Final Report from the University of Delaware to the AIChE Committee*, 345 East 47 Street, New York, American Institute of Chemical Engineers, 1958.
16. ———, ———, ———, and ———: *Tray Efficiencies in Distillation Columns—Final Report from the University of Delaware to the AIChE Committee*, 345 East 47 Street, New York, American Institute of Chemical Engineers, 1958.
17. E. R. Gilliland and T. K. Sherwood: "Diffusion of Vapors into Air Streams," *Ind. Eng. Chem.* **26**(5): 516 (1934).
18. J. P. Graham, J. W. Fulton, M. S. Kuk, and C. D. Holland: "Predictive Methods for the Determination of Vaporization Efficiencies," *Chem. Sci.*, **28**: 473 (1973).
19. R. Higbie: "The Rate of Absorption of a Pure Gas Into a Still Liquid During Short Periods of Exposure," *Trans. AIChE*, **31**: 365 (1935).
20. J. O. Hirschfelder, R. B. Bird, and E. L. Spotz: "The Transport Properties of Gases and Gaseous Mixtures II," *Chem. Rev.*, **44**: 205 (1949).
21. C. D. Holland and N. E. Welch: "Steam Batch Distillation Calculation," *Petroleum Refiner*, **36**(5): 251 (1957).
22. ———: *Fundamentals and Modeling of Separation Processes, Absorption, Distillation, Evaporation, and Extraction*, Prentice-Hall, Inc., Englewood Cliffs, N.J., 1975.
23. ———, A. E. Hutton, and G. P. Pendon, "Prediction of Vaporization Efficiencies for Multicomponent Mixtures by Use of Existing Correlations for Vapor and Liquid Film Coefficients," *Chem. Eng. Sci.*, **26**: 1729 (1971).
24. ——— and K. S. McMahon: "Comparison of Vaporization Efficiencies with Murphree-type Efficiencies in Distillation—I," *Chem. Eng. Sci.*, **25**: 431 (1970).

25. M. Leva: "Flow through Irrigated Dumped Packings, Pressure Drop, Loading Flooding," *Chem. Eng. Prog. Symp. Ser.*, **50**(10): 51 (1954).
26. W. K. Lewis and W. G. Whitman: "Principles of Gas Absorption," *Ind. Eng. Chem.*, **16**: 1215 (1924).
27. W. E. Lobo, L. Friend, F. Haohmall, and F. Zenz, "Limiting Capacity of Dumped Tower Packing," *Trans. AIChE*, **41**: 693 (1945).
28. W. Nernst: "Theorie der Reaktions ges chwindigkeit in heterogenen Systemem," *Z. Phys. Chem.*, **47**: 52 (1904).
29. T. K. Sherwood, R. L. Pigford, and C. R. Wilke: *Mass Transfer*, McGraw-Hill Book Company, New York, 1975.
30. ———, G. H. Shipley, and F. A. L. Holloway: "Flooding Velocities in Packed Columns," *Trans. AIChE*, **30**: 765 (1938).
31. B. D. Smith: *Design of Equilibrium Stage Processes*, McGraw-Hill Book Company, New York, 1963.
32. Y. P. Tang: "Diffusion of Carbon Dioxide in Liquids—Concentration Dependence of Diffusion Coefficient and Diffusion in Mixed Solvents," Ph.D. Dissertation, University of Texas, 1963.
33. D. L. Taylor, Parke Davis, and C. D. Holland: "Determination of Plate Efficiencies from Operation Data," *AIChE J.*, **10**: 864 (1964).
34. L. H. Tung and H. G. Drickamer: "Diffusion Through an Interface—Binary System," *J. Chem. Phys.*, **20**: 6 (1952).
35. W. H. Walker, W. K. Lewis, W. H. McAdams, and E. R. Gilliland: *Principles of Chemical Engineering*, 3d ed., McGraw-Hill Book Company, New York, 1937.
36. J. E. Vivian and C. J. King: "The Mechanism of Liquid-Phase Resistance to Gas Absorption in a Packed Column," *AIChE J.*, **10**: 221 (1964).
37. W. G. Whitman: "The Two-Film Theory of Gas Absorption," *Chem. Met. Eng.*, **29**: 146 (1923).
38. C. R. Wilke: "Diffusional Properties of Multicomponent Gases," *Chem. Eng. Prog.*, **46**: 95 (1950).
39. ——— and D. Chang: "Correlation of Diffusion Coefficients in Dilute Solutions," *AIChE J.*, **1**: 264 (1955).
40. ———: "A Viscosity Equation for Gas Mixtures," *J. Chem. Phys.*, **18**: 517 (1950).

FOURTEEN

THERMODYNAMIC RELATIONSHIPS
FOR MULTICOMPONENT MIXTURES†

To solve distillation problems involving multicomponent mixtures, vapor-liquid equilibrium data and enthalpy data are needed. The methods used to obtain these data may be classified as follows: (1) the use of a single equation of state and (2) the use of multiple equations of state and/or correlations for the prediction of the liquid and vapor parts of the K values and the enthalpies. This classification was suggested by Adler et al.[2] in an excellent paper on the industrial uses of equations of state. Although the first approach, the use of a single equation of state, is the more desirable, many industrial problems are encountered in which this approach is too inaccurate and the second approach is used.

In Sec. 14-1 the fundamental thermodynamic relationships for multicomponent mixtures are presented. In Sec. 14-2, the use of the thermodynamic relationships developed in Sec. 14-1 and an equation of state to predict the K values and enthalpies is demonstrated for several equations of state. The equations of state which appear to be the most popular choice of industry are treated in detail.

Sections 14-3 through 14-5 are devoted to the use of several equations of state and/or correlations used in the calculation of the K values and enthalpies which are needed in a distillation calculation. A variety of methods are presented for the calculation of ideal solution K values, and several methods for computing activity coefficients such as the Wilson equation and the UNIFAC method are presented.

† This chapter was written in collaboration with Dr. N. J. Tetlow of Dow Chemical Company.

14-1 FUNDAMENTAL PRINCIPLES OF THERMODYNAMICS NEEDED IN THE CALCULATION OF VAPOR-LIQUID EQUILIBRIA AND ENTHALPIES OF MULTICOMPONENT MIXTURES

A knowledge of the thermodynamics of pure components is assumed. In view of this assumption, the general development moves rapidly from the equations for pure components to the corresponding equations for multicomponent mixtures.

The First and Second Laws of Thermodynamics

The first law of thermodynamics (the energy of the universe is constant) implies that the energy function U is independent of path. As a consequence of the second law (heat does not of itself flow from a body of lower temperature to one of higher temperature), the entropy function S can be shown to be independent of path. For a pure component or a mixture at constant composition, both the first and second laws are contained in the expression

$$dU = T\,dS - P\,dV \tag{14-1}$$

The statement that the functions U and S are independent of path has the physical significance that the functions U and S depend on the state of a system as described by its temperature, pressure, and composition, and they are independent of all previous states of the system. That is, such functions are independent of the history of the system, and they are known by a variety of names such as "potential functions," "exact differentials," and "point functions."

The function W (work) and the function Q (heat) which appear in one statement of the first law of thermodynamics ($\Delta U = Q - W$) are "path functions." These functions depend upon the path followed in going from one state to another.

From the standpoint of thermodynamics, the most significant mathematical properties of state functions are summarized in the following statement. The necessary and sufficient condition that the line integral

$$I = \int_{C} M(x, y)\,dx + N(x, y)\,dy \tag{14-2}$$

be independent of the path described by curve C is that

$$\frac{\partial M}{\partial y} = \frac{\partial N}{\partial x} \tag{14-3}$$

This statement means two things. First, if the line integral I is independent of path, then $\partial M/\partial y = \partial N/\partial x$. Second, if $\partial M/\partial y = \partial N/\partial x$, then the line integral I is independent of path. The equality given by Eq. (14-3) is commonly called the Maxwell relationship.

Since the internal energy U is independent of path, it follows from Eqs. (14-1) through (14-3) that

$$\left(\frac{\partial T}{\partial V}\right)_S = -\left(\frac{\partial P}{\partial S}\right)_V \tag{14-4}$$

This is one of the well-known "Maxwell relationships." Another useful property of functions whose line integrals are independent of path is the following one. If the line integral I is independent of path, a function $f(x, y)$ exists such that

$$\frac{\partial f(x, y)}{\partial x} = M(x, y) \quad \text{and} \quad \frac{\partial f(x, y)}{\partial y} = N(x, y) \tag{14-5}$$

Equation (14-1) implies that U is a function of the independent variables S and V, that is

$$U = U(S, V)$$

Thus

$$dU = \left(\frac{\partial U}{\partial S}\right)_V dS + \left(\frac{\partial U}{\partial V}\right)_S dV \tag{14-6}$$

Comparison of Eqs. (14-1) and (14-6) shows that

$$\left(\frac{\partial U}{\partial S}\right)_V = T \quad \text{and} \quad \left(\frac{\partial U}{\partial V}\right)_S = -P$$

In addition to the functions U and S, other thermodynamic functions in common use are as follows: the enthalpy H, the work function A, and the free energy function G. These are also state functions, and they are defined as follows

$$H = U + PV \tag{14-7}$$

$$A = U - TS \tag{14-8}$$

$$G = H - TS \tag{14-9}$$

Combination of these definitions with Eq. (14-1) gives the following set of equations for a pure component where the total number of moles is held fixed (or a mixture in which the moles of each component is held fixed)

$$dH = T\, dS + V\, dP \tag{14-10}$$

$$dA = -S\, dT - P\, dV \tag{14-11}$$

$$dG = -S\, dT + V\, dP \tag{14-12}$$

Each of these is an equivalent statement of the first and second laws of thermodynamics for a pure component or a mixture in which the number of moles of each component is held fixed. These equations are also said to describe closed systems in that there is no exchange of mass between the system and the surroundings.

Consider now the free energy function G. As indicated by Eq. (14-12), G is a function of P and T. Thus

$$\left(\frac{\partial G}{\partial T}\right)_P = -S \quad \text{and} \quad \left(\frac{\partial G}{\partial P}\right)_T = V \tag{14-13}$$

and

$$\left(\frac{\partial S}{\partial P}\right)_T = -\left(\frac{\partial V}{\partial T}\right)_P \tag{14-14}$$

For a multicomponent mixture in which the composition is free to vary, the total free energy is a function not only of P and T but also the moles n_i of each component present in the mixture, that is,

$$G = G(P, T, n_1, n_2, \ldots, n_c)$$

and

$$dG = \left(\frac{\partial G}{\partial T}\right)_{P, n_i} dT + \left(\frac{\partial G}{\partial P}\right)_{T, n_i} dP + \sum_{i=1}^{c} \left(\frac{\partial G}{\partial n_i}\right)_{P, T, n_{j \neq i}} dn_i \tag{14-15}$$

In the case of the first two partial derivatives, the subscript n_i means that the moles of all components ($i = 1, 2, \ldots, c$) are held fixed. The subscript $n_{j \neq i}$ which appears on the derivatives included in the summation means that the moles of all components except i are held fixed. The partial derivative $(\partial G/\partial n_i)_{P, T, n_{j \neq i}}$ is called the partial molar free energy, and denoted by $\bar{G}_i$. The partial molar free energy $\bar{G}_i$ was introduced by Gibbs who called it μ_i, the chemical potential. Thus

$$\left(\frac{\partial G}{\partial n_i}\right)_{T, P, n_{j \neq i}} = \bar{G}_i = \mu_i = \mu_i(P, T, n_1, n_2, \ldots, n_c) \tag{14-16}$$

Since the moles of each component of the mixture are to be held fixed in computing the first two partial derivatives of Eq. (14-15), it follows from Eq. (14-12) that

$$\left(\frac{\partial G}{\partial T}\right)_{P, n_i} = -S \quad \text{and} \quad \left(\frac{\partial G}{\partial P}\right)_{T, n_i} = V \tag{14-17}$$

Thus, Eq. (14-15) may be restated as follows

$$dG = -S\, dT + V\, dP + \sum_{i=1}^{c} \mu_i\, dn_i \tag{14-18}$$

which is a statement of the first and second laws of thermodynamics for multicomponent mixtures. When Eq. (14-18) is combined with the expressions obtained by taking the total differentials of Eqs. (14-7) through (14-9), the following equivalent forms for the first and second laws are obtained

$$dU = T\,dS - P\,dV + \sum_{i=1}^{c} \mu_i\,dn_i \tag{14-19}$$

$$dH = T\,dS + V\,dP + \sum_{i=1}^{c} \mu_i\,dn_i \tag{14-20}$$

$$dA = -S\,dT - P\,dV + \sum_{i=1}^{c} \mu_i\,dn_i \tag{14-21}$$

Since $U = U(S, V, n_1, n_2, \ldots, n_c)$, $H = H(S, P, n_1, n_2, \ldots, n_c)$, and $A = A(T, V, n_1, n_2, \ldots, n_c)$, these functions may be expanded in a form analogous to Eq. (14-15) and the coefficients of dS, dP, dT, and dV determined in the same manner as shown for Eq. (14-18). Comparison of the expressions so obtained with Eqs. (14-19) through (14-21) shows that

$$\mu_i = \left(\frac{\partial U}{\partial n_i}\right)_{S, V, n_j \neq i} \qquad \mu_i = \left(\frac{\partial H}{\partial n_i}\right)_{S, P, n_j \neq i} \qquad \mu_i = \left(\frac{\partial A}{\partial n_i}\right)_{V, T, n_j \neq i} \tag{14-22}$$

IDEAL AND NONIDEAL SOLUTIONS

Several equivalent definitions for ideal solutions exist; that is, any one of these may be taken as the definition and each of the others will follow as a consequence. Thus, the choice of a particular definition is largely a matter of personal preference. The definition taken here was selected because it describes a physical phenomenon which is easily visualized.

An ideal solution is defined simply as one which obeys Amagat's law of additive volumes. That is, the volume of an ideal solution at a given temperature and pressure is equal to the sum of the volumes of its pure constituents at the same temperature and pressure

$$V = n_1 v_1 + n_2 v_2 + \cdots + n_c v_c \tag{14-23}$$

where n_i = moles of component i; the total moles of solution is denoted by n;
$$n = \sum_{i=1}^{c} n_i = \sum_i n_i$$
v_i = volume of one mole of pure component i at the temperature T and pressure P of the solution
V = volume of n moles of solution at P and T

By this definition, volume is conserved when components that form an ideal solution are mixed. For the above definition to be of significant practical use, it is necessary that Eq. (14-23) hold over a reasonable range of temperatures, pressures, and compositions.

Nonideal solutions may be represented by a formula which is symmetrical to Eq. (14-23). This formula is obtained by use of Euler's theorem (Appendix A, Theorem A-6) for homogeneous functions. These extensive thermodynamic functions V, U, S, H, A, and G are all homogeneous functions of degree 1. This degree is assigned to that class of functions having the common property that

$$V(\lambda n_1, \lambda n_2, \ldots, \lambda n_c) = \lambda V(n_1, n_2, \ldots, n_c) \tag{14-24}$$

where λ is any real number. For example, if the moles of each component in a mixture are doubled (at constant temperature and pressure), the volume of the resulting mixture is twice the initial volume. For such functions Euler's formula[60] is

$$V = n_1 \bar{V}_1 + n_2 \bar{V}_2 + \cdots + n_c \bar{V}_c \tag{14-25}$$

where $\bar{V}_i = (\partial V/\partial n_i)_{P, T, n_j \neq i} = $ partial molar volume of component i; evaluated at the pressure, temperature, and composition of the mixture. Note that, $\bar{V}_i$ is not homogeneous in the intensive variables P and T, but it is a function of these variables.

THE FUGACITY

In the analysis of thermodynamic systems, it has been found that a new variable called the fugacity is more convenient to use than is the free energy. The fugacity of pure component i at the temperature T and pressure P composed of pure component i is defined by

$$d\mu = RTd \ln f \quad \text{(at constant } T) \tag{14-26}$$

$$\lim_{P \to 0} \frac{f}{P} = 1 \tag{14-27}$$

where the chemical potential μ is equal to the free energy per mole of component i, that is,

$$\mu = \mu(P, T) = \left(\frac{\partial G}{\partial n}\right)_{P, T} = \left[\frac{\partial(ng)}{\partial n}\right]_{P, T} = g = g(P, T) \tag{14-28}$$

The second condition, Eq. (14–27), is equivalent to supposition fact that all actual gases and substances approach a perfect gas in the limit as the pressure goes to zero.

The fugacity of component i in a mixture of c components at the temperature T and pressure P is defined by

$$d\mu_i = RTd \ln \hat{f}_i \quad \text{(at constant } T) \tag{14-29}$$

$$\lim_{P \to 0} \frac{\hat{f}_i}{p_i} = 1 \tag{14-30}$$

where $\hat{f}_i = \hat{f}_i(P, T, n_1, n_2, \ldots, n_c) = $ fugacity of component i in the mixture
$\mu_i = \mu_i(P, T, n_1, n_2, \ldots, n_c) = (\partial G/\partial n_i)_{P, T, n_j \neq i} = \bar{G}_i$
$p_i = $ partial pressure of component i; $p_i = P y_i$

PHYSICAL EQUILIBRIUM

The necessary conditions for any mixture (ideal or nonideal) of components in each of two phases to be in equilibrium may be taken as follows

$$T^V = T^L$$

$$P^V = P^L \tag{14-31}$$

$$\mu_i^V = \mu_i^L \qquad (i = 1, 2, \ldots, c)$$

Beginning with equations of this type, the Gibbs phase rule may be deduced

$$\mathscr{P} + \Phi = c + 2 \tag{14-32}$$

where $\mathscr{P}$ = number of phases
Φ = number of degrees of freedom
c = number of components

As a consequence of the last expression of Eq. (14-31), it can be shown that

$$\hat{f}_i^V = \hat{f}_i^L \tag{14-33}$$

Since the fugacity $\hat{f}_i^V$ is a function of the temperature T, pressure P, and composition $\{y_i\}$ of the vapor phase, it follows that $\hat{f}_i^V$ may be restated in the following functional form

$$\hat{f}_i^V = \gamma_i^V f_i^V y_i \tag{14-34}$$

where $\gamma_i^V = \gamma_i^V(P, T, \{y_i\})$, the activity coefficient of pure component i in the vapor phase
f_i^V = the fugacity of pure component i in the vapor state at T and P of the vapor mixture

Similarly, for component i in the liquid phase

$$\hat{f}_i^L = \gamma_i^L f_i^L x_i \tag{14-35}$$

where $\gamma_i^L = \gamma_i^L(P, T, \{x_i\})$, the activity coefficient of component i in the liquid phase
f_i^L = the fugacity of pure component i in the liquid state at T and P of the liquid mixture

Then, Eq. (14-33) may be restated as follows

$$y_i = \left(\frac{\gamma_i^L f_i^L}{\gamma_i^V f_i^V} \right) x_i \tag{14-36}$$

The vapor-liquid distribution coefficient, $K_i = y_i/x_i$, is commonly called the K value of component i. Thus

$$K_i = \frac{\gamma_i^L f_i^L}{\gamma_i^V f_i^V} \tag{14-37}$$

Instead of showing the activity for component i in the vapor phase explicitly, it has become customary to restate Eq. (14-37) in the following form

$$K_i = \frac{\gamma_i^L f_i^L}{(\gamma_i^V f_i^V/P)P} = \frac{\gamma_i^L f_i^L}{\hat{\phi}_i^V P} = \frac{\gamma_i^L \phi_i^L}{\hat{\phi}_i^V} \tag{14-38}$$

where $\phi_i^L = f_i^L/P$†
$\phi_i^V = f_i^V/P$
$\hat{\phi}_i^V = \gamma_i^V f_i^V/P = \hat{f}_i^V/(Py_i)$

(Recognition of the fact that $\hat{\phi}_i^V = \hat{f}_i^V/Py_i$ is useful in the application of the equations of state for the calculation of $\hat{\phi}_i^V$.)

In the development of the calculational procedures presented in other chapters of this text, the symmetrical form of Eq. (14-36) is used and the ratio f_i^L/f_i^V is denoted by K_i and called the *ideal solution K value*. In the present treatment (and only in this chapter), the ideal solution K value is denoted by K_i^I. Equation (14-36) may be restated in terms of the ideal solution K value as follows

$$y_i = K_i x_i = \frac{\gamma_i^L}{\gamma_i^V} K_i^I x_i = \left(\frac{\gamma_i^L}{\hat{\phi}_i^V/\phi_i^V}\right) K_i^I x_i \tag{14-39}$$

This form of the equilibrium relationship is used because the behavior of a large number of mixtures can be approximated by use of ideal solutions. For the case where both the vapor and liquid phase form ideal solutions, it is shown in the next section that $\gamma_i^L = 1$ and $\gamma_i^V = 1$. Consequently, for ideal solutions, Eq. (14-39) reduces to

$$y_i = K_i x_i = K_i^I x_i \tag{14-40}$$

Useful Thermodynamic Relationships

In order to develop the relationship between the chemical potential $\mu_i^V(P, T, \{y_i\})$ of component i in a mixture and the chemical potential $\mu_i^V(P, T)$ of pure component i at the same pressure P and temperature T as the mixture, the change in chemical potential over the path shown in Fig. 14-1 (with $P_0 = P$) is computed.

For step 1

$$\mu_i^V(P_2, T) - \mu_i^V(P, T) = RT \int_{f_{i, P_0}^V}^{f_{i, P_2}^V} \frac{df_i^V}{f_i^V} = RT \ln \frac{f_{i, P_2}^V}{f_i^V} \tag{14-41}$$

For step 2

$$\mu_i^V(P, T, \{y_i\}) - \mu_i^V(P_2, T) = RT \int_{f_{i, P_2}^V}^{\hat{f}_i^V} \frac{df_i^V}{f_i^V} = RT \ln \frac{\hat{f}_i^V}{f_{i, P_2}^V} = 0 \tag{14-42}$$

† Numerous authors (Refs. 12, 25, 51) denote the liquid fugacity coefficient by v_i rather than ϕ_i^L.

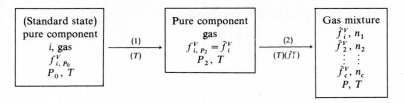

Note: P_2 is that pressure at which f_i^V of the pure component is equal to its fugacity $\hat{f}_i^V$ in the mixture.

Figure 14-1 Path for computing the difference in free energy of a component in a mixture and in its standard state at the same temperature.

Addition of these two equations gives the desired result

$$\bar{G}_i^V = g_i^V + RT \ln \frac{\hat{f}_i^V}{f_i^V} \qquad (14\text{-}43)$$

where $\bar{G}_i^V = \mu_i^V(P, T, \{y_i\}) =$ partial molar free energy or chemical potential of component i evaluated at P, T, and $\{y_i\}$ of the vapor phase
$g_i^V = \mu_i^V(P, T) =$ free energy or chemical potential per mole of pure component i in the vapor phase at P and T

A similar development may be used for the liquid phase to obtain

$$\bar{G}_i^L = g_i^L + RT \ln \frac{\hat{f}_i^L}{f_i^L} \qquad (14\text{-}44)$$

where $\bar{G}_i^L = \mu_i^L(P, T, \{x_i\}) =$ partial molar free energy or chemical potential of component i evaluated at P, T, and $\{x_i\}$ of the liquid phase
$g_i^L = \mu_i^L(P, T) =$ free energy or chemical potential per mole of pure component i in the liquid phase at P and T

When $\hat{f}_i^V$ and $\hat{f}_i^L$ in Eqs. (14-43) and (14-44) are replaced by their equivalents as given by Eqs. (14-34) and (14-35) respectively, one obtains

$$\bar{G}_i^V = g_i^V + RT \ln \gamma_i^V y_i \qquad (14\text{-}45)$$

and

$$\bar{G}_i^L = g_i^L + RT \ln \gamma_i^L x_i \qquad (14\text{-}46)$$

THE LEWIS AND RANDALL RULE

From Eqs. (14-3) and (14-18), it follows that

$$\left[\frac{\partial \mu_i^V(P, T, \{y_i\})}{\partial P} \right]_{P, n_i} = \left(\frac{\partial V}{\partial n_i} \right)_{P, T, n_j \neq i} = \bar{V}_i^V$$

Similarly, for a pure component, Eq. (14-12) gives

$$\left(\frac{\partial G}{\partial P}\right)_T = \left[\frac{\partial(ng_i^V)}{\partial P}\right]_T = n\left(\frac{\partial g_i^V}{\partial P}\right)_T = nv_i^V \tag{14-47}$$

or

$$\left(\frac{\partial g_i^V}{\partial P}\right)_T = v_i^V \tag{14-48}$$

Thus, partial differentiation of Eq. (14-43) with respect to P gives

$$\bar{V}^V - v_i^V = RT\left(\frac{\partial \ln \hat{f}_i^V/f_i^V}{\partial P}\right)_{T,\, n_i} \tag{14-49}$$

If, over the interval from 0 to P, the partial molar volume $\bar{V}_i^V$ for component i is equal to the molar volume v_i^V for pure component i (that is $\bar{V}_i^V = v_i^V$), then the Lewis and Randall rule holds for component i over the pressure interval 0 to P. (Note that for the ideal solution behavior to exist over the interval 0 to P, it is necessary that the Lewis-Randall rule hold for each component of the mixture over the interval 0 to P.) Thus, if $\bar{V}_i^V = v_i^V$, then Eq. (14-49) reduces to

$$\left(\frac{\partial \ln \hat{f}_i^V/f_i^V}{\partial P}\right)_{T,\, n_i} = 0 \tag{14-50}$$

This equation shows that when the temperature and composition of a mixture are held fixed, the ratio $\hat{f}_i^V/f_i^V$ is independent of pressure. At any given temperature, the general solution of Eq. (14-50) is

$$\hat{f}_i^V/f_i^V = \eta_i(y_1, y_2, \ldots, y_c) \tag{14-51}$$

A more precise expression for η_i is given by recalling that the limit of $\hat{f}_i^V/f_i^V$ as P goes to zero is y_i, that is,

$$\lim_{P \to 0} \hat{f}_i^V/f_i^V = y_i = \lim_{P \to 0} \eta_i$$

But since $\hat{f}_i^V/f_i^V$ and η_i are independent of P, the following relationship, called the Lewis and Randall fugacity rule, holds at all pressures in the interval 0 to P

$$\hat{f}_i^V = f_i^V y_i \tag{14-52}$$

Similarly, if over the interval 0 to P, $\bar{V}_i^L = v_i^L$ for component i in the liquid phase, then it follows as a consequence that the Lewis and Randall rule holds for component i over the pressure interval 0 to P

$$\hat{f}_i^L = f_i^L x_i \tag{14-53}$$

Thus, by replacing $\hat{f}_i^V$ and $\hat{f}_i^L$ in Eqs. (14-43) and (14-44) by the expressions given by Eqs. (14-52) and (14-53), the following expressions are obtained for ideal solutions

$$\bar{G}_i^V = g_i^V + RT \ln y_i \qquad (i = 1, 2, \ldots, c) \tag{14-54}$$

and

$$\bar{G}_i^L = g_i^L + RT \ln x_i \qquad (i = 1, 2, \ldots, c) \qquad (14\text{-}55)$$

Also, a comparison of Eqs. (14-52) and (14-53) with (14-34) and (14-35) shows that for ideal solutions

$$\gamma_i^L = 1 \qquad \gamma_i^V = 1 \qquad (14\text{-}56)$$

for all i.

RELATIONSHIPS BETWEEN THE PARTIAL MOLAR QUANTITIES

Since Eq. (14-9) applies for both a mixture and a pure component, it may be differentiated termwise with respect to n_i with P, T, and the remaining mole numbers held fixed to give

$$\bar{G}_i = \bar{H}_i - T\bar{S}_i \qquad (14\text{-}57)$$

The quantities $\bar{G}_i$, $\bar{H}_i$, and $\bar{S}_i$ are defined as follows

$$\bar{G}_i = \left(\frac{\partial G}{\partial n_i}\right)_{P, T, n_{j \neq i}} \qquad \text{partial molar free energy} \qquad (14\text{-}58)$$

$$\bar{H}_i = \left(\frac{\partial H}{\partial n_i}\right)_{P, T, n_{j \neq i}} \qquad \text{partial molar enthalpy} \qquad (14\text{-}59)$$

$$\bar{S}_i = \left(\frac{\partial S}{\partial n_i}\right)_{P, T, n_{j \neq i}} \qquad \text{partial molar entropy} \qquad (14\text{-}60)$$

Important thermodynamic relationships involving the partial molar quantities are as follows

$$-\bar{S}_i = \left(\frac{\partial \bar{G}_i}{\partial T}\right)_{P, n_i} = \left(\frac{\partial \mu_i}{\partial T}\right)_{P, n_i} \qquad (14\text{-}61)$$

$$\left(\frac{\partial \bar{G}_i/T}{\partial T}\right)_{P, n_i} = -\frac{\bar{H}_i}{T^2} \qquad (14\text{-}62)$$

For ideal vapor and liquid solutions

$$\bar{H}_i^V = h_i^V \qquad (14\text{-}63)$$

$$\bar{H}_i^L = h_i^L \qquad (14\text{-}64)$$

Proofs of the above relationships are left as problems for the student; see Probs. 14-7 and 14-8.

THE VIRTUAL VALUES OF THE PARTIAL MOLAR ENTHALPIES[27, 28]

Use of the virtual values of the partial molar enthalpies makes it unnecessary to compute the partial molar enthalpies in order to compute the correct enthalpy of a mixture. The virtual values of the partial molar enthalpies are defined as follows

$$\hat{H}_i = h_i^\circ + \Omega \qquad (14\text{-}65)$$

where $\hat{H}_i = H_i^V(P, T, \{n_i\}) = $ virtual value of the partial molar enthalpy of component i in a mixture at temperature T and pressure P

$h_i^\circ = $ enthalpy of one mole of component i in the perfect gas state at the temperature T and the pressure $P = 1$ atmosphere

$\Omega = h(P, T, \{n_i\}) - h^\circ(1, T, \{n_i\})$, the enthalpy of one mole of mixture at the temperature T and pressure P minus the enthalpy of one mole of the same mixture in the perfect gas state at T and at $P = 1$ atmosphere; Ω is called the enthalpy departure function

$h^\circ = \sum_i (n_i/n)h_i^\circ$

Although the virtual values of the partial molar enthalpies are generally unequal to the partial molar enthalpies, they may be used to compute the correct enthalpy of the mixture, that is,

$$nh(P, T, \{n_i\}) = \sum_i n_i \bar{H}_i = \sum_i n_i \hat{H}_i \qquad (14\text{-}66)$$

The equality given by this equation is readily established by beginning with the fact that the enthalpy of one mole of any mixture may be expressed in terms of h° and the departure function Ω as follows

$$h = h^\circ + \Omega \qquad (14\text{-}67)$$

[Equation (14-67) is also seen to follow immediately from the definition given above for Ω.] Then the enthalpy of n moles $(n_1 + n_2 + \cdots + n_c = n)$ of a mixture is given by

$$nh = nh^\circ + n\Omega \qquad (14\text{-}68)$$

Termwise differentiation of Eq. (14-68) with respect to n_k at constant pressure, temperature, and with all of the n_i's held fixed except n_k yields

$$\bar{H}_k = \left[\frac{\partial(nh)}{\partial n_k}\right]_{P, T, n_{j \neq k}} = \frac{\partial}{\partial n_k}\left(\sum_i n_i h_i^\circ\right) + \left(\frac{\partial n}{\partial n_k}\right)\Omega + n\left(\frac{\partial\Omega}{\partial n_k}\right)_{P, T, n_{j \neq k}} \qquad (14\text{-}69)$$

which reduces to

$$\bar{H}_k = h_k^\circ + \Omega + n\left(\frac{\partial\Omega}{\partial n_k}\right)_{P, T, n_{j \neq k}} \qquad (14\text{-}70)$$

For convenience in making subsequent comparisons, let the subscript k in the above equation be replaced by the subscript i. Multiplication of the resulting expression by n_i followed by the summation over all component yields

$$nh = \sum_i n_i \bar{H}_i = \sum_i n_i h_i^\circ + n\Omega + n\sum_i n_i\left(\frac{\partial\Omega}{\partial n_i}\right)_{P, T, n_{j \neq i}} \qquad (14\text{-}71)$$

Since Ω is a homogeneous function of degree zero in the n_i's, it follows from

Euler's theorem (Appendix A, Theorem A-6) that

$$\sum_i n_i \left(\frac{\partial \Omega}{\partial n_i} \right)_{P, T, n_{j \neq i}} = 0$$

Thus, Eq. (14-71) reduces to

$$nh = \sum_i n_i \bar{H}_i = \sum_i n_i h_i^{\circ} + n\Omega \qquad (14\text{-}72)$$

Now if one begins with the definition of $\hat{H}_i$ as given by Eq. (14-65) and multiplies each term of this equation by n_i and then sums the resulting expression over all components, one obtains

$$\sum_i n_i \hat{H}_i = \sum_i n_i h_i^{\circ} + n\Omega \qquad (14\text{-}73)$$

Comparison of Eqs. (14-72) and (14-73) establishes the validity of Eq. (14-66). Consequently, the virtual values of the partial molar enthalpies may be employed in the calculation of the correct enthalpy of a mixture even though the virtual values of the partial molar enthalpies are unequal to the partial molar enthalpies. [The relationship between $\bar{H}_i$ and $\hat{H}_i$ follows immediately from Eqs. (14-65) and (14-70)].

The departure function Ω (which is needed in the evaluation of the virtual values of the partial molar enthalpies [Eq. (14-65)]) may be evaluated through the use of an equation of state for the mixture.

Next, the equations needed to evaluate the thermodynamic functions for a given equation of state are presented.

Equations for the Evaluation of the Thermodynamic Functions in Terms of the Independent Variables, Temperature, Pressure, Volume and Composition

First, paths for the evaluation of the thermodynamic functions of pure components in terms of the independent variables $\{P, T\}$ are presented. Then the paths are presented for the evaluation of the thermodynamic functions for mixtures in terms of the independent variables $(P, T, \{n_i\})$.

The state of any pure component is uniquely determined by fixing any two variables. Consider first the case where pressure and temperature are taken to be the independent variables. The expressions for the thermodynamic functions are obtained by use of the path shown in Fig. 14-2. The standard state is taken to be a perfect gas at P_0 and T_0. The gas is taken from its standard state to its final state P and T by the following path.

1. Change the temperature from T_0 to T with the pressure held constant at P_0.
2. Change the pressure from P_0 to P^* with the temperature held constant at T.
3. Change the pressure from P^* to P with the temperature held constant at T.

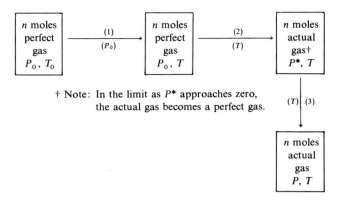

† Note: In the limit as P^* approaches zero, the actual gas becomes a perfect gas.

Figure 14-2 Path used to develop the thermodynamic functions for pure gases in terms of temperature and pressure.

The third container may be regarded as a "converter" because a perfect gas enters and an actual gas leaves. The equations involving the third container [see steps (2) and (3) of Fig. 14-2] become correct in the limit as the pressure P^* is allowed to go to zero. The actual gas leaving the third container approaches a perfect gas as the pressure approaches zero. This path follows closely the one originally proposed by Beattie.[5] The resulting thermodynamic expressions obtained by use of the path shown in Fig. 14-2 are given in Table 14-1.

The path used for evaluation of the thermodynamic functions of mixtures in terms of temperature, pressure, and composition is shown in Fig. 14-3. The standard state of each component is taken to be the pure component in the perfect gas state T_0 and P_0. Again, in the limit as P^* is allowed to approach zero, the container at P^* and T (see Fig. 14-3) becomes a "convertor" in the sense that a perfect gas mixture enters and an actual gas mixture leaves. The resulting expressions for the thermodynamic functions are presented in Table 14-1.

The path used to evaluate the thermodynamic functions of pure components as functions of V and T is shown in Fig. 14-4. The standard state is V_0 and T_0 and the final state is V and T. The gas is taken to its final state along the path shown in Fig. 14-4. In the limit as V^* is allowed to approach infinity, the second container becomes a "convertor" as described for the previous case. The resulting expression obtained for the thermodynamic functions are presented in Table 14-2.

In Fig. 14-5, the path used to evaluate the thermodynamic functions of mixtures in terms of V, T, and $\{n_i\}$ is shown. Again, the standard state of each component is taken to be the perfect gas state at the temperature and volume, T_0 and V_0. After the components have been mixed isothermally, the mixture is taken along the path shown in Fig. 14-5 to the final state V and T. The corresponding equations obtained for thermodynamic functions are given in Table 14-2.

Table 14-1 Equations for pure components as functions of P and T and mixtures as functions of P, T, and $\{n_i\}$

1. Pure components

$$H(P, T) = \int_0^P \left[V - T\left(\frac{\partial V}{\partial T}\right)_P \right] dP + nh°$$

$$S(P, T) = \int_0^P \left[\frac{nR}{P} - \left(\frac{\partial V}{\partial T}\right)_P \right] dP - nR \ln P + ns°$$

$$A(P, T) = \int_0^P \left(V - \frac{nRT}{P} \right) dP + nRT \ln P - PV + n(h° - Ts°)$$

$$RT \ln f = \int_0^P \left(\frac{V}{n} - \frac{RT}{P} \right) dP + RT \ln P$$

2. Mixtures

$$H(P, T, \{n_i\}) = \int_0^P \left[V - T\left(\frac{\partial V}{\partial T}\right)_{P, n_i} \right] dP + \sum_i n_i h_i°$$

$$S(P, T, \{n_i\}) = \int_0^P \left[\frac{\sum_i n_i R}{P} - \left(\frac{\partial V}{\partial T}\right)_{P, n_i} \right] dP - \sum_i n_i R \ln Py_i + \sum_i n_i s_i°$$

$$A(P, T, \{n_i\}) = \int_0^P \left(V - \frac{\sum_i n_i RT}{P} \right) dP + \sum_i n_i RT \ln Py_i - PV + \sum_i n_i(h_i° - Ts_i°)$$

$$RT \ln \hat{f}_i = \int_0^P \left[\left(\frac{\partial V}{\partial n_i}\right)_{P, T, n_i} - \frac{RT}{P} \right] dP + RT \ln Py_i$$

where

$$h_i° = h_i°(P_0, T_0) + \int_{T_0}^T c_{pi}° \, dT$$

$$s_i° = s_i°(P_0, T_0) + \int_{T_0}^T \frac{c_{pi}°}{T} \, dT$$

$h_i°(P_0, T_0) =$ enthalpy at the standard state of a perfect gas at P_0 and T_0

$s_i°(P_0, T_0) =$ entropy at the standard state of a perfect gas at P_0 and T_0

RESTATEMENT OF THE THERMODYNAMIC EQUATIONS IN TERMS OF THE RESIDUAL WORK FUNCTION, $\tilde{A}(\rho, T)$

The restatement of the equations given in Table 14-2 in terms of the residual work function is presented in Table 14-3. The use of the reformulated form of the equations for the evaluation of the thermodynamic functions is a useful technique which appears to have been suggested first by Benedict et al.[6]

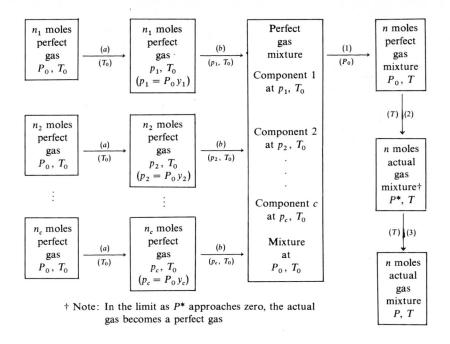

Figure 14-3 Path used to develop the thermodynamic functions for gas mixtures in terms of temperature, pressure, and composition.

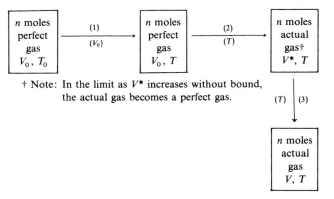

Figure 14-4 Path used to develop the thermodynamic functions for pure gases in terms of temperature and volume.

Table 14-2 Equations for pure components as functions of V and T and mixtures as functions of V, T, and $\{n_i\}$

1. Pure components

$$H(V, T) = \int_V^\infty \left[P - T\left(\frac{\partial P}{\partial T}\right)_V \right] dV + PV + nu^\circ$$

$$S(V, T) = \int_V^\infty \left[\frac{nR}{V} - \left(\frac{\partial P}{\partial T}\right)_V \right] dV + nR \ln \frac{V}{nRT} + ns^\circ$$

$$A(V, T) = \int_V^\infty \left(P - \frac{nRT}{V} \right) dV - nRT \ln \frac{V}{nRT} + n(u^\circ - Ts^\circ)$$

$$RT \ln f = \int_V^\infty \left[\left(\frac{\partial P}{\partial n}\right)_{V, T} - \frac{RT}{V} \right] dV - RT \ln \frac{V}{nRT}$$

2. Mixtures

$$H(V, T, \{n_i\}) = \int_V^\infty \left[P - T\left(\frac{\partial P}{\partial T}\right)_{V, n_i} \right] dV + PV + \sum_i n_i u_i^\circ$$

$$S(V, T, \{n_i\}) = \int_V^\infty \left[\frac{\sum_i n_i R}{V} - \left(\frac{\partial P}{\partial T}\right)_{V, n_i} \right] dV + \sum_i n_i R \ln \frac{V}{n_i RT} + \sum_i n_i s_i^\circ$$

$$A(V, T, \{n_i\}) = \int_V^\infty \left(P - \frac{\sum_i n_i RT}{V} \right) dV - \sum_i n_i RT \ln \frac{V}{n_i RT} + \sum_i n_i(u_i^\circ - Ts_i^\circ)$$

$$RT \ln \hat{f}_i = \int_V^\infty \left[\left(\frac{\partial P}{\partial n_i}\right)_{V, T, n_{j \neq i}} - \frac{RT}{V} \right] dV - RT \ln \frac{V}{n_i RT}$$

where

$$u_i^\circ = u_i^\circ(V_0, T_0) + \int_{T_0}^T c_{vi}^\circ \, dT$$

$$s_i^\circ = s_i^\circ(V_0, T_0) + \int_{T_0}^T \frac{c_{vi}^\circ}{T} \, dT + R \ln \frac{T}{T_0}$$

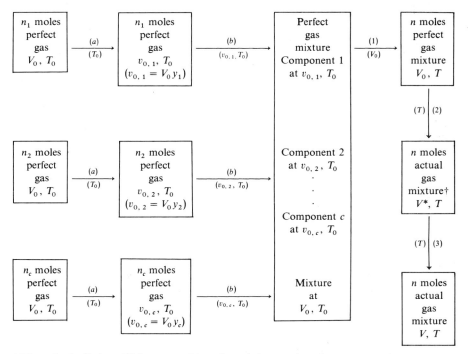

† Note: In the limit as V^* increases without bound the actual gas becomes a perfect gas.

Figure 14-5 Path used to develop the thermodynamic functions for gas mixtures in terms of temperature, volume, and composition.

Generally, in the solution of problems involving the thermodynamic functions U, H, G, A, μ, and f, the values of the functions at the system pressure and temperature are needed. However, most of the equations of state are usually stated in the form of P as a function of V and T which makes the integrals appearing in Table 14-2 easier to evaluate than those in Table 14-1. Also, instead of evaluating all of the integrals given in Table 14-2, one needs to evaluate only one of these because all of the functions may be restated in terms of this integral. The integral commonly selected for evaluation by use of an equation of state is the one appearing in the expression for $A(V, T)$. A summary of these results is presented in Table 14-3. Since $U(V, T) = U(P, T)$, $H(V, T) = H(P, T)$, ..., $\mu(V, T) = \mu(P, T)$, and $f(V, T) = f(P, T)$, the expressions shown in Table 14-3 may be used to compute the values of the thermodynamic functions at the systems P and T. The common integral, called the "residual work content" after Benedict et al.,[6] is defined as follows

$$\tilde{A}(V, T) = \frac{1}{n} \int_V^\infty \left(P - \frac{nRT}{V} \right) dV \qquad (14\text{-}74)$$

Table 14-3 Statement of the thermodynamic functions of Table 14-2 in terms of the residual work content $\tilde{A}$

---

1. Pure components

$$\tilde{A} = \frac{1}{n} \int_V^{\infty} \left(P - \frac{nRT}{V} \right) dV = \int_0^{\rho} \left(\frac{P - \rho RT}{\rho^2} \right) d\rho$$

$$H = n\tilde{A} - nT \left(\frac{\partial \tilde{A}}{\partial T} \right)_V + n \left(\frac{P}{\rho} - RT \right) + nh^{\circ}$$

$$S = -n \left(\frac{\partial \tilde{A}}{\partial T} \right)_V - nR \ln \rho RT + ns^{\circ}$$

$$A = n\tilde{A} + nRT \ln \rho RT + n(u^{\circ} - Ts^{\circ})$$

$$RT \ln \frac{f}{P} = \tilde{A} + RT \ln \frac{\rho RT}{P} + \left(\frac{P}{\rho} - RT \right)$$

---

2. Mixtures

$$\tilde{A} = \frac{1}{n} \int_V^{\infty} \left(P - \frac{\sum_i nRT}{V} \right) dV = \int_0^{\rho} \left(\frac{P - \rho RT}{\rho^2} \right) d\rho$$

$$H = n\tilde{A} - nT \left(\frac{\partial \tilde{A}}{\partial T} \right)_{V, n_i} + n \left(\frac{P}{\rho} - RT \right) + \sum_i n_i h_i^{\circ}$$

$$S = -n \left(\frac{\partial \tilde{A}}{\partial T} \right)_{V, n_i} + \sum_i n_i R \ln \frac{V}{n_i RT} + \sum_i n_i s_i^{\circ}$$

$$A = n\tilde{A} - \sum_i n_i RT \ln \frac{V}{n_i RT} + \sum_i n_i (u_i^{\circ} - Ts_i^{\circ})$$

$$RT \ln \frac{\hat{f}_i^V}{P y_i} = \left[\frac{\partial (n\tilde{A})}{\partial n_i} \right]_{\rho, T, n_{j \neq i}} + \left(\frac{P}{\rho} - RT \right) + RT \ln \frac{\rho RT}{P}$$

Let the molar density be denoted by ρ. Then $\rho = n/V$, and the above integral may be restated in terms of ρ and T to give

$$\tilde{A}(V, T) = \tilde{A}(\rho, T) = \int_0^{\rho} \left(\frac{P - \rho RT}{\rho^2} \right) d\rho \qquad (14\text{-}75)$$

Thus, the expression given in Table 14-2 for $A(V, T)$ reduces to

$$A(V, T) = n\tilde{A}(V, T) + nRT \ln \rho RT + n(u^{\circ} - Ts^{\circ}) \qquad (14\text{-}76)$$

which is seen to be the expression given in Table 14-3 for $A(V, T)$. Once an equation of state has been specified, the integral called the residual work content may be evaluated.

14-2 EQUATIONS OF STATE AND THEIR USE IN THE PREDICTION OF K VALUES AND ENTHALPIES

The equations of state which appear to be finding the widest application for distillation calculations consist of the modified versions of the Benedict-Webb-Rubin equation of state, the Soave-Redlich-Kwong equation of state, and the Peng-Robinson equation of state. All of the equations needed to compute the K values and enthalpies by use of these equations of state are presented in this section. In addition an outline of the development of the equations needed to compute the K values and enthalpies by use of the original Benedict-Webb-Rubin equation of state is presented. Also, an outline of the development of the equations needed to calculate fugacities and enthalpies by the original Redlich-Kwong equation of state is presented. Unlike the original Benedict-Webb-Rubin equation of state which is recommended for computing the properties of both vapor and liquid phases, the Redlich-Kwong equation is recommended for computing properties of only the vapor phase. The Soave-Redlich-Kwong and the Peng-Robinson equations of state are, however, recommended for computing the properties of both phases for certain pure components and mixtures.

The Benedict-Webb-Rubin Equation of State[6, 7]

The Benedict-Webb-Rubin (BWR) equation of state is given by

$$P = RT\rho + \left(B_0 RT - A_0 - \frac{C_0}{T^2}\right)\rho^2 + (bRT - a)\rho^3$$

$$+ a\alpha\rho^6 + \frac{c\rho^3}{T^2}(1 + \gamma\rho^2)\exp(-\gamma\rho^2) \quad (14\text{-}77)$$

where A_0, B_0, C_0, a, b, c, α, and γ are empirical constants. In some correlations, some of these constants are taken to be functions of temperature. The mixing rules commonly employed are given in Table 14-4.

In the solution of distillation problems, the only functions generally required are the enthalpies and fugacities of the pure components and the enthalpies of mixtures as well as the fugacities of the components in a mixture. Thus, only the expressions for these quantities are given in tables for the equations of state.

The first step in the development of the expressions for these functions is the development of the expression for the residual work content $\tilde{A}$. For the BWR equation of state, the expression for $\tilde{A}$ is obtained by carrying out the integration implied by Eq. (14-75) through the use of Eq. (14-77). The resulting expression obtained for $\tilde{A}$ is given in Table 14-4.

The expression for the enthalpy of one mole of a pure component is obtained by use of the expression given by h in Table 14-3 and the expression given in Table 14-4 for $\tilde{A}$. Similarly the expression for the fugacity of a pure component is found by use of the expression given in Table 14-3 for f and the expression for $\tilde{A}$. The resulting expressions for h and f are given in item 3 of Table 14-4.

Table 14-4 The Benedict-Webb-Rubin equation of state[6, 7]

1. Values of the Constants of the BWR equation for 12 Hydrocarbons[7]

Units: $P = \text{lb/in}^2$ abs
$\rho = \text{lb mol/ft}^3$
$T = $ degree Rankine (°F + 459.63)
$R = 10.7335 \text{ lb ft}^3/(\text{in}^2 \text{ lb mol °R})$

Substance	Methane	Ethylene	Ethane	Propylene	Propane
Mol wt	16.031	28.031	30.047	42.047	44.062
B_0	0.682401	0.89198	1.00554	1.36263	1.55884
A_0	6,995.25	12,593.6	15,670.7	23,049.2	25,913.4
$C_0 \times 10^{-6}$	275.763	1,602.28	2,194.27	5,365.97	6,209.93
b	0.867325	2.20678	2.85393	4.79997	5.77355
a	2,984.12	15,645.5	20,850.2	46,758.6	57,248.0
$c \times 10^{-6}$	498.106	4,133.14	6,413.14	20,083.0	25,247.8
$\alpha \times 10^3$	511.172	731.661	1,000.44	1,873.12	2,495.77
$\gamma \times 10^2$	153.961	236.844	302.790	469.325	464.524
$A_0^{1/2}$	83.6376	112.221	125.183	151.820	160.983
$C_0^{1/2} \times 10^{-3}$	16.6061	40.0285	46.8430	73.2528	78.8031
$b^{1/3}$	0.953644	1.30193	1.41845	1.68686	1.79396
$a^{1/3}$	14.3970	25.0109	27.5235	36.0264	38.5408
$C^{1/3} \times 10^{-2}$	7.92684	16.0486	18.5788	27.1813	29.3360
$\alpha^{1/3} \times 10$	7.99567	9.01094	10.0015	12.3270	13.5644
$\gamma^{1/2} \times 10$	12.4082	15.3897	17.4009	21.6640	23.7597

Substance	i-Butane	i-Butylene	n-Butane	i-Pentane
Mol wt	58.078	56.062	58.078	72.094
B_0	2.20329	1.85858	1.99211	2.56386
A_0	38,587.4	33,762.9	38,029.6	4,825.36
$C_0 \times 10^{-6}$	10,384.7	11,329.6	12,130.5	21,336.7
b	10.8890	8.93375	10.2636	17.1441
a	117,047.0	102,251.0	113,705.0	226,902.0
$c \times 10^{-6}$	55,977.7	53,807.2	61,925.6	136,025.0
$\alpha \times 10^3$	4,414.96	3,744.17	4,526.93	6,987.77
$\gamma \times 10^2$	872.447	759.401	872.447	1,188.07
$A^{1/2}$	196.437	183.747	195.012	219.667
$C_0^{1/2} \times 10^{-3}$	101.905	106.441	110.139	146.071
$b^{1/3}$	2.21647	2.07497	2.17320	2.57853
$a^{1/3}$	48.9165	46.7620	48.4464	60.9929
$C^{1/3} \times 10^{-2}$	38.2530	37.7521	39.5625	51.4281
$\alpha^{1/3} \times 10$	16.4050	15.5281	16.5425	19.1181
$\gamma^{1/3} \times 10$	29.5372	27.5573	29.5372	34.4683

Table 14-4 (*continued*)

Substance	n-Pentane	n-Hexane	n-Pentane
Mol wt	77.094	86.109	100.125
B_0	2.51096	2.84835	3.18782
A_0	45,928.8	5,443.4	66,070.6
$C_0 \times 10^{-6}$	25,917.2	40,556.2	57,984.0
b	17.1441	28.0032	38.9917
a	246,148.0	429,901.0	626,106.0
$c \times 10^{-6}$	161,306.0	296,077.0	483,427.0
$\alpha \times 10^3$	7,439.92	11,553.9	17,905.6
$\gamma \times 10^2$	1,218.86	1,711.15	2,309.42
$A_0^{1/2}$	214.310	233.331	257.042
$C_0^{1/2} \times 10^{-3}$	160.988	201.386	240.799
$b^{1/3}$	2.57853	3.03671	3.39097
$a^{1/3}$	62.6707	75.4726	85.5494
$C^{1/3} \times 10^{-2}$	54.4349	66.6492	78.4822
$\alpha^{1/3} \times 10$	19.5220	22.6070	26.1616
$\gamma^{1/2} \times 10$	34.9122	41.3660	48.0564

2. Residual work content

$$\tilde{A} = \left(B_0 RT - A_0 - \frac{C_0}{T^2} \right)\rho + \frac{(bRT - a)}{2}\rho^2 + \frac{a\alpha\rho^5}{5} + \frac{c\rho^2}{T^2}\left[\frac{1}{\nu\rho^2} - \left(\frac{1}{\nu\rho^2} + \frac{1}{2} \right)\exp\left(-\gamma\rho^2 \right) \right]$$

3. Pure components

$$h - h^\circ = \left(B_0 RT - 2A_0 - \frac{4C_0}{T^2} \right)\rho + \left(bRT - \frac{3a}{2} \right)\rho^2 + \frac{6a\alpha\rho^5}{5} + \frac{c\rho^2}{T^2}\left[3\left(\frac{1 - e^{-\gamma\rho^2}}{\gamma\rho^2} \right) + \left(\gamma\rho^2 - \frac{1}{2} \right)e^{-\gamma\rho^2} \right]$$

$$+ \rho T \frac{dA_0}{dT} + \frac{\rho}{T}\frac{dC_0}{dT} - \frac{RT^2\rho^2}{2}\frac{db}{dT} + \frac{c}{\gamma^2 T}\left[1 - \left(1 + \gamma\rho^2 + \frac{\gamma^2\rho^4}{2} \right)\exp\left(-\gamma\rho^2 \right) \right]\frac{d\gamma}{dT}$$

$$RT \ln \frac{f}{P} = RT \ln \frac{\rho RT}{P} + 2\left(B_0 RT - A_0 - \frac{C_0}{T^2} \right)\rho$$

$$+ \frac{3}{2}(bRT - a)\rho^2 + \frac{6}{5}a\alpha\rho^5 + \frac{c\rho^2}{T^2}\left[\frac{1 - e^{-\gamma\rho^2}}{\gamma\rho^2} + \left(\frac{1}{2} + \gamma\rho^2 \right)e^{-\gamma\rho^2} \right]$$

4. Mixtures

The mixing rules are as follows:

$$A_0 = \left(\sum_i y_i A_{0i}^{1/2} \right)^2 \qquad a = \left(\sum_i y_i a_i^{1/3} \right)^3 \qquad \alpha = \left(\sum_i y_i \alpha_i^{1/3} \right)^3$$

$$B_0 = \sum_i y_i B_{0i} \qquad b = \left(\sum_i y_i b_i^{1/3} \right)^3 \qquad \gamma = \left(\sum_i y_i \gamma_i^{1/2} \right)^2$$

$$C_0 = \left(\sum_i y_i C_{0i}^{1/2} \right)^2 \qquad c = \left(\sum_i y_i c_i^{1/3} \right)^3$$

(Continued on page 514.)

Table 14-4 (*continued*)

$$\Omega = h - h° = \left(B_0 RT - 2A_0 - \frac{4C_0}{T^2}\right)\rho + \left(bRT - \frac{3a}{2}\right)\rho^2 + \frac{6a\alpha\rho^5}{5}$$

$$+ \frac{c\rho^2}{T^2}\left[3\left(\frac{1 - e^{-\gamma\rho^2}}{\gamma\rho^2}\right) + \left(\gamma\rho^2 - \frac{1}{2}\right)e^{-\gamma\rho^2}\right] + \rho T \frac{\partial A_0}{\partial T} + \frac{\rho}{T}\frac{\partial C_0}{\partial T}$$

$$- \frac{RT^2\rho^2}{2}\frac{\partial b}{\partial T} - \frac{c}{\gamma^2 T}\left[1 - \left(1 + \gamma\rho^2 + \frac{\gamma^2\rho^4}{2}\right)\exp\left(-\gamma\rho^2\right)\right]\frac{\partial\gamma}{\partial T}$$

$$RT \ln \frac{\hat{f}^V}{Py_i} = RT \ln \frac{\rho RT}{P} + [(B_0 + B_{0i})RT - 2(A_0 A_{0i})^{1/2} - 2(C_0 C_{0i})^{1/2}/T^2]\rho$$

$$+ 3/2[RT(b^2 b_i)^{1/3} - (a^2 a_i)^{1/2}]\rho^2 + 3/5[a(\alpha^2\alpha_i)^{1/3} + \alpha(a^2 a_i)^{1/3}]\rho^5$$

$$+ \frac{3\rho^2(c^2 c_i)^{1/3}}{T^2}\left(\frac{1 - e^{-\gamma\rho^2}}{\gamma\rho^2} - \frac{e^{-\gamma\rho^2}}{2}\right) - \frac{2\rho^2 c}{T^2}\left(\frac{\gamma_i}{\gamma}\right)^{1/2}\left(\frac{1 - e^{-\gamma\rho^2}}{\gamma\rho^2} - e^{-\gamma\rho^2} - \frac{\gamma\rho^2 e^{-\gamma\rho^2}}{2}\right)$$

$RT \ln \hat{f}_i^L/Px_i$ = an expression of the same form as shown for the vapor phase. In this case, however, the liquid mole fractions and the liquid density are used in the evaluation of the right-hand side of the above expression instead of the vapor density and vapor mole fractions $\{y_i\}$.

The enthalpy of a mixture is found by use of the expression given in Table 14-3 for the enthalpy and the expression given by Eq. (14-75) for $\tilde{A}$. The expression for the fugacity $\hat{f}_i$ of a component in a mixture is found in a similar manner. Expressions for these quantities for the BWR equation of state are given in item 4 of Table 14-4.

For a given P, T, and sets of vapor and liquid compositions, the equation of state [Eq. (14-77)] is solved for the vapor and liquid densities ρ which satisfy this equation. On the basis of these values of P, T, composition, and ρ, the expressions given in Table 14-4 for the functions may be evaluated.

In the calculation of the properties for the vapor phase, the vapor density and the vapor mole fractions $\{y_i\}$ are used. In the calculation of the properties for the liquid phase the liquid density and the liquid mole fractions $\{x_i\}$ are used. For example, in the calculation of the K value for a given component, the right-hand side of the expression given in Table 14-7 for $\hat{f}_i^V/Py_i$ is evaluated on the basis of P, T, the vapor density, and the vapor mole fractions $\{y_i\}$. Similarly, for the calculation of $\hat{f}_i^L/Px_i$, the right-hand side of the expression shown in Table 14-4 is evaluated on the basis of P, T, the liquid density, and the liquid mole fractions $\{x_i\}$. The resulting values obtained for $\hat{f}_i^V/Py_i$ and $\hat{f}_i^L/Px_i$ are then employed to compute the K values for component i as follows

$$\frac{\hat{f}_i^L/Px_i}{\hat{f}_i^V/Py_i} = K_i \tag{14-78}$$

CONSTANTS AND MODIFICATIONS OF THE BENEDICT-WEBB-RUBIN EQUATION OF STATE

Compilations of constants for the BWR equation of state have been given by several authors.[14, 35] The pure component constants given by Benedict et al.[7] for the hydrocarbons (methane through n-heptane) yield accurate values of the properties at densities up to about 1.8 times the critical density. For temperatures well below the normal boiling points, constants are given by Orye.[46] Also included in this compilation are constants for H_2, N_2, CO_2, and H_2S. For temperatures 0.6 times the critical, Reid et al.[55] recommend the constants of Cooper and Goldfrank,[14] Orye,[46] and Kaufman.[35] For temperatures less than $0.6T_c$, the constants of Orye and Kaufman are recommended. Viswanath and Su[70] generalized the BWR equation of state based on a modified theorem of corresponding states and determined a set of constants for densities up to twice the critical in the reduced pressure range of 0 to 40 and the reduced temperature range of 1.0 to 15.0. Modifications and a variety of applications of the BWR equation of state have been proposed by several authors (Refs. 39, 45, 62, 63, 64, 65, and 77).

The constants proposed by Starling for mixtures are presented in Ref. 65. With the advantage of increased accuracy achieved by use of the 11-constant equation proposed by Starling[65] came the disadvantage that the constants could no longer be obtained from pure component properties. Experimental equilibrium data, density data, and enthalpy data for mixtures had to be included in the regression procedure for obtaining the constants.[2]

Cubic Equations of State

Martin[44] initiated an analysis of the cubic equations by use of the following generalized form of the cubic equation of state

$$P = \frac{RT}{v} - \frac{-\alpha(T)}{(v + \beta)(v + \gamma)} + \frac{-\delta(T)}{v(v + \beta)(v + \gamma)} \qquad (14\text{-}79)$$

where α and δ are functions of temperature and β and γ are constants. By making suitable choices of the constants and functions and performing appropriate rearrangements and translations, Martin was able to obtain other well-known cubic equations of state such as the Redlich-Kwong and the Peng-Robinson from Eq. (14-79). Tests by Martin showed that the best two-term cubic equation of state is the Martin equation which is presented in Table 14-9.

The Redlich-Kwong Equation of State[53]

The Redlich-Kwong equation of state is given by

$$P = \frac{RT}{v - b} - \frac{a/T^{1/2}}{v(v + b)} \qquad (14\text{-}80)$$

where a and b are constants which depend upon the critical temperature and pressure as shown below

$$v = V/n = 1/\rho = ZRT/P = \text{molar volume} \tag{14-81}$$

The combining rules for mixtures are as follows

$$a = \left(\sum_i y_i a_i^{1/2} \right)^2 \tag{14-82}$$

$$b = \sum_i y_i b_i \tag{14-83}$$

where $a_i^{1/2} = (0.4278R^2 T_{ci}^{2.5}/P_{ci})^{1/2}$
$\quad b_i = 0.0867RT_{ci}/P_{ci}$
$\quad P_{ci} = $ critical pressure of component i
$\quad T_{ci} = $ critical temperature of component i

For convenience, let

$$A = \sum_i y_i A_i$$
$$\tag{14-84}$$
$$B = \sum_i y_i B_i$$

where

$$A_i = \frac{a_i^{1/2}}{RT^{1.25}} \qquad B_i = \frac{b_i}{RT}$$

Then

$$A^2 = \left(\sum_i y_i A_i \right)^2 = \frac{1}{R^2 T^{2.25}} \left(\sum_i y_i a_i^{1/2} \right)^2 = \frac{a}{R^2 T^{2.5}}$$
$$\tag{14-85}$$
$$B = \sum_i y_i B_i = \frac{1}{RT} \sum_i y_i b_i = \frac{b}{RT}$$

and thus

$$A^2 R^2 T^{2.5} = a \tag{14-86}$$

$$BRT = b \tag{14-87}$$

Use of the expressions given by Eqs. (14-86) and (14-87) to eliminate a and b from Eq. (14-80) yields

$$P = \frac{\rho RT}{1 - \rho RTB} - \frac{A^2 \rho^2 (RT)^2}{(1 + \rho RTB)} \tag{14-88}$$

ENTHALPIES AND FUGACITIES

Expressions for the enthalpies and fugacities are developed in a manner analogous to that demonstrated for the BWR equation of state. The expression given by Eq. (14-75) for $\tilde{A}$ and the expressions given in Table 14-3 are used to obtain the formulas presented in Table 14-5. The enthalpy of a mixture is found by

Table 14-5 The Redlich-Kwong equation of state[53]

1. Equations of state and constants

The equation of state is given by Eq. (14-84), and the constants are given by Eq. (14-85).

2. Residual work content

$$\tilde{A} = -RT \left[\ln \left(1 - \frac{BP}{Z} \right) + \frac{A^2}{B} \ln \left(1 + \frac{BP}{Z} \right) \right]$$

3. Pure component

$$h - h° = RT(Z - 1) - \frac{3}{2} \frac{RTA^2}{B} \ln \left(1 + \frac{BP}{Z} \right)$$

$$\log \frac{f}{P} = Z - 1 - \ln (Z - BP) - \frac{A^2}{B} \ln \left(1 + \frac{BP}{Z} \right)$$

4. Mixtures

The mixing rules and constants are given by Eqs. (14-85) through (14-87).

$$\Omega = h - h° = RT(Z - 1) + \frac{RTA^2}{B} \left(\frac{2T}{A^2} \sum_i y_i \frac{\bar{A}_i^2}{\alpha_i^{1/2}} \frac{d\alpha_i^{1/2}}{dT} - 1 \right) \ln \left(1 + \frac{BP}{Z} \right)$$

$$\ln \frac{\hat{f}_i^V}{Py_i} = \frac{B_i}{B} (Z - 1) - \ln (Z - BP) - \frac{A^2}{B} \left(\frac{2A_i}{A} - \frac{B_i}{B} \right) \ln \left(1 + \frac{BP}{Z} \right)$$

where

$$A_i = \frac{a_i^{1/2} \alpha_i^{1/2}}{RT}$$

$$\bar{A}_i^2 = A_i \sum_j y_j A_j (1 - k_{ij})$$

$$\ln \frac{\hat{f}_i^L}{Px_i} = \text{(See note for this expression in Table 14-4.)}$$

evaluating A and B by use of the combining rules given by Eqs. (14-82) through (14-84). The first step in the application of these equations is the calculation of the liquid and vapor densities which satisfy Eq. (14-88) for a given temperature, pressure, and sets of vapor and liquid compositions. Restatement of Eq. (14-88) in terms of the compressibility factor yields

$$Z^3 - Z^2 + [A^2P - BP(1 + BP)]Z - A^2PBP = 0 \qquad (14\text{-}89)$$

This cubic equation in Z has either three real roots or one real root and two conjugate imaginary roots.[60] For the case where three real roots exist, the smallest root corresponds to the liquid phase and is called the "liquid root," and the largest root corresponds to the vapor phase and is called the "vapor root." The intermediate root has no physical meaning. For the case where only one real root exists, it may be either a vapor or a liquid root. For $T < T_c$, the single real root is a vapor root if the corresponding density is less than the critical density $(\rho < \rho_c)$ and a liquid root if $\rho > \rho_c$. If $T > T_c$, the single real root is a vapor root. Methods for the determination of the real roots are readily available;[60] see also Adler et al.[2] for the retrograde region. After Eq. (14-89) has been solved for the real roots, the enthalpies and fugacities are found by use of the expressions given in Table 14-5.

The Redlich-Kwong equation of state is commonly regarded as one of the best of the two-parameter equations of state. It may be used to predict vapor-phase properties above the critical temperature for any pressure.[53] This equation and modifications of it have been recommended for all nonpolar compounds. However, the fugacities and enthalpies predicted by the Redlich-Kwong equation for the liquid phase are commonly regarded as inferior to those predicted by the modified version of the Redlich-Kwong equation which was proposed by Soave.[59] Although a similar modification has been proposed by Wilson,[73, 74] only the modification proposed by Soave[59] is presented.

The Soave-Redlich-Kwong Equation of State[59]

While the Redlich-Kwong equation is said to give volumetric and thermal properties of pure components and of mixtures with good accuracy, the vapor-liquid-equilibrium data predicted by this equation often gives poor results.[59] To improve this equation for the prediction of vapor-liquid-equilibrium data, Soave[59] proposed the following modified form of the Redlich-Kwong equation of state

$$P = \frac{RT}{v - b} - \frac{a\alpha}{v(v + b)} \tag{14-90}$$

where the constants a, b, and α are given in Table 14-6. The quantity α is a function of the reduced temperature and a parameter called the acentric factor ω_i, which is defined as follows

$$\omega_i = -\log_{10} \left. \frac{P_i}{P_{ci}} \right|_{T_r = 0.7} - 1 \tag{14-91}$$

The acentric factor ω_i was introduced by Pitzer et al.[49] and has been used in a number of equations of state. In the modified equation, the parameter a of the original equation is taken to be temperature-dependent. With the application of mixing rules given in Table 14-6, the proposed equation can be extended successfully to vapor-liquid-equilibrium calculations of multicomponent mixtures of nonpolar compounds, with the exclusion of carbon dioxide. Less accurate results are obtained for mixtures containing hydrogen.[59] The constants, mixing rules, and the expressions for the fugacities and enthalpies are given in Table 14-6.

The Peng-Robinson Equation of State[47]

Peng and Robinson[47] pointed out that the Redlich-Kwong and the Soave modification of the Redlich-Kwong equation of state have the common failure of not being able to predict accurate liquid densities even though the vapor densities computed by these equations are satisfactory. In order to effect this desired improvement, Peng and Robinson proposed the following modified form of the Redlich-Kwong equation of state

$$P = \frac{RT}{v - b} - \frac{\alpha a}{v(v + b) + b(v - b)} \tag{14-92}$$

The equation performs as well as or better than the Soave-Redlich-Kwong in all cases tested and shows its greatest advantages in the prediction of liquid phase densities.[47] The constants, mixing rules, and the expressions for the fugacities and enthalpies for this equation of state are given in Table 14-7.

The Virial Equation of State

Most all of the equations of state which have been proposed are essentially of an empirical nature. Two of the best known exceptions are the van der Waals equation of state and the virial equation of state. The equations for the virial equation of state are given in Table 14-8.

Both theoretical knowledge and experimental data are available for the evaluation of the second virial coefficient B_{ij}, but little is known about the third- and higher-order virial coefficients. If the third- and higher-order virial coefficients are neglected, the resulting equations may be used for low to moderate pressures and densities, say densities up to about one-half the critical density.[55]

Other Equations of State

A number of other equations of state have been proposed. Some of these are presented in Table 14-9. Lee et al.[38] proposed an equation of state in which a procedure similar to the Chao-Seader method (described below) was used for the calculation of K values. Lee et al. state that the K values obtained by this method are more accurate than those given by the Chao-Seader method,[12] particularly at low temperatures.

Sugie and Lu[66] presented an equation of state (see Table 14-9) which consists of a modification of the Redlich-Kwong equation through the use of a deviation function and Pitzer's tables.[49] The authors state that the resulting equation is expected to be applicable in the region $0.56 < T_r < 1.0$ and suitable for predicting the vapor and the liquid compressibility factors as well as the vapor-phase fugacities of nonpolar substances and mixtures.

In the formulation of the Barner-Adler equation of state,[3] the authors placed primary emphasis on obtaining a generalized set of constants for the region of temperature and pressure most frequently encountered by the process

Table 14-6 The Soave-Redlich-Kwong equations of state[59]

1. Constants of the Soave-Redlich-Kwong equation of state

$$a_i = 0.42747 \frac{R^2 T_{ci}^2}{P_{ci}} \qquad A_i = \frac{a_i^{1/2} \alpha_i^{1/2}}{RT} \qquad \alpha_i^{1/2} = 1 + m_i(1 - T_{ri}^{1/2}) \qquad B_i = \frac{b_i}{RT}$$

$$b_i = 0.08664 \frac{RT_{ci}}{P_{ci}} \qquad m_i = 0.480 + 1.574\omega_i - 0.176\omega_i^2$$

ω_i = acentric factor for component i.

2. Residual work content

$$\bar{A} = -RT\left[\ln\left(1 - \frac{BP}{Z}\right) + \frac{A^2}{B}\ln\left(1 + \frac{BP}{Z}\right)\right]$$

3. Pure components

$$h - h° = RT(Z - 1) + \frac{RTA^2}{B}\left(\frac{T}{\alpha}\frac{d\alpha}{dT} - 1\right)\ln\left(1 + \frac{BP}{Z}\right)$$

$$\ln\frac{f}{P} = Z - 1 - \ln(Z - BP) - \frac{A^2}{B}\ln\left(1 + \frac{BP}{Z}\right)$$

4. Mixtures

The mixing rules are as follows

$$A = \sum_i y_i A_i$$

$$A^2 = \left(\sum_i y_i A_i\right)^2$$

$$B = \sum_i y_i B_i$$

For systems containing CO_2, H_2S, and polar compounds, Soave[59] suggested that the empirical factor k_{ij} be included in the expression for A^2 as follows

$$A^2 = \sum_i \sum_j y_i y_j A_i A_j(1 - k_{ij})$$

and in the formula which follows

$$\bar{A}_i^2 = A_i \sum_j y_j A_j(1 - k_{ij})$$

Table 14-6 (*continued*)

The following values of k_{ij} for the Soave-Redlich-Kwong equation were provided by Professor Erbar[20]

Compound	Carbon dioxide	Hydrogen sulfide	Nitrogen	Carbon monoxide
Methane	0.12	0.08	0.02	−0.02
Ethylene	0.15	0.07	0.04	
Ethane	0.15	0.07	0.06	
Propylene	0.08	0.07	0.06	
Propane	0.15	0.07	0.08	
Isobutane	0.15	0.06	0.08	
n-Butane	0.15	0.06	0.08	
Isopentane	0.15	0.06	0.08	
n-Pentane	0.15	0.06	0.08	
n-Hexane	0.15	0.05	0.08	
n-Heptane	0.15	0.04	0.08	
n-Octane	0.15	0.04	0.08	
n-Nonane	0.15	0.03	0.08	
n-Decane	0.15	0.03	0.08	
n-Undecane	0.15	0.03	0.08	
Carbon dioxide	....	0.12	....	−0.04
Cyclohexane	0.15	0.03	0.08	
Methyl cyclohexane	0.15	0.03	0.08	
Benzene	0.15	0.03	0.08	
Toluene	0.15	0.03	0.08	
o-Xylene	0.15	0.03	0.08	
m-Xylene	0.15	0.03	0.08	
p-Xylene	0.15	0.03	0.08	
Ethylebenzene	0.15	0.03	0.08	

$$\Omega = h - h^\circ = RT(Z - 1) + \frac{RTA^2}{B}\left[\frac{2T}{A^2}\sum_i y_i \bar{A}_i^2\left(\frac{1}{\alpha_i^{1/2}}\frac{d\alpha_i^{1/2}}{dT}\right) - 1\right]\ln\left(1 + \frac{BP}{Z}\right)$$

$$\ln\frac{\hat{f}_i^V}{Py_i} = \frac{B_i}{B}(Z - 1) - \frac{A^2}{B}\left(\frac{2\bar{A}_i^2}{A^2} - \frac{B_i}{B}\right)\ln\left(1 + \frac{BP}{Z}\right)$$

$$\ln\frac{\hat{f}_i^L}{Px_i} = \text{(See note for this expression in Table 14-4.)}$$

Table 14-7 The Peng-Robinson equation of state

---

1. Equation of state and constants

$$P = \frac{RT}{v - b} - \frac{\alpha a}{v(v + b) + b(v - b)}$$

For any component i, the constants a and b are given by

$$a_i = 0.45724 R^2 T_{ci}^2 / P_{ci}$$

$$b_i = 0.07780 RT_{ci} / P_{ci}$$

$$\alpha_i^{1/2} = 1 + m_i(1 - T_{ri}^{1/2})$$

$$m_i = 0.37464 + 1.54226\omega_i - 0.26992\omega_i^2$$

$$A_i = \frac{a_i^{1/2}\alpha_i^{1/2}}{RT}$$

$$B_i = \frac{b_i}{RT}$$

---

2. Residual work content

$$\tilde{A} = -RT\left[\ln\left(1 - \frac{BP}{Z}\right) + \frac{A^2}{2\sqrt{2}B}\ln\frac{Z + (1 + \sqrt{2})BP}{Z + (1 - \sqrt{2})BP}\right]$$

---

3. Pure components

$$h - h° = RT(Z - 1) + \frac{RTA^2}{2\sqrt{2}B}\left(\frac{T}{\alpha}\frac{d\alpha}{dT} - 1\right)\left[\ln\frac{Z + (1 + \sqrt{2})BP}{Z + (1 - \sqrt{2})BP}\right]$$

$$\ln\frac{f}{P} = Z - 1 - \ln(Z - BP) - \frac{A^2}{2\sqrt{2}B}\ln\frac{Z + (1 + \sqrt{2})BP}{Z + (1 - \sqrt{2})BP}$$

---

4. Mixtures

The mixing rules are as follows

$$A = \sum_i y_i A_i \qquad A^2 = \left(\sum_i y_i A_i\right)^2 = \sum_i \sum_j y_i y_j A_i A_j$$

$$B = \sum_i y_i B_i$$

When the interaction coefficient k_{ij} is included in the definition of a, one obtains

$$A^2 = \sum_i \sum_j y_i y_j A_i A_j (1 - k_{ij})$$

$$\Omega = h - h° = RT(Z - 1) + \frac{RTA^2}{2\sqrt{2}B}\left[\frac{2T}{A^2}\sum_i y_i \bar{A}_i^2\left(\frac{1}{\alpha_i^{1/2}}\frac{d\alpha_i^{1/2}}{dT}\right) - 1\right]\ln\frac{Z + (1 + \sqrt{2})BP}{Z + (1 - \sqrt{2})BP}$$

Table 14-7 (*continued*)

where

$$\bar{A}_i^2 = A_i \sum_j y_j A_j (1 - k_{ij})$$

$$\ln \frac{\hat{f}_i^V}{P y_i} = \frac{B_i}{B}(Z - 1) - \ln(Z - BP) - \frac{A^2}{2\sqrt{2}\,B}\left(\frac{2\bar{A}_i^2}{A^2} - \frac{B_i}{B}\right) \ln \frac{Z + (1 + \sqrt{2})BP}{Z + (1 - \sqrt{2})BP}$$

where

$$\bar{A}_i^2 = A_i \sum_j y_j A_j (1 - k_{ij})$$

$$\ln \frac{\hat{f}_i^L}{P x_i} = (\text{See note for this expression in Table 14-4.})$$

Table 14-8 The virial equation of state

1. Equation of state and constants

$$Z = \frac{Pv}{RT} = 1 + \frac{B}{v} + \frac{C}{v^2} + \cdots$$

where B and C are the second and third virial coefficients. Since values of C are generally unknown, the truncated form of this equation of state is most commonly used

$$Z = 1 + \frac{B}{v}$$

2. Residual work content

$$\tilde{A} = BRT\rho + \frac{CRT\rho^2}{2} + \cdots$$

$$\ln \frac{f}{P} = 2B\rho + \frac{3}{2}C\rho^2 + \cdots - \ln Z$$

3. Mixtures

The mixing rules are as follows

$$B = \sum_i \sum_j y_i y_j B_{ij} \qquad \text{the second virial coefficient}$$

$$C = \sum_i \sum_j \sum_k y_i y_j y_k C_{ijk} \qquad \text{the third virial coefficient}$$

$$\ln \frac{\hat{f}_i^V}{P y_i} = 2\rho \sum_j y_j B_{ij} + \frac{3\rho^2}{2} \sum_j \sum_k y_i y_k C_{ijk} - \ln Z$$

Table 14-9 Equations of state

---

1. The Lee-Erbar-Edmister equation of state[38]

$$P = \frac{RT}{v-b} - \frac{a}{v(v-b)} + \frac{bc}{v(v-b)(v+b)}$$

where a, b, and c are constants

---

2. The Sugie-Lu equation of state[66]

$$P = \frac{RT}{v-b+c} - \frac{aT^{-0.5}}{(v+c)(v+b+c)} + \sum_{j=1}^{10} \frac{d_j T + e_j T^{-0.5}}{v^{j+1}}$$

where a, b, c, $\{d_j\}$, and $\{e_j\}$ are constants

---

3. The Barner-Adler equation of state[3]

$$P = \frac{RT}{v-b} - \frac{af_a}{v(v-b)} + \frac{cf_c}{v(v-b)^2} - \frac{df_d}{v(v-b)^3} + \frac{ef_e}{v(v-b)^4}$$

where a, c, d, and e are constants and f_a, f_c, f_d, and f_e are functions defined by Barner and Adler.[3]

---

4. The Martin equation of state[44]

$$P_r = \frac{T_r}{Z_c v_r - B} - \frac{A(T_r)}{(Z_c v_r + C)^2}$$

where B and C are constants, Z_c is the experimental critical compressibility factor and $A(T_r)$ is a parameter which depends upon the reduced temperature, T_r.

---

engineers—the dew-point vapor region. Reasonably good results are said to be obtained near the critical point and at reduced volumes $v_r(v_r = v/v_c, \rho = 1/v)$ as low as 0.6 in the reduced temperature range of 1.0 to 1.5. The equation is not applicable to the liquid phase, nor is it recommended for reduced temperatures exceeding 1.5. Reliable results can be expected for gas-phase density, fugacity, and enthalpies of pure, nonpolar compounds and slightly nonpolar compounds.[3]

In addition to the equations of state listed in Table 14-9, other useful equations of state have been proposed. In order to cover the whole range of reduced temperatures T_r and reduced pressures P_r of practical interest in hydrocarbon processing, Lee and Kesler[39] proposed the use of two equations of state similar in form to the BWR equation of state, one for a simple fluid and one for a reference fluid. The Lee-Kesler correlation also makes use of the acentric factor.

14-3 USE OF MULTIPLE EQUATIONS OF STATE FOR THE CALCULATION OF K VALUES

The Chao-Seader method[12] is an example of the use of multiple equations of state for the calculation of K values. The Redlich-Kwong equation of state is used to compute the vapor-phase fugacity coefficient $\hat{\phi}_i^V$, the Hildebrand equation for the calculation of the liquid-phase activity coefficient γ_i^L, and an extension of Pitzer's modified form of the principle of corresponding states for the calculation of the liquid-phase fugacity coefficient ϕ_i^L.

A number of methods are presented for the prediction of K values. First, methods are presented for the calculation of K values for mixtures which form ideal solutions in both the vapor and liquid phases. Then the convergence-pressure method, the Chao-Seader, and the Grayson-Streed modification of the Chao-Seader method are presented.

Calculation of K Values for Mixtures Whose Vapor and Liquid Phases Form Ideal Solutions

For the case where both the vapor and the liquid phases form ideal solutions, the general expression for the K value given by Eqs. (14-37) and (14-38) reduces to

$$K_i = \frac{\phi_i^L}{\phi_i^V} = \frac{f_i^L}{f_i^V} = K_i^I \tag{14-93}$$

where f_i^L and f_i^V are the fugacities of the pure component i, evaluated at the temperature T and pressure P of the mixture.

Thus, to compute K values for ideal solutions, it is necessary only to evaluate the fugacities f_i^V and f_i^L at the temperature T and pressure P of the mixture. Except for the special case where the vapor forms a perfect gas which is treated below, the calculation of the K values is divided into two parts: (1) the calculation of the vapor fugacities for pure components, and (2) the calculation of the liquid fugacities for pure components.

Consider the next special case where both phases form ideal solutions and the vapor phase obeys the perfect gas law $(Pv = RT)$.

K VALUES FOR SYSTEMS IN WHICH THE VAPOR PHASE IS A PERFECT GAS

For nonpolar compounds, the perfect gas state is approached at relatively low pressures and temperatures, and f_i^V approaches P. As proposed by Lewis and Kay[45] the fugacity f_i^L is evaluated at the vapor pressure of component i at the temperature T, that is,

$$f_{i,\,P}^L \cong f_{i,\,P_i}^L = f_{i,\,P_i}^V \tag{14-94}$$

Again for nonpolar compounds at relatively low pressures and high temperatures, the vapor at P and T approaches a perfect gas ($f^V_{i,\,P_i} = P_i$) and Eq. (14-93) reduces to

$$K^l_i = \frac{P_i}{P} \qquad (14\text{-}95)$$

This equation has limited application because many liquids (such as polar compounds) do not form ideal solutions at low pressures.

CALCULATION OF THE VAPOR PHASE FUGACITY OF A PURE COMPONENT

Vapor fugacities for pure components may be computed by use of the equations of state which have been presented as well as others which have not been presented. Unless extensive testing has been carried out, the range of values of the variables given by the developers should not be exceeded.

To apply an equation of state, it is first solved for the vapor density or vapor compressibility factor at the given P and T, and then the value so obtained is used in the equation of state; see Tables 14-4 through 14-9.

Alternatively if compressibility factor charts are available, these Z factors may be used to compute the fugacity of pure component i by use of the following equation which is readily obtained by commencing with Eqs. (14-12) and (14-26)

$$\ln \frac{f^V_i}{P} = \int_0^{P_r} (Z - 1) \frac{dP_r}{P_r} \qquad (14\text{-}96)$$

where Z is evaluated at P and T, and P_r is of course equal to P/P_{ci}.

CALCULATION OF THE LIQUID PHASE FUGACITY OF A PURE COMPONENT

A formula for computing the variation of the fugacity of pure component i in the liquid phase with pressure at a given temperature is found by first restating Eq. (14-26) in the following form

$$\left(\frac{\partial \mu^L_i}{\partial P}\right)_T = RT\left(\frac{\partial \ln f^L_i}{\partial P}\right)_T \qquad (14\text{-}97)$$

The following relationship follows from Eq. (14-5) and (14-18)

$$\left(\frac{\partial \mu^L_i}{\partial P}\right)_T = \frac{1}{n}\left(\frac{\partial g^L}{\partial P}\right)_T = \frac{V}{n} = v^L_i \qquad (14\text{-}98)$$

Elimination of $(\partial \mu^L_i / \partial P)_T$ from Eqs. (14-97) and (14-98) followed by integration of the result so obtained from the vapor pressure P_i of pure component i at the temperature T to the total pressure P gives

$$f_i^L = f_{i,\,P_i}^L \exp \int_{P_i}^{P} \frac{v_i^L}{RT} \, dP \tag{14-99}$$

When two phases composed of the pure component i are in equilibrium at the temperature T and the corresponding saturation pressure P_i

$$f_{i,\,P_i}^V = f_{i,\,P_i}^L \tag{14-100}$$

Thus, Eq. (14-99) may be restated in the following form

$$f_i^L = \phi_i^s P_i \exp \int_{P_i}^{P} \frac{v_i^L}{RT} \, dP \tag{14-101}$$

where

$$\phi_i^s = f_{i,\,P_i}^V / P_i$$

The fugacity coefficient ϕ_i^s may be calculated by use of an equation of state or other methods for computing vapor fugacities as described above. (Note, in this case, however, f_i^V / P is replaced by $f_{i,\,P_i}^V / P_i$.) If the liquid state exists at all pressures between P_i and the total pressure P of the mixture, the fugacity of the liquid is found by integrating Eq. (14-101) which gives

$$f_i^L = \phi_i^s P_i \exp \left[\frac{v_i^L}{RT} (P - P_i) \right] \tag{14-102}$$

In the integration of the exponential term (called the Poynting factor), the effect of pressure on the molar volume of the liquid was neglected. Equations (14-96) and (14-102) represent direct methods for the calculation of the vapor and liquid fugacities by use of Z factor charts or equations of state, provided that both the vapor and liquid phases exist. There are, however, many sets of P and T for which either the vapor or the liquid phase does not exist for the pure component, and extrapolation procedures are required to find the corresponding fugacities. For the following sets of P and T, and extrapolation procedure is needed to compute either the vapor or the liquid fugacity:

1. If $T < T_{ci}$ and $P > P_i$, the vapor phase does not exist.
2. If $T < T_{ci}$ and $P < P_i$, the liquid phase does not exist.
3. If $T > T_{ci}$, the liquid phase does not exist at any P.

A number of methods have been used to estimate the fugacities for the hypothetical phases. Some of those suggested are enumerated below.

Case 1 $T < T_{ci}$ **and** $P > P_i$ Since the vapor phase does not exist, the fugacity of the vapor must be determined by extrapolation. Extrapolations may be carried out either isothermally or isobarically. For example, if the pressure is held fixed, then the values of the vapor fugacity coefficient ϕ_i^V may be evaluated at each of several values of T for which the vapor phase exists. A plot of ϕ_i^V versus T may then be extrapolated to the hypothetical vapor state at the specified P and T. A numerical example using this approach is presented in Ref. 29. A variety of other methods of extrapolation have been suggested.[24, 40, 61] For example Lewis and Kay[40] extrapolated plots of f_i^V versus $f_{i, P_i}^V / K_i$ while Souders et al.[61] extrapolated plots of log K versus log P.

Case 2 $T < T_{ci}$ **and** $P < P_i$ When the liquid phase does not exist, the liquid fugacity at the specified P and T may be estimated by use of an extrapolation at either constant temperature or constant pressure from the vapor to the hypothetical liquid phase. An example is given in Ref. 29 in which the extrapolation is carried out isobarically.

Case 3 $T > T_{ci}$ In this case, the liquid phase does not exist at any pressure. Lewis and Kay[40] suggested the extrapolation of a plot of log f_{i, P_i}^V or log P_i vs. $1/T_r$ while Souders et al.[61] used experimental K values and heats of solution data in their extrapolation procedure. Adler et al.[2] suggested the use of the Chao-Seader (or Grayson-Streed) equations or back-calculation from experimental vapor-liquid equilibrium data.

Calculation of K Values of the Aliphatic Hydrocarbons by Use of Convergence Pressure

The concept of convergence pressure has either been recognized or employed in the correlation of K data by a variety of investigators.[9, 23, 61] It appears to be an outgrowth by the plots of log K vs. log P suggested in 1932 by Souders et al.[61] The pressure at which all of the K's of a multicomponent mixture appear to approach unity on a log K vs. log P plot at constant temperature has come to be known as the convergence pressure. The possibility of predicting the convergence pressure of a multicomponent mixture by means of an equivalent binary mixture was suggested by Katz and Kurata.[34] Eventually such a method was developed by Hadden.[30, 31] The development of this method depends upon the phase behavior of binary mixtures, a discussion of which follows.

Consider the hypothetical system consisting of components A and B. Figure 14-6 was constructed on the basis of the general characteristics of the experimental results of several investigators[4, 36] and the discussions of others;[9, 18, 56, 58] it represents the phase behavior of a typical binary mixture. With respect to composition, the limiting conditions of the mixture are represented by the single curves for pure A and B. These are the vapor pressure curves for pure A and pure B, and they are seen to terminate at the critical temperature and pressure for each component. Each envelope contains all of the two-phase mixtures ranging from bubble-point liquid to dew-point vapor. The left- and right-hand boundaries of each envelope represent the bubble-point and dew-point curves, respectively.

The intersection of two envelopes gives the composition of the vapor and the liquid at the temperature and pressure that locate the point of intersection. For

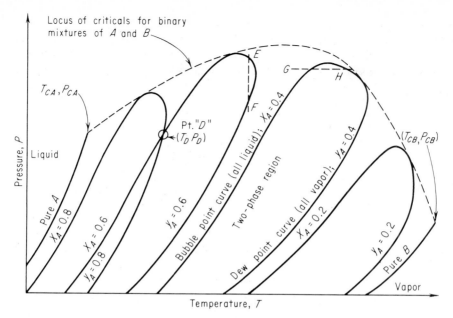

Figure 14-6 Phase behavior and locus of critical temperatures and pressures for binary mixtures of components A and B.

example, point D of Fig. 14-6 defines the equilibrium state

$$P = P_D \qquad T = T_D$$

$$y_A = 0.8 \qquad x_A = 0.6$$

$$K_A = 4/3 \qquad K_B = 0.2/0.4 = 1/2$$

At the critical temperature and pressure the liquid and vapor phases are indistinguishable. One such point exists for each envelope. The curve connecting the points for all envelopes is called the "locus of criticals."

The phenomenon of "retrograde condensation" is also illustrated by Fig. 14-6. As pointed out by Brown,[9] if either the bubble-point or dew-point curve is crossed twice while passing through the two-phase region by either isobaric or isothermal paths, retrograde condensation will occur. Lines EF and GH represent isothermal and isobaric paths, respectively, which would produce retrograde condensation.

CONVERGENCE PRESSURE

According to the Gibbs phase rule, Eq. (14-32), the number of degrees of freedom for a two-phase system is equal to the number of components. Thus for a binary there are two degrees of freedom so long as two phases persist. The

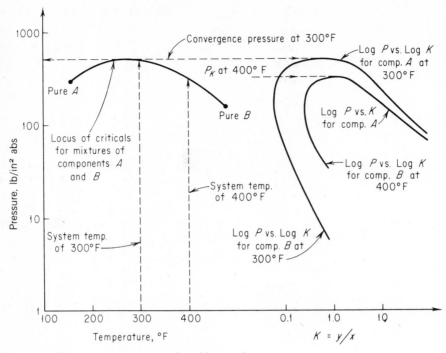

Figure 14-7 Convergence pressures for a binary mixture.

fixing of any two variables, such as the system pressure and the system temperature denoted by T_s and P_s, completely determines the properties of the system. That is both of the K_i's, as well as the vapor and liquid compositions, are determined along with convergence pressure, which is the ordinate of the point of intersection of the line $T_s = T_c$ and the locus of criticals as illustrated in Fig. 14-7. This type of graph was used Hadden.[31] It consists of a combined graph of the locus of criticals and the associated log P vs. log K isotherm.

For a ternary system with two phases present, there are three degrees of freedom, and three variables must be fixed in order to completely determine the system. Carter et al.[11] elected to fix T, P, and the composition parameter C_2, which is defined as follows

$$C_2 = x_2/(x_2 + x_3) \tag{14-103}$$

where the components are numbered in the order of decreasing volatility. For each set of values for P, T, and C_2, the convergence pressure is uniquely determined as illustrated in Fig. 14-8. The log P vs. log K plots were constructed on the basis of one choice of C_2, namely, $C_2 = 0.4$. This figure illustrates the fact that only the specification of C_2 and of the system temperature determines the convergence pressure and the log P vs. log K isotherm. Also, this figure demonstrates that a ternary system may be regarded as three binaries; components 1 and 2, components 1 and 3, and components 2 and 3. The broken line in Fig. 14-8 represents the locus of criticals for the ternary mixture with $C_2 = 0.4$.

Hadden[31] found that better correlations were obtained by use of a composition parameter based on mass (lb-mass or gm-mass) fractions rather than mole fractions. For a ternary mixture the mass-composition parameter is defined by

$$M_2 = m_2/(m_2 + m_3) \tag{14-104}$$

where the mass fractions are denoted by the m_i's.

Hadden found that the critical temperature of a mixture of components 2 and 3 in a ternary mixture at a specified value of M_2 could be computed with sufficient accuracy for use in subsequent correlations by use of the simple proportionality

$$\frac{t_{c3} - t_{c\overline{2}}}{M_2} = \frac{t_{c3} - t_{c2}}{1} \tag{14-105}$$

where $t_{c\overline{2}}$ = critical temperature (°F) of a mixture of components 2 and 3 at the specified value of M_2. The lower case t is used to emphasize the fact that the temperatures are in °F

t_{c2}, t_{c3} = critical temperatures of components 2 and 3. (Note that for developments in which c is used to identify the critical temperature and pressure, n is used to represent the total number of components)

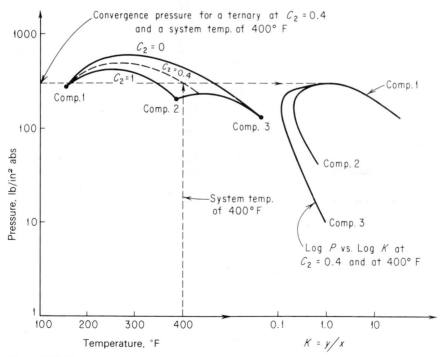

Figure 14-8 Convergence pressure for ternary systems.

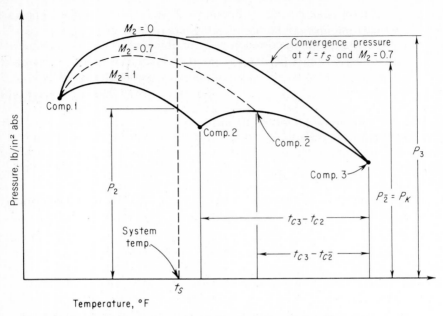

Figure 14-9 Convergence pressure and critical temperature relationships.

These temperature differences are shown in Fig. 14-9. Substitution of the definition of M_2 into Eq. (14-105), followed by rearrangement yields

$$t_{c\bar{2}} = \frac{m_2 t_{c2} + m_3 t_{c3}}{m_2 + m_3} \tag{14-106}$$

which is recognized as the weight average of the critical temperatures of components 2 and 3.

Calculation of the convergence pressure for a ternary mixture may be accomplished by the following formula, proposed by Winn,[75]

$$(P_{\bar{2}} - P_3)/M_2 = (P_2 - P_3)/1 \tag{14-107}$$

where P_2, P_3 = convergence pressure for the binaries composed of components (1, 2) and (1, 3), respectively, at the system temperature t_s. If $t_s > T_{c2}$ then P_2 is taken equal to the critical pressure for component 2, that is $P_2 = P_{c2}$

$P_K = P_{\bar{2}}$ = convergence pressure for the ternary mixture at the specified values of M_2 and t_s

These pressure differences are likewise depicted in Fig. 14-9, and in view of the definition of M_2, Eq. (14-107) reduces to

$$P_K = P_{\bar{2}} = \frac{m_2 P_2 + m_3 P_3}{m_2 + m_3} \tag{14-108}$$

Equations (14-106) and (14-108) and Fig. 14-9 strongly suggest the representation of a ternary mixture (at M_2 fixed) as an equivalent binary composed of component 1 and the pseudoheavy component $\bar{2}$. Equation (14-106) serves to define the pseudoheavy as that hypothetical component having a critical temperature $t_{c\bar{2}}$ (the weight average temperature of components 2 and 3 at the specified value of M_2). The binary composed of components 1 and $\bar{2}$ has a convergence pressure $P_K = P_{\bar{2}}$, which is also of convergence pressure of the ternary mixture. Thus, for the determination of convergence pressure, the ternary may be replaced by an equivalent binary mixture.

MULTICOMPONENT MIXTURES

The above concepts may be extended as demonstrated elsewhere.[26] The critical temperature of the pseudoheavy component is given by the following extension of Eq. (14-106), namely,

$$
t_{c\bar{2}} = \frac{\sum\limits_{i=2}^{n} m_i t_{ci}}{\sum\limits_{i=2}^{n} m_i} \tag{14-109}
$$

where component 1 is taken to be the lightest component of the mixture.

The convergence pressure is given by an extension of Eq. (14-108), namely,

$$
P_K = P_{c\bar{2}} = \frac{\sum\limits_{i=2}^{n} m_i P_{ci}}{\sum\limits_{i=2}^{n} m_i} \tag{14-110}
$$

Of these formulas, only the one for $t_{c\bar{2}}$ given by Eq. (14-109) is used in the correlation which follows.

The GPSA (Gas Processors Suppliers Association) charts as well as other correlations are based on the postulate that there exists one and only one log P vs. log K_i chart for each choice of $\{P_s, t_s, t_{c\bar{2}}\}$ or $\{P_s, t_s, P_K\}$, that is

$$
K_i = K_i(P_s, t_s, t_{c\bar{2}}) = K_i(P_s, t_s, P_K) \tag{14-111}
$$

THE GPSA (GAS PROCESSORS SUPPLIERS ASSOCIATION) CHARTS

Charts containing plots of log K vs. log P for each of several convergence pressures are presented in the *Engineering Data Book*.[19] Charts are presented for nitrogen, methane, ethylene, ethane, propylene, propane, *i*-butane, *n*-butane, *i*-pentane, *n*-pentane, hexane, octane, decane, hydrogen sulfide, selected binaries, and the normal boiling fractions.

In the use of these charts, the pseudoheavy component of the mixture is determined by use of Eq. (14-109). The intersection of the line $t = t_{c\bar{2}}$ with the locus of criticals determines the critical pressure for component $\bar{2}$. If component

$\bar{2}$ does not happen to correspond to a pure component, a symmetrical locus is sketched as shown in Fig. 14-10.

The intersection of this curve with the line $t = t_s$ determines the convergence pressure P_K. The K value for any particular component at the system pressure and temperature is then found by use of the appropriate log K vs. log P plot as demonstrated in the lower diagram of Fig. 14-10.

In the determination of the bubble-point temperature corresponding to a specified system pressure and liquid-phase composition, the procedure described might appear to lead to a trial determination of both the system temperature and the convergence pressure. However, because of the relatively small dependence of the K_i's on the convergence pressure, the trial-and-error procedure is reduced primarily to the finding of the bubble-point temperature. This characteristic relationship between the K_i's and P_K permits the use of a single convergence pressure over either the entire column or several plates.

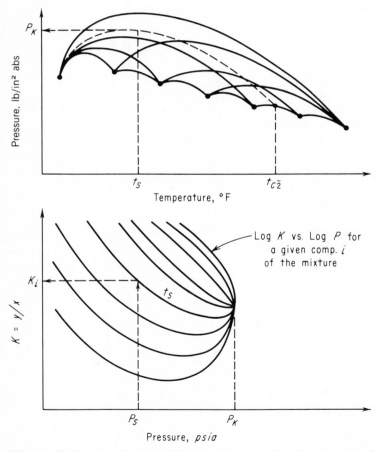

Figure 14-10 Sketches of typical GPSA (Gas Processors Suppliers Association) charts.

If the system temperature is less than the critical temperature of the lightest component of the mixture, a convergence pressure for the mixture does not exist. However, in order to use the charts, one must be chosen. Following Hadden,[31] this pressure is called the "quasi-convergence" pressure. Both Hadden[31] and Winn[75] recommend that for such systems the convergence pressure be taken equal to the critical pressure of the lightest component. Another choice has been recommended by Cajander et al.[10]

Another problem which sometimes arises is the choice of the light component of the equivalent binary mixture for multicomponent mixtures that are dilute solutions of the relatively light components. Hadden[31] recommended that the light component of the equivalent binary be selected as the lightest component of the mixture having a mole fraction (in the liquid phase) equal to or greater than 0.001.

Work continues in the area of graphical correlations as is witnessed by the recent correlation of Kesler et al.[37] The authors present a single nomograph for the hydrocarbons boiling above 210°F.

The Chao-Seader Method for the Calculation of K Values[12]

The Chao-Seader method uses the Redlich-Kwong equation of state for the calculation of $\hat{\phi}_i^V$, Hildebrand's equation for the calculation of the liquid activity coefficient γ_i^L, and an extension of Pitzer's modified form of the principle of corresponding states for the calculation of the liquid fugacity ratio ϕ_i^L.

To compute $\hat{\phi}_i^V$, one first determines the vapor root of the Redlich-Kwong equation. Then $\hat{\phi}_i^V$ is computed by use of the expression given for $\ln \hat{f}_i^V / P y_i$ in Table 14-5.

The liquid-phase fugacity coefficient $\phi_i^L = f_i^L/P$ may be calculated from a generalized correlation in terms of reduced temperature and pressure such as those of Lydersen et al.[42] and Curl and Pitzer.[15] Chao and Seader used a modified form of the Curl and Pitzer correlation. The correlation was modified by use of experimental data such that appropriate values of ϕ_i^L could be computed for the case where a component does not exist as a liquid and for the case of low temperatures. The following expression was proposed for the calculation of the fugacity coefficient for any component i in the liquid phase

$$\log_{10} \phi^L = \log_{10} \phi^{(0)} + \omega \log_{10} \phi^{(1)} \qquad (14\text{-}112)$$

The quantity $\phi^{(0)}$ is given by

$$\log_{10} \phi^{(0)} = A_0 + A_1 T_r + A_2 T_r + A_3 T_r^2 + A_4 T_r^3$$
$$+ (A_5 + A_6 T_r + A_7 T_r^2)P_r + (A_8 + A_9 T_r)P_r^2 - \log_{10} P_r \qquad (14\text{-}113)$$

The quantity $\phi^{(1)}$ is given by

$$\log_{10} \phi^{(1)} = -4.23893 + 8.65808 T_r - 1.22060/T_r$$
$$- 3.15224 T_r^3 - 0.025(P_r - 0.6) \qquad (14\text{-}114)$$

The coefficients for Eq. (14-113) which were given by Chao and Seader are listed in Table 14-10. Special coefficients for the A's of Eq. (14-113) for methane and hydrogen, are given in Table 14-10 for these components. When these A's are used, the acentric factor ω is set equal to zero.

Three constants must be known for each component in order to determine its liquid-fugacity coefficient ϕ_i^L. These are the critical temperature, critical pressure, and the acentric factor ω. Values of ω presented by Chao and Seader[12] are listed in Table 14-10.

The activity coefficients γ_i^L were predicted by use of Hildebrand's equation

$$\ln \gamma_i^L = \frac{v_i^L(\delta_i - \delta)^2}{RT} \tag{14-115}$$

To use this equation, two constants must be known for each component, the solubility parameter δ_i and the molar volume v_i^L. The average value of the solubility parameter δ for the solution is computed as follows

$$\delta = \frac{\sum_i x_i v_i^L \delta_i}{\sum_i x_i v_i^L} \tag{14-116}$$

A list of values of δ_i and v_i^L for several components given by Chao and Seader are reproduced in Table 14-10.

The following restrictions were stated by Chao and Seader.[12]

(a) For hydrocarbons (except methane)
Reduced temperature range: 0.5 to 1.3 (based on the pure component critical temperature).
Pressure: pressures up to 3,000 lb/in^2 abs but not to exceed 0.8 of the critical pressure of the system.

(b) For light gases—hydrogen and methane
Temperature range: $-100°F$ to 0.93 of the pseudo reduced temperature of the equilibrium mixture but not to exceed 500°F. The pseudo reduced temperature is based on the mole fraction average of the critical temperatures of the components.
Pressures: up to about 8,000 lb/in^2 abs.
Concentration: up to about 20 mole percent of other dissolved gases in the liquid.

The Grayson-Streed Modification of the Chao-Seader Method for Calculation of K Values[25]

Experimental vapor-liquid equilibrium data obtained by Grayson and Streed[25] were used to extend the temperature range of the Chao-Seader K-value method

Table 14-10 Constants for use in the Chao-Seader correlation

1. Coefficients for the Chao-Seader correlation of the liquid fugacity coefficient[12]

Coefficient	Simple fluid	Methane	Hydrogen
A_0	5.75748	2.43840	1.96718
A_1	-3.01761	-2.24550	1.02972
A_2	-4.98500	-0.34084	-0.054009
A_3	2.02299	0.00212	0.0005288
A_4	0.0	-0.00223	0.0
A_5	0.08427	0.10486	0.008585
A_6	0.26667	-0.03691	0.0
A_7	-0.31138	0.0	0.0
A_8	-0.02655	0.0	0.0
A_9	0.02883	0.0	0.0

2. Pure-component constants used by Chao-Seader[12] and Grayson-Streed[25] for the Hildebrand equation

Substance	T_c† °R	P_c† lb/in² abs	Modified ω	δ (cal/cm³)$^{1/2}$	v^L cm³/g mol
Hydrogen	60.2	....	....	3.25	32
Paraffins					
Methane	343.9	673.1		5.68	52
Ethane	550.0	709.8	0.1064	6.05	68
Propane	665.7	617.4	0.1538	6.40	84
i-butane	734.7	529.1	0.1825	6.73	105.5
n-butane	765.3	550.7	0.1953	6.73	101.4
i-pentane	829.8	483.0	0.2104	7.02	117.4
n-pentane	845.6	489.5	0.2387	7.02	116.1
neo-pentane	780.3	464.0	(0.195)	7.02	123.3
n-hexane	914.2	440.0	0.2927	7.27	131.6
n-heptane	972.3	396.8	0.3403	7.430	147.5
n-octane	1024.9	362.1	0.3992	7.551	163.5
n-nonane	1073.0	332	0.4439	7.65	179.6
n-decane	1114	304	0.4869	7.72	196.0
n-undecane	1154	282	0.5210	7.79	212.2
n-dodecane	1188	262	0.5610	7.84	228.6
n-tridecane	1220	250	0.6002	7.89	244.9
n-tetradecane	1251	235	0.6399	7.92	261.9
n-pentadecane	1278	220	0.6743	7.96	277.8
n-hexadecane	1303	206	0.7078	7.99	294.1
n-heptadecane	1328	191	0.7327	8.03	310.4
Olefins					
Ethylene	509.5	742.1	0.0949	6.08	61
Propylene	657.2	667.0	0.1451	6.43	79
1-butene	755.3	583.0	0.2085	6.76	95.3
cis-2-butene	770	600.0	0.2575	6.76	91.2

(continued on page 538.)

Table 14-10 (*continued*)

Substance	T_c† °R	P_c† lb/in² abs	Modified ω	δ (cal/cm³)$^{1/2}$	v^L cm³/g mol
Olefins					
trans-2-butene	770	600.0	0.2230	6.76	93.8
i-butene	752.2	579.8	0.1975	6.76	95.4
1, 3-butadiene	765	628	0.2028	6.94	88.0
1-pentene	853	586	0.2198	7.05	110.4
cis-2-pentene	(866)	(518.8)	(0.206)	7.05	107.8
trans-2-pentane	(863)	(515.8)	(0.209)	7.05	109.0
2-methyl-1-butene	(850)	(514.4)	(0.200)	7.05	108.7
3-methyl-1-butene	(831)	(507)	(0.149)	7.05	112.8
2-methyl-2-butene	(870)	(527.6)	(0.212)	7.05	106.7
1-hexane	(920)	471.7	0.2463	(7.40)	125.8
Naphthenes					
Cyclopentane	921.2	655.0	0.2051	8.11	94.7
Methylcyclopentane	959.0	599.0	0.2346	7.85	113.1
Cyclohexane	977.7	561.0	0.2032	8.20	108.7
Methylcyclohexane	1030.2	504.4	0.2421	7.83	128.3
Aromatics					
Benzene	1012.7	714	0.2130	9.16	89.4
Toluene	1069.2	590	0.2591	8.92	106.8
o-xylene	1138.3	530	0.2904	8.99	121.2
m-xylene	1114.9	510	0.3045	8.82	123.5
p-xylene	1113.1	500	0.2969	8.77	124.0
Ethylbenzene	1115.8	540	0.2936	8.79	123.1

† Taken from Grayson and Streed.[25]

from 500 to 800°F at pressures up to 3,000 lb/in² abs. New equations for computing the liquid-fugacity coefficient for the heavier hydrocarbons were developed. The K values for the heavier hydrocarbons were correlated in terms of normal boiling fractions (125–150°F fraction, 150–200°F fraction, 200–250°F fraction, 650–700°F fraction).

For all components, the vapor-fugacity coefficients $\hat{\phi}_i^V$ were computed by use of the Redlich-Kwong equation of state as originally proposed by Chao and Seader. The fugacity coefficient ϕ_i^L ($\phi_i^L = f_i^L/P$) for any component i in the liquid phase was again correlated by use of Eq. (14-112). In this case, however, the new data obtained by Grayson and Streed were used to obtain an improved set of coefficients in the following correlation

$$\log_{10} \phi^{(0)} = A_0 + A_1/T_r + A_2 T_r + A_3 T_r^2 + A_4 T_r^3$$
$$+ (A_5 + A_6 T_r + A_7 T_r^2)P_r + (A_8 + A_9 T_r)P_r^2 - \log_{10} P_r \quad (14\text{-}117)$$

These coefficients obtained by Grayson and Streed are presented in Table 14-11. When Eq. (14-112) is applied to methane and hydrogen, ω is set equal to zero. The quantity $\phi^{(1)}$ is calculated by use of Eq. (14-114). In this case, however, the limits of this equation were altered in order to fit the high-temperature data. In particular, for values of $T_r > 1.0$, the values of $\phi^{(1)}$ were set equal to the value at $T_r = 1.0$.

For the 45 pure components, the values of δ_i and v_i^L are the same as those given by Chao and Seader (see Table 14-10). The value of δ for the fraction were computed by Grayson and Streed, by use of the following relationship for each component in the liquid state

$$\delta = \left(\frac{\Delta H - RT}{v^L} \right)^{1/2} \tag{14-118}$$

where ΔH is the latent heat of vaporization in cal per g-mol. For the pure components shown in Table 14-11, ω was computed by use of Eq. (14-91) and for the fractions, the value of ω given in Table 14-11 was computed by use of the following formula

$$\omega = \frac{3}{7} \left[\frac{\log_{10} (P_c/14.7)}{T_c/T_b - 1} \right] - 1 \tag{14-119}$$

where T_b denotes the normal boiling-point temperature in °R, T_c is critical temperature in °R, and P_c is the critical pressure in lb/in² abs.

Calculation of Enthalpies for Mixtures of Nonpolar to Slightly Polar Compounds

For computing the enthalpy departures of pure gases and gas mixtures, the equations of state of Sec. 14-2 may be used for both the vapor and liquid phases: (1) modified versions of the BWR equation of state, such as the one proposed by Starling, (2) the Soave-Redlich-Kwong, and (3) the Peng-Robinson equation of state. Another method which has been tested extensively and found to give accurate results even for polar gas mixtures is the Yen-Alexander correlation.[78]

For hydrocarbon gas mixtures (including the light gases such as N_2, CO_2, and H_2S), Reid et al.[55] recommend the Soave modification of the Redlich-Kwong equation, the Lee-Kesler equation, or the Lee-Erbar-Edmister equation. If the mixture contains hydrogen, the Soave equation should be used. For gas mixtures containing nonhydrocarbons, Reid et al.[55] have recommended the Lee-Kesler correlation, the Yen-Alexander correlation, the Barner-Adler equation of state, or the Sugie-Lu equation of state.

Enthalpy departures for liquid mixtures of hydrocarbons which do not contain hydrogen may be calculated by use of the Soave modification of the Redlich-Kwong equation of state.

Table 14-11 Constants for use in the Grayson-Streed correlation[25]

1. Coefficients for use in the Grayson-Streed correlation of the liquid fugacity coefficient

Constant	Simple fluid	Methane	Hydrogen
A_0	2.05135	1.36822	1.50709
A_1	−2.10899	−1.54831	2.74283
A_2	0.0	0.0	−0.02110
A_3	−0.19396	0.02889	0.00011
A_4	0.02282	−0.01076	0.0
A_5	0.08852	0.10486	0.008585
A_6	0.0	−0.02529	0.0
A_7	−0.00872	0.0	0.0
A_8	−0.00353	0.0	0.0
A_9	0.00203	0.0	0.0

2. Constants for use in the Hildebrand equation for the petroleum fractions

Petroleum fraction	T_c °R	P_c lb/in² abs	ω	δ (cal/cm³)$^{1/2}$	v^L cm³/g mol
125–150	903	477	0.2759	7.45	127.7
150–200	945	458	0.3177	7.53	135.3
200–250	994	440	0.3573	7.70	142.9
250–300 (A)	1043	420	0.3910	7.87	152.9
250–300 (R)	1050	430	0.4020	7.96	151.4
300–350 (D)	1100	393	0.4275	8.01	167.5
300–350 (C)	1113	400	0.4042	8.12	164.7
350–400	1159	371	0.4536	8.18	181.4
400–450 (A)	1216	343	0.4926	8.27	200.9
400–450 (R)	1210	330	0.4941	8.17	205.3
450–500 (A)	1260	320	0.5332	8.28	219.8
450–500 (R)	1260	290	0.5645	8.13	236.8
500–550 (A)	1305	288	0.5844	8.26	245.3
500–550 (R)	1290	265	0.5888	8.06	260.2
550–600 (A)	1350	262	0.6441	8.24	27˙.8
550–600 (R)	1325	240	0.6668	8.02	288.2
600–650 (A)	1403	234	0.6682	8.20	307.1
600–650 (R)	1370	210	0.7291	7.94	325.4
650–700 (A)	1442	213	0.7659	8.17	336.0
650–700 (R)	1415	180	0.8090	7.84	377.6
700 + (A)	1580	147	1.2127	7.98	473.5
700 + (R)	1620	100	1.6848	7.67	647.7

Note: Code letters refer to different gas oils and different hydrocracking conversion levels used in the experiments:

A = heavy gas oils at 40–90% conversion
C = raffinate from propane deasphalting of heavy gas oils, both at a low conversion level (15%)
D = heavy gas oils at a somewhat higher conversion level (25%)
R = raffinate from propane deasphalting at 40–90% conversion

14-4 ESTIMATION OF THE FUGACITY COEFFICIENTS FOR THE VAPOR PHASE OF POLAR GAS MIXTURES

The use of the virial equation of state and the correlations of Pitzer and Curl[15, 48, 49] for the prediction of fugacity coefficients for nonpolar mixtures is presented first and then the method is extended to polar gas mixtures.

Prediction of the Fugacities Coefficients for Polar and Nonpolar Mixtures

In order to compute f_i^V/P or $\hat{f}_i^V/Py_i$, by the truncated virial equation of state (see Table 14-8), it is necessary to evaluate the second virial coefficient B. There follows a summary of the formulas needed to evaluate the second virial coefficient for both polar and nonpolar compounds.

ESTIMATION OF THE SECOND VIRIAL COEFFICIENT B_{ii} FOR PURE NONPOLAR GASES

Pitzer and Curl[48, 49] developed the following correlation for B_{ii}, namely,

$$\frac{P_{ci}B_{ii}}{RT_{ci}} = f^{(0)}(T_r) + \omega_i f^{(1)}(T_r) \tag{14-120}$$

where $T_r = T/T_{ci}$. The acentric factor ω_i is defined by Eq. (14-91). The functional forms for $f^{(0)}$ and $f^{(1)}$ which were originally proposed by Pitzer and Curl[48] were modified by Tsonopoulos[67] to give

$$f^{(0)}(T_r) = 0.1445 - \frac{0.330}{T_r} - \frac{0.1385}{T_r^2} - \frac{0.0121}{T_r^3} - \frac{0.000607}{T_r^8} \tag{14-121}$$

$$f^{(1)}(T_r) = 0.0637 + \frac{0.331}{T_r^2} - \frac{0.423}{T_r^3} - \frac{0.008}{T_r^8} \tag{14-122}$$

ESTIMATION OF THE SECOND VIRIAL COEFFICIENT FOR PURE POLAR GASES

The following correlations were given by Tsonopoulos[67] for two broad classes of polar compounds, namely, nonhydrogen bonding and hydrogen bonding compounds.

Tsonopoulos recommends the following correlations

$$\frac{P_{ci}B_{ii}}{RT_{ci}} = f^{(0)}(T_r) + \omega_i f^{(1)}(T_r) + f^{(2)}(T_r) \tag{14-123}$$

The function $f^{(2)}(T_r)$ is defined as follows for the nonhydrogen bonding polar gases

$$f^{(2)}(T_r) = \frac{a}{T_r^6} \tag{14-124}$$

For ketones

$$a = -0.00020483\mu_r \tag{14-125}$$

For ethers

$$\ln(-a) = -12.63147 + 2.09681 \ln \mu_r \tag{14-126}$$

The reduced dipole moment is defined by

$$\mu_r = \frac{10^5 \mu_i^2 P_{ci}}{T_{ci}^2} \tag{14-127}$$

where the dipole moment μ_i is in Debye, the critical pressure is in atmospheres, and the critical temperature is in degrees Kelvin.

For hydrogen-bonding polar gases, the following form of the function $f^{(2)}(T_r)$ of Eq. (14-124) is recommended

$$f^{(2)}(T_r) = \frac{a}{T_r^6} - \frac{b}{T_r^8} \tag{14-128}$$

For straight-chain 1-alkanols

$$a = 0.0878$$

$$b = 0.00908 + 0.0006957\mu_r \tag{14-129}$$

Estimation of the Cross-Coefficient $B_{ij} (i \neq j)$

Expressions recommended by Tsonopoulos[67] for the cross coefficients are obtained from Eqs. (14-121) through (14-124) and (14-127) by replacing B_{ii} by B_{ij}, P_{ci} by P_{cij}, T_{ci} by P_{cij}, ω_i by ω_{ij}, and T_r is now equal to T/T_{cij}

$$T_{cij} = (T_{ci} T_{cj})^{1/2}(1 - k_{ij}) \tag{14-130}$$

$$\omega_{ij} = 1/2(\omega_i + \omega_j)$$

$$P_{cij} = 4T_{cij} \frac{(P_{ci} v_{ci}/T_{ci} + P_{cj} v_{cj}/T_{cj})}{(v_{ci}^{1/3} + v_{cj}^{1/3})^3}$$

where v_c is the molar critical volume and k_{ij} is a characteristic constant for each binary.

The expressions given by Eq. (14-130) suffice for the description of nonpolar-nonpolar binaries. Characteristics constants for binary constants for nonpolar systems have been tabulated by Chuch and Prausnitz[13] and a correlation for k_{ij} for the gases such as H_2, He, etc. has been given by Hiza and Duncan.[33] Tabulations of values of k_{ij} for various types of binary mixtures are also given by Tsonopoulos.

For polar-nonpolar binaries, B_{ij} is assumed to have no polar term, and

$$a_{ij} = 0 \tag{14-131}$$

$$b_{ij} = 0 \tag{14-132}$$

For polar-polar binaries, the polar contribution to B_{ij} is calculated by assuming that

$$a_{ij} = 0.5(a_i + a_j) \tag{14-133}$$

$$b_{ij} = 0.5(b_i + b_j) \tag{14-134}$$

For a given temperature, pressure, and composition, the corresponding density and Z factor given by the equation of state is needed. The vapor root is found by solving the truncated virial equation of state (see Table 14-8) for Z, and for small values of BP/RT, the expression so obtained reduces to $Z = 1 + BP/RT$, that is,

$$Z = \frac{1 + \sqrt{1 + 4BP/RT}}{2} \cong \frac{1}{2} + \frac{1}{2}\left[1 + \frac{1}{2}\left(\frac{4BP}{RT}\right)\right] = 1 + \frac{BP}{RT} \tag{14-135}$$

The value of Z so obtained is used to compute the corresponding value of the density which is needed in the evaluation of the expressions for the fugacities.

14-5 CALCULATION OF LIQUID-PHASE ACTIVITY COEFFICIENTS

The composition of the liquid phase can have a profound effect on the K value for a given component of a mixture. In fact, experimental evidence shows that large variations in the K's result from changes in the liquid compositions at pressures so low that the vapor not only forms an ideal solution but may also behave as a perfect gas mixture. Thus, for many systems, Eq. (14-37) reduces to

$$K_i = \frac{\gamma_i^L f_i^L}{\gamma_i^V f_i^V} \cong \frac{\gamma_i^L f_i^L}{P} \cong \frac{\gamma_i^L P_i}{P} \tag{14-136}$$

In spite of the simplicity of this relationship, the expressions for computing the activity coefficients for multicomponent mixtures are quite cumbersome. Since the equations of Wohl,[76] Van Laar,[68] Margules,[43] Scatchard and Hamer,[57] Redlich and Kister,[52] and others are well documented in the literature, a restatement of these equations is not presented. Instead, a brief introduction to the newer methods including the Wilson equation, the NRTL, UNIQUAC, ASOG, and the UNIFAC methods is presented.

There follows a presentation of certain fundamental relationships needed in the prediction of activity coefficients for the liquid phase.

Thermodynamic Relationships for Fugacities and Activity Coefficients for the Liquid Phase

This section is divided into three parts. First, condensable components and then noncondensable components are considered. Then the use of the excess free energy function in the calculation of the activity coefficient is presented.

CONDENSABLE COMPONENTS

A condensable component (or subcritical component) is one whose critical temperature is greater than the temperature of the mixture. The expression proposed by Prausnitz et al.[50] for computing the fugacity $\hat{f}_i^L$ of a component in a liquid mixture at the pressure P and temperature T may be obtained as follows. From the definition of the fugacity of a component in a liquid mixture [Eq. (14-29)], the following expression is obtained

$$\left(\frac{\partial \mu_i^L}{\partial P}\right)_{T,\,n_i} = RT\left(\frac{\partial \ln \hat{f}_i^L}{\partial P}\right)_{T,\,n_i} \tag{14-137}$$

From Eqs. (14-3) and (14-18), $(\partial \mu_i^L/\partial P)_{T,\,n_i} = \bar{V}_i^L$, and consequently

$$\bar{V}_i^L = RT\left(\frac{\partial \ln \hat{f}_i^L}{\partial P}\right)_{T,\,n_i} \tag{14-138}$$

Thus

$$\int_{P_0}^{P}\left(\frac{\partial \ln \hat{f}_i^L}{\partial P}\right)_{T,\,n_i} dP = \int_{P_0}^{P}\frac{\bar{V}_i^L}{RT}\,dP \tag{14-139}$$

and integration yields

$$\hat{f}_i^L = \hat{f}_i^{OL} \exp \int_{P_0}^{P}\frac{\bar{V}_i^L}{RT}\,dP \tag{14-140}$$

where $\hat{f}_i^{OL}$ is the fugacity of component i in the liquid mixture at the temperature T and the standard state pressure P_0. Use of Eq. (14-35) gives

$$\hat{f}_i^{OL} = \gamma_i^{OL} f_i^{OL} x_i \tag{14-141}$$

where γ_i^{OL} is the activity coefficient of component i in the mixture at the temperature T and pressure P_0, and f_i^{OL} is the fugacity of pure component i in the liquid phase at the temperature T and the standard state pressure P_0. The expression suggested by Prausnitz is obtained by combining Eqs. (14-140) and (14-141), namely,

$$\hat{f}_i^L = \gamma_i^{OL} f_i^{OL} x_i \exp \int_{P_0}^{P}\frac{\bar{V}_i^L}{RT}\,dP \tag{14-142}$$

When the solution is far from its critical conditions, $\bar{V}_i^L$ does not usually differ appreciably from v_i^l and may be replaced by it in Eq. (14-142); otherwise, $\bar{V}_i^L$ may be evaluated by use of an equation of state or correlations given by Prausnitz.[50] The activity coefficient may be adjusted to the pressure P_0 by use of Eq. (14-149). The fugacity f_i^{OL} of the pure component i may be calculated from Eq. (14-99) by replacing the upper limit P by P_0 to give

$$f_i^{OL} = f_{i,\,P_i}^L \exp \int_{P_i}^{P_0}\frac{v_i^L}{RT}\,dP = \phi_i^s P_i \exp \int_{P_i}^{P_0}\frac{v_i^L}{RT}\,dP$$

$$\cong \phi_i^s P_i v_i^L \exp\,(P_0 - P_i) \tag{14-143}$$

where P_i = vapor pressure of pure component i at the temperature T

f^L_{i,P_i} = fugacity of pure component i in the liquid state at a pressure equal to its vapor pressure P_i at the temperature T

$\phi^s_i = f^L_{i,P_i}/P_i = f^V_{i,P_i}/P_i$ = fugacity coefficient of component i at its vapor pressure P_i at the temperature T

If the vapor pressure P_i is well below the critical pressure, then ϕ^s_i may be computed by use of an equation of state. If the pressure P_0 is allowed to go to zero, the quantity f^{0L}_i computed by use of Eq. (14-143) becomes the "liquid fugacity extrapolated to zero pressure."

NONCONDENSABLE COMPONENTS

A noncondensable component i is defined as one which is a constituent of a mixture which is at a temperature very much above the critical temperature of component i.

For a noncondensable component, the standard state fugacity may be defined as Henry's constant for that component in a pure reference solvent R which is a constituent of the solvent mixture. As the composition of the solution approaches that of the pure solvent R ($x_R \to 1$, $x_i \to 0$), the solution approximates an ideal solution and $\gamma^L_i \to 1$. Thus

$$\lim_{\substack{x_i \to 0 \\ x_R \to 1}} \frac{\hat{f}^L_i}{x_i} = \lim_{\substack{x_i \to 0 \\ x_R \to 1}} \gamma^{0L}_i f^{0L}_i = f^{0L}_i = \mathcal{H}^0_i \qquad (14\text{-}144)$$

Thus, for a noncondensable component, Eq. (14-142) is applicable where the standard state fugacity f^{0L}_i is equal to the Henry law constant H^0_i (evaluated at zero pressure). Henry law constants are found experimentally at the vapor pressure P_R of the reference solvent R. To correct the Henry law constant to zero pressure, Eq. (14-142) may be applied to give

$$\mathcal{H}^0_i = \mathcal{H}_i \bigg|_{P_R} \exp \int_{P_R}^0 \frac{\bar{V}^\infty_{i,R}}{RT} dP \qquad (14\text{-}145)$$

Where $\bar{V}^\infty_{i,R}$ is the molar volume (or to be more precise the partial molar volume) which component i would have in a solution which is infinitely dilute in the solvent R. Correlations of the partial molar volumes for noncondensable components have been given by Prausnitz et al.[50]

CALCULATION OF ACTIVITY COEFFICIENTS

In the above equations, the activity coefficient adjusted to pressure P_0 appears. In order to develop the formula needed to adjust the activity coefficient from the system pressure P to the reference pressure P_0, one may begin with an equation for the liquid phase which is of the same form as Eq. (14-44), namely,

$$\bar{G}^L_i(P, T, \{n_i\}) - g^L_i(P, T) = RT \ln \frac{\hat{f}^L_i}{f^L_i} \qquad (14\text{-}146)$$

Partial differentiation of each member of Eq. (14-46) with respect to pressure yields

$$\bar{V}_i^L - v_i^L = RT\left(\frac{\partial \ln \hat{f}_i^L/f_i^L}{\partial P}\right)_{T, n_i} \tag{14-147}$$

From Eq. (14-34), it follows that

$$\bar{V}_i^L - v_i^L = RT\left(\frac{\partial \ln \gamma_i^L}{\partial P}\right)_{T, n_i} \tag{14-148}$$

Integration of this expression with respect to pressure from P to P_0 yields the following result upon rearrangement

$$\gamma_i^{0L} = \gamma_i^L \exp \int_P^{P_0} \frac{\bar{V}_i^L - v_i^L}{RT} dP \tag{14-149}$$

Thus, if the activity coefficient γ_i^L is either known or can be calculated by use of a correlation at any pressure P, it may be adjusted to the reference pressure P_0 through the use of Eq. (14-149).

Activity coefficients may be estimated by use of a number of models such as those enumerated in the introduction of Sec. 14-5. These models are generally stated in terms of the excess free energy per mole of solution. The excess free energy g^E is the difference in free energy of one mole of the actual mixture minus the free energy which the mixture would have if it were an ideal solution, that is,

$$ng^E = \sum_i n_i \mu_i^L(P, T, \{n_i\}) - \sum_i n_i \mu_i^L(P, T) = \sum_i n_i(\bar{G}_i^L - g_i^L) \tag{14-150}$$

When this expression is combined with Eq. (14-146), one obtains

$$ng^E = \sum_i n_i RT \ln \gamma_i^L \tag{14-151}$$

Then

$$\left[\frac{\partial(ng^E)}{\partial n_i}\right]_{P, T, n_{j \neq i}} = \sum_i n_i RT\left(\frac{\partial \ln \gamma_i^L}{\partial n_i}\right)_{P, T, n_{j \neq i}} + RT \ln \gamma_i^L \tag{14-152}$$

Since the summation on the right-hand side of Eq. (14-152) is recognized as the Gibbs-Duhem equation (Ref. 51) which has the value zero, Eq. (14-152) reduces to

$$\left[\frac{\partial(ng^E)}{\partial n_i}\right]_{P, T, n_{j \neq i}} = RT \ln \gamma_i^L \tag{14-153}$$

This relationship is used in the development of the expressions given in Table 14-12 for the activity coefficients.

The Wilson equation (suggested by Wilson[72] in 1964) is suitable for a large variety of nonideal mixtures. This equation, a semiempirical generalization of the Flory-Huggins equation, is presented in Table 14-12. The term $(\lambda_{ij} - \lambda_{ii})$ is an empirically determined energy term which is closely related to the difference in cohesive energy between an i-j pair and an i-i pair. Over small temperature ranges, the variation of the energy difference $(\lambda_{ij} - \lambda_{ii})$ with temperature may be neglected with good accuracy. By use of Eq. (14-153) and the expression given for g^E, the formula given in Table 14-12 for the activity coefficient is obtained.

The NRTL equation was developed by Renon and Prausnitz.[54] The expression for the excess energy function g^E (see Table 14-12) was developed on the basis of a "two-liquid" theory. In the theoretical development, the λ_{ij}'s are the Gibbs interaction energies and the α_{ij}'s are the reciprocals of the lattice coordination numbers. In practice, the quantities $(\lambda_{ji} - \lambda_{ii})$, $(\lambda_{ij} - \lambda_{jj})$, and α_{ij} are taken to be three adjustable parameters per binary pair in the mixture and are obtained by a regression of the equilibrium data. The NRTL equation has been used extensively. It is superior to the Wilson equation in that it can be used to represent liquid-liquid systems $(\alpha_{ij} < 0.426)$. However, the NRTL equation requires three parameters per binary pair, whereas the Wilson equation requires only two parameters per binary pair.

On the basis of a quasi-chemical theory, Abrams and Prausnitz[1] developed

Table 14-12 A summary of selected methods for computing activity coefficients

1. Wilson equation

$$\frac{g^E}{RT} = -\sum_i x_i \ln\left(\sum_j \Lambda_{ij} x_j\right)$$

$$\ln \gamma_i^L = 1 - \ln\left(\sum_j x_j \Lambda_{ij}\right) - \sum_k \left(\frac{x_k \Lambda_{ki}}{\sum_j x_j \Lambda_{kj}}\right)$$

where

$$\Lambda_{ij} = \frac{v_j^L}{v_i^L} \exp\left[-\frac{(\lambda_{ij} - \lambda_{ii})}{RT}\right]$$

$$\lambda_{ji} = \lambda_{ij} \qquad \Lambda_{ji} \neq \Lambda_{ij}$$

2. NRTL equation

$$\frac{g^E}{RT} = \sum_i x_i \left(\frac{\sum_j x_j \tau_{ji} \Lambda_{ji}}{\sum_l x_l \Lambda_{li}}\right)$$

$$\ln \gamma_k = \sum_j \left(\frac{x_j \tau_{jk} \Lambda_{jk}}{\sum_l x_l \Lambda_{lk}}\right) + \sum_i \left(\frac{x_i \Lambda_{ki}}{\sum_l x_l \Lambda_{li}}\right)\left[\tau_{ki} - \sum_j \left(\frac{x_j \tau_{ji} \Lambda_{ji}}{\sum_l x_l \Lambda_{li}}\right)\right]$$

where

$$\tau_{ji} = \frac{\lambda_{ji} - \lambda_{ii}}{RT} \qquad \Lambda_{ji} = \exp\left(-\alpha_{ji} \tau_{ji}\right)$$

$$\tau_{ij} = \frac{\lambda_{ij} - \lambda_{jj}}{RT} \qquad \Lambda_{ij} = \exp\left(-\alpha_{ij} \tau_{ij}\right) \qquad \alpha_{ij} = \alpha_{ji}$$

(continued on page 548.)

Table 14-12 (*continued*)

---

3. UNIQUAC equation

$$g^E = g^C + g^R$$

$$\frac{g^C}{RT} = \sum_i x_i \ln \frac{\Phi_i}{x_i} + \frac{z}{2} \sum_i x_i q_i \ln \frac{\theta_i}{\Phi_i}$$

$$\frac{g^R}{RT} = -\sum_i \left(x_i q_i \ln \sum_j \theta_j \Lambda_{ji} \right)$$

$$\ln \gamma_i = \ln \gamma_i^C + \ln \gamma_i^R$$

where

$$\ln \gamma_i^C = 1 - \frac{\Phi_i}{x_i} + \ln \frac{\Phi_i}{x_i} - \frac{z}{2} q_i \left(1 - \frac{\Phi_i}{\theta_i} + \ln \frac{\Phi_i}{\theta_i} \right)$$

$$\ln \gamma_i^R = q_i \left[1 - \ln \sum_i \theta_j \Lambda_{ji} - \sum_k \left(\frac{\theta_k \Lambda_{ik}}{\sum_j \theta_j \Lambda_{jk}} \right) \right]$$

$$\Lambda_{ji} = \exp \left[-\frac{(\lambda_{ji} - \lambda_{ii})}{RT} \right] \qquad \Phi_i = \frac{x_i r_i}{\sum_j x_j r_j}$$

$$\Lambda_{ij} = \exp \left[-\frac{(\lambda_{ij} - \lambda_{jj})}{RT} \right] \qquad \theta_i = \frac{x_i q_i}{\sum_j x_j q_j}$$

---

z = lattice coordination number

an equation for the excess free energy per mole of mixture which is known as the UNIQUAC equation (see Table 14-12). The excess free energy function is taken to be the sum of g^C, the combinatorial part, and g^R, the residual part of the excess free energy. The activity coefficient is relatively insensitive to the choice of the coordination number—provided that a reasonable value for the lattice co-ordination number z is chosen.[1] Abrams and Prausnitz[1] used the value $z = 10$. The parameters r_i and q_i are the van der Walls volume and area, respectively, of the molecule relative to those of a standard segment of lattice structure of the quasi-chemical theory. The pure component constants may be computed from bond angles and distances as shown by Abrams and Prausnitz.[1] In practice the differences in interaction energies $(\lambda_{ji} - \lambda_{ii})$ and $(\lambda_{ij} - \lambda_{jj})$ are taken to be two adjustable parameters per binary pair in the mixture. These differences in inter-action energies must be obtained from experimental results. The UNIQUAC equation is suitable for both miscible and immiscible systems and requires only two adjustable parameters per binary pair. Abrams and Prausnitz state that it is about as satisfactory as the Wilson equation but no better for miscible systems. For liquid-liquid systems, the authors state that the UNIQUAC equation is superior to the NRTL equation. By choice of appropriate simplifying assumptions, the UNIQUAC equation may be reduced to other equations such as the Wilson, NRTL, Margules, and Van Laar as shown by Abrams and Prausnitz.[1]

Group Contribution Method for Computing Activity Coefficients

When experimental data for compounds or mixtures are not available, the group contribution methods may be used to estimate the properties of the pure compounds and mixtures. As observed by Fredenslund et al.,[21] group contribution methods necessarily involve a compromise between accuracy and the amount of data required.

An objective of the group contribution methods for the calculation of activity coefficients is the reduction of the amount of data required for the binary mixture parameters. Thousands of compounds are of interest and the number of binary pairs of these is astronomical. However, these compounds may be constructed from a relatively small number of molecular groups, and the binary pairs existing between the groups become manageable. The data required is reduced to that needed to determine the interaction parameter for the group pairs. The compromise in accuracy occurs because it is assumed that each group behaves in the same manner in all of the different molecules. In spite of this compromise in accuracy, group methods appear to be the best available method when specific data are lacking.

THE ASOG METHOD FOR PREDICTING ACTIVITY COEFFICIENTS

The analytical-solution-of-groups (ASOG) method was developed by Derr and Deal[17] and Wilson,[72] and a compilation of parameters has been prepared by Kojima and Tochigi.[41] In this method, the activity coefficient of component i in a mixture is computed as follows

$$\ln \gamma_i = \ln \gamma_i^S + \ln \gamma_i^G \qquad (14\text{-}154)$$

where the activity coefficient γ_i^S depends only upon the size of the group and γ_i^G depends upon the type of group such as CH_2, CO, or OH. The activity coefficient γ_i^S is computed by use of the following formula which is based on the Flory-Huggins theory for athermal mixtures of unequal-sized molecules

$$\ln \gamma_i^S = 1 - \mathcal{R}_i + \ln \mathcal{R}_i \qquad (14\text{-}155)$$

where $\mathcal{R}_i = S_i / \sum_j S_j x_j$

$\quad x_j$ = mole fraction of component j
$\quad S_j$ = number of size groups in molecule j

The activity coefficient γ_i^G is computed by use of the equation

$$\ln \gamma_i^G = \sum_k v_{ki}(\ln \Gamma_k - \ln \Gamma_{ki}) \qquad (14\text{-}156)$$

where the summation is taken over all groups present in the mixture, and
$\quad \Gamma_k$ = activity coefficient of group k in the mixture
$\quad \Gamma_{ki}$ = activity coefficient of group k in the standard state of pure component i
$\quad v_{ki}$ = number of interaction group k in molecule i

The activity coefficient Γ_k is found by use of the following form of the Wilson equation

$$\ln \Gamma_k = 1 - \ln \left(\sum_l X_l \Lambda_{kl} \right) - \sum_l \left(\frac{X_l \Lambda_{lk}}{\sum_m X_m \Lambda_{lm}} \right) \qquad (14\text{-}157)$$

where the group mole fraction X_k for the particular group k is defined by

$$X_k = \frac{\sum_j x_j v_{kj}}{\sum_j x_j \sum_k v_{kj}} \qquad (14\text{-}158)$$

Also, $\ln \Gamma_{ki}$ is computed as follows

$$\ln \Gamma_{ki} = 1 - \ln \sum_l X_{li} \Gamma_{kl} - \sum_l \left(\frac{X_{li} \Lambda_{lk}}{\sum_j X_{ji} \Lambda_{lj}} \right) \qquad (14\text{-}159)$$

where X_{li} is the mole fraction of group l in pure component i, and it is calculated as follows

$$X_{li} = \frac{v_{li}}{\sum_k v_{ki}} \qquad (14\text{-}160)$$

THE UNIFAC METHOD FOR PREDICTING ACTIVITY COEFFICIENTS

The UNIFAC method, developed by Fredenslund et al.[21, 22] is similar in concept to the ASOG method, but it is based on the UNIQUAC equation

$$\ln \gamma_i = \ln \gamma_i^C + \ln \gamma_i^R$$

where the superscript C refers to the combinatorial part of the activity coefficient and the superscript R refers to the residual part of the activity coefficient.

The combinatorial part of the activity coefficient is computed by use of the following expression

$$\ln \gamma_i^C = \ln \frac{\Phi_i}{x_i} + \frac{z}{2} q_i \ln \frac{\theta_i}{\Phi_i} + l_i - \frac{\Phi_i}{x_i} \sum_j x_j l_j \qquad (14\text{-}161)$$

where $l_i = (z/2)(r_i - q_i) - (r_i - 1)$ $\qquad q_i = \sum_k v_{ki} Q_k$
$\quad r_i = \sum_k v_{ki} R_k$ $\qquad \theta_i = x_i q_i / \sum_j x_j q_j$ $\qquad \Phi_i = x_i r_i / \sum_j x_j r_j$
$\quad x_i =$ mole fraction of component i in the mixture
$\quad v_{ki} =$ count of the kth group in component i
$\quad R_k =$ a measure of Van der Waals volume ratio for group k
$\quad Q_k =$ a measure of Van der Waals area ratio for group k

Group parameters R_k and Q_k are obtained from Van der Waals' group volume V_{wk} and surface areas A_{wk} given by Bondi[8]

$$R_k = \frac{V_{wk}}{15.17} \quad \text{and} \quad Q_k = \frac{A_{wk}}{2.5 \times 10^9} \tag{14-162}$$

The numbers 15.17 and 2.5×10^9 are normalization factors recommended by Abrams and Prausnitz.[1] Values of R_k and Q_k are given in Table 14-13.

The residual part of the activity coefficient is computed as follows

$$\ln \gamma_i^R = \sum_k v_{ki}(\ln \Gamma_k - \ln \Gamma_{k,i}) \tag{14-163}$$

where the summations are over all groups, and

Γ_k = activity coefficient for group k in the mixture
$\Gamma_{k,i}$ = activity coefficient for group k in pure component i

The activity coefficient Γ_k is computed as follows

$$\ln \Gamma_k = Q_k \left[1 - \ln \left(\sum_m \Theta_m \Psi_{mk} \right) - \sum_m \left(\frac{\Theta_m \Psi_{km}}{\sum_n \Theta_n \Psi_{nm}} \right) \right] \tag{14-164}$$

where the summations are over all groups, and

$\Theta_m = X_m Q_m / \sum_n X_n Q_n$, area fraction of group m
$X_m = \sum_j x_j v_{mj} / \sum_n \sum_j x_j v_{nj}$, mole fraction of group m in the mixture
$\Psi_{mn} = \exp(-a_{mn}/T)$
$\{a_{mn}\}$ = set of group interaction parameters, $a_{mn} \neq a_{nm}$

The activity coefficient $\ln \Gamma_{ki}$ is computed as follows

$$\ln \Gamma_{ki} = Q_k \left[1 - \ln \left(\sum_m \Theta_{mi} \Psi_{mk} \right) - \sum_m \left(\frac{\Theta_{mi} \Psi_{km}}{\sum_n \Theta_{ni} \Psi_{nm}} \right) \right] \tag{14-165}$$

where the summations are over all groups in component i, and

$\Theta_{mi} = X_{mi} Q_m / \sum_n X_{ni} Q_n$, area fraction of group m in component i
$X_{mi} = v_{mi} / \sum_n v_{ni}$, mole fraction of group m in component i

Values for the group-interaction parameters a_{mn} must be evaluated from experimental phase-equilibrium data. Sets of values of a_{nm} and a_{mn} taken from Ref. 21 are presented in Table 14-14.

Table 14-13 Group volume and group area parameters[21]

Group	Subgroup no.	Subgroup	Name	R_k	Q_k	Sample group assignment
1	....	CH_2	Alkane group			
	1A	CH_3	End group of hydrocarbon chain	0.9011	0.848	Ethane: 2 CH_3
	1B	CH_2	Middle group in hydrocarbon chain	0.6744	0.540	n-butane: 2 CH_3, 2 CH_2
	1C	CH	Middle group in hydrocarbon chain	0.4469	0.228	Isobutane: 3 CH_3, 1 CH
2	....	C=C	Olefin group, α-olefin only	1.3454	1.176	α-butene: 1 C=C, 1 CH_2, 1 CH_3
3	....	ACH	Aromatic carbon group	0.5313	0.400	Benzene: 6 ACH
4	....	$ACCH_2$	Aromatic carbon-alkane group			
	4A	$ACCH_2$	General case	1.0396	0.660	Ethylbenzene: 5 ACH, 1 $ACCH_2$, 1 CH_3
	4B	$ACCH_3$	Toluene group	1.2663	0.968	Toluene: 5 ACH, 1 $ACCH_3$
5	....	COH	Alcohol group, includes nearest CH_2			
	5A	COH	General case	1.2044	1.124	Ethanol: 1 CH_3, 1 COH
	5B	MCOH	Methanol	1.4311	1.432	Methanol: 1 MCOH
	5C	CHOH	Secondary alcohol	0.9769	0.812	Isopropanol: 2 CH_3, 1 CHOH
6	....	H_2O	Water	0.9200	1.400	Water: 1 H_2O
7	....	ACOH	Aromatic carbon-alcohol group	0.8952	0.680	Phenol: 5 ACH, 1 ACOH
8	....	CO	Carbonyl group	0.7713	0.640	Acetone: 2 CH_3, 1 CO

Table 14-13 (*continued*)

Group	Subgroup no.	Subgroup	Name	R_k	Q_k	Sample group assignment
9	....	CHO	Aldehyde group	0.9980	0.948	Propionaldehyde: 1 CH_3, 1 CH_2, 1 CHO
10	....	COO	Ester group	1.0020	0.880	Methyl acetate: 2 CH_3, 1 COO
11	....	O	Ether group	0.2439	0.240	Diethyl ether: 2 CH_3, 2 CH_2, 1 O
12			Primary amine group, includes nearest CH_2			
	12A	CNH_2	General case	1.3692	1.236	n-Propylamine: 1 CH_3, 1 CH_2, 1 CNH_2
	12B	$MCNH_2$	Methylamine	1.5959	1.544	Methylamine: 1 $MCNH_3$
13	....	NH	Secondary amine group	0.5326	0.396	Diethylamine: 2 CH_3, 2 CH_2, 1 NH
14	....	$ACNH_2$	Aromatic carbon-amine group	1.0600	0.816	Aniline: 5 ACH, 1 $ACNH_2$
15			Nitrile group, includes nearest CH_2			
	15A	MCCN	Acetonitrile	1.8701	1.724	Acetonitrile: 1 MCCN
	15B	CCN	General case	1.6434	1.416	Propionitrile: 1 CCN, 1 CH_3
16			Chloride group			
	16A	Cl-1	Cl on end carbon	0.7660	0.720	1,2-Dichlorethane: 2 CH_2, 2 Cl-1
	16B	Cl-2	Cl on middle carbon	0.8069	0.728	1,2,3-Trichloropropane: 2 CH_2, 1 CH, 2 Cl-1, 1 Cl-2
17	....	$CHCl_2$	Dichloride group, end group only	2.0672	1.684	1,1-Dichlorethane: 1 CH_3, 1 $CHCl_2$
18	....	ACCl	Aromatic carbon-chloride group	1.1562	0.844	Chlorobenzene: 5 ACH, 1 ACCl

553

Table 14-14 Group interaction parameters[21]

a_{nm} and a_{mn}

Group	CH$_2$	C=C	ACH	ACCH$_2$	COH	H$_2$O	ACOH	CO	CHO	Group
CH$_2$	0	−200.0	32.08	26.78	931.2	1452	1860	1565	685.9	1
C=C	2520	0	651.6	1490	943.3	578.3	×	1400	×	2
ACH	15.26	−144.3	0	167.0	705.9	860.7	1310	651.1	×	3
ACCH$_2$	−15.84	−309.2	−146.8	0	856.2	3000	740.0	3000	×	4
COH	169.7	254.2	83.5	92.61	0	−320.8	×	462.3	480.0	5
H$_2$O	657.7	485.4	361.5	385.0	287.5	0	462.6	470.8	234.5	6
ACOH	3000	×	3000	3000	×	−558.2	0	×	×	7
CO	3000	3000	101.8	75.00	−106.5	−532.6	×	0	−49.24	8
CHO	343.2	×	×	×	3000	−226.4	×	39.47	0	9
COO	348.0	×	325.5	3000	167.5	×	−254.1	333.6	×	10
O	2160	×	−75.50	3000	−13.44	×	×	−39.81	×	11
CNH$_2$	−16.74	90.37	−38.64	×	−109.8	−527.7	×	×	×	12
NH	3000	8.922	37.94	×	−700.0	−882.7	×	×	×	13
ACNH$_2$	3000	×	3000	3000	×	236.6	×	×	×	14
CCN	27.31	43.03	−66.44	−150.0	337.9	227.0	×	447.7	×	15
Cl	−119.6	242.1	−90.43	52.69	357.0	618.2	×	62.00	×	16
CHCl$_2$	31.06	−72.88	×	×	×	467.0	×	37.63	×	17
ACCl	121.1	×	1000	×	586.3	1472	×	×	×	18
Group	1	2	3	4	5	6	7	8	9	

Table 14-14 (*continued*)

Group	COO	O	CNH$_2$	NH	ACNH$_2$	CCN	Cl	CHCl$_2$	ACCl	Group
CH$_2$	687.5	472.6	422.1	800.0	1330	601.6	523.2	60.45	194.2	1
C=C	×	×	349.9	515.2	×	691.3	253.8	259.5	×	2
ACH	159.1	37.24	179.7	487.2	680.0	290.1	124.0	×	−99.9	3
ACCH$_2$	110.0	680.0	×	×	640.0	3000	33.84	×	×	4
COH	174.3	−204.6	−166.8	3000	×	79.85	194.6	×	69.97	5
H$_2$O	×	×	385.3	743.8	−314.6	118.5	158.4	247.2	190.6	6
ACOH	−470.2	×	×	×	×	×	×	×	×	7
CO	−180.1	475.5	×	×	×	−307.4	628.0	874.5	×	8
CHO	×	×	×	×	×	×	×	×	×	9
COO	0	−26.15	×	×	×	×	×	×	×	10
O	−290.0	0	×	×	×	×	×	×	×	11
CNH$_2$	×	×	0	×	×	×	×	×	−10.0	12
NH	×	×	×	0	×	×	×	×	−60.0	13
ACNH$_2$	×	×	×	×	0	×	×	×	3000	14
CCN	×	×	×	×	×	0	−100.0	×	25.0	15
Cl	×	×	×	×	×	100.0	0	−308.5	×	16
CHCl$_2$	×	×	×	×	×	×	790.0	0	×	17
ACCl	×	×	3000	3000	110.0	3000	×	×	0	18
Group	10	11	12	13	14	15	16	17	18	

NOTATION

A	total work function; energy units
$\bar{A}_i$	partial molar value; $\bar{A}_i = (\partial A/\partial n_i)_{P,\,T,\,n_{j \neq i}}$, energy units per mole
$\tilde{A}$	residual work function; defined by Eq. (14-69)
B_{ii}	second virial coefficient for pure component i, volume per mole
B_{ij}	cross virial coefficient for components i and j, volume per mole
f	the fugacity; defined by Eq. (14-26), atm
f_i^V, f_i^L	fugacities of pure component i in the vapor and liquid states, respectively at the temperature T and total pressure P of the mixture, atm
$\hat{f}_i^V, \hat{f}_i^L$	fugacities of component i in vapor and liquid phase mixtures, respectively; evaluated at the temperature, pressure, and compositions of the respective phases, atm
$f_{i,\,P_i}^V$	fugacity of pure component i in the vapor state; evaluated at its vapor pressure at the temperature T (Note $f_{i,\,P_i}^V = f_{i,\,P_i}^L$.)
f_i^{0L}	fugacity of pure component i in the liquid state; evaluated at the temperature T and pressure P_0, atm
$f^{(0)}(T_r)$	a function of reduced temperature; see Eq. (14-121)
$f^{(1)}(T_r)$	a function of reduced temperature; see Eq. (14-122)
$f^{(2)}(T_r)$	a function of the reduced temperature and the reduced dipole moment; see Eq. (14-124)
g	free energy per mole, energy units per mole
G	total free energy, energy units
$\bar{G}_i$	partial molar free energy; defined below Eq. (14-16), energy units per mole
h	enthalpy per mole, energy units per mole
H	total enthalpy, energy units
$\bar{H}_i$	partial molar enthalpy; defined by Eq. (14-59), energy units per mole
$\hat{H}_i$	virtual value of the partial molar free energy; defined by Eq. (14-65)
K_i	vapor-liquid distribution coefficient, $K_i \equiv y_i/x_i$; defined by Eqs. (14-36) and (14-37)
K_i^I	ideal solution K value, $K_i^I = f_i^L/f_i^V$ (In the remaining chapters of this book, the ideal solution K value is denoted by K_i.)
m_i	mass of component i in a mixture, mass units
M	mass composition parameter; defined by Eq. (14-104)
n	total number of moles; $n_i =$ moles of component i
p	partial pressure of component i, $p_i = Py_i$, atm
P	total pressure, atm
P_i	vapor pressure of pure component i at the temperature T, atm
P_r	reduced pressure, $P_r = P/P_c$, where P_c is the critical pressure
P_k	convergence pressure, atm
R	gas constant in consistent units

s	entropy per mole, energy units per degree Kelvin (or Rankine) per mole
S	total entropy, energy units per degree Kelvin (or Rankine)
$\bar{S}_i$	partial molar entropy; defined by Eq. (14-60); energy units per mole per degree Kelvin (or Rankine)
t	temperature, °F
T	absolute temperature, in degrees Kelvin or Rankine
T_r	reduced temperature, $T_r = T/T_c$, where T_c is the critical temperature
u	internal energy per mole, energy units per mole
U	total internal energy, energy units
v	molar volume, volume units per mole
v_r	reduced molar volume; $v_r = v/v_c$, where v_c is the critical volume
V	total volume, volume units
$\bar{V}_i$	partial molar volume of component i; defined below Eq. (14-25)
x_i, y_i	mole fractions of component i in the liquid and vapor phases, respectively

Greek Letters

γ_i^V, γ_i^L	activity coefficients of component i in the vapor and liquid phases, respectively
δ_i	solubility parameter in Hildebrand's equation for the activity coefficient; see Eq. (14-115)
$\bar{\delta}$	average value of the solubility parameter; defined by Eq. (14-116)
μ	chemical potential; defined by Eq. (14-26), energy units per mole
μ_r	reduced dipole moment; defined by Eq. (14-127)
ρ	molar density, moles per unit volume
ϕ_i^L	fugacity coefficient for pure component i in the liquid state at the pressure P and temperature T; $\phi_i^L = f_i^L/P$ (This liquid fugacity coefficient is commonly denoted by v_i in the literature.)
ϕ_i^V	fugacity coefficient for pure component i in the vapor state at the pressure P and temperature T; $\phi_i^V = f_i^V/P$
ϕ_i^s	fugacity coefficient for pure component i, evaluated at its saturation pressure P_i at the temperature T; $\phi_i^s = f_{i, P_i}^V/P_i = f_{i, P_i}^L/P_i$
$\hat{\phi}_i^L$	$\hat{f}_i^L/(PX_i) = \gamma_i^L \phi_i^L$
$\hat{\phi}_i^V$	$\hat{f}_i^V/(Py_i) = \gamma_i^V \phi_i^V$
Ω	an enthalpy departure function; defined by Eq. (14-67)

Subscripts

i, j, k, l, m, n	counting integers

Superscripts

L	liquid state
V	vapor state

PROBLEMS

14-1 Use the path shown in Fig. 14-2 for pure components to develop the expressions for the thermodynamic expressions given in Table 14-1 and for the functions, U, G, and μ.

14-2 Use the path shown in Fig. 14-3 for mixtures to develop the expressions for the thermodynamic expressions given in Table 14-1 and for the functions U, G, and μ_i.

14-3 Repeat Prob. 14-1 for the path shown in Fig. 14-4 and the expressions given in Table 14-2 for pure components.

14-4 Repeat Prob. 14-2 for the path shown in Fig. 14-5 and the expressions given in Table 14-2 for mixtures.

14-5 Use the definition given by Eq. (14-77) for the residual work content to obtain the thermodynamic expressions for pure components given in Table 14-3 as well as those for U, G, and μ.

14-6 Repeat Prob. 14-5 for mixtures.

14-7 Develop the relationships given by Eqs. (14-61) and (14-62).

14-8 Develop the relationships given by Eqs (14-63) and (14-64) for ideal solutions.

14-9 Obtain the expression given in Table 14-4 for the residual work content A for the BWR equation of state.

14-10 Repeat Prob. 14-9 by obtaining the expression given in Table 14-5 for the Redlich-Kwong equation of state.

14-11 Use the expressions of Table 14-5 which are needed to obtain the expressions for $h - h°$ and $\ln f/P$ which are given in Table 14-3 for pure components which obey the Redlich-Kwong equation of state.

14-12 Use the expressions of Table 14-3 which are needed to obtain the expressions for Ω and $\ln \hat{f}_i^V/Py_i$ which are given in Table 14-5 for mixtures which obey the Redlich-Kwong equation of state.

14-13 By use of the relationship given by Eq. (14-153) and the Wilson excess free energy function given in Table 14-12, obtain the corresponding expression given in this table for $\ln \gamma_i^L$.

14-14 The UNIFAC method may be used to find the activity coefficient for n-propanol (component 1) in the presence of n-heptane (component 2) at 333 K by using the equations given in the text in the order in which the answers to the intermediate steps are listed. The mole fraction of n-propanol is $x_1 = 0.1$ and the mole fraction of n-heptane is $x_2 = 0.9$.

(a) Verify the following information by use of Table 14-13.

n-propanol ($i = 1$) contains the following groups

1CH$_3$ group:	$k = 1A$	$v_{1A,1} = 1$	$R_{1A} = 0.9011$
			$Q_{1A} = 0.848$
1CH$_2$ group:	$k = 1B$	$v_{1B} = 1$	$R_{1B} = 0.6744$
			$Q_{1B} = 0.540$
1COH group:	$k = 5$	$v_{5,1} = 1$	$R_5 = 1.2044$
			$Q_5 = 1.124$

n-heptane ($i = 2$)

2CH$_3$ groups:	$= 1A$	$v_{1A,2} = 2$	$R_{1A} = 0.9011$
			$Q_{1A} = 0.848$
5CH$_2$ groups:	$k = 1B$	$v_{1B,2} = 5$	$R_{1B} = 0.6744$
			$Q_{1B} = 0.540$

(b) Use the expressions given below Eq. (14-161) to show that

$r_1 = 2.7799 \quad q_1 = 2.512 \quad r_2 = 5.1742 \quad q_2 = 4.295$
$\Phi_1 = 0.05633 \quad \Phi_2 = 0.9437 \quad \theta_1 = 0.0597 \quad \theta_2 = 0.9403 \quad l_1 = -0.4404 \quad l_2 = -0.2832$

(c) Use the results found above to show that Eq. (14-161) gives

$$\ln \gamma_1^C = -0.1163$$

(d) To compute the activity coefficients $\{\Gamma_{ki}\}$ for group k in pure component i by use of Eq. (14-165), first observe that groups $CH_3 = 1A$, $CH_2 = 1B$ belong to group 1 and COH belongs to group 5, and then use Table 14-14

$$a_{1,5} = 931.2 \text{ K}$$
$$a_{5,1} = 169.7 \text{ K}$$

and show that

$$\Psi_{1,5} = 0.06103$$
$$\Psi_{5,1} = 0.6007$$

(e) In the calculation of $\Gamma_{1A,1}$ for pure n-propanol ($i = 1$), observe that three groups are involved: $1CH_3$, $1CH_2$, and $1COH$. Then show that

$$X_{1A,1} = 1/3 \qquad X_{1B} = 1/3 \qquad X_{COH,1} = 1/3$$

Next use the expression beneath Eq. (14-164) to show that for component 1, n-propanol

$$\Theta_{1A,1} = 0.3376 \qquad \Theta_{1B,1} = 0.2150 \qquad \Theta_{5,1} = 0.4475$$

(f) Use the results found in part (e) to show that by use of Eq. (14-165), the following result is obtained for pure n-propanol

$$\ln \Gamma_{1A,1} = 0.3962$$

(g) Repeat part (f) and show that for pure n-propanol

$$\ln \Gamma_{1B,1} = 0.2523$$
$$\ln \Gamma_{5,1} = 0.2523$$

(h) To compute Γ_{1A} for the mixture, first show that for the mixture

$$X_{1A} = 0.2879$$
$$X_{1B} = 0.6970$$
$$X_5 = 0.01515$$

(i) Then for the mixture, show that

$$\Theta_{1A} = 0.3829 \qquad \Theta_{1B} = 0.5904 \qquad \Theta_5 = 0.02671$$

(j) For the mixture, use Eq. (14-164) to show that

$$\ln \Gamma_{1A} = 0.006776$$
$$\ln \Gamma_{1B} = 0.004315$$
$$\ln \Gamma_5 = 2.867$$

(k) Use the above results to show that for component 1, Eq. (14-163) gives

$$\ln \gamma_1^R = 1.783$$

Then show that the UNIQUAC equation gives

$$\ln \gamma_1 = 1.6667 \qquad \gamma_1 = 5.29$$

(The experimental value found by Van Ness et al.[69] was $\gamma_1 = 5.38$.)

REFERENCES

1. D. S. Abrams and J. M. Prausnitz: "Statistical Thermodynamics of Liquid Mixtures: A New Expression for the Excess Gibbs Energy of Partly or Completely Miscible Systems," *AIChE J.*, **21**: 116 (1975).
2. S. B. Adler, C. F. Spencer, H. Ozkardesh, and C.-M. Kuo: "Industrial Uses of Equations of State: A State-of-the-Art-Review," *Phase Equilibria and Fluid Properties in the Chemical Industry*, T. S. Storvick and S. I. Sandler (eds.), ACS Symposium Series 60, American Chemical Society, Washington, D.C., 1977.
3. H. E. Barner and S. B. Adler: "Three-Parameter Formulation of the Joffe Equation of State," *Ind. Eng. Chem. Fundam.*, **9**(4): 521 (1970).
4. F. H. Barr-David: "Notes on Phase Relations of Binary Mixtures in the Region of the Critical Point," *AIChE J.*, **2**: 426 (1956).

5. J. A. Beattie: "The Computation of the Thermodynamic Properties of Real Gases and Mixtures of Real Gases," *Chem. Rev.*, **44**:141 (1949).
6. M. Benedict, G. B. Webb, and L. C. Rubin: "An Empirical Equation for the Thermodynamic Properties of Light Hydrocarbons and Their Mixtures," *J. Chem. Phys.*, **8**:334 (1940).
7. ———, ———, and ———: "An Empirical Equation for Thermodynamic Properties of Light Hydrocarbons and Their Mixtures—Constants for Twelve Hydrocarbons," *Chem. Eng. Prog.*, **47**(8):419 (1951).
8. A. Bondi: *Physical Properties of Molecular Crystals, Liquids and Gases*, John Wiley & Sons, New York, 1968.
9. G. G. Brown: "Vapor-Liquid Phase Equilibria in Hydrocarbon Systems," *Pet. Eng.*, **11**(8):25 (1940).
10. B. C. Cajander, H. G. Hipkin, and J. M. Lenoir: "Prediction of Equilibrium Ratios from Nomograms of Improved Accuracy," *J. Chem. Eng. Data*, **5**(3):251 (1960).
11. R. T. Carter, B. H. Sage, and W. N. Lacey: "Phase Behavior in Methane-Propane-*n*-Pentane Systems," *Am. Inst. Min. Met. Eng. Tech. Pub.*, 1250 (October 1940).
12. K. C. Chao and J. D. Seader: "A General Correlation of Vapor-Liquid Equilibria in Hydrocarbon Mixtures," *AIChE J.*, **7**(4):598 (1961).
13. P. L. Chuch and J. M. Prausnitz: "Vapor-Phase Fugacity Coefficients in Nonpolar and Quantum-Gas Mixtures," *Ind. Eng. Chem. Fundam.*, **6**:192 (1967).
14. H. W. Cooper and J. C. Goldfrank: "BWR Constants and New Correlations," *Hydrocarbon Process.*, **46**(12):141 (1967).
15. R. F. Curl, Jr., and K. S. Pitzer: "Volumetric and Thermodynamic Properties of Fluids-Enthalpy, Free Energy and Entropy," *Ind. Eng. Chem.*, **50**:265 (1958).
16. C. H. Deal, E. L. Derr, and M. N. Papadopoulos: "Activity Coefficients and Molecular Structure," *Ind. Eng. Chem. Fundam.*, **1**:17 (1962).
17. E. L. Derr and C. H. Deal: "Analytical Solutions of Groups: Correlation of Activity Coefficients through Structural Groups Parameters," *Inst. Chem. Eng. Symp. Ser.* (Institute of Chemical Engineers, London) **32**(3):40 (1969).
18. B. F. Dodge: *Chemical Engineering Thermodynamics*, McGraw-Hill Book Company, New York, 1944.
19. *Engineering Data Book* compiled and edited by the Gas Processors Suppliers Association (1812 First Place, Tulsa, Oklahoma 74103) and the Gas Products Association, 9th ed., 1972.
20. J. H. Erbar: Oklahoma State University, Personal Communication.
21. A. Fredenslund, R. L. Jones, and J. M. Prausnitz: "Group Contribution Estimation of Activity Coefficients in Nonideal Liquid Mixtures," *AIChE J.*, **21**:1086 (1975).
22. ———, J. Gmehling, and P. Rasmussen: *Vapor-Liquid Equilibria Using UNIFAC*, Elsevier Scientific Publishing Company, New York, 1977.
23. E. R. Gilliland and H. W. Scheeline: "High Pressure Vapor-Liquid Equilibria: Propylene-Isobutane and Propane-Hydrogen Sulfide," *Ind. Eng. Chem.*, **32**:48 (1940).
24. B. W. Gamson and K. M. Watson: "High Pressure Vapor-Liquid Equilibria," *Nat. Pet. News*, **36**(R):623 (September 6, 1944).
25. H. G. Grayson and C. W. Streed: "Vapor-Liquid Equilibria for High Temperature, High Pressure Hydrogen-Hydrocarbon Systems," Paper No. 20 PD7, Sixth World Petroleum Conference, Frankfurt, Germany, 1963.
26. C. D. Holland: *Multicomponent Distillation*, Prentice-Hall, Englewood Cliffs, N.J., 1963.
27. ——— and P. T. Eubank: "Solve More Distillation Problems: Part 2—Partial Molar Enthalpies Calculated," *Hydrocarbon Process.*, **53**(11):176 (1974).
28. ———: "Energy Balances for Systems Involving Nonideal Solutions," *Ind. Eng. Chem. Fundam.*, **16**(1):143 (1977).
29. ——— and R. G. Anthony: *Fundamentals of Chemical Reaction Engineering*, Prentice-Hall, Englewood Cliffs, N.J., 1979.
30. S. T. Hadden: "Vapor-Liquid Equilibria in Hydrocarbon Systems," *Chem. Eng. Prog.*, **44**:37 (1948).

31. ———: "Convergence Pressure in Hydrocarbon Vapor-Liquid Equilibria," *Chem. Eng. Prog. Symp. Ser.*, **49**(7):53 (1953).

32. ——— and H. C. Grayson: "New Charts for Hydrocarbon Vapor-Liquid Equilibria," *Pet. Refiner*, **40**(9):207 (1961).

33. M. J. Hiza and A. G. Duncan: "A Correlation for the Prediction of Interaction Energy Parameters for Mixtures of Small Molecules," *AIChE J.*, **16**:733 (1970).

34. D. L. Katz and F. Kurata: "Retrograde Condensation," *Ind. Eng. Chem.*, **32**:817 (1940).

35. T. G. Kaufman: "A Method for Phase Equilibria Calculations Based on Generalized Benedict-Webb-Rubin Constants," *Ind. Eng. Chem. Fundam.*, **7**:115 (1968).

36. W. B. Kay: "Liquid-Vapor Phase Equilibrium Relations in the Ethane-*n*-Heptane System," *Ind. Eng. Chem.*, **30**:459 (1938).

37. M. G. Kesler, B. I. Lee, M. J. Fish, and S. T. Hadden: "Correlation Improves K-Value Predictions," *Hydrocarbon Process.*, **56**:257 (1977).

38. B. I. Lee, J. H. Erbar, and W. C. Edmister: "Properties for Low Temperature Hydrocarbon Process Calculations," *AIChE J.*, **19**(2):349 (1973).

39. ——— and M. G. Kesler: "A Generalized Thermodynamic Correlation Based on Three-Parameter Corresponding States," *AIChE J.*, **21**:510 (1975).

40. W. K. Lewis and W. C. Kay: "Fugacity of Various Hydrocarbons Above Their Vapor Pressure and Below Their Critical Temperature," *Oil Gas J.*, **32**(45):40 (1934).

41. K. Kojima and K. Tochigi, *Prediction of Vapor-Liquid Equilibria by ASOG Method*, New York, Elsevier Scientific Publishing Co., 1979.

42. A. L. Lydersen, R. A. Greenkorn, and O. A. Hougen: "Generalized Thermodynamic Properties of Pure Fluids," *Univ. Wis., Coll. Eng., Eng. Stn. Rep. 4*, Madison, 1955.

43. M. Margules: *Sitzungsber, Akad. Wiss. Wien, Math. Naturwiss. Kl.*, **104**(2)1243 (1895).

44. Joseph J. Martin: "Cubic Equations of State—Which?", *Ind. Eng. Chem. Fundam.*, **18**(2):81 (1979).

45. J. B. Opfell, B. H. Sage, and K. S. Pitzer: "Application of Benedict Equation to Theorem of Corresponding States," *Ind. Eng. Chem.*, **48**:2069 (1956).

46. R. V. Orye: "Prediction and Correlation of Phase Equilibria and Thermal Properties with the BWR Equation of State," *Ind. Eng. Chem. Process. Des. Dev.*, **8**:579 (1969).

47. D.-Y. Peng and D. B. Robinson: "A New Two-Constant Equation of State," *Ind. Eng. Chem. Fundam.*, **15**(1):59 (1976).

48. K. S. Pitzer and R. F. Curl, Jr: "The Volumetric and Thermodynamic Properties of Fluids. III. Empirical Equation for the Second Virial Coefficient," *J. Am. Chem. Soc.*, **79**:2369 (1957).

49. ———, D. Z. Lippmann, R. F. Curl, Jr., C. M. Huggins, and D. E. Peterson: "The Volumetric and Thermodynamic Properties of Fluids. II. Compressibility Factor, Vapor Pressure and Entropy of Vaporization," *J. Am. Chem. Soc.*, **77**:3433 (1955).

50. J. M. Prausnitz, C. A. Eckert, R. V. Orye, and J. P. O'Connell: *Computer Calculations for Multicomponent Vapor-Liquid Equilibria*, Prentice-Hall, Englewood Cliffs, N.J., 1967.

51. ———: *Molecular Thermodynamics of Fluid Phase Equilibria*, Prentice-Hall, Englewood Cliffs, N.J., 1969.

52. O. Redlich and A. T. Kister: "Algebraic Representation of Thermodynamic Properties and Classification of Solutions," *Ind. Eng. Chem.*, **40**:345 (1948).

53. ——— and J. N. S. Kwong: "On the Thermodynamics of Solutions. V. An Equation of State. Fugacities of Gaseous Solutions," *Chem. Rev.*, **44**:233 (1949).

54. H. Renon and J. M. Prausnitz: "Local Compositions in Thermodynamic Excess Functions for Liquid Mixtures," *AIChE J.*, **14**:135 (1968).

55. R. C. Reid, J. M. Prausnitz, and T. K. Sherwood: *The Properties of Gases and Liquids*, 3d ed., McGraw-Hill Book Company, New York, 1977.

56. C. S. Robinson and E. R. Gilliland: *Elements of Fractional Distillation*, 4th ed., McGraw-Hill Book Company, New York, 1950.

57. G. Scatchard and W. J. Hamer: "The Application of Equations for the Chemical Potentials to Partly Miscible Solutions," *J. Am. Chem. Soc.*, **57**:1805 (1935).

58. J. M. Smith and H. C. Van Ness: *Introduction to Chemical Engineering Thermodynamics,* McGraw-Hill Book Company, New York, 1959.

59. G. Soave: "Equilibrium Constants from a Modified Redlich-Kwong Equation of State," *Chem. Eng. Sci.,* **27**(6): 1197 (1971).

60. I. S. Sokolnikoff and E. S. Sokolnikoff: *Higher Mathematics for Engineers and Physicists,* 2d ed., McGraw-Hill Book Company, New York, 1941.

61. M. Souders, Jr., W. W. Selheimer, and G. G. Brown: "Equilibria Between Liquid and Vapor Solutions of Paraffin Hydrocarbons," *Ind. Eng. Chem.,* **24**: 517 (1932).

62. K. E. Starling and J. E. Powers: "Enthalpy of Mixtures by Modified BWR Equation," *Ind. Eng. Chem. Fundam.,* **9**(4): 531 (1970).

63. ————: "Thermo Data Refined for LPG, Part 1: Equation of State and Computer Prediction," **50**: 101 (1971).

64. ———— and M. S. Han: "Thermo Data Refined for LPG, Part 14: Mixtures," *Hydrocarbon Process.,* **51**: 101 (1971).

65. ————: *Fluid Thermodynamics for Light Petroleum Systems,* Gulf Publishing Co., Houston, 1973.

66. H. Sugie and B. C.-Y. Lu: "Generalized Equation of State for Vapors and Liquids," *AIChE J.,* **17**(5): 1068 (1971).

67. C. Tsonopoulos: "An Empirical Correlation of Second Virial Coefficients," *AIChE J.,* **20**(2): 263 (1974).

68. J. J. Van Laar: "Zur Theorie der Dampfspannugen, von binären Gemishen," *Z. Phys. Chem.,* **83**: 599 (1913).

69. H. C. Van Ness, C. A. Soczek, G. L. Peloquin, and R. L. Machado: "Thermodynamic Excess Properties of Three Alcohol-Hydrocarbon Systems," *J. Chem. Eng. Data,* **10**: 163 (1965).

70. D. S. Viswanath and G.-J. Su: Generalized Thermodynamic Properties of Real Gases, "Part II. Generalized Benedict-Webb-Rubin Equation of State for Real Gases," *AIChE J.,* **11**(2): 205 (1965).

71. G. M. Wilson and C. H. Deal: "Activity Coefficients and Molecular Structure," *Ind. Eng. Chem. Fundam.,* **1**: 20 (1962).

72. ————: "Vapor-Liquid Equilibrium. XI. A New Expression for the Excess Free Energy of Mixing," *J. Am. Chem. Soc.,* **86**: 127 (1964).

73. ————: "Vapor-Liquid Equilibria, Correlation by Means of a Modified Redlich-Kwong Equation of State," *Adv. Cryog. Eng.,* **19**: 168 (1964).

74. ————: "Calculation of Enthalpy Data from a Modified Redlich-Kwong Equation of State," *Adv. Cryog. Eng.,* **11**: 392 (1966).

75. F. W. Winn: "Simplified Nomographic Presentation: Hydrocarbon Vapor-Liquid Equilibria," *Chem. Eng. Prog. Symp. Ser.,* **48**(2): 121 (1952).

76. K. Wohl: "Thermodynamic Evaluation of Binary and Ternary Liquid Systems," *Trans. Am. Inst. Chem. Eng.,* **42**: 215 (1946).

77. T. Yamada: "An Improved Generalized Equation of State," *AIChE J.,* **19**: 286 (1973).

78. L. C. Yen and R. E. Alexander: "Estimation of Vapor and Liquid Enthalpies," *AIChE J.,* **11**: 334 (1965).

SELECTED TOPICS IN MATRIX OPERATIONS AND NUMERICAL METHODS FOR SOLVING MULTIVARIABLE PROBLEMS

In this chapter, topics in matrix operation and numerical methods for solving multivariable problems are presented and examined. In Sec. 15-1, the use of linked lists in the storage of large sparse matrices is presented, and the scaling of matrices is presented in Sec. 15-2. The presentation and examination of selected numerical methods are presented in Sec. 15-3.

15-1 STORAGE OF LARGE SPARSE MATRICES

Large sparse matrices such as those encountered in Chap. 5 are commonly stored in computers in *packed form*, a term used to mean that only the nonzero elements are stored. Storage in packed form is almost essential because of the large amount of internal computer storage which would be required to store all of the zero elements.

In order to avoid the storage of the zeros and to locate the nonzero elements in storage, the use of *linked lists* has become popular.[7] The following information must be available for this type of storage: the address of the first nonzero element of a given row, the value of the element, the column location of the element, and the location of the next element of the row. Packed storage through the use of a linked list will be illustrated by the storage of an original matrix and the matrix as modified by each row operation of an **LU** factorization.

Consider first the storage of the matrix **A**,

$$
\mathbf{A} = \begin{bmatrix} 1 & 3 & 1 & 0 \\ 6 & 2 & 0 & 0 \\ 0 & 5 & 0 & 3 \\ 4 & 0 & 2 & 9 \end{bmatrix} \tag{15-1}
$$

and let **IPV** = vector storing pivot order during factorization

RIP = vector pointing to storage location of first nonzero element in a row

VE = vector of the nonzero elements of **A** as well as a predetermined number of zeros for the storage of numbers resulting from fill anticipated from subsequent row operations

CI = vector containing the column numbers in which the elements of **A** appear

NSL = vector containing the location of the next nonzero element of the row

NASL = points to the next empty location in **VE**

The storage of the matrix given by Eq. (15-1) is shown in Fig. 15-1. That the linked list shown constitutes the storage of matrix **A** is readily demonstrated as follows. The element 1 of **IPV** and the corresponding element 1 of **RIP** means that the first nonzero element of row 1 is stored in location 4. Go to location 4, and observe that the elements 1 of **VE**, 1 of **CI**, and 7 of **NSL** mean that the value of the first nonzero element of row 1 is 1, it is a member of column 1, and the next nonzero element of row 1 is in location 7. Go to location 7, and observe that the elements 3 of **VE**, 2 of **CI**, and 12 of **NSL** mean that the next nonzero element of row 1 is 3, it is a member of column 2, and the next nonzero element of row 1 is in location 12. Go to location 12, and observe that the elements 1 of **VE**, 3 of **CI** and -1 of **NSL** mean that the next nonzero element of row 1 has the value of 1, it is a member of column 3, and the -1 is used to signify that this is the last nonzero element of row 1. Thus, return to **IPV** and take the next element 2 which corresponds to row 2. The corresponding element 2 of **RIP** means that the first nonzero element of row 2 is in location 18. Repeat the sequence of steps described above for rows 2, 3, and 4. The **NASL** = 10 is the location of the next empty location. The **NSL** = 13 of location 10 points to the next potential empty location. After location 10 has been filled, **NASL** is reset to 13, the next empty location. This procedure is repeated as the storage is filled and an analogous procedure is used as the storage is emptied.

The steps in the **LU** factorization of **A** by use of row operations as demonstrated in Chap. 4 follow together with the vector containing the pivot row information. First, the original matrix with its corresponding **IPV** vector is

$$
\textbf{IPV} \qquad
\begin{array}{|c|}
\hline
1 \\ \hline
2 \\ \hline
3 \\ \hline
4 \\ \hline
\end{array}
\begin{bmatrix}
1 & 3 & 1 & 0 \\
6 & 2 & 0 & 0 \\
0 & 5 & 0 & 3 \\
4 & 0 & 2 & 9
\end{bmatrix} = \textbf{A}
\qquad (15\text{-}2)
$$

Select row 2 as the first pivot row, and use it to eliminate the other elements from column 1. The resulting matrix with the negative of the multipliers stored in parentheses in the place of the corresponding zero elements and the corresponding **IPV** vector follow.

Location	1	2	3	4	5	6	7	8	9	10	11	12	13	14	15	16	17	18	19	20	21	22	23	24
VE	5			1			3	4	2		3	1		2	9			6						
CI	2			1			2	1	3		4	3		2	4			1						
NSL	11	19	5	7	6	16	12	9	15	13	−1	−1	2	−1	−1	20	21	14	3	17	22	24	0	23

RIP

4
18
1
8

IPV

1
2
3
4

NASL = 10 (Initial value)

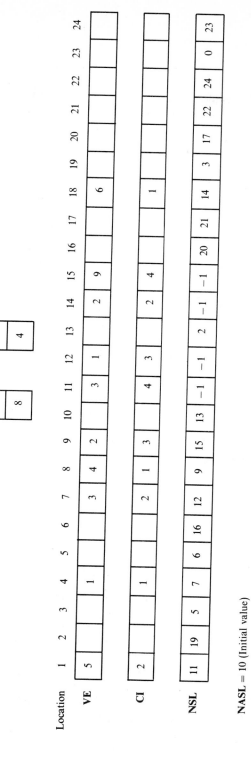

Figure 15-1 Storage of matrix A [Eq. (15-1)] as a linked list.

565

IPV

$$
\begin{array}{c}
\boxed{2} \\
\boxed{1} \\
\boxed{3} \\
\boxed{4}
\end{array}
\begin{bmatrix}
(1/6) & 16/6 & 1 & 0 \\
6 & 2 & 0 & 0 \\
0 & 5 & 0 & 3 \\
(4/6) & -8/6 & 2 & 9
\end{bmatrix} = A_1
$$

first pivot row and 6 is the pivot element

The elements of **IPV** vector denote the reordering of the rows of **A** in the transformed matrix. Row 2 of **A** is now the new row 1 and row 1 of **A** is now new row 2 of the transformed matrix. Rows 3 and 4 of **A** are also rows 3 and 4 of the transformed matrix.

Next row 3 of **A** is selected as the second pivot row, and the corresponding row operations are performed to give

IPV

$$
\begin{array}{c}
\boxed{2} \\
\boxed{3} \\
\boxed{1} \\
\boxed{4}
\end{array}
\begin{bmatrix}
(1/6) & (16/30) & 1 & -48/30 \\
6 & 2 & 0 & 0 \\
0 & 5 & 0 & 3 \\
(4/6) & (-8/30) & 2 & 294/30
\end{bmatrix} = A_2
$$

first pivot row
second pivot row and 5 is the pivot element

The elements of **IPV** mean that row 2 of the original matrix **A** is row 1 of the transformed matrix, row 3 of matrix **A** is row 2 of the transformed matrix, row 1 of **A** is row 3 of the transformed in a matrix, and row 4 of **A** is also row 4 of the transformed matrix.

Next row 4 of **A** is picked as the third pivot row, and the associated row operations yield

IPV

$$
\begin{array}{c}
\boxed{2} \\
\boxed{3} \\
\boxed{4} \\
\boxed{1}
\end{array}
\begin{bmatrix}
(1/6) & (16/30) & (1/2) & -195/30 \\
6 & 2 & 0 & 0 \\
0 & 5 & 0 & 3 \\
(4/6) & (-8/30) & 2 & 294/30
\end{bmatrix} = A_3
$$

first pivot row
second pivot row
third pivot row and 2 is the pivot element

The **IPV** vector contains the information that rows 2, 3, 4, and 1 of **A** are rows 1, 2, 3, and 4 of the transformed matrix.

The three transformed matrices A_1, A_2, and A_3 are stored as linked lists in Fig. 15-2. To verify the fact that the linked list in Fig. 15-2 contains the elements of the transformed matrices A_1, A_2, and A_3, an analysis analogous to that used to confirm the fact that the linked list in Fig. 15-2 contains the elements of **A** [Eq. (15-1)] may be performed. By comparison of Figs. 15-1 and 15-2, the fill resulting from the row operations is readily determined. The above formulation of a linked list storage procedure follows closely that used by Gallun.[4]

Matrix A_1

IPV:

2	
1	
3	
4	

RIP:

4
18
1
8

Location	1	2	3	4	5	6	7	8	9	10	11	12	13	14	15	16	17	18	19	20	21	22	23	24
VE	5			$\frac{1}{6}$			$\frac{16}{6}$	$\frac{4}{6}$	2	$-\frac{8}{6}$	3	1	2	2	9			6						
CI	2			1			2	1	3	2	4	3	4	2	4			1						
NSL	11	19	5	7	6	16	12	10	15	9	-1	-1	2	-1	-1	20	21	14	3	17	22	24	0	23

NASL = 13

Matrix A_2

IPV:

2	
3	
1	
4	

RIP:

4
18
1
8

Location	1	2	3	4	5	6	7	8	9	10	11	12	13	14	15	16	17	18	19	20	21	22	23	24
VE	5			$\frac{1}{6}$			$\frac{16}{30}$	$\frac{4}{6}$	2	$-\frac{8}{30}$	3	1	$-\frac{48}{30}$	2	$\frac{294}{30}$			6						
CI	2			1			2	1	3	2	4	3	4	2	4			1						
NSL	11	19	5	7	6	16	12	10	15	9	-1	13	-1	-1	-1	20	21	14	3	17	22	24	0	23

NASL = 2

Matrix A_3

IPV:

2	
3	
4	
1	

RIP:

4
18
1
8

Location	1	2	3	4	5	6	7	8	9	10	11	12	13	14	15	16	17	18	19	20	21	22	23	24
VE	5			$\frac{1}{6}$			$\frac{16}{30}$	$\frac{4}{6}$	2	$-\frac{8}{6}$	3	$\frac{1}{2}$	$-\frac{195}{30}$	2	$\frac{294}{30}$			6						
CI	2			1			2	1	3	2	4	3	4	2	4			1						
NSL	11	19	5	7	6	16	12	10	15	9	-1	13	-1	-1	-1	20	21	14	3	17	22	24	0	23

NASL = 2

Figure 15-2 Storage of matrices A_1, A_2, and A_3 as linked lists.

15-2 SCALING OF MATRICES

Two of the many possible methods for scaling matrices are presented, which are called (1) *variable scaling and row scaling* and (2) *column scaling and row scaling*. The first of these consists of scaling the variables and functions. This type of scaling is frequently effected in the formulation of a problem by making suitable choices of the variables and functions. In terms of matrix operations, it consists of variable scaling followed by row scaling. This well-known procedure was used by Burdett[3] in the analysis of a system of 17 evaporators.

In the second method, row scaling is preceded by column scaling. Neither of these scaling procedures yields an optimum solution to the scaling problem as pointed out by Tewarson[7] who also provides several references to more advanced scaling methods.

Variable Scaling and Row Scaling

This procedure is developed for the case where the problem is to be solved by use of the Newton-Raphson method. For the general case of n independent equations in n unknowns, n equations are obtained upon application of the Newton-Raphson method. The equations so obtained may be represented by the following matrix equation

$$\mathbf{B}_k \Delta \mathbf{X}_k = -\mathbf{f}_k \tag{15-3}$$

where $\mathbf{B}_k$ is the square jacobian matrix of order n and $\Delta \mathbf{X}_k$ and $\mathbf{f}_k$ are conformable column vectors

$$\mathbf{B}(\mathbf{X}_k) = \begin{bmatrix} \dfrac{\partial f_1}{\partial x_1} & \dfrac{\partial f_1}{\partial x_2} & \cdots & \dfrac{\partial f_1}{\partial x_n} \\ \vdots & \vdots & & \vdots \\ \dfrac{\partial f_n}{\partial x_1} & \dfrac{\partial f_n}{\partial x_2} & & \dfrac{\partial f_n}{\partial x_n} \end{bmatrix}$$

$$\Delta \mathbf{X}_k = \begin{bmatrix} \Delta x_1 & \Delta x_2 & \cdots & \Delta x_n \end{bmatrix}^T \qquad \Delta x_j = x_{j, k+1} - x_{jk}$$

Also,

$$\Delta \mathbf{X}_k = \mathbf{X}_{k+1} - \mathbf{X}_k \qquad \mathbf{X}_k = \begin{bmatrix} x_{1k} & x_{2k} & \cdots & x_{nk} \end{bmatrix}^T$$

$$\mathbf{f}_k = \begin{bmatrix} f_{1k} & f_{2k} & \cdots & f_{nk} \end{bmatrix}^T$$

The notation $\mathbf{B}(\mathbf{X}_k)$ is used to emphasize that the jacobian matrix $\mathbf{B}_k$ is to be evaluated at $\mathbf{X} = \mathbf{X}_k$.

Let $\mathbf{R}_k$ be a diagonal matrix whose elements r_{ii} are just greater than the absolute value of the corresponding elements of $\mathbf{X}_k$, that is,

$$r_{11} > |x_{1k}| \qquad r_{22} > |x_{2k}|, \ldots, r_{nn} > |x_{nk}| \tag{15-4}$$

[Except for the restriction that r_{ii} must never be set up equal to zero, the inequality given by Eq. (15-4) need not be adhered to precisely in practice.] The elements of $\mathbf{R}_k$ should be close, however, in absolute value to the corresponding elements of $\mathbf{X}_k$. Equation (15-3) may be restated in terms of the matrix $\mathbf{R}_k$ in the following manner

$$\mathbf{D}_k \Delta \mathbf{Y}_k = -\mathbf{f}_k \tag{15-5}$$

where

$$\Delta \mathbf{Y}_k = \mathbf{R}_k^{-1} \Delta \mathbf{X}_k \tag{15-6}$$

$$\mathbf{D}_k = \mathbf{B}_k \mathbf{R}_k \tag{15-7}$$

After the multiplication implied by Eq. (15-7) has been carried out, form the diagonal matrix $\mathbf{M}_k$ whose elements m_{ii} are selected such that for each row

$$m_{ii} = \text{maximum of } |d_{ij}| \text{ over all elements of row } i \tag{15-8}$$

Premultiplication of each side of Eq. (15-5) by $\mathbf{M}_k^{-1}$ yields

$$(\mathbf{M}_k^{-1} \mathbf{D}_k) \Delta \mathbf{Y}_k = -(\mathbf{M}_k^{-1} \mathbf{f}_k) \tag{15-9}$$

Next, compute the values of the elements of $\mathbf{E}_k$ and $\mathbf{F}_k$, which are defined as follows

$$\mathbf{E}_k = \mathbf{M}_k^{-1} \mathbf{D}_k \qquad \mathbf{F}_k = \mathbf{M}_k^{-1} \mathbf{f}_k \tag{15-10}$$

Then

$$\Delta \mathbf{Y}_k = -\mathbf{E}_k^{-1} \mathbf{F}_k \tag{15-11}$$

Column Scaling and Row Scaling

A jacobian matrix may be scaled by first dividing each column by that element of the row which is largest in absolute value. Let $\mathbf{D}_k$ denote the diagonal matrix which contains the reciprocals of the elements of the respective columns which are largest in absolute value, and let $\{a_{ij}\}$ denote the elements of $\mathbf{J}_k$. The elements $\{d_{ij}\}$ of $\mathbf{D}_k$ are as follows

$$d_{11} = 1/[\text{maximum } |a_{i1}| \text{ of column 1 of matrix } \mathbf{J}_k]$$

$$d_{22} = 1/[\text{maximum } |a_{i2}| \text{ of column 2 of matrix } \mathbf{J}_k]$$

$$\vdots \tag{15-12}$$

$$d_{nn} = 1/[\text{maximum } |a_{in}| \text{ of column } n \text{ of matrix } \mathbf{J}_k]$$

The column scaling is represented by

$$(\mathbf{J}_k \mathbf{D}_k)(\mathbf{D}_k^{-1} \Delta \mathbf{X}_k) = -\mathbf{f}_k \tag{15-13}$$

Next row scaling is performed on the matrix $\mathbf{J}_k \mathbf{D}_k$. Let $\mathbf{E}_k$ denote the diagonal matrix which contains the reciprocals of the elements of the respective rows which are largest in absolute value, and let $\{b_{ij}\}$ denote the elements of

$\mathbf{J}_k \mathbf{D}_k$. The elements $\{e_{ii}\}$ of $\mathbf{M}_k$ are as follows

$$e_{11} = 1/[\text{maximum } |b_{1j}| \text{ of row 1 of matrix } \mathbf{J}_k \mathbf{D}_k]$$

$$e_{22} = 1/[\text{maximum } |b_{2j}| \text{ of row 2 of matrix } \mathbf{J}_k \mathbf{D}_k]$$

$$\vdots \tag{15-14}$$

$$e_{nn} = 1/[\text{maximum } |b_{nj}| \text{ of row } n \text{ of matrix } \mathbf{J}_k \mathbf{D}_k]$$

The row scaling is represented by

$$(\mathbf{E}_k \mathbf{J}_k \mathbf{D}_k)(\mathbf{D}_k^{-1} \Delta \mathbf{X}_k) = -\mathbf{E}_k \mathbf{f}_k \tag{15-15}$$

Thus

$$\Delta \mathbf{X}_k = -\mathbf{D}_k (\mathbf{E}_k \mathbf{J}_k \mathbf{D}_k)^{-1} \mathbf{E}_k \mathbf{f}_k \tag{15-16}$$

15-3 NUMERICAL METHODS FOR SOLVING MULTIVARIABLE PROBLEMS

Three well-known numerical methods for solving multivariable problems are presented as well as their convergence characteristics. The methods considered are direct iteration, the Newton-Raphson method, and Broyden's method.

Solution of Multivariable Problems by Use of the Method of Direct Iteration

Any iterative procedure in which the calculated values of the variables are used without alteration to make the next trial calculation is referred to herein as direct iteration.

A variety of names are used in the literature to describe calculational procedures for solving sets of linear and nonlinear algebraic equations such as *iteration, successive iteration,* and *successive substitution.*

Let it be required to find the desired solution to a nonlinear set of algebraic equations, which may be represented as follows

$$f_1(x_1, x_2, \ldots, x_n) = 0$$

$$f_2(x_1, x_2, \ldots, x_n) = 0$$

$$\vdots \tag{15-17}$$

$$f_n(x_1, x_2, \ldots, x_n) = 0$$

Let the trial vector for the kth trial be denoted by

$$\mathbf{X}_k = [x_{1k} \ x_{2k} \ \cdots \ x_{nk}]^T$$

For convenience let each function of Eq. (15-17) be represented by

$$f_i(\mathbf{X}) = 0 \qquad (i = 1, 2, \ldots, n) \tag{15-18}$$

and let

$$\mathbf{f}(\mathbf{X}) = [f_1(\mathbf{X}) \; f_2(\mathbf{X}) \; \cdots \; f_n(\mathbf{X})]^T \tag{15-19}$$

Then in matrix notation, Eq. (15-17) may be represented

$$\mathbf{f}(\mathbf{X}) = \mathbf{0} \tag{15-20}$$

Now, let the solution set of values of the variables of Eq. (15-17) be denoted by **a**, where

$$\mathbf{a} = [a_1 \; a_2 \; \cdots \; a_n]^T \tag{15-21}$$

The equations employed in the iterative procedure may be obtained by expressing each of the equations given by Eq. (15-17) in terms of the variables in a manner such that one expression is obtained for each x_i, namely,

$$x_i = F_i(\mathbf{X}) \qquad (i = 1, 2, \ldots, n) \tag{15-22}$$

It should be observed that the functions $F_i(\mathbf{X})$ may be formed from any suitable combination of the functions $f_i(\mathbf{X})$. That is, F_i may be any function for which $F_i(\mathbf{X})$ approaches a_i as $\mathbf{X}$ approaches **a** for each i from $i = 1$ through $i = n$.

In the Jacobi method of iteration, a new set $\mathbf{X}_k$ of values of the variables is computed on the basis of an assumed set $\mathbf{X}_{k-1}$ as follows

$$x_{ik} = F_i(\mathbf{X}_{k-1}) \qquad (i = 1, 2, \ldots, n) \tag{15-23}$$

The set $\mathbf{X}_k$ so obtained is used to compute the next set in a similar manner.

Now let R denote the set of all vectors $\mathbf{X}$ whose elements satisfy the condition $|x_j - a_j| \le h$ $(j = 1, 2, \ldots, n)$ for which

$$\sum_{j=1}^{n} \left| \frac{\partial F_i(\mathbf{X})}{\partial x_j} \right| < \mu < 1 \qquad (i = 1, 2, \ldots, n) \tag{15-24}$$

for each i. If the starting vector $\mathbf{X}_0$ is an element of R, then

$$\lim_{N \to \infty} \mathbf{X}_N = \mathbf{a} \tag{15-25}$$

A method for initiating the proof of this proposition is given in Prob. 15-6.

In the procedure which is analogous to the Gauss-Seidel solution of linear algebraic equations, the most recent value of each variable is used at each point in the calculational procedure. The Gauss-Seidel type of iteration differs from the Jacobi type of iteration in that instead of computing each element of $\mathbf{X}_k$ on the basis of an assumed vector $\mathbf{X}_{k-1}$, as indicated by Eq. (15-23), the following procedure is employed

$$\begin{aligned}
x_{1k} &= F_1(x_{1, k-1}, x_{2, k-1}, x_{3, k-1}, \ldots, x_{n-1, k}, x_{n, k-1}) \\
x_{2k} &= F_2(x_{1k}, x_{2, k-1}, x_{3, k-1}, \ldots, x_{n-1, k-1}, x_{n, k-1}) \\
x_{3k} &= F_3(x_{1k}, x_{2k}, x_{3, k-1}, \ldots, x_{n-1, k-1}, x_{n, k-1}) \\
&\;\;\vdots \\
x_{nk} &= F_n(x_{1k}, x_{2k}, x_{3k}, \ldots, x_{n-1, k}, x_{n, k-1})
\end{aligned} \tag{15-26}$$

When the conditions given by Eqs. (15-24) and (15-25) are satisfied, convergence of the Gauss-Seidel method can be assured; see Prob. 15-7.

Another sufficient condition for the convergence of the Jacobi method of iteration follows. Let R be the set of all starting vectors X_0 for which the largest Hilbert norm of any matrix B generated by the iterative process has the property that

$$\|B\|_{\text{III}} = \sqrt{\Lambda_1} < 1 \tag{15-27}$$

where

$$B = \begin{bmatrix} \dfrac{\partial F_1}{\partial x_1} & \dfrac{\partial F_1}{\partial x_2} & \cdots & \dfrac{\partial F_1}{\partial x_n} \\[2mm] \dfrac{\partial F_2}{\partial x_1} & \dfrac{\partial F_2}{\partial x_2} & \cdots & \dfrac{\partial F_2}{\partial x_n} \\[1mm] \cdots\cdots\cdots\cdots\cdots\cdots\cdots\cdots \\[1mm] \dfrac{\partial F_n}{\partial x_1} & \dfrac{\partial F_n}{\partial x_2} & \cdots & \dfrac{\partial F_n}{\partial x_n} \end{bmatrix}$$

The norm $\|B\|_{\text{III}}$ is defined in App. A.

The Newton-Raphson method for the solution of n equations in n unknowns takes the form given by Eq. (15-3). In the application of this method, it is recommended that the convergence characteristics be checked by solving a wide variety for examples. The use of different initial sets of values for the variables should also be investigated. Also, if only positive roots of the functions are desired, provisions should be made for an alternative selection of variables for the next trial when one or more negative values are computed by an intermediate trial.

In carrying out the following analysis, it is convenient to make use of the Newton-Raphson equations stated in the form given by Eq. (15-3). Also, the notation $B(X_k)$ is used to emphasize the fact that the jacobian matrix B_k is to be evaluated at X_k.

Let the solution set of the independent variables be represented by the column vector α

$$\alpha = [a_1 \ a_2 \ \cdots \ \alpha_n]^T \tag{15-28}$$

As the name of this set of values of the variables implies

$$f_i(\alpha) = 0 \qquad (i = 1, 2, \ldots, n)$$

or

$$f(\alpha) = 0 \tag{15-29}$$

In the analysis that follows, it is shown that if X_0 is any starting vector which is "close enough" [explained below Eq. (15-43)] to the solution vector α, if the elements of $f(X_0)$ and $B(X_0)$ are continuous, and if the determinant of $B(X_0)$ is nonzero ($|B(X_0)| \neq 0$), then the Newton-Raphson procedure converges to the solution set α. The general approach used in the construction of the proof follows that of Carnahan et al.[1]

To initiate this analysis, it is supposed that the elements of f and B are continuous over all starting vectors X_0 and all vectors X_k generated therefrom by use of the Newton-Raphson method. It is further supposed that the determinant $|B| \neq 0$ for all X_0 and X_k.

Since $|B(X_k)| \neq 0$, the inverse $B^{-1}(X_k)$ exists. Then Eq. (15-3) may be solved for ΔX_k to give

$$\Delta X_k = -B^{-1}(X_k)f(X_k) \tag{15-30}$$

This equation may be restated in the following form by use of the result given by Eq. (15-29)

$$\Delta X_k = B^{-1}(X_k)[f(\alpha) - f(X_k)] \tag{15-31}$$

The difference $[f(\alpha) - f(X_k)]$ of the two column vectors may be replaced by its equivalent as given by the *mean value theorem of differential calculus for multivariable functions* (Theorem A-7), since the continuity requirements of the theorem are satisfied by suppositions stated above. Then, each element i of $[f(X_k) - f(\alpha)]$ may be stated as follows

$$f_i(X_k) - f_i(\alpha) = \sum_{j=1}^{n} \frac{\partial f_i(\xi_{ik})}{\partial x_j}(x_{ij} - \alpha_j) \qquad (i = 1, 2, \ldots, n) \tag{15-32}$$

The elements of ξ_{ik} consist of $\alpha_1 + \beta_{ik}(x_{1k} - \alpha_1)$, $\alpha_2 + \beta_{ik}(x_{2k} - \alpha_2)$, $\ldots$, $\alpha_n + \beta_{ik}(x_{nk} - \alpha_n)$, where $0 < \beta_{ik} < 1$. Thus

$$f(X_k) - f(\alpha) = B(\xi_{ik})E_k \tag{15-33}$$

The notation $B(\xi_{ik})$ is used to imply that the first row of functions in B is to be evaluated at ξ_{1k}, the second row at ξ_{2k}, and finally the nth row at ξ_{nk}. The column vector E_k has as its elements the error of each variable at the end of the kth trial, that is,

$$E_k = X_k - \alpha \tag{15-34}$$

Now observe that

$$E_{k+1} = X_{k+1} - \alpha = X_k - \alpha + \Delta X_k = E_k + \Delta X_k \tag{15-35}$$

Use of Eqs. (15-31) and (15-33) permits Eq. (15-35) to be restated as follows

$$E_{k+1} = B^{-1}(X_k)D_k E_k \tag{15-36}$$

where D_k is defined by

$$D_k = B(X_k) - B(\xi_{ik})$$

Equation (15-36) is seen to be a recurrence formula which gives the error vector at the end of the $k + 1$st iteration in terms of the error vector for the kth iteration. By repeated substitution of the vector E_k into the expression for E_{k+1}, it is readily shown that the error vector at the end of the first $N + 1$ iterations is given by

$$\mathbf{E}_{N+1} = \mathbf{B}^{-1}(\mathbf{X}_N)\mathbf{D}_N B^{-1}(\mathbf{X}_{N-1})\mathbf{D}_{N-1}, \ldots, \mathbf{B}^{-1}(\mathbf{X}_0)\mathbf{D}_0 \mathbf{E}_0 \qquad (15\text{-}37)$$

Let $\|\mathbf{B}^{-1}\|_{\mathrm{III}}$ and $\|\mathbf{D}\|_{\mathrm{III}}$ denote the largest values of $\|\mathbf{B}^{-1}(\mathbf{X}_l)\|_{\mathrm{III}}$ and $\|\mathbf{D}_l\|_{\mathrm{III}}$, respectively, over all l from 1 through N. Application of vector and matrix norms (presented in App. A) to Eq. (15-37) yields

$$\|\mathbf{E}_{N+1}\|_{\mathrm{III}} \leq [\|\mathbf{B}^{-1}\|_{\mathrm{III}}\|\mathbf{D}\|_{\mathrm{III}}]^{N+1}\|\mathbf{E}_0\|_{\mathrm{III}} \qquad (15\text{-}38)$$

The Hilbert norms of matrices $\mathbf{D}$ and $\mathbf{B}^{-1}$ are given by

$$\|\mathbf{D}\|_{\mathrm{III}} = \sqrt{\Lambda_1} \qquad (15\text{-}39)$$

$$\|\mathbf{B}^{-1}\|_{\mathrm{III}} = \sqrt{\delta_1} \qquad (15\text{-}40)$$

where Λ_1 is the largest eigenvalue of $\mathbf{D}^T\mathbf{D}$ and δ_1 is the largest eigenvalue of $(\mathbf{B}^{-1})^T\mathbf{B}^{-1}$. Thus

$$\|\mathbf{E}_{N+1}\| \leq [\sqrt{\delta_1 \Lambda_1}]^{N+1}\|\mathbf{E}_0\|_{\mathrm{III}} \qquad (15\text{-}41)$$

Hence, it is evident that if

$$\sqrt{\delta_1 \Lambda_1} < 1 \qquad (15\text{-}42)$$

then

$$\lim_{N \to \infty} \|\mathbf{E}_{N+1}\|_{\mathrm{III}} = 0 \qquad (15\text{-}43)$$

A starting vector $\mathbf{X}_0$ is said to be "close enough" to α if the Newton-Raphson iterations lead to a value of $\sqrt{\delta_1 \Lambda_1} < 1$, where δ_1 and Λ_1 are defined by Eqs. (15-39) and (15-40). Let R be the set of all starting vectors $\mathbf{X}_0$ for which the maximum Hilbert norms of $\mathbf{B}^{-1}$ and $\mathbf{D}$ over all iterations by the Newton-Raphson method satisfy the condition given by Eq. (15-42). Observe that the solution vector α is an element of R, since $\mathbf{D}(\alpha) = \mathbf{0}$.

From the definition of the vectors $\mathbf{X}_0$ of R, observe that each vector of the sequence $\mathbf{X}_1, \mathbf{X}_2, \ldots, \mathbf{X}_k, \mathbf{X}_{k+1}, \ldots$ generated by the Newton-Raphson method on the basis of any $\mathbf{X}_0$ of R is also a member of R.

Since it has not shown that the set of R contains any elements other than the solution vector α, it will now be shown that under certain conditions, the set R contains infinitely many starting vectors. Let S be the set of all starting vectors $\mathbf{X}$ and all vectors $\mathbf{X}_k$ generated therefrom (by use of the Newton-Raphson method) for which the corresponding Hilbert norms of $\mathbf{D}$ are less than unity. Since this condition is satisfied by each element of R, then each element of R is an element of S, and R is a subset of (belongs to) S. No other restriction was placed on the value of the Hilbert norm of $\mathbf{B}^{-1}$ (boundedness of this norm is assured, however, by the continuity conditions imposed on the elements of $\mathbf{B}$ and the condition that $|B| \neq 0$). The condition given by Eq. (15-42) is not necessarily satisfied by each element of S, and consequently S may contain elements which do not belong to R. It will now be shown that if the elements of $\mathbf{f}(\mathbf{X})$ and $\mathbf{B}(\mathbf{X})$ are continuous over all $\mathbf{X}$ of S and if $|\mathbf{B}(\mathbf{X})| \neq 0$ over all $\mathbf{X}$ of S, then there exists an infinite subset P of S which is also a subset of R.

Since $|\mathbf{B}(\mathbf{X})| \neq 0$, it follows that $\mathbf{B}^{-1}(\mathbf{X})$ exists and the Hilbert norm of $\mathbf{B}^{-1}$ is bounded. Let $\sqrt{\delta_{1,b}}$ denote the least upper bound of the Hilbert norm of $\mathbf{B}^{-1}$; that is, for any $\mathbf{X}$ of S

$$\|\mathbf{B}^{-1}(\mathbf{X})\|_{\text{III}} \leq \sqrt{\delta_{1,b}}$$

From the definition of $\mathbf{D}_k$, element j of row i of this matrix is given by

$$\frac{\partial f_i(\mathbf{X}_k)}{\partial x_j} - \frac{\partial f_i(\xi_{ik})}{\partial x_j}$$

This difference may be made arbitrarily small by choosing the starting vector $\mathbf{X}$ sufficiently close to the solution vector $\boldsymbol{\alpha}$. Since, by Theorem A-1-2, the largest eigenvalue of a real, nonsingular, symmetric matrix can never exceed the largest sum of the absolute value of the elements of any row (or column) of $\mathbf{D}^T\mathbf{D}$, it follows that Λ_1 may be made arbitrarily small by choosing the elements of $\mathbf{X}$ suitably close to the elements of $\boldsymbol{\alpha}$. Thus, there exist infinitely many starting vectors $\mathbf{X}_P$ of S such that the corresponding Hilbert norm $\sqrt{\Lambda_{1,P}}$ satisfies the following inequality

$$\delta_{1,b}\Lambda_{1,P} < 1$$

Let the set of all such vectors be denoted by P. Since the Hilbert norm $\sqrt{\delta_{1,P}}$ corresponding to $\sqrt{\Lambda_{1,P}}$ is the largest one generated by the Newton-Raphson method when $\mathbf{X}_P$ is used as a starting vector, it follows that

$$\delta_{1,P}\Lambda_{1,P} \leq \delta_{1,b}\Lambda_{1,P} < 1 \tag{15-44}$$

Since each vector $\mathbf{X}_P$ of the set P satisfies the inequality given by Eq. (15-44) it is also an element of R. Since each element of P is an element of R, it follows that P is subset of R.

The Broyden Modification of the Newton-Raphson Method

There follows the development of a class of methods proposed by Broyden[2] for the purpose of overcoming some of the disadvantages of the Newton-Raphson method. One serious disadvantage of the Newton-Raphson method is the time required to evaluate all of the elements of the jacobian matrix $\mathbf{B}_k$. Even if the functions f_i are sufficiently simple for their partial derivatives to be obtained analytically, the amount of labor required to evaluate all n^2 of these may be excessive. If the partial derivatives of the f_i's are approximated numerically, the amount of labor required constitutes a serious disadvantage.

The second disadvantage of the Newton-Raphson method is that it may not converge unless the initial estimate of the solution is sufficiently good.

Despite the disadvantages of the Newton-Raphson method, it is easy to apply, has a sound theoretical basis, and for many problems its convergence is rapid. Furthermore, according to Broyden,[2] competitors such as those based upon the minimization of some norm of the vector **f** by the steepest descent,[8] or some similar method, tend to converge slowly in some instances, although a new method recently described by Kizner[6] may ultimately prove satisfactory.

In the class of methods proposed by Broyden, the partial derivatives $\partial f_i / \partial x_j$ in the jacobian matrix are evaluated only once. In each successive trial the elements of the inverse of the jacobian matrix are corrected by use of computed values of the functions f_i. Throughout the development which follows, it is supposed that the functions f_i are real variable functions of real variables and that the functions are continuous and differentiable. If the jacobian matrix $\mathbf{B}_k$ in the Newton-Raphson equation [Eq. (15-3)] is nonsingular, then $\mathbf{B}_k^{-1}$ exists and

$$\mathbf{X}_{k+1} = \mathbf{X}_k - \mathbf{B}_k^{-1}\mathbf{f}_k \qquad (15\text{-}45)$$

where $\Delta \mathbf{X}_k = \mathbf{X}_{k+1} - \mathbf{X}_k$.

Let $\mathbf{P}_k$ be defined as follows

$$\mathbf{P}_k = -\mathbf{A}_k \mathbf{f}_k \qquad (15\text{-}46)$$

where $\mathbf{A}_k$ is some approximation of $\mathbf{B}_k$. Then a simple modification of the Newton-Raphson method consists of

$$\mathbf{X}_{k+1} = \mathbf{X}_k + s_k \mathbf{P}_k \qquad (15\text{-}47)$$

where s_k is a scalar which is picked as described below.

Now let the vector **X** be defined by

$$\mathbf{X} = \mathbf{X}_k + s\mathbf{P}_k \qquad (15\text{-}48)$$

where s is arbitrary. Each function f_i may now be regarded as a function of the single variable s. Since the partial derivatives $\partial f_i / \partial x_j$ are assumed to exist, each derivative with respect to s is given by

$$\frac{df_i}{ds} = \sum_{j=1}^{n} \frac{\partial f_i}{\partial x_j} \frac{dx_j}{ds} \qquad (i = 1, 2, \ldots, n) \qquad (15\text{-}49)$$

It will now be shown that an approximation to the derivatives given by Eq. (15-49) may be used to improve the approximate matrix $\mathbf{A}_k$. By use of the relationship obtained by differentiating the members of Eq. (15-47) with respect to s, $d\mathbf{X}/ds = \mathbf{P}_k$, it is possible to restate Eq. (15-49) in the following form

$$\frac{d\mathbf{f}}{ds} = \mathbf{B}\mathbf{P}_k \qquad (15\text{-}50)$$

Thus, if an accurate estimate of $d\mathbf{f}/ds$ were available, it could be used with Eq. (15-50) to estimate a condition that any approximation to the jacobian matrix must satisfy. Suppose, however, that the partial derivatives are evaluated numerically. Since the value of the jacobian matrix at the point $\mathbf{X}_{k+1}$ is desired, the derivative $d\mathbf{f}/ds$ at the point $\mathbf{X}_{k+1}$ may be evaluated numerically. The numerical approximation of the derivative df_i/ds may be computed as follows

$$\frac{df_i}{ds} = \frac{f_i(\mathbf{X}_k + s_k\,\mathbf{P}_k) - f_i[\mathbf{X}_k + (s_k - \varepsilon)\mathbf{P}_k]}{\varepsilon} \tag{15-51}$$

or in simpler notation

$$-\frac{df_i}{ds} \cong \frac{f_i(s_k - \varepsilon) - f_{i,\,k+1}}{\varepsilon}$$

Thus, the complete set of derivatives may be represented as follows

$$\mathbf{f}(s_k - \varepsilon) \cong \mathbf{f}_{k+1} - \varepsilon\,\frac{d\mathbf{f}}{ds} \tag{15-52}$$

Elimination of $d\mathbf{f}/ds$ from Eqs. (15-50) and (15-52) yields

$$\mathbf{f}_{k+1} - \mathbf{f}(s_k - \varepsilon_k) \cong \varepsilon_k\,\mathbf{BP}_k \tag{15-53}$$

In the class of methods proposed by Broyden, the improved approximation $\mathbf{A}_{k+1}$ of $\mathbf{A}_k$ is selected such that the following equation is satisfied

$$\mathbf{f}_{k+1} - \mathbf{f}(s_k - \varepsilon_k) = \varepsilon_k\,\mathbf{A}_{k+1}\,\mathbf{P}_k \tag{15-54}$$

Next, let the square matrix $\mathbf{H}_k$ of order n be defined by

$$\mathbf{H}_k = -\mathbf{A}_k^{-1} \tag{15-55}$$

and the column vector $\mathbf{Y}_k$ of order n by

$$\mathbf{Y}_k = \mathbf{f}_{k+1} - \mathbf{f}(s_k - \varepsilon_k) \tag{15-56}$$

When stated in terms of these matrices, Eqs. (15-46) and (15-54) become

$$\mathbf{P}_k = \mathbf{H}_k\,\mathbf{f}_k \tag{15-57}$$

$$\mathbf{H}_{k+1}\mathbf{Y}_k = -\varepsilon_k\,\mathbf{P}_k \tag{15-58}$$

respectively. Also, Eqs. (15-54) and (15-56) may be combined to give

$$\mathbf{Y}_k = \varepsilon_k\,\mathbf{A}_{k+1}\,\mathbf{P}_k \tag{15-59}$$

which relates the change $\mathbf{Y}_k$ in the vector function to a change of $\mathbf{X}$ in the direction $\mathbf{P}_k$. In the procedure called "Method 1" by Broyden, $\mathbf{A}_{k+1}$ is chosen so that the change in $\mathbf{f}$ predicted by $\mathbf{A}_{k+1}$ in a direction $\mathbf{Q}_k$ orthogonal to $\mathbf{P}_k$ is the same as would be predicted by $\mathbf{A}_k$, that is,

$$\mathbf{A}_{k+1}\mathbf{Q}_k = \mathbf{A}_k\mathbf{Q}_k \qquad \mathbf{Q}_k^T\mathbf{P}_k = 0 \tag{15-60}$$

When these two relationships are combined with Eq. (15-59), the following formula for $\mathbf{A}_{k+1}$ is obtained (see Prob. 15-2)

$$\mathbf{A}_{k+1} = \mathbf{A}_k + \frac{(\mathbf{Y}_k - \varepsilon_k\mathbf{A}_k\mathbf{P}_k)\mathbf{P}_k^T}{\varepsilon_k\,\mathbf{P}_k^T\mathbf{P}_k} \tag{15-61}$$

In the solution of problems by use of a computer, it is preferable to store $\mathbf{H}_k$ rather than $\mathbf{A}_k$. To obtain $\mathbf{H}_{k+1}$ from $\mathbf{A}_{k+1}$, Householder's formula is used. If $\mathbf{A}$ is a nonsingular matrix and $\mathbf{X}$ and $\mathbf{Y}$ are vectors, all of order n, and if $(\mathbf{A} + \mathbf{XY}^T)$ is nonsingular, then Householder's formula

$$(\mathbf{A} + \mathbf{XY}^T)^{-1} = \mathbf{A}^{-1} - \frac{\mathbf{A}^{-1}\mathbf{XY}^T\mathbf{A}^{-1}}{1 + \mathbf{Y}^T\mathbf{A}^{-1}\mathbf{X}}$$

applies. When this formula is applied to Eq. (15-61), one obtains

$$\mathbf{H}_{k+1} = \mathbf{H}_k - \frac{(\varepsilon_k\mathbf{P}_k + \mathbf{H}_k\mathbf{Y}_k)\mathbf{P}_k^T\mathbf{H}_k}{\mathbf{P}_k^T\mathbf{H}_k\mathbf{Y}_k} \tag{15-62}$$

(see Prob. 15-3).

In the application of these relationships, ε_k is set equal to s_k, which amounts to using a full step size to approximate the total derivative; see Eq. (15-51). The numerical value of s_k is picked such that convergence is promoted; that is, for the kth trial, s_k is picked such that the euclidean norm (see App. A) of $\mathbf{f}_{k+1}$ is less than the euclidean norm of $\mathbf{f}_k$, that is,

$$\left[\sum_{i=1}^n f_i^2(s_k)\right]^{1/2} < \left[\sum_{i=1}^n f_i^2(s_{k-1})\right]^{1/2} \tag{15-63}$$

The first value of s_k (denoted by the second subscript) is taken as unity

$$s_{k,\,1} = 1$$

If this value of s_k satisfies Eq. (15-63), it is used; otherwise, a second value is computed by use of the following formula which was developed by Broyden

$$s_{k,\,2} = \frac{(1 + 6\eta)^{1/2} - 1}{3\eta} \tag{15-64}$$

where

$$\eta = \frac{\left[\sum_{i=1}^n f_i^2(1)\right]}{\left[\sum_{i=1}^n f_i^2(0)\right]}$$

As pointed out by Broyden, other methods for picking s_k may be used. For example, s_k may be picked such that the euclidean norm of $\mathbf{f}$ is minimized.

The steps of a calculational procedure proposed by Broyden are presented in Chap. 4.

PROBLEMS

15-1 Broyden's derivation of Eq. (15-64) is initiated as follows. Let $\phi(x)$ denote the square of the euclidean norm, and let the ideal quadratic of this norm be defined as one which has the properties that $\phi_q(1) = 0$ and $\phi_q'(1) = 0$.

(a) Show that the ideal quadratic has the following representation

$$\phi_q(x) = (1 - 2x + x^2)\phi(0)$$

(b) In practice, the square of the euclidean norm at $x = 1$ is very rarely equal to zero. If the deviation of the square of the norm from zero at $x = 1$ is due to the contribution of a single cubic term, show that the ideal cubic approximation of ϕ has the following representation

$$\phi_c(x) = (1 - 2x + x^2)\phi(0) + x^3\phi(1)$$

(c) By setting $d\phi_c/dx = 0$ and solving for the positive root lying between zero and unity, show that the expression given by Eq. (15-64) is obtained.

15-2 This problem consists of a suggested outline for the development of the relationship given by Eq. (15-61). In this suggested development, the trial subscript k which appears in Eqs. (15-57) through (15-60) is dropped. Let the given trial vector $\mathbf{P}_k$ of order n which appears in the text be denoted by $\mathbf{P}_1$. Now pick $n - 1$ other vectors, $\mathbf{P}_2, \mathbf{P}_3, \ldots, \mathbf{P}_n$, which are mutually orthogonal and orthogonal to $\mathbf{P}_1$, that is,

$$\begin{aligned} \mathbf{P}_i^T\mathbf{P}_j &= 0 \qquad \text{if } i \neq j \\ \mathbf{P}_i^T\mathbf{P}_j &\neq 0 \qquad \text{if } i = j \end{aligned} \tag{A}$$

[Note that the vector $\mathbf{Q}_k$ which appears in Eq. (15-60) may be selected from the set $\mathbf{P}_2, \ldots, \mathbf{P}_n$.]

(a) Let the matrix formed by these column vectors be denoted by

$$\mathbf{C} = [\mathbf{P}_1 \ \mathbf{P}_2 \ \cdots \ \mathbf{P}_n] = \begin{bmatrix} P_{11} & P_{12} & \cdots & P_{1n} \\ P_{21} & P_{22} & \cdots & P_{2n} \\ \vdots & \vdots & & \vdots \\ P_{n1} & P_{n2} & \cdots & P_{nn} \end{bmatrix} \tag{B}$$

and let $\mathbf{D}$ denote the matrix formed by the transpose of each column vector

$$\mathbf{D} = [\mathbf{P}_1^T \ \mathbf{P}_2^T \ \cdots \ \mathbf{P}_n^T] = \begin{bmatrix} P_{11} & P_{21} & \cdots & P_{n1} \\ P_{12} & P_{22} & \cdots & P_{n2} \\ \vdots & \vdots & & \vdots \\ P_{1n} & P_{2n} & \cdots & P_{nn} \end{bmatrix} \tag{C}$$

Show that

$$\mathbf{DC} = \begin{bmatrix} \mathbf{P}_1^T\mathbf{P}_1 & 0 & 0 & 0 \\ 0 & \mathbf{P}_2^T\mathbf{P}_2 & 0 & 0 \\ \hdashline & & & \\ 0 & & 0 & \mathbf{P}_n^T\mathbf{P}_n \end{bmatrix} \tag{D}$$

(b) Let the diagonal matrix on the right-hand side of Eq. (D) be denoted by $[\mathbf{P}_i^T\mathbf{P}_i]$ and its inverse by $[1/\mathbf{P}_i^T\mathbf{P}_i]$, that is,

$$\left[\frac{1}{\mathbf{P}_i^T\mathbf{P}_i}\right] = [\mathbf{P}_i^T\mathbf{P}_i]^{-1} = \begin{bmatrix} \dfrac{1}{\mathbf{P}_1^T\mathbf{P}_1} & 0 & 0 & 0 \\ 0 & \dfrac{1}{\mathbf{P}_2^T\mathbf{P}_2} & 0 & 0 \\ \hdashline & & & \\ 0 & \cdots & 0 & \dfrac{1}{\mathbf{P}_n^T\mathbf{P}_n} \end{bmatrix} \tag{E}$$

Show that

$$
C^{-1} = \left[\frac{1}{P_i^T P_i} \right] D = \begin{bmatrix} \dfrac{P_1^T}{P_1^T P_1} \\ \dfrac{P_2^T}{P_2^T P_2} \\ \vdots \\ \dfrac{P_n^T}{P_n^T P_n} \end{bmatrix} \tag{F}
$$

(c) Next, let Eq. (15-58) and the first expression of Eq. (15-55) be restated in the following respective forms

$$
BP_1 = \frac{Y}{\varepsilon} \tag{G}
$$

$$
BP_j = AP_j \qquad (j \geq 2) \tag{H}
$$

where A and B are nonsingular square matrices of order n and Y is a compatible column vector. (Note B corresponds to A_{k+1}, A to A_k, Y to Y_k and ε to ε_k in the text.) Show that

$$
BC = A \left[\frac{A^{-1}Y}{\varepsilon} \quad P_2 \quad P_3 \quad \cdots \quad P_n \right]
$$

and that

$$
B = A \left[\frac{A^{-1}Y}{\varepsilon} \quad P_2 \quad P_3 \quad \cdots \quad P_n \right] C^{-1}
$$

By partitioning the multiplication implied by this expression, show that

$$
B = \frac{YP_1^T}{\varepsilon P_1^T P_1} + A[P_2 \quad P_3 \quad \cdots \quad P_n] \begin{bmatrix} \dfrac{P_2^T}{P_2^T P_2} \\ \vdots \\ \dfrac{P_n^T}{P_n^T P_n} \end{bmatrix} \tag{I}
$$

(d) Begin with the relationship

$$
I = CC^{-1}
$$

and show that

$$
A = \frac{AP_1 P_1^T}{P_1^T P_1} + A[P_2 \quad P_3 \quad \cdots \quad P_n] \begin{bmatrix} \dfrac{P_2^T}{P_2^T P_2} \\ \vdots \\ \dfrac{P_n^T}{P_n^T P_n} \end{bmatrix} \tag{J}
$$

(e) By use of the results given by Eqs. (I) and (J), obtain the desired formula, Eq. (15-61).

15-3 Begin with Eq. (15-61) and show that when Householder's formula is applied one obtains the result given by Eq. (15-63).

Hint:

Let

$$
b = \varepsilon_k P_k^T P_k
$$

$$
X = \frac{Y_k - \varepsilon_k A_k P_k}{b}
$$

Then Eq. (15-49) becomes

$$\mathbf{A}_{k+1} = \mathbf{A}_k + \mathbf{X}\mathbf{P}_k^T$$

Observe that

$$\mathbf{A}_{k+1}^{-1} = [\mathbf{A}_k + \mathbf{X}\mathbf{P}_k^T]^{-1} \tag{A}$$

and that the right-hand side of the expression may be evaluated by use of Householder's formula.

15-4 (a) Solve the following set of equations by use of the Newton-Raphson method. For the starting vector, take $x_k = y_k = z_k = 0$.

$$f_1(x, y, z) = x + 4y + z + 1$$

$$f_2(x, y, z) = 2x + 3y + 2z + 2$$

$$f_3(x, y, z) = 3x + 10y + 4z - 7$$

Answer: $x = -11$, $y = 0$, $z = 10$

(b) Explain why only one trial is required to obtain the solution.

15-5 Find the positive roots of the following equations by use of the Newton-Raphson method. For the starting vector, take $x_k = 2$, $y_k = 3$

$$f_1(x, y) = x^2 - y^2 - 2$$

$$f_2(x, y) = 2x^2 - 3y^2 + 6$$

Answer: $x = \sqrt{12}$, $y = \sqrt{10}$

15-6 Prove the proposition whose statement is given by Eq. (15-24) and (15-25).
Hints:

(1) First observe that

$$x_{ik} - a_i = F_i(\mathbf{X}_{k-1}) - F_i(\mathbf{a}) \qquad (1 \le i \le n) \tag{A}$$

(2) *Next* make use of *the mean value theorem of differential calculus for multivariable functions* (A-6), namely,

$$x_{ik} - a_i = \sum_{j=1}^{n} \frac{\partial F_i(\boldsymbol{\xi}_{i, k-1})}{\partial x_j}(x_{j, k-1} - a_j) \qquad (1 \le i \le n) \tag{B}$$

where

$$\boldsymbol{\xi}_{i, k-1} = [\xi_{i, 1, k-1} \ \xi_{i, 2, k-1} \ \cdots \ \xi_{i, n, k-1}]^T$$

$$\xi_{i, l, k-1} = a_l + \beta_{i, k-1}(x_{l, k-1} - a_l)$$

where

$$0 < \beta_{i, k-1} < 1$$

15-7 Show that the conditions given by Eqs. (15-24) and (15-25) are sufficient to assure convergence for the Gauss-Seidel method of iteration.
Hint:

Make use of the fact that the calculations of x_{1k} may be represented by Eq. (B) of Prob. 15-6. Since $\mathbf{X}_{k-1}$ is assumed to belong to R, $\boldsymbol{\varepsilon}_{1, k-1}$ belongs to R.

15-8 When Eq. (B) of Prob. 15-6 is stated in matrix notation, one obtains

$$\mathbf{E}_k = \mathbf{B}_{k-1} \mathbf{E}_{k-1} \tag{A}$$

where **B** is defined below Eq. (15-27), and

$$\mathbf{E}_k = \mathbf{X}_k - \mathbf{a} \tag{B}$$

Show that if the conditions given by Eq. (15-27) are satisfied, then convergence can be assured for the Jacobi method of iteration.

15-9 Show that if the F_i's are linear, a necessary and sufficient condition for convergence is that each eigenvalue of **B** is less than unity.

Hints:

(1) By repeated substitution, show that Eqs. (A) and (B) of Prob. 15-8 give

$$\mathbf{E}_N = \mathbf{B}^N \mathbf{E}_0 \tag{A}$$

(2) If B is a nonsingular matrix of order n with n distinct eigenvalues, then

$$\mathbf{Q}^{-1}\mathbf{B}\mathbf{Q} = \Lambda$$

where Λ is the diagonal matrix which has as its elements the n eigenvalues of **B**.

15-10 The real symmetric matrix **H** is defined by

$$\mathbf{H} = \mathbf{A}^T \mathbf{A}$$

where **A** is real, nonsingular, and of order n. Show that the eigenvalues of **H** are all real.

15-11 Show that all of the roots of the real symmetric matrix **H** (defined in Prob. 15-10) are positive.

15-12 If the eigenvalues of **H** (defined in Prob. 15-10) are all distinct, then the eigenvectors corresponding to eigenvalues are mutually orthogonal.

15-13 Show that the vectors forming an orthonormal system are linearly independent.

REFERENCES

1. Brice Carnahan, H. A. Luther, and J. O. Wilkes: *Applied Numerical Methods*, John Wiley and Sons, New York, 1969.
2. C. G. Broyden: "A Class of Methods for Solving Nonlinear Simultaneous Equations," *Mathematics of Computation*, **19**: 577 (1956).
3. J. W. Burdett: "Prediction of Steady State and Unsteady State Response Behavior of a Multiple Effect Evaporator System," Ph.D. Dissertation, Texas A&M University, College Station, Texas 1969.
4. S. E. Gallun: Personal Communication (1979).
5. C. D. Holland: *Multicomponent Distillation*, Prentice-Hall, Englewood Cliffs, N.J., 1963.
6. Leon Kizner: "A Numerical Method for Finding Solutions of Nonlinear Equations," *J. Soc. Ind. Appl. Math.*, **12**: 424 (1964).
7. R. P. Tewarson: *Sparse Matrices*, Academic Press, New York, 1973.
8. D. J. Wilde and C. S. Beightler, *Foundations of Optimization*, Prentice-Hall, Englewood Cliffs, N.J., 1967.

MATRIX NORMS AND THEOREMS

In Sec. A-1, an abbreviated treatment of matrix and vector norms is presented, and in Sec. A-2, mathematical theorems used in the text are presented.

A-1 MATRIX AND VECTOR NORMS

A brief presentation of certain matrix relationships which are useful in the analysis of a series of matrices follows. More extensive treatments of matrices are available; see for example Refs. 1, 2, 4, 5, 6, and 8. For a treatment of the general subject of matrices, see Amundson,[1] and for a brief summary of some of the properties, see Ref. 7.

In the following development, it is supposed that the reader is familiar with the matrix operations of addition, subtraction, multiplication, and inversion. When any of these operations are indicated, it is of course supposed that the matrices involved are conformable in the operations indicated. Also, it should be noted that any matrix other than a column or row matrix is considered to be a square matrix.

Unless otherwise noted, it is supposed throughout the following developments that the elements of each matrix are real numbers. A square matrix $\mathbf{A}$ of order n is represented as follows:

$$\mathbf{A} = \begin{bmatrix} a_{11} & a_{12} & \cdots & a_{1n} \\ a_{21} & a_{22} & \cdots & a_{2n} \\ \cdots\cdots\cdots\cdots\cdots\cdots\cdots \\ a_{n1} & a_{n2} & \cdots & a_{nn} \end{bmatrix} \tag{A-1}$$

A square matrix is said to be *symmetric* if $a_{ij} = a_{ji}$.

A *column matrix* or *column vector* of order n has the following meaning.

$$\mathbf{X} = \begin{bmatrix} x_1 \\ x_2 \\ \cdots \\ x_n \end{bmatrix} \tag{A-2}$$

The transpose of $\mathbf{A}$, denoted by $\mathbf{A}^T$, is formed by exchanging the corresponding rows and columns of $\mathbf{A}$, that is,

$$\mathbf{A}^T = \begin{bmatrix} a_{11} & a_{21} & \cdots & a_{n1} \\ a_{12} & a_{22} & \cdots & a_{n2} \\ \cdots\cdots\cdots\cdots\cdots\cdots \\ a_{1n} & a_{2n} & \cdots & a_{nn} \end{bmatrix} \tag{A-3}$$

If $\mathbf{A}$ is symmetric, then $\mathbf{A}^T = \mathbf{A}$.

The transpose of the column vector $\mathbf{X}$ is seen to be the following *row matrix* or *row vector*

$$\mathbf{X}^T = [x_1 \ x_2 \ \cdots \ x_n] \tag{A-4}$$

Throughout the following development, a vector will always be represented by a column vector, and the transpose of a vector will always be represented by a row vector. The transpose of a product of two matrices (which are conformable in multiplication) is equal to the product of the transposes taken in reverse order

$$(\mathbf{AB})^T = \mathbf{B}^T\mathbf{A}^T \tag{A-5}$$

The validity of this relationship is readily demonstrated by use of two square matrices of order 2. Multiplication of $\mathbf{A}$ by $\mathbf{B}$ gives

$$\mathbf{AB} = \begin{bmatrix} a_{11} & a_{12} \\ a_{21} & a_{22} \end{bmatrix}\begin{bmatrix} b_{11} & b_{12} \\ b_{21} & b_{22} \end{bmatrix}$$

$$= \begin{bmatrix} (a_{11}b_{11} + a_{12}b_{21}) & (a_{11}b_{12} + a_{12}b_{22}) \\ (a_{21}b_{11} + a_{22}b_{21}) & (a_{21}b_{12} + a_{22}b_{22}) \end{bmatrix} = \mathbf{C} \tag{A-6}$$

Then from Eq. (A-3), the transpose of $\mathbf{C}$ is given by

$$(\mathbf{AB})^T = \mathbf{C}^T = \begin{bmatrix} (a_{11}b_{11} + a_{12}b_{21}) & (a_{21}b_{11} + a_{22}b_{21}) \\ (a_{11}b_{12} + a_{12}b_{22}) & (a_{21}b_{12} + a_{22}b_{22}) \end{bmatrix} \tag{A-7}$$

The right-hand side of Eq. (A-5) has the following representation

$$\mathbf{B}^T\mathbf{A}^T = \begin{bmatrix} b_{11} & b_{21} \\ b_{12} & b_{22} \end{bmatrix}\begin{bmatrix} a_{11} & a_{21} \\ a_{12} & a_{22} \end{bmatrix} \tag{A-8}$$

which upon multiplication yields the transpose of $\mathbf{C}$ [Eq. (A-7)],

$$\mathbf{B}^T\mathbf{A}^T = \mathbf{C}^T \tag{A-9}$$

Since $(\mathbf{AB})^T = \mathbf{C}^T$ and $\mathbf{B}^T\mathbf{A}^T = \mathbf{C}^T$, it follows that $(\mathbf{AB})^T = \mathbf{B}^T\mathbf{A}^T$, [Eq. (A-5)].

The determinant of $\mathbf{A}$ is commonly denoted by $|\mathbf{A}|$ or det $\mathbf{A}$. If $|\mathbf{A}| = 0$, the matrix $\mathbf{A}$ is said to be *singular*; and if $|\mathbf{A}| \neq 0$, the matrix $\mathbf{A}$ is said to be *nonsingular*. One rule for the multiplication of two determinants of the same order is the same rule as that for the multiplication of two square matrices of the same order. Thus

$$|\mathbf{A}\,\|\,\mathbf{B}| = |\mathbf{C}| \tag{A-10}$$

and since the elements of $\mathbf{C}$ are identically the same as those of the matrix product $\mathbf{AB}$, it follows that

$$|\mathbf{AB}| = |\mathbf{A}\,\|\,\mathbf{B}| \tag{A-11}$$

Thus, the determinant of the product of two matrices is equal to the product of their determinants.

The vector product $\mathbf{X}^T\mathbf{X}$ is seen to be a scalar, namely,

$$\mathbf{X}^T\mathbf{X} = \sum_{i=1}^{n} x_i^2 \tag{A-12}$$

and the square root of this scalar is called the *euclidean norm* of an n-dimensional vector space, as discussed in a subsequent section. If each element of the vector $\mathbf{X}$ is divided by k $(k^2 = x_1^2 + x_2^2 + \cdots + x_n^2)$, then the vector $\mathbf{X}$ is said to have been normalized. That is, let

$$\mathbf{Y} = \frac{1}{k}\mathbf{X} \tag{A-13}$$

where the elements of $\mathbf{Y}$ are related to those of $\mathbf{X}$ by

$$y_i = \frac{x_i}{\left(\sum\limits_{i=1}^{n} x_i^2\right)^{1/2}} \tag{A-14}$$

Thus, it follows that the normalized vector $\mathbf{Y}$ has the property that

$$\mathbf{Y}^T\mathbf{Y} = 1 \tag{A-15}$$

For any two vectors $\mathbf{X}$ and $\mathbf{Y}$ of the same order, the products $\mathbf{X}^T\mathbf{Y}$ and $\mathbf{Y}^T\mathbf{X}$ are equal to the same scalar

$$\mathbf{X}^T\mathbf{Y} = \sum_{i=1}^{n} x_i y_i = \mathbf{Y}^T\mathbf{X} \tag{A-16}$$

Two vectors $\mathbf{X}$ and $\mathbf{Y}$ are said to be *orthogonal* if

$$\mathbf{X}^T\mathbf{Y} = 0 \tag{A-17}$$

For example, the vectors $\mathbf{X}_1$ and $\mathbf{X}_2$ are seen to be orthogonal.

$$\mathbf{X}_1^T\mathbf{X}_2 = \begin{bmatrix} 1 & -2 & 1 \end{bmatrix} \begin{bmatrix} 2 \\ 1 \\ 0 \end{bmatrix} = 0$$

The corresponding normalized vectors $\mathbf{Y}_1$ and $\mathbf{Y}_2$ are also orthogonal

$$\mathbf{Y}_1^T \mathbf{Y}_2 = [1/3 \ -2/3 \ 2/3] \begin{bmatrix} 2/\sqrt{5} \\ 1/\sqrt{5} \\ 0 \end{bmatrix} = 0$$

and

$$\mathbf{Y}_1^T \mathbf{Y}_1 = [1/3 \ -2/3 \ 2/3] \begin{bmatrix} 1/3 \\ -2/3 \\ 2/3 \end{bmatrix} = 1$$

$$\mathbf{Y}_2^T \mathbf{Y}_2 = [2/\sqrt{5} \ 1/\sqrt{5} \ 0] \begin{bmatrix} 2/\sqrt{5} \\ 1/\sqrt{5} \\ 0 \end{bmatrix} = 1$$

Orthogonal vectors $\mathbf{Y}_1$ and $\mathbf{Y}_2$ which have been normalized are said to be *orthonormal*. A system of vectors is said to be orthonormal if the normalized vectors of the system are mutually orthogonal.

If the elements x_i of $\mathbf{X}$ are complex and $\bar{\mathbf{X}}$ is a column vector that contains the corresponding complex conjugates $\bar{x}_i$ of x_i, then the product $\bar{\mathbf{X}}^T \mathbf{X}$ is a real valued scalar

$$\bar{\mathbf{X}}^T \mathbf{X} = \sum_{i=1}^{n} \bar{x}_i x_i = \sum_{i=1}^{n} |x_i|^2 \tag{A-18}$$

where $|x_i| =$ the modulus or absolute value of the complex number. For example if $x_i = \alpha_i + j\beta_i$, then $|x_i| = \sqrt{\alpha_i^2 + \beta_i^2}$, where α_i and β_i are real numbers, and j is the complex unit with property $j^2 = -1$.

For real matrices, the product $\mathbf{X}^T \mathbf{A}\mathbf{X}$ is also a scalar, whose value is given by

$$\mathbf{X}^T \mathbf{A}\mathbf{X} = \sum_{i=1}^{n} \sum_{j=1}^{n} x_i a_{ij} x_j \tag{A-19}$$

The matrix $\mathbf{A}$ is said to be *positive definite* if $\mathbf{X}^T \mathbf{A}\mathbf{X} > 0$ for all conformable vectors $\mathbf{X} \neq \mathbf{0}$. If $\mathbf{X}^T \mathbf{A}\mathbf{X} \geq 0$ for all conformable vectors $\mathbf{X} \neq 0$, then $\mathbf{A}$ is said to be *positive semidefinite*.

The vectors $\mathbf{X}_1, \mathbf{X}_2, \ldots, \mathbf{X}_n$ (all of the same order) are said to be *linearly independent* if and only if the one set of scalars $c_1, c_2, \ldots, c_n$ that satisfies the equation

$$c_1 \mathbf{X}_1 + c_2 \mathbf{X}_2 + c_3 \mathbf{X}_3 + \cdots + c_n \mathbf{X}_n = \mathbf{0} \tag{A-20}$$

is the set $c_1 = c_2 = \cdots = c_n = 0$. An example of a set of independent vectors is the set of unit vectors for three-dimensional space

$$\mathbf{E}_1 = \begin{bmatrix} 1 \\ 0 \\ 0 \end{bmatrix} \qquad \mathbf{E}_2 = \begin{bmatrix} 0 \\ 1 \\ 0 \end{bmatrix} \qquad \mathbf{E}_3 = \begin{bmatrix} 0 \\ 0 \\ 1 \end{bmatrix}$$

Note that if a set of vectors is linearly independent, no one vector of the set can be formed from a linear combination of the remaining vectors of the set. A set of vectors is said to be *linearly dependent* if there exists some set of scalars c_1, $c_2, \ldots, c_n$ not all zero that satisfy Eq. (A-20).

The following well-known theorem, which is stated without proof, is needed in subsequent developments.

Theorem A-1-1 If $\mathbf{A}$ is a square matrix of order n and $\mathbf{X}$ is a conformable column vector, the equations $\mathbf{AX} = \mathbf{0}$ have a nontrivial solution if and only if $|\mathbf{A}| = 0$.

Any solution $\mathbf{X} \neq \mathbf{0}$ is called a nontrivial solution of the equations $\mathbf{AX} = \mathbf{0}$. Note that any solution $\mathbf{X}$ is not unique because $k\mathbf{X}$ (where k is any nonzero scalar) is also a solution.

The *eigenvalues* of the matrix $\mathbf{A}$ are the scalars λ that make

$$|\mathbf{A} - \lambda\mathbf{I}| = 0 \tag{A-21}$$

The development leading to this definition follows. Consider the matrix equation

$$\mathbf{AX} = \lambda\mathbf{X}$$

Then

$$(\mathbf{A} - \lambda\mathbf{I})\mathbf{X} = \mathbf{0} \tag{A-22}$$

By Theorem A-1-1 there exists a nontrivial solution of Eq. (A-22) if and only if $|\mathbf{A} - \lambda\mathbf{I}| = 0$.

For each eigenvalue λ_i that makes $|\mathbf{A} - \lambda_i\mathbf{I}| = 0$, there exists a corresponding *eigenvector* (not unique) given by

$$\mathbf{AX}_i = \lambda_i\mathbf{X}_i \tag{A-23}$$

Properties of the Real Symmetric Matrix $\mathbf{H} = \mathbf{A}^T\mathbf{A}$ ($\mathbf{A}$ is Real, Nonsingular, and of Order n)

Since $\mathbf{A}$ is real, then the transpose $\mathbf{A}^T$ is real, and consequently $\mathbf{H}$ is real. Since $|\mathbf{A}| \neq 0$ and $|\mathbf{A}^T| \neq 0$, it follows from Eq. (A-11) that $|\mathbf{H}| \neq 0$. Thus, $\mathbf{H}$ is nonsingular. That the matrix $\mathbf{H}$ is symmetric follows immediately

$$\mathbf{H}^T = (\mathbf{A}^T\mathbf{A})^T = \mathbf{A}^T(\mathbf{A}^T)^T = \mathbf{A}^T\mathbf{A} = \mathbf{H}$$

If $\mathbf{A}$ is real and nonsingular, then for any arbitrary real vector $\mathbf{X} \neq \mathbf{0}$, $\mathbf{H}$ is positive definite. The truth of this statement is established as follows

$$\mathbf{X}^T\mathbf{HX} = \mathbf{X}^T\mathbf{A}^T\mathbf{AX} = (\mathbf{AX})^T(\mathbf{AX}) \tag{A-24}$$

Since $\mathbf{A}$ is nonsingular, it follows from Theorem A-1-1 that $\mathbf{AX} \neq \mathbf{0}$. Thus, $\mathbf{AX}$ is equal to some nonzero vector $\mathbf{Y}$, and Eq. (A-24) reduces to

$$\mathbf{X}^T\mathbf{HX} = \mathbf{Y}^T\mathbf{Y} > 0 \tag{A-25}$$

Consequently, $\mathbf{H}$ is positive definite.

It can be shown that the eigenvalues of the real symmetric matrix $\mathbf{H}(\mathbf{H} = \mathbf{A}^T\mathbf{A}$, where $\mathbf{A}$ is real and nonsingular) are real and positive. Also, if the eigenvalues of $\mathbf{H}$ are all distinct, then the eigenvectors corresponding to the different eigenvalues are mutually orthogonal.

A system of mutually orthogonal vectors is readily converted to an orthonormal system by dividing each vector of the system by the square root of the sum of the squares of its elements $[(\mathbf{X}^T\mathbf{X})^{1/2}]$. It can be shown that the vectors forming an orthonormal system are linearly independent. Thus, if the n eigenvalues of the real symmetric matrix $\mathbf{H}$ are all distinct, there exist n orthonormal and linearly independent eigenvectors. This result may be generalized because it can be shown that to each eigenvalue of a real symmetric matrix there corresponds as many linearly independent orthonormal eigenvectors as the multiplicity of the eigenvalue; see for example Hadley.[6] That is, if an eigenvalue is repeated k times, there exists k linearly independent eigenvectors corresponding to this repeated eigenvalue. Thus, for the n eigenvalues $\lambda_1 \geq \lambda_2 \geq \lambda_3 \geq \cdots \geq \lambda_n > 0$ of the real symmetric matrix $\mathbf{H}$, there exists n orthonormal and linearly independent eigenvectors.

Norms of Vectors and Matrices

In the investigation of the characteristics of series of vectors and matrices, the concept of norms of vectors and matrices proves to be quite useful. The norm of a vector $\mathbf{X}$ of order n is a nonnegative number $\|\mathbf{X}\|$ satisfying the following conditions:

1. $\|\mathbf{X}\| > 0$ for $\mathbf{X} \neq \mathbf{0}$ and $\|\mathbf{0}\| = 0$.
2. $\|c\mathbf{X}\| = |c|\,\|\mathbf{X}\|$ for any numerical multiplier c.
3. $\|\mathbf{X} + \mathbf{Y}\| \leq \|\mathbf{X}\| + \|\mathbf{Y}\|$ (the triangle inequality).

The most common norms that satisfy these three conditions are as follows:

   I. The maximum norm $\|\mathbf{X}\|_{\mathrm{I}} = \max_i |x_i|$, which means that the norm is equal to the absolute value of that element x_i whose absolute value is equal to or greater than that of any other element of $\mathbf{X}$.
  II. The absolute norm $\|\mathbf{X}\|_{\mathrm{II}} = \sum_{i=1}^n |x_i|$.
 III. The euclidean norm $\|\mathbf{X}\|_{\mathrm{III}} = (\sum_{i=1}^n |x_i|^2)^{1/2}$.

The euclidean norm is seen to be the length of the vector $\mathbf{X}$. For $\mathbf{X}$ real

$$\|\mathbf{X}\|_{\mathrm{III}} = (\mathbf{X}^T\mathbf{X})^{1/2} = \left(\sum_{i=1}^n x_i^2\right)^{1/2} \tag{A-26}$$

The euclidean norm is used exclusively in the analysis presented in the text. Furthermore, since the elements of the vectors and matrices corresponding to the processes considered in the text are real, it is supposed throughout the following development that all of the numbers are real.

It will now be shown that the euclidean norm satisfies conditions 1, 2, and 3. For $\mathbf{X} \neq \mathbf{0}$, it is evident that

$$\|\mathbf{X}\|_{\mathrm{III}} = \left(\sum_{i=1}^{n} x_i^2 \right)^{1/2} > 0$$

If $\mathbf{X} = \mathbf{0}$, then $(\sum_{i=1}^{n} 0^2)^{1/2} = 0$, and condition 1 is satisfied. The euclidean norm of $c\mathbf{X}$ is given by

$$\|c\mathbf{X}\|_{\mathrm{III}} = \left[\sum_{i=1}^{n} (cx_i)^2 \right]^{1/2} = |c| \left(\sum_{i=1}^{n} x_i^2 \right)^{1/2}$$

and condition 2 is satisfied. To show that the euclidean norm satisfies the triangle inequality, one may commence with the Schwarz inequality

$$\left(\sum_{i=1}^{n} x_i^2 \right)^{1/2} \left(\sum_{i=1}^{n} y_i^2 \right)^{1/2} \geq \sum_{i=1}^{n} x_i y_i \tag{A-27}$$

Multiplication of both sides of Eq. (A-27) by 2, followed by the addition of $\sum_{i=1}^{n} x_i^2$ and $\sum_{i=1}^{n} y_i^2$ to both sides yields an expression which is readily reduced to give

$$\left[\sum_{i=1}^{n} (x_i + y_i)^2 \right]^{1/2} \leq \left(\sum_{i=1}^{n} x_i^2 \right)^{1/2} + \left(\sum_{i=1}^{n} y_i^2 \right)^{1/2} \tag{A-28}$$

This result is recognized as the triangle inequality for the euclidean norm.

In an analogous manner, the norm of a square matrix $\mathbf{A}$ is defined as a nonnegative number $\|\mathbf{A}\|$ satisfying the following conditions:

1. $\|\mathbf{A}\| > 0$ if $\mathbf{A} \neq \mathbf{0}$ and $\|\mathbf{0}\| = 0$.
2. $\|c\mathbf{A}\| = |c| \|\mathbf{A}\|$ for any numerical multiplier c.
3. $\|\mathbf{A} + \mathbf{B}\| \leq \|\mathbf{A}\| + \|\mathbf{B}\|$.
4. $\|\mathbf{AB}\| \leq \|\mathbf{A}\| \|\mathbf{B}\|$.

Since the matrix $\mathbf{A}$ and the vector $\mathbf{X}$ generally appear together as $\mathbf{AX}$, it is found desirable from the standpoint of application to pick norms in pairs such that the following inequality is satisfied

$$\|\mathbf{AX}\| \leq \|\mathbf{A}\| \|\mathbf{X}\| \tag{A-29}$$

The norm of a matrix is said to be *compatible* with a given vector norm provided Eq. (A-29) is satisfied for any matrix $\mathbf{A}$ and any conformable vector $\mathbf{X}$. The following procedure is employed for the construction of the matrix norm such that it is compatible with a given vector norm. First, observe that the lengths of all vectors in any n-dimensional vector space span the set of all real numbers, and that the process of normalization of each vector throws the lengths of all vectors into a space containing the single element unity. The norm of the matrix **A** is taken to be the maximum of the norms of the vector **AX**, where **X** is picked from the set of all vectors having a norm of unity. That is

$$\|\mathbf{A}\| = \max_{\|\mathbf{X}\|=1} \|\mathbf{A}\mathbf{X}\| \tag{A-30}$$

First it will be shown that for any real nonsingular matrix **A** of order n, there exists a particular vector $\mathbf{X}_1$ such that $\|\mathbf{X}_1\| = 1$ and $\|\mathbf{A}\mathbf{X}_1\| = \|\mathbf{A}\|$. The matrix norm so constructed is said to be *subordinate* to the given vector norm. Then it will be shown that the matrix norm so constructed satisfies the four conditions required of a matrix norm as well as the compatibility relationship. Only the matrix norm subordinate to the euclidean vector norm $\|\mathbf{X}\|_{\text{III}}$ given by Eq. (A-26) is considered. Now consider the vector **AX**, whose vector norm as given by Eq. (A-26), is

$$\|\mathbf{A}\mathbf{X}\|_{\text{III}}^2 = (\mathbf{A}\mathbf{X})^T \mathbf{A}\mathbf{X} = \mathbf{X}^T \mathbf{A}^T \mathbf{A}\mathbf{X} = \mathbf{X}^T \mathbf{H}\mathbf{X} \tag{A-31}$$

Since **A** is real and nonsingular, it follows from the previous section that **H** is real, nonsingular, symmetric, and has eigenvalues which may be ordered as follows

$$\lambda_1 \geq \lambda_2 \geq \cdots \geq \lambda_n > 0$$

Let the corresponding orthonormal system of eigenvectors be denoted by

$$\mathbf{X}_1, \mathbf{X}_2, \ldots, \mathbf{X}_n$$

Since these eigenvectors are linearly independent, any arbitrary vector **X** with a euclidean norm of unity may be stated as a linear combination of this orthonormal set of eigenvectors

$$\mathbf{X} = c_1 \mathbf{X}_1 + c_2 \mathbf{X}_2 + \cdots + c_n \mathbf{X}_n \tag{A-32}$$

Since the norm of **X** is unity, it follows that

$$\|\mathbf{X}\|_{\text{III}}^2 = \mathbf{X}^T \mathbf{X} = c_1^2 + c_2^2 + \cdots + c_n^2 = 1$$

The factors $\mathbf{X}^T$ and **HX** in Eq. (A-31) are given by

$$\mathbf{X}^T = c_1 \mathbf{X}_1^T + c_2 \mathbf{X}_2^T + \cdots + c_n \mathbf{X}_n^T$$

and

$$\mathbf{H}\mathbf{X} = c_1 \lambda_1 \mathbf{X}_1 + c_2 \lambda_2 \mathbf{X}_2 + \cdots + c_n \lambda_n \mathbf{X}_n$$

Since the eigenvectors $\mathbf{X}_1, \mathbf{X}_2, \ldots, \mathbf{X}_n$ are orthonormal, it follows that Eq. (A-31) reduces to

$$\|\mathbf{A}\mathbf{X}\|_{\text{III}}^2 = \mathbf{X}^T \mathbf{H}\mathbf{X} = \lambda_1 c_1^2 + \lambda_2 c_2^2 + \cdots + \lambda_n c_n^2$$
$$\leq \lambda_1 (c_1^2 + c_2^2 + \cdots + c_n^2) = \lambda_1$$

The maximum value of $\|\mathbf{AX}\|_{\text{III}}$ occurs when $\mathbf{X} = \mathbf{X}_1$. For this case

$$\|\mathbf{AX}_1\|_{\text{III}}^2 = \mathbf{X}_1^T \mathbf{HX}_1 = \lambda_1 \mathbf{X}_1^T \mathbf{X}_1 = \lambda_1$$

Thus, the norm of the matrix $\mathbf{A}$ is given by

$$\|\mathbf{A}\|_{\text{III}} = \max_{\|\mathbf{X}_1\|=1} \|\mathbf{AX}_1\|_{\text{III}} = \sqrt{\lambda_1} \tag{A-33}$$

The matrix norm $\|\mathbf{A}\|_{\text{III}} = \sqrt{\lambda_1}$, where λ_1 is the largest eigenvalue of $\mathbf{H} = \mathbf{A}^T\mathbf{A}$, which is commonly referred to as the *Hilbert* or *spectral* norm of matrix $\mathbf{A}$.

Now it will be shown that the Hilbert matrix norm and the euclidean vector norm satisfy the four requirements as well as the compatibility relationship. To show that the first condition is satisfied, let $\mathbf{A}$ be any real nonsingular, nonzero matrix ($\mathbf{A} \neq \mathbf{0}$). Then there exists a vector $\mathbf{X}$ such that $\|\mathbf{X}\|_{\text{III}} = 1$ and such that the euclidean norm of the vector $\mathbf{AX}$ is a maximum and nonzero. Moreover, the particular vector $\mathbf{X}$ that satisfies these conditions is the eigenvector $\mathbf{X}_1$ corresponding to the largest eigenvalue λ_1. That is

$$\|\mathbf{A}\|_{\text{III}} = \|\mathbf{AX}_1\|_{\text{III}} = \sqrt{\lambda_1} > 0$$

If $\mathbf{A} = \mathbf{0}$, then certainly there exists a vector $\mathbf{X}$ such that

$$\|\mathbf{A}\|_{\text{III}} = \max_{\|\mathbf{X}\|=1} \|\mathbf{0X}\|_{\text{III}} = 0$$

and consequently condition 1 is satisfied. To show that condition 2 is satisfied, let c be any real number multiplier of the real nonsingular matrix $\mathbf{A}$. By definition of the euclidean vector norm, it follows that

$$\|c\mathbf{AX}\|_{\text{III}} = [c^2(\mathbf{AX})^T(\mathbf{AX})]^{1/2} = |c|[(\mathbf{AX})^T(\mathbf{AX})]^{1/2}$$
$$= |c|\|\mathbf{AX}\|_{\text{III}}$$

Since there exists an $\mathbf{X}$ having a euclidean norm of unity for which the euclidean vector norm of $\mathbf{AX}$ is a maximum, it follows that

$$\|c\mathbf{A}\|_{\text{III}} = \max_{\|\mathbf{X}\|=1} |c|\|\mathbf{AX}\|_{\text{III}} = |c|\|\mathbf{A}\|_{\text{III}}$$

and thus condition 2 is satisfied. To verify the compatibility condition, let $\mathbf{Y}$ be any real nonzero vector and let $\mathbf{X}$ denote the vector obtained by normalization of $\mathbf{Y}$. That is

$$\mathbf{X} = \frac{\mathbf{Y}}{\|\mathbf{Y}\|_{\text{III}}}$$

Now let $\mathbf{A}$ be any real nonsingular matrix of the same order as $\mathbf{Y}$. Then from the properties of the vector norm, it follows that

$$\|\mathbf{AY}\|_{\text{III}} = \|\mathbf{AX}(\|\mathbf{Y}\|)\|_{\text{III}} = \|\mathbf{Y}\|_{\text{III}}\|\mathbf{AX}\|_{\text{III}}$$

Since the vector norm of $\mathbf{AX}$ is equal to or less than the vector norm of $\mathbf{AX}_1$ (the Hilbert norm of $\mathbf{A} = \|\mathbf{A}\|_{\text{III}} = \|\mathbf{AX}_1\|_{\text{III}} = \sqrt{\lambda_1}$), it follows that

$$\|\mathbf{AY}\|_{\text{III}} \leq \|\mathbf{A}\|_{\text{III}}\|\mathbf{Y}\|_{\text{III}}$$

To verify the third condition, consider first any two square matrices $\mathbf{A}$ and $\mathbf{B}$ which are real and nonsingular. Now pick a vector $\mathbf{X}_0$ such that $\|\mathbf{X}_0\|_{\mathrm{III}} = 1$ and such that the vector norm $\|(\mathbf{A} + \mathbf{B})\mathbf{X}_0\|_{\mathrm{III}}$ is a maximum. Then the triangle inequality (condition 3 for vector norms) may be used to show that

$$\|\mathbf{A} + \mathbf{B}\|_{\mathrm{III}} = \max_{\|\mathbf{X}_0\| = 1} \|(\mathbf{A} + \mathbf{B})\mathbf{X}_0\|_{\mathrm{III}} \leq \|\mathbf{A}\mathbf{X}_0\|_{\mathrm{III}} + \|\mathbf{B}\mathbf{X}_0\|_{\mathrm{III}}$$

$$\leq \|\mathbf{A}\|_{\mathrm{III}} + \|\mathbf{B}\|_{\mathrm{III}}$$

and thus the third condition is satisfied.

To show that the condition 4 is satisfied, let $\mathbf{A}$ and $\mathbf{B}$ be any two real nonsingular matrices of the same order. Let $\mathbf{X}_0$ denote a vector with a euclidean norm of unity for which the vector norm $\|\mathbf{A}\mathbf{B}\mathbf{X}_0\|_{\mathrm{III}}$ takes on its maximum value. Then

$$\|\mathbf{A}\mathbf{B}\|_{\mathrm{III}} = \max_{\|\mathbf{X}_0\| = 1} \|\mathbf{A}\mathbf{B}\mathbf{X}_0\|_{\mathrm{III}} = \|\mathbf{A}(\mathbf{B}\mathbf{X}_0)\|_{\mathrm{III}}$$

But

$$\|\mathbf{A}(\mathbf{B}\mathbf{X}_0)\|_{\mathrm{III}} \leq \|\mathbf{A}\|_{\mathrm{III}} \|\mathbf{B}\mathbf{X}_0\|_{\mathrm{III}} \leq \|\mathbf{A}\|_{\mathrm{III}} \|\mathbf{B}\|_{\mathrm{III}}$$

Thus

$$\|\mathbf{A}\mathbf{B}\|_{\mathrm{III}} \leq \|\mathbf{A}\|_{\mathrm{III}} \|\mathbf{B}\|_{\mathrm{III}}$$

An Upper Bound of the Largest Eigenvalue of the Real, Symmetric Matrix H

Since the Hilbert norm of the matrix $\mathbf{H}$ ($\mathbf{H} = \mathbf{A}^T\mathbf{A}$, where $\mathbf{A}$ is real and nonsingular) is equal to $\sqrt{\lambda_1}$, a method is needed for approximating λ_1, the largest eigenvalue of $\mathbf{H}$. One approximation of λ_1 is given by Gerschgorin's theorem (see Smith[9]). For the special case of the symmetric matrix $\mathbf{H}$, Gerschgorin's theorem may be stated as follows:

Theorem A-1-2 The largest eigenvalue λ_1 of the real symmetric matrix $\mathbf{H}$ ($\mathbf{H} = \mathbf{A}^T\mathbf{A}$, where $\mathbf{A}$ is real and nonsingular) is equal to or less than the largest sum of absolute values of the elements of $\mathbf{H}$ along any row (or column), that is,

$$\lambda_1 \leq \max_i \sum_{j=1}^{n} |h_{ij}| \tag{A-34}$$

where λ_1 is the largest eigenvalue of $\mathbf{H}$.

PROOF Let $\mathbf{X}_1$ be the eigenvector with elements $(x_1, x_2, \ldots, x_n)$ corresponding to the eigenvalue λ_1. Let x_s denote that element of $\mathbf{X}_1$ that has the largest absolute value. The matrix equation

$$\mathbf{H}\mathbf{X}_1 = \lambda_1\mathbf{X}_1$$

may be displayed as follows

$$h_{11}x_1 + h_{12}x_2 + \cdots + h_{1n}x_n = \lambda_1 x_1$$
$$h_{21}x_1 + h_{22}x_2 + \cdots + h_{2n}x_n = \lambda_1 x_2$$
$$\cdots\cdots\cdots\cdots\cdots\cdots\cdots\cdots\cdots\cdots$$
$$h_{s1}x_1 + h_{s2}x_2 + \cdots + h_{sn}x_n = \lambda_1 x_s$$
$$h_{n1}x_1 + h_{n2}x_2 + \cdots + h_{nn}x_n = \lambda_1 x_n$$

Division of each member of the nth equation by x_s gives

$$\lambda_1 = h_{s1}\left(\frac{x_1}{x_s}\right) + h_{s2}\left(\frac{x_2}{x_s}\right) + \cdots + h_{sn}\left(\frac{x_n}{x_s}\right)$$

Since by selection

$$\left(\frac{x_i}{x_s}\right) \leq 1 \qquad (i = 1, 2, \ldots, n) \tag{A-35}$$

it follows that

$$\lambda_1 \leq |h_{s1}| + |h_{s2}| + \cdots + |h_{sn}| \tag{A-36}$$

Since **H** is symmetric, it follows that the sum of the absolute values of the elements of row s is equal to the sum of the absolute values of the elements of column s. If the sum of the absolute values of the elements of some other row is greater than the sum for row s, then this sum is also greater than λ_1. Thus, to find an upper bound of λ_1, one needs only to locate that row having the largest sum of absolute values of its elements.

A-2 THEOREMS

Although the following definitions and theorems are to be found in most texts dealing with functional analysis, they are repeated here for the convenience of the reader.

Definition A-2-1 *Continuity of $f(x)$ at x_0* The function of $f(x)$ is said to be continuous at the point x_0 if, for every positive number ε, there exists a δ_ε depending upon ε such that for all x of the domain for which

$$|x - x_0| < \delta_\varepsilon$$

then

$$|f(x) - f(x_0)| < \varepsilon$$

Definition A-2-2 *Continuity of $f(x)$ in an interval* A function which is continuous at each point in an interval is said to be continuous in the interval.

Theorem A-2-1 If the function $f(x)$ is continuous in the interval $a \leq x \leq b$ and $f(a) \lessgtr k \lessgtr f(b)$, then there exists a number c in the interval $a < c < b$ such that

$$f(c) = k$$

Theorem A-2-2 *Mean value theorem of differential calculus* If the function $f(x)$ is continuous in the interval $a \leq x \leq b$ and differentiable at every point of the interval $a < x < b$, then there exists at least one value ξ such that

$$f(b) = f(a) + (b - a)f'[a + \xi(b - a)]$$

where $0 < \xi < 1$.

Theorem A-2-3 *Mean value theorem of integral calculus* If the function $f(x)$ is continuous in the interval $a \leq x \leq b$, then

$$\int_a^b f(x)\, dx = f(\xi)(b - a)$$

where $a \leq \xi \leq b$.

Theorem A-2-4 *Taylor's theorem* If the functions $f(x)$, $f'(x)$, ..., $f^{(n)}(x)$ are continuous for each x in the interval $a \leq x \leq b$, and $f^{(n+1)}(x)$ exists for each x in the interval $a < x < b$, then there exists a ξ in the interval $a < x < b$ such that

$$f(a + h) = f(a) + hf'(a) + \frac{h^2}{2!} f^{(2)}(a) + \frac{h^3}{3!} f^{(3)}(a) + \cdots + \frac{h^n}{n!} f^{(n)}(a) + R_n$$

where $h = b - a$, and the remainder R_n is given by the formula

$$R_n = \frac{h^{n+1}}{(n+1)!} f^{(n+1)}(\xi) \qquad (a < \xi < b)$$

Theorem A-2-5 *Weirstrass M test* Let $f_1(x) + f_2(x) + \cdots + f_n(x) + \cdots$ be a series of functions of x defined in the interval $a < x < b$. If there exists a convergent series of positive constants

$$M_1 + M_2 + \cdots + M_n + \cdots$$

such that

$$|f_i(x)| \leq M_i \qquad \text{(for all } i)$$

for all x in the open interval $a < x < b$, then the series of functions is uniformly and absolutely convergent for $a < x < b$.

Definition A-2-3 A function $f(x_1, x_2, \ldots, x_n)$ of n variables $x_1, x_2, \ldots, x_n$ is said to be homogeneous of degree m if the function is multiplied by λ^m when the arguments $x_1, x_2, \ldots, x_n$ are replaced by $\lambda x_1, \lambda x_2, \ldots, \lambda x_n$, respectively.

That is, if $f(x_1, x_2, \ldots, x_n)$ is homogeneous of degree m, then

$$f(\lambda x_1, \lambda x_2, \ldots, \lambda x_n) = \lambda^m f(x_1, x_2, \ldots, x_n)$$

Theorem A-2-6 *Euler's theorem* If the function $f(x_1, x_2, \ldots, x_n)$ is homogeneous of degree m and has continuous first partial derivatives, then

$$x_1 \frac{\partial f}{\partial x_1} + x_2 \frac{\partial f}{\partial x_2} + \cdots + x_n \frac{\partial f}{\partial x_n} = mf(x_1, x_2, \ldots, x_n)$$

Theorem A-2-7 *Mean value theorem of differential calculus for multivariable functions* Let $f(x, y, z)$ be continuous and have continuous first partial derivatives in a domain D. Furthermore, let (x_0, y_0, z_0) and $(x_0 + h, y_0 + k, z_0 + l)$ be points in D such that the line segment joining these points lies in D. Then

$$f(x_0 + h, y_0 + k, z_0 + l) - f(x_0, y_0, z_0) = h\frac{\partial f}{\partial x} + k\frac{\partial f}{\partial y} + l\frac{\partial f}{\partial z}$$

where each partial is evaluated at the point

$$(x_0 + \alpha h, y_0 + \alpha k, z_0 + \alpha l) \qquad 0 < \alpha < 1$$

BIBLIOGRAPHY

1. N. R. Amundson: *Mathematical Methods in Chemical Engineering—Matrices and Their Applications*, Prentice-Hall, Englewood Cliffs, N.J., 1966.
2. Brice Carnahan, H. A. Luther, and J. O. Wilkes: *Applied Numerical Methods*, John Wiley and Sons, New York, 1969.
3. V. N. Faddeeva: *Computational Methods of Linear Algebra*, Dover, New York, 1959.
4. Carl-Erik Froberg: *Introduction to Numerical Analysis*, Addison-Wesley, Reading, Mass., 1965.
5. B. A. Frazer, W. J. Duncan, and A. R. Collar: *Elementary Matrices and Some Applications to Dynamics and Differential Equations*, Cambridge University Press, New York, 1938.
6. G. Hadley: *Linear Algebra*, Addison-Wesley, Reading, Mass., 1961.
7. C. D. Holland: *Unsteady State Processes with Applications in Multicomponent Distillation*, Prentice-Hall, Englewood Cliffs, N.J., 1966.
8. Leon Lapidus: *Digital Computation for Chemical Engineers*, McGraw-Hill Book Company, New York, 1962.
9. G. D. Smith: *Numerical Solution of Partial Differential Equations*, Oxford University Press, New York, 1965.

EQUILIBRIUM AND ENTHALPY DATA

Table B-1 Equilibrium data†

$P = 300 \text{ lb/in}^2 \text{ abs}, (K_i/T)^{1/3} = a_{1i} + a_{2i}T + a_{3i}T^2 + a_{4i}T^3 \ (T \text{ in } °R)$

Component	$a_1 \times 10^2$	$a_2 \times 10^5$	$a_3 \times 10^8$	$a_4 \times 10^{12}$
CH_4	32.718139	−9.6951405	6.9229334	−47.361298
C_2H_4	−5.177995	62.124576	−37.562082	8.0145501
C_2H_6	−9.8400210	67.545943	−37.459290	−9.0732459
C_3H_6	−25.098770	102.39287	−75.221710	153.84709
C_3H_8	−14.512474	53.638924	−5.3051604	−173.58329
$i\text{-}C_4H_8$	−10.104481	21.400418	38.564266	−353.65419
$i\text{-}C_4H_{10}$	−18.967651	61.239667	−17.891649	−90.855512
$n\text{-}C_4H_{10}$	−14.181715	36.866353	16.521412	−248.23843
$i\text{-}C_5H_{12}$	−7.5488400	3.2623631	58.507340	−414.92323
$n\text{-}C_5H_{12}$	−7.5435390	2.0584231	59.138344	−413.12409
$n\text{-}C_6H_{14}$	1.1506919	−33.885839	97.795401	−542.35941
$n\text{-}C_7H_{16}$	5.5692758	−50.705967	112.17338	−574.89350
$n\text{-}C_8H_{18}$	7.1714400	−52.608530	103.72034	−496.46551
400	2.5278960	−17.311330	33.502879	−126.25039
500	3.3123291	−16.652384	24.310911	−64.148982

† Taken from S. T. Hadden: "Vapor-Liquid Equilibria in Hydrocarbon System," *Chem. Eng. Prog.*, **44**:37 (1948).

Table B-2 Enthalpy data†

$P = 300 \text{ lb/in}^2 \text{ abs}; (h_i)^{1/2} = c_{1i} + c_{2i} T + c_{3i} T^2 (T \text{ in } °\text{R});$
$(H_i)^{1/2} = e_{1i} + e_{2i} T + e_{3i} T^2 (T \text{ in } °\text{R}), \text{Btu/lb mol}$

Component	c_1	$c_2 \times 10$	$c_2 \times 10^4$	e_1	$e_2 \times 10^4$	$e_3 \times 10^6$
CH_4	−17.899210	1.7395763	−3.7596114	44.445874	501.04559	7.3207219
C_2H_4	−7.2915000	1.5411962	−1.6088376	56.79638	615.93154	2.4088730
C_2H_6	−8.4857000	1.6286636	−1.9498601	61.334520	588.75430	11.948654
C_3H_6	−12.427900	1.8834652	−2.4839140	71.828480	658.55130	11.299585
C_3H_8	−14.500060	1.9802223	−2.9048837	81.795910	389.81919	36.470900
$i\text{-}C_4H_8$	−16.553450	2.161865	−3.1476209	139.17444	−822.39488	120.39298
$i\text{-}C_4H_{10}$	−16.5534050	2.1618650	−3.1476209	147.65414	−1185.2942	152.87778
$n\text{-}C_4H_{10}$	−20.298110	2.3005743	−3.8663417	152.66798	−1153.4842	146.64125
$i\text{-}C_5H_{12}$	−23.356460	2.5017453	−4.3917897	130.96679	−197.98604	82.549947
$n\text{-}C_5H_{12}$	−24.371540	2.5636200	−4.6499694	128.90152	2.0509603	64.501496
$n\text{-}C_6H_{10}$	−23.870410	2.6768089	−4.4197793	85.834950	1522.3917	−34.018595
$n\text{-}C_7H_{16}$	−25.314530	2.8246389	−4.5418718	94.682620	1479.5387	−19.105299
$n\text{-}C_8H_{18}$	−22.235050	2.8478429	−3.8850819	106.32806	1328.3949	1.6230737
400	−203.32192	6.3932857	−21.611909	72.328160	1893.3822	−59.003314
500	1.9205300	3.0179232	−2.2183809	138.49658	1497.8171	18.641269

† Based on data taken from J. B. Maxwell: *Data Book on Hydrocarbons*, D. Van Nostrand Company, New York, 1955.

Table B-3 Equilibrium data†

I. Vapor Pressures: Constants for the Antoine equation

$$\ln P \text{ (mmHg)} = A - \frac{B}{C + T \text{ (K)}}$$

Component	A	B	C
Toluene	16.01365	3096.516	−53.6680
Ethylbenzene	16.01951	3279.456	−59.9440
Styrene	16.01933	3328.571	−63.7200
Isopropylbenzene	15.97224	3363.593	−65.3730
1-methyl-3-ethylbenzene	16.15452	3521.067	−64.6410
α-methylstyrene	13.88532	2216.659	−132.9380
cis-1-propylbenzene	15.22500	3056.589	−95.6551

† *Selected Values of Properties of Hydrocarbon and Related Compounds*, American Research Project 44, Thermodynamics Research Center, Chemical Engineering Division, Texas Engineering Experiment Station.

Table B-4 Enthalpy data†

II. Vapor Enthalpies: Curve fit constants for the enthalpy function

$$\frac{H - H_0^\circ}{T} = A + BT + CT^2 + DT^3$$

where H is the enthalpy in the standard state of a perfect gas at the temperature T (K) in cal per gm mole, and H_0° is the enthalpy of the perfect gas at 0 K.

Component	A	B	C	D
Toluene	-1.658389	0.7856237×10^{-1}	$-0.1114677 \times 10^{-3}$	0.9747316×10^{-7}
Ethylbenzene	7.417126	0.2327917×10^{-1}	0.515253×10^{-4}	$-0.3961877 \times 10^{-7}$
Styrene	18.09991	$-0.6840097 \times 10^{-1}$	$0.28509830 \times 10^{-3}$	$-0.2388318 \times 10^{-6}$
Isopropyl-benzene	21.29170	$-0.8031827 \times 10^{-1}$	0.3468338×10^{-3}	$-0.2915282 \times 10^{-6}$
1-methyl-3-ethylbenzene	14.81064	$-0.1858342 \times 10^{-1}$	0.1802130×10^{-3}	$-0.1501775 \times 10^{-6}$
α-methyl-styrene	149.5145	-1.079530	0.2885290×10^{-2}	$-0.2408575 \times 10^{-5}$
cis-1-propyl-benzene	149.5145	-1.079530	0.2885290×10^{-2}	$-0.2408575 \times 10^{-5}$

III. Liquid Enthalpies:

$$h - H_0^\circ = H - H_0^\circ - \lambda$$

By use of the Clausius-Clapeyron equation

$$\frac{d \ln P}{dP} = \frac{\lambda}{RT^2}$$

and the Antoine equation of part I, the following expression is obtained for the latent heat of vaporization λ

$$\lambda = RT^2 \left[\frac{B}{(C + T)^2} \right]$$

† *Selected Values of Properties of Hydrocarbon and Related Compounds*, Thermodynamics Research Center, Chemical Engineering Division, Texas Engineering Experiment Division.

Table B-5 Curve fit parameters for the enthalpies used for absorbers

I. Liquid enthalpies† used for packed absorbers for the temperature range of $-25°F$ to $40°F$ at $P = 800$ lb/in² abs

Component	b_{1i}	b_{2i}	b_{3i}	b_{4i}
CO_2	0.22524075×10^4	0.54462643×10^1	$0.27910080 \times 10^{-1}$	$-0.18765335 \times 10^{-4}$
N_2	0.15837112×10^4	0.37315121×10^1	$0.17655857 \times 10^{-1}$	$-0.14662071 \times 10^{-4}$
CH_4	0.81635181×10^3	0.72064600×10^1	$0.15354034 \times 10^{-1}$	$-0.84406456 \times 10^{-5}$
C_2H_6	0.97404712×10^3	0.11454294×10^2	$0.79399534 \times 10^{-2}$	$-0.42183183 \times 10^{-6}$
C_3H_8	0.21237510×10^4	0.46383524×10^1	$0.31726830 \times 10^{-1}$	$-0.12580301 \times 10^{-4}$
$i\text{-}C_4H_{10}$	0.17543628×10^4	0.92456856×10^1	$0.30206113 \times 10^{-1}$	$-0.89584664 \times 10^{-5}$
$n\text{-}C_4H_{10}$	0.32309192×10^4	0.66175545×10^1	$0.38262386 \times 10^{-1}$	$-0.16110935 \times 10^{-4}$
$i\text{-}C_5H_{12}$	0.33611663×10^4	0.39552670×10^1	$0.54925647 \times 10^{-1}$	$-0.25869682 \times 10^{-4}$
$n\text{-}C_5H_{12}$	0.43454375×10^4	0.10596339×10^2	$0.43731511 \times 10^{-1}$	$-0.19637475 \times 10^{-4}$
$n\text{-}C_6H_{14}$	-0.44150469×10^4	0.70354599×10^2	$-0.67470074 \times 10^{-1}$	$0.60245657 \times 10^{-4}$
$n\text{-}C_7H_{16}$	0.66707016×10^2	0.18159073×10^2	$0.38164884 \times 10^{-1}$	$-0.42837073 \times 10^{-5}$
$n\text{-}C_8H_{18}$	-0.10632578×10^2	0.19229950×10^2	$0.40186413 \times 10^{-1}$	$-0.70521889 \times 10^{-6}$
$n\text{-}C_9H_{20}$	-0.79141992×10^4	0.81615143×10^2	$-0.79501927 \times 10^{-1}$	$0.83943509 \times 10^{-4}$
$n\text{-}C_{10}H_{22}$	-0.67810352×10^4	0.74108551×10^2	$-0.58315706 \times 10^{-1}$	$0.75087155 \times 10^{-4}$

† $h_i = b_{1i} + b_{2i}T + b_{3i}T^2 + b_{4i}T^3$ (T in °R), Btu/lb mol; based on data provided by the Exxon Company, Baytown, Texas.

(Continued on page 600.)

Table B-5 (*Continued*)

II. Vapor Enthalpies‡ Used for Packed Absorbers for the Temperature Range of $-25°F$ to $40°F$ at $P = 800 \text{ lb/in}^2$ abs

Component	c_{1i}	c_{2i}	c_{3i}	c_{4i}
CO_2	0.13978977×10^5	-0.96359463×10^1	$0.38228422 \times 10^{-1}$	$-0.26870170 \times 10^{-4}$
N_2	0.48638672×10^4	-0.21227379×10^1	$0.17565668 \times 10^{-1}$	$-0.11367006 \times 10^{-4}$
CH_4	0.63255430×10^4	-0.20747757×10^1	$0.18532634 \times 10^{-1}$	$-0.10630416 \times 10^{-4}$
C_2H_6	0.10628934×10^5	-0.28718834×10^1	$0.24877094 \times 10^{-1}$	$-0.13233222 \times 10^{-4}$
C_3H_8	0.13954383×10^5	-0.41930256×10^1	$0.32614145 \times 10^{-1}$	$-0.15483340 \times 10^{-4}$
$i\text{-}C_4H_{10}$	0.94088984×10^4	0.39262680×10^2	$-0.55596594 \times 10^{-1}$	$0.51507392 \times 10^{-4}$
$n\text{-}C_4H_{10}$	0.57302344×10^4	0.75117737×10^2	-0.13120884×10^0	$0.10517908 \times 10^{-3}$
$i\text{-}C_5H_{12}$	0.83081953×10^4	0.75267792×10^2	-0.12945843×10^0	$0.10845697 \times 10^{-3}$
$n\text{-}C_5H_{12}$	0.12804211×10^5	0.61654007×10^2	$-0.97365201 \times 10^{-1}$	$0.84398722 \times 10^{-4}$
$n\text{-}C_6H_{14}$	0.23001684×10^5	0.27744919×10^2	$-0.31545494 \times 10^{-1}$	$0.49981289 \times 10^{-4}$
$n\text{-}C_7H_{16}$	0.14876816×10^5	0.59342438×10^2	$-0.81853271 \times 10^{-1}$	$0.81429855 \times 10^{-4}$
$n\text{-}C_8H_{18}$	0.32793215×10^5	-0.35040283×10^2	0.11162955×10^0	$-0.42647429 \times 10^{-4}$
$n\text{-}C_9H_{20}$	0.47024656×10^5	-0.95395035×10^2	0.24547529×10^0	$-0.13209638 \times 10^{-3}$
$n\text{-}C_{10}H_{22}$	0.55238211×10^5	-0.13195618×10^3	0.32518369×10^0	$-0.18188384 \times 10^{-3}$

‡ $H_i = c_{1i} + c_{2i}T + c_{3i}T^2 + c_{4i}T^3$ (T in °R), Btu/lb mol; based on data provided by the Exxon Company, Baytown, Texas.

Table B-6 Curve fit parameters for K values used for absorbers

I. K values† used for the packed absorber for the temperature range of $-25°F$ to $40°F$ and at a pressure of 800 lb/in² abs

Component	a_{1i}	a_{2i}	a_{3i}	a_{4i}
CO_2	$-0.62822223 \times 10^{-1}$	$0.30688802 \times 10^{-3}$	$0.39996468 \times 10^{-6}$	$-0.57899830 \times 10^{-9}$
N_2	0.50596821	$-0.43488364 \times 10^{-3}$	$-0.15009991 \times 10^{-5}$	$0.34494154 \times 10^{-8}$
CH_4	0.15584934	$-0.15205775 \times 10^{-3}$	$0.50349212 \times 10^{-6}$	$-0.17713546 \times 10^{-9}$
C_2H_6	$0.91486037 \times 10^{-1}$	$-0.16355944 \times 10^{-3}$	$0.33741924 \times 10^{-6}$	$0.14797150 \times 10^{-9}$
C_3H_8	$0.37769508 \times 10^{-1}$	$-0.64491702 \times 10^{-4}$	$0.29233627 \times 10^{-6}$	$-0.48597680 \times 10^{-11}$
$i\text{-}C_4H_{10}$	$0.36708355 \times 10^{-1}$	$-0.94310963 \times 10^{-4}$	$0.28026648 \times 10^{-6}$	$0.10462797 \times 10^{-10}$
$n\text{-}C_4H_{10}$	$0.37231278 \times 10^{-1}$	$-0.13635085 \times 10^{-3}$	$0.37584653 \times 10^{-6}$	$-0.69237741 \times 10^{-10}$
$i\text{-}C_5H_{12}$	0.1947034×10^{-1}	$-0.40284984 \times 10^{-4}$	$0.14439195 \times 10^{-6}$	$0.56656790 \times 10^{-10}$
$n\text{-}C_5H_{12}$	$0.15414596 \times 10^{-1}$	$-0.34736106 \times 10^{-4}$	$0.12591028 \times 10^{-6}$	$0.73157133 \times 10^{-10}$
$n\text{-}C_6H_{14}$	$0.88765752 \times 10^{-3}$	$0.37082646 \times 10^{-4}$	$-0.40746951 \times 10^{-7}$	$0.15187203 \times 10^{-9}$
$n\text{-}C_7H_{16}$	$0.63677356 \times 10^{-2}$	$-0.64409760 \times 10^{-5}$	$0.31793974 \times 10^{-7}$	$0.78284379 \times 10^{-10}$
$n\text{-}C_8H_{18}$	$0.99674799 \times 10^{-2}$	$-0.34673591 \times 10^{-4}$	$0.82305291 \times 10^{-7}$	$0.21022392 \times 10^{-10}$
$n\text{-}C_9H_{20}$	$0.78793392 \times 10^{-2}$	$-0.23886125 \times 10^{-4}$	$0.62435951 \times 10^{-7}$	$0.25793478 \times 10^{-10}$
$n\text{-}C_{10}H_{22}$	$0.64146556 \times 10^{-2}$	$-0.16131104 \times 10^{-4}$	$0.30005250 \times 10^{-7}$	$0.30266026 \times 10^{-10}$

† $\sqrt[3]{K_i/T} = a_{1i} + a_{2i}T + a_{3i}T^2 + a_{4i}T^3$ (T in °R), Btu/lb mol; based on data provided by the Exxon Company, Baytown, Texas.

Table B-7 Equilibrium data†

$P = 50 \text{ lb/in}^2 \text{ abs}; (K_i/T)^{1/3} = a_1 + a_2 T + a_3 T^2 + a_4 T^3 \ (T \text{ in } °R).$

Component	$a_1 \times 10$	$a_2 \times 10^3$	$a_3 \times 10^6$	$a_4 \times 10^9$
CH_4	5.097584	0.2407971	−0.5376841	0.2354444
C_2H_6	−7.578061	3.602315	−3.955079	1.456571
C_3H_8	−0.1246870	4.932274	−5.430016	2.036879
$n\text{-}C_4H_{10}$	−6.460362	2.319527	−2.058817	0.6341839
$n\text{-}C_5H_{12}$	−8.381815	2.952740	−2.949674	1.053882
$n\text{-}C_6H_{14}$	−2.634813	−0.5820756	0.0990418	−0.2293738
$n\text{-}C_7H_{16}$	−0.4456271	0.2688671	1.065180	−0.5817661
500	9.139924	−3.573887	4.539999	−1.810713

† Based on data taken from *Equilibrium Ratio Data Book*, published by Natural Gasoline Association of America, 421 Kennedy Building, Tulsa, Oklahoma (1957).

‡ Taken from F. E. Hess, Ph.D. Dissertation, Texas A&M University, 1977. For steam, $i = 35$, the data were taken from Hadden and Grayson, "New Charts for Hydrocarbon Vapor-Liquid Equilibria," *Hydrocarbon Process.*, **40**(9): 207 (1961).

§ Water for all stages except the condenser-accumulator.

¶ For the pure water in the condenser, $K_i = P_i/P$. The vapor pressure data were taken from API Research Project 44, *Selected Values of Properties of Hydrocarbon and Related Compounds*, Thermodynamics Research Center, Chemical Engineering Division, Texas Engineering Experiment Station.

Table B-8 Vapor-liquid equilibrium data for the pseudo components in the pipestill example[‡]

$(K_i/T)^{1/3} = a_{1i} + a_{2i} T + a_{3i} T^2 + a_{4i} T^3$; T = temperature in °F/400;
Range 100°F $\leq$ Temp. °F $\leq$ 700°F

Component	$a_{1i} \times 10^4$	$a_{2i} \times 10^4$	$a_{3i} \times 10^3$	$a_{4i} \times 10^3$
1	84,942.239	−19,887.558	−738.57033	439.25069
2	37,663.443	45,078.397	−5,433.5339	1,605.0239
3	22,981.107	37,241.402	−3,296.8558	861.6139
4	14,534.533	27,932.276	−1,454.6345	236.77591
5	7,554.5532	29,336.869	−1,316.8658	184.44471
6	3,962.6526	26,418.903	−932.23369	97.713149
7	2,309.7991	22,918.657	−718.28079	98.482722
8	2,040.0548	19,313.782	−321.45931	−19.950431
9	2,158.6454	12,104.426	496.89309	−281.05582
10	2,032.9867	9,064.7138	794.28601	−364.73093
11	1,924.5697	6,357.7303	1,026.3977	−421.25578
12	1,741.0456	2,028.4179	1,335.7317	−483.56342
13	1,578.0995	788.79938	1,376.6444	−476.85528
14	1,427.6863	−226.54712	1,387.0991	−461.36825
15	1,414.31	−1,744.1716	1,473.7543	−474.69622
16	1,154.433	−1,595.3605	1,347.2038	−415.66348
17	994.05834	−2,229.3182	1,329.3031	−396.72015
18	922.63432	−2,649.8906	1,277.2033	−365.17737
19	833.49285	−2,464.3903	1,206.3332	−335.30425
20	829.69445	−3,034.0565	1,177.8056	−313.66712
21	733.63979	−2,951.5385	1,071.923	−267.74846
22	702.12854	−3,294.9733	990.60792	−229.31727
23	628.16242	−3,213.7041	865.85971	−179.49571
24	598.69948	−3,284.5679	777.50048	−143.821
25	561.07935	−3,212.9363	658.30091	−101.77001
26	406.66694	−2,385.7242	438.28074	−30.639625
27	315.52522	−1,594.1901	225.65800	30.828092
28	92.231454	−377.8281	153.61966	88.445616
29	−58.394251	637.07327	−148.87792	129.47591
30	−244.44226	1,620.172	−264.5609	143.78666
31	78.385574	−336.40966	14.851827	27.154798
32	28.736876	−110.61544	3.4640033	1.0991692
33	−20.244529	151.27163	−26.103035	14.875848
34	1.4685537	6.3780247	−2.2125512	2.8874915
35[‡]	51,841.823	834.21498	−1,070.65	278.80417

For water in the condenser-accumulator[¶]

$$a_{1i} = 0.1384708 \qquad a_{2i} = 11.25389$$

$$a_{3i} = -38.06869 \qquad a_{4i} = 62.49521$$

Valid for 68.02°F $\leq$ Temp. °F $\leq$ 122.05°F

Table B-9 Vapor enthalpies for the pipestill example†

$(H_i/10^5)^{1/2} = c_{1i} + c_{2i} T + c_{3i} T^2$

T = temperature °F/400;

H_i = vapor enthalpy, Btu/lb mol;

Component	$c_{1i} \times 10$	$c_{2i} \times 10^2$	$c_{3i} \times 10^5$
1	2.1988716	7.4948894	−6.778551
2	2.9833282	7.6319279	821.31019
3	3.4742411	9.8454002	715.69776
4	3.9110834	11.631621	665.87499
5	4.3247089	13.029787	732.4235
6	4.7046199	13.958863	908.22069
7	4.9473952	14.312364	1006.8035
8	5.0650597	14.528556	1050.5139
9	5.2397923	14.928689	1108.9134
10	5.3571892	15.290289	1141.2631
11	5.4736208	15.674217	1171.2766
12	5.7001889	16.349937	1235.8147
13	5.8109603	16.6436	1269.5199
14	5.9235975	17.001623	1300.351
15	6.0358735	17.35331	1330.9679
16	6.1487622	17.715556	1361.9732
17	6.26219	18.09599	1391.0851
18	6.3771977	18.515026	1417.6289
19	6.4341315	18.703899	1431.0889
20	6.5506671	19.132183	1463.9682
21	6.6689498	19.609297	1492.5246
22	6.8485831	20.369453	1530.375
23	7.0300771	21.133566	1571.4875
24	7.2136842	21.902457	1613.4967
25	7.4612264	22.948864	1666.4161
26	7.8086943	24.458663	1744.3925
27	8.2921778	26.847078	1841.4273
28	8.7716315	29.193879	1936.6453
29	9.2331601	31.936614	2016.6142
30	9.7683089	35.84527	2096.5584
31	10.152315	39.383389	2124.0245
32	10.172237	41.523173	2020.2658
33	10.132643	44.949763	839.92281
34	10.132643	44.949763	839.92281
35	4.0730043	8.4736299	−1886.8246

† Taken from F. E. Hess, Ph.D. Dissertation, Texas A&M University, 1977. For steam, $i = 35$, the data were taken from the steam tables given by J. M. Smith and H. C. Van Ness, *Introduction to Chemical Engineering Thermodynamics*, McGraw-Hill Book Company, New York (1959).

Table B-10 Liquid enthalpies for the pipestill example†

$(h_i/10^5)^{1/2} = b_{1i} + b_{2i}T + b_{3i}T^2$;
T = temperature °F/400;
h_i = liquid enthalpy, Btu/lb mol

Component	$b_{1i} \times 10^2$	$b_{2i} \times 10$	$b_{3i} \times 10^2$
1	16.852671	1.3185405	− 1.7392434
2	18.921665	1.7881551	− 1.9871128
3	22.863145	1.8596551	− 1.1429565
4	24.269822	2.4522686	− 2.7301188
5	26.53429	2.6857005	− 2.8230318
6	28.625662	2.6705425	− 1.7891758
7	29.788779	2.787255	− 1.8455584
8	30.366562	2.8459005	− 1.8739535
9	31.266992	2.936742	− 1.9161221
10	31.932373	3.0035441	− 1.9494004
11	32.603256	3.0712096	− 1.9828791
12	33.837528	3.1969409	− 2.0393011
13	34.403084	3.2551934	− 2.0627634
14	35.018279	3.3186895	− 2.0907043
15	35.6222662	3.381028	− 2.1167825
16	36.232888	3.4443953	− 2.1439566
17	36.852055	3.508743	− 2.1707096
18	37.502862	3.5759313	− 2.1988534
19	37.806915	3.6075768	− 2.2108511
20	38.467918	3.6764037	− 2.2400138
21	39.16724	3.74865	− 2.27022
22	40.243367	3.8605474	− 2.3181668
23	41.318941	3.9732338	− 2.3661674
24	42.396584	4.08584	− 2.4080059
25	43.830823	4.237531	− 2.466202
26	45.856446	4.4531328	− 2.5451993
27	48.822658	4.7701528	− 2.6528091
28	51.64276	5.0796201	− 2.7508439
29	54.586766	5.4092751	− 2.8452967
30	58.221254	5.8343151	− 2.9275716
31	60.846498	6.173097	− 2.9394836
32	60.990181	6.2911368	− 2.9264184
33	62.130268	6.2461345	− 2.5728939
34	62.130268	6.2461345	− 2.5728939
35	5.0942051	2.4548845	− 6.7540262

† Taken from F. E. Hess, Ph.D. Dissertation, Texas A&M University, 1977. For steam, $i = 35$, the data were taken from the steam tables given by J. M. Smith and H. C. Van Ness, *Introduction to Chemical Engineering Thermodynamics*, McGraw-Hill Book Company, New York (1959).

Table B-11 Liquid enthalpies[†]

$h_i = a_i + b_i T + c_i T^2$ (T in °R), Btu/lb mol

Component	a_i	b_i	c_i
Methanol	-0.3119436×10^4	-0.4145198×10	0.2131106×10^{-1}
Acetone	-0.115334×10^5	0.1770348×10^2	0.1166435×10^{-1}
Ethanol	0.4046348×10^3	-0.2410286×10^2	0.4728230×10^{-1}
Water	-0.87838059×10^4	0.1758450×10^2	0.3651369×10^{-3}
Methyl acetate	-0.1275019×10^5	0.1831477×10^2	0.1544864×10^{-1}
Benzene	-0.1048661×10^5	0.1197147×10^2	0.1898900×10^{-1}
Chloroform	-0.1061259×10^5	0.1661526×10^2	0.1018617×10^{-1}
Methylcyclohexane	-0.5748773×10^4	0.1415027×10^2	0.3715891×10^{-1}
Toluene	-0.5920285×10^4	0.1550792×10^2	0.2143342×10^{-1}
Phenol	-0.2226400×10^5	0.4839999×10^2	0.0

[†] Taken from an M.S. Thesis by S. E. Gallun, Texas A&M University, 1975.

Table B-12 Ideal gas and pure component vapor enthalpies†

$H_i = a_i + b_i T + c_i T^2 + d_i T^3 + e_i T^4$ (T in °R), Btu/lb mol

Component	a_i	b_i	c_i	d_i	e_i
Methanol	0.1174119×10^5	0.7121495×10	0.5579442×10^{-2}	$-0.4506170 \times 10^{-6}$	$-0.2091904 \times 10^{-10}$
Acetone	0.867332×10^4	0.4735799×10	0.1452860×10^{-1}	$-0.1121397 \times 10^{-5}$	$-0.2018173 \times 10^{-9}$
Ethanol	0.106486×10^5	0.7515997×10	0.1151360×10^{-1}	$-0.1682096 \times 10^{-5}$	$0.9036333 \times 10^{-10}$
Water	0.1545871×10^5	0.8022526×10	$-0.4745722 \times 10^{-3}$	0.6878047×10^{-6}	$-0.1439752 \times 10^{-9}$
Methyl acetate	0.855164×10^4	0.3952399×10	0.1489360×10^{-1}	$-0.1066871 \times 10^{-5}$	$-0.2983536 \times 10^{-9}$
Benzene	0.1266990×10^5	-0.8743198×10	0.3191165×10^{-1}	$-0.7611106 \times 10^{-5}$	0.7677461×10^{-9}
Chloroform	0.8893305×10^4	0.5674799×10	0.1125665×10^{-1}	$-0.3292178 \times 10^{-5}$	0.3790719×10^{-9}
Methylcyclomexane	0.1805154×10^5	-1.072998	0.3568761×10^{-1}	0.0	0.0
Toluene	0.1628525×10^5	0.6948364×10	0.1784207×10^{-1}	0.0	0.0
Phenol	-0.4000000×10	0.4839999×10^2	0.0	0.0	0.0

† Based on the correlation of Rihani and Doraiswany's on pp. 182 through 186 of *The Properties of Gases and Liquids*, by R. C. Reid and T. K. Sherwood, 2d ed., McGraw-Hill Book Company, New York, 1966.

Table B-13 Antoine constants†

$$\log_{10} P_i = A_i - \left(\frac{B_i}{C_i + T}\right) \qquad (P_i \text{ in mmHg}, T \text{ in } °C)$$

Component	A_i	$B_i \times 10^{-3}$	$C_i \times 10^{-2}$
Methanol	7.87863	1.47311	2.3000
Acetone	7.02447	1.16000	2.24000
Ethanol	8.04494	1.55430	2.22650
Water	7.96681	1.66821	2.28000
Methyl acetate	7.20211	1.232830	2.28000
Benzene	6.90565	1.211033	2.20790
Chloroform	6.90328	1.163030	2.27400
Methylcyclohexane	6.826890	1.276864	2.21630
Toluene	6.953340	1.343943	2.19377
Phenol	8.232267	2.251778	2.39831

† Taken from M. J. Holmes and J. Van Winkle, "Predictions of Ternary Vapor-Liquid Equilibria in Miscible Systems from Binary Data," *Ind. Eng. Chem.*, **62**(1):21 (1970).

Table B-14 Molar volume constants†

$$\alpha_i = a_i + b_i T + c_i T^2 \quad (\alpha_i \text{ in cc/g mol}, T \text{ in } °R)$$

Component	a_i	b_i	c_i
Methyl alcohol	0.6451094×10^2	-0.1095359	0.1195526×10^{-3}
Acetone	0.5686523×10^2	0.468039×10^{-2}	0.5094978×10^{-4}
Ethanol	0.5370027×10^2	$-0.1728176 \times 10^{-1}$	0.4938200×10^{-4}
Water	0.2288676×10^2	$-0.2023121 \times 10^{-1}$	0.2115899×10^{-4}
Methyl acetate	0.1359977×10^3	-0.2594739	0.2845980×10^{-3}
Benzene	0.7084254×10^2	0.8357935×10^{-2}	0.4894158×10^{-4}
Chloroform	0.6106564×10^2	0.1681204×10^{-1}	0.3676007×10^{-4}

† Taken from M. J. Holmes and M. Van Winkle, "Predictions of Ternary Vapor-Liquid Equilibria in Miscible Systems from Binary Data," *Ind. Eng. Chem.*, **62**(1):21 (1970).

Table B-15 Wilson parameters[†]

$\lambda_{ij} - \lambda_{ii}$ (cal/g mol)

Component	1 (methanol)	2 (acetone)	3 (methyl acetate)	4 (benzene)	5 (chloroform)
1 (methanol)	0.0	0.66429×10^3	0.84583×10^3	0.167946×10^4	0.170257×10^4
2 (acetone)	-0.21501×10^3	0.0	-0.65210×10^2	0.4949199×10^3	-0.73150×10^2
3 (methyl acetate)	-0.98420×10^2	0.16126×10^3	0.0	0.101700×10^2	0.791000×10^2
4 (benzene)	0.21613×10^3	-0.16790×10^3	0.20054×10^3	0.0	0.14844×10^3
5 (chloroform)	-0.37225×10^3	-0.31310×10^3	-0.43141×10^3	-0.20850×10^3	0.0

[†] Taken from J. W. Hudson and M. Van Winkle, "Multicomponent Vapor-Liquid Equilibria in Miscible Systems from Binary Parameters," *Ind. Eng. Chem.*, **9**(3): 466 (1970).

Table B-16 Wilson parameters†

$\lambda_{ij} - \lambda_{ii}$ (cal/g mol)

Component	1 (methanol)	2 (acetone)	3 (ethanol)	4 (water)
1 (methanol)	0.0	0.66408×10^3	0.59844×10^3	0.20530×10^3
2 (acetone)	-0.21495×10^3	0.0	0.38170×10^2	0.43964×10^3
3 (ethanol)	0.51139×10^3	0.41896×10^3	0.0	0.38230×10^3
4 (water)	0.48216×10^3	0.140549×10^3	0.95549×10^3	0.0

† Taken from M. J. Holmes and M. Van Winkle, "Predictions of Ternary Vapor-Liquid Equilibria in Miscible Systems from Binary Data," *Ind. Eng. Chem.*, **62**(1):21 (1970).

Table B-17 Physical constants†

Component	T_c K	P_c atm	v_c cm^3/g mol	ω	ω_H	μ Debye	η
Methanol	513.2	78.5	118.0	0.557	0.105	1.66	1.21
Acetone	508.7	46.6	213.5	0.309	0.187	2.88	0.00
Ethanol	516.0	63.0	161.3	0.637	0.152	1.69	1.10
Water	647.4	218.3	55.2	0.344	0.010	1.83	0.00
Methyl acetate	506.9	46.3	228.0	0.326	0.215	1.72	0.62
Benzene	562.0	48.6	260.1	0.211			
Chloroform	536.6	54.0	276.0	0.214	0.187	1.02	0.28

T_c = critical temperature ω_H = acentric factor of the homograph of the component
P_c = critical pressure μ = dipole moment
v_c = critical volume η = self-interaction parameter
ω = acentric factor

† Taken from J. M. Prausnitz, C. A. Eckert, R. V. Orye, and J. P. O. O'Connell, *Computer Calculations for Multicomponent Vapor-Liquid Equilibria*, Prentice-Hall, Englewood Cliffs, N.J. (1967).

Table B-18 Equilibrium and enthalpy relationships used in the solution of Examples 5-1 and 5-2†

1. *Equilibrium relationship*

The form of the equilibrium relationship used herein

$$\gamma_i^V f_i^V y_i = \gamma_i^L f_i^L x_i \tag{1}$$

may be restated in the form of Eq. (4) on p. 5 of Prausnitz, et al.†

$$\phi_i y_i P = \gamma_i x_i f_i^{OL} \tag{2}$$

which may be restated in the following form for plate j

$$y_{ji} \phi_{ji} P_j = \gamma_{ji} P_{ji}^S \phi_{ji}^{OL} \tag{3}$$

where $\phi_{ji}^{OL} = f_{ji}^{OL}/P_{ji}^S$

P_{ji}^S = saturation pressure of component i at the temperature of the mixture

$x_{ji} = l_{ji}/\sum_{i=1}^{c} l_{ji}$

$y_{ji} = v_{ji}/\sum_{i=1}^{c} v_{ji}$

When these definitions are substituted into Prausnitz's equation[18] of Chap. 4, Eq. (4) is obtained for the calculation of γ_{ji}, namely,

$$\gamma_{ji} = \left\{ \exp\left[1 - \sum_{k=1}^{c} \left(\frac{l_{jk}\Lambda_{ki}}{\sum_{m=1}^{c} l_{jk}\Lambda_{km}} \right) \right] \right\} \frac{\sum_{m=1}^{c} l_{jm}}{\sum_{m=1}^{c} l_{jm}\Lambda_{im}} \tag{4}$$

where

$$\Lambda_{ki} = \frac{v_{ji}^L}{v_{jk}^L} \exp\left[-\frac{(\lambda_{ki} - \lambda_{kk})}{RT_j} \right]$$

The Antoine equation is used to calculate P_{ji}^S, and ϕ_{ji}^{OL} was computed by use of the equations given below which were taken from RSTATE of Prausnitz, et al. Although ϕ_{ji}^{OL} should be differentiated with respect to temperature in the Newton-Raphson procedure, it was found that this step could be eliminated with the saving of considerable computational effort in the solution of Examples 5-1 and 5-2.

The following equations were used to compute ϕ_{ji}^{OL}

$\phi_{ji}^S = \exp f_1 + \omega_i f_2$

$f_1 = -3.5021358 + T_r\{5.6085595 + T_r[-3.076574 + T_r(0.57335015)]\}$

$f_2 = -3.7690418 + T_r(4.3538729 + T_r(0.3166137 + T_r(0.12666184 + T_r(-1.1662283$

$\quad + T_r(-0.10665730 + T_r(0.12147436 + T_r(0.18927037 + T_r(0.14936906 + T_r(0.024364816$

$\quad + T_r(-0.068603516 + T_r(-0.015172164 + T_r(0.012089114)))))))))))))$

$T_r = T_j/T_{ci}$

ω_i = acentric factor; see Table B-17

$\phi_{ji}^{OL} = \phi_{ji}^S \exp\left[-v_{ji}^L P_{ji}^S/(RT_j) \right] \tag{6}$

where $R = 1.987$ cal/gm mol K.

Table B-18 (*continued*)

The vapor phase fugacity coefficient ϕ_{ji} was calculated exactly as described in Chap. 3, in Prausnitz et al., and Eqs. (10) through (23) were used directly.

2. Enthalpy deviations

The enthalpy H_i of pure component i is related to its fugacity f_i by the well known thermodynamic relationship given in Eq. (7). The fugacity can be related to an equation of state through Eq. (8). The equation of state used in the Prausnitz monograph is given by Eq. (9) where Z has the usual meaning as defined by Eq. (10).

$$H_i - H_i^\circ = -RT^2\left(\frac{\partial \ln f_i}{\partial T}\right)_P \tag{7}$$

$$\ln\left(\frac{f_i}{P}\right) = \int_0^P \left(\frac{Z-1}{P}\right) dP \tag{8}$$

$$Z = 1 + B/v \tag{9}$$

$$Z = Pv/RT \tag{10}$$

Equations (9) and (10) can be used to integrate Eq. (8) and the result substituted into Eq. (7) gives Eq. (11). Equation (12) is the result of eliminating v between Eqs. (9) and (10).

$$H_i - H_i^\circ = -RT^2\left(2 - \frac{1}{Z}\right)\left(\frac{\partial Z}{\partial T}\right)_P \tag{11}$$

$$Z = \frac{1 + \sqrt{1 + 4BP/RT}}{2} \tag{12}$$

In the calculation of ϕ_{ji}^V, a mixture virial B is calculated using mixing rules described in Chap. 3 of the monograph. If B is substituted into Eqs. (11) and (12), the result is the virtual value of the partial molar enthalpy, Ω. Rigorous application of the Almost Band Algorithm requires that the derivative of Ω be calculated with respect to temperature for use in the convergence procedure. This was not done in the solution of Examples 5-1 and 5-2.

---

† J. M. Prausnitz, C. A. Eckert, R. V. Orye, and J. P. O'Connell, *Computer Calculations for Multicomponent Vapor-Liquid Equilibria*, Prentice-Hall, Englewood Cliffs, N.J. (1967).

Table B-19 Vapor-liquid equilibrium data

$y_i = \gamma_i K_i x_i$

1. K values†

Component	Vapor-liquid equilibrium ratio
Acetic acid (1)	$K_1 = 0.001 \quad t \leq 74.45°C$ $K_1 = 2.25 \times 10^{-2}t - 1.666 \quad t > 74.45°C$
Ethanol (2)	$\log_{10} K_2 = -2.3 \times 10^3/T + 6.58825,\ T$ in K
Water (3)	$\log_{10} K_3 = -2.3 \times 10^3/T + 6.48351,\ T$ in K
Ethyl acetate (4)	$\log_{10} K_4 = -2.3 \times 10^3/T + 6.74151,\ T$ in K

2. Activity coefficients‡

$$\log_{10} \gamma_1 = A_1 x_2^2 + A_2 x_3^2 + A_3 x_4^2 + A_4 x_2 x_3 + A_5 x_2 x_4 + A_6 x_3 x_4$$
$$+ A_7 x_1 x_2^2 + A_8 x_1 x_3^2 + A_9 x_1 x_4^2 + A_{10} x_1 x_2 x_3 + A_{11} x_2 x_3 x_4$$
$$+ A_{12} x_3 x_4 x_1 + A_{13} x_4 x_1 x_2 + A_{14} x_2 x_3^2 + A_{15} x_2 x_4^2 + A_{16} x_3 x_4^2$$

A_i	Acetic acid (1)	Ethanol (2)	Water (3)	Ethyl acetate (4)
1	−0.554296	0.581778	0.688636	−0.0601361
2	−0.324357	0.209245	0.0243031	0.229575
3	−0.103685	−0.257329	0.375534	1.86575
4	−0.705455	−0.562636	1.27548	0.355191
5	−2.01335	−0.314853	1.77863	0.468416
6	−2.25362	0.451732	0.696279	1.5110
7	0.837926	−0.115411	0.936722	−0.0599682
8	0.523760	0.069531	0.449357	0.0673994
9	0.434061	0.0740529	0.717790	−3.15997
10	−0.534056	0.187010	1.44979	0.941858
11	−3.25231	−0.369985	−2.11099	−1.92225
12	5.90329	−0.082339	0.746905	−0.755731
13	3.35400	−0.409472	1.12914	1.03791
14	0.197296	1.09247	0.120436	0.365254
15	−0.452660	0.192416	−1.64268	−1.36587
16	0.014715	−0.172565	0.330018	−2.13818

The remaining activity coefficients are obtained by rotating the subscripts on the x's as follows:

$$1 \to 2 \to 3 \to 4 \to 1$$

† M. Hirata and H. Komatsu, "Vapor-Liquid Equilibrium Relations Accompanied with Chemical Reaction," Kagaku Kogaku Abr. Ed., **5**(1): 143 (1967).

‡ I. Suzuki, H. Komatsu, and M. Hirata, "Formulation and Prediction of Quaternary Vapor-Liquid Equilibria Accompanied by Esterification," *J. Chem. Eng. Jpn.*, **4**(3): 26 (1971).

Table B-20 Enthalpies of perfect gases and latent heats of vaporization†

1. Enthalpy data for the perfect gas state

Component	$H_i^\circ = \Delta H_{F298,i}^\circ + \int_{298}^{T} c_p^\circ \, dT$ (cal/g mol)
Acetic acid	$H^\circ = -106.607.89 + 1.156T + 3.0435 \times 10^{-2}T^2 - 1.3956 \times 10^{-5}T^3 + 2.955 \times 10^{-9}T^4$
Ethanol	$H^\circ = -58855.738 + 2.153T + 2.5565 \times 10^{-2}T^2 - 6.68 \times 10^{-6}T^3 + 8.2 \times 10^{-11}T^4$
Water	$H^\circ = 60135.845 + 7.701T + 2.2975 \times 10^{-4}T^2 + 8.4 \times 10^{-7}T^3 - 2.15 \times 10^{-10}T^4$
Ethyl acetate	$H^\circ = -110238.9 + 1.728T + 4.8625T^2 - 1.665 \times 10^{-5}T^3 + 1.729 \times 10^{-9}T^4$, where T is in degrees Kelvin

2. Latent heats of vaporization‡

Component	λ_i at normal boiling point (cal/g mol)	T_B Normal boiling point temperature	T_c Critical temperature
Acetic acid	5600	391.1	594.4
Ethanol	9260	351.5	516.2
Water	9717	373.2	647.3
Ethyl acetate	7700	350.3	523.2

$$\lambda_{i,2} = \lambda_{i,1}\left(\frac{1 - T_{r,2}}{1 - T_{r,1}}\right)^{0.38} \qquad T_r = T/T_c$$

† *Selected Values of Properties of Hydrocarbon and Related Compounds,* Thermodynamics Research Center, Chemical Engineering Division, Texas Engineering Experiment Station.

‡ R. C. Reid and T. K. Sherwood, *The Properties of Gases and Liquids,* McGraw-Hill Book Company, New York, 1977.

‡ D. S. Viswanath and Gowd-Den Su, "Generalized Thermodynamic Properties of Real Gases," *AIChE J.,* **11**(2):202 (1965).

§ H. J. Arnikar, T. S. Rao, and A. A. Bodne, "A Gas Chromatographic Study of the Kinetics of Uncatalysed Esterification of Acetic Acid by Ethanol," *J. Chromatogr.,* **47**:265 (1970).

Table B-21 Vapor and liquid enthalpies

The vapor and liquid enthalpies were calculated by use of the following values of the virtual values of the partial molar enthalpies

$$\hat{H}_i = \hat{H}_i^\circ + \Omega$$

$$\hat{h}_i = \hat{H}_i - \lambda_i(T)$$

where $\Omega = -RT^2(\partial \ln f/\partial T)_{P, M}$

The following expression based on the BWR equation of state was used for computing Ω

$$\Omega = (B_0 RT - 2A_0 - 4C_0/T^2)\rho + \left(\frac{2bRT - 3a}{2}\right)\rho^2 + \frac{6a\alpha\rho^5}{5}$$

$$+ \frac{c\rho^2}{T}\left[3\frac{1 - \exp\left(-\gamma\rho^2\right)}{\gamma\rho^2} + \left(-\frac{1}{2} + \gamma\rho^2\right)\exp\left(-\gamma\rho^2\right)\right]$$

where P is in lb/in^2 abs, T is in °R, and ρ is the molar density.
The following coefficients were used for the pure components.

1. Coefficients for the BWR equation for state‡

Coefficient	Acetic acid	Ethanol	Ethyl acetate	Water
B_0	1.04651	0.81653	1.39142	0.29786
A_0	3,804.33	26,001.91	4.452×10^4	1.178×10^4
C_0	3.83×10^{10}	1.97×10^{10}	3.47×10^{10}	1.40×10^{10}
b	6.96508	4.31489	12.31281	0.5643
a	94,930.17	51,071.14	14,7709.5	8,368.41
c	1.593×10^{11}	6.46×10^{10}	1.92×10^{11}	1.66×10^{10}
αa	6.2865×10^5	1.6491×10^5	2.991×10^6	1.2778×10^3
γ	11.2486	6.9685	19.885	0.91127

2. Mixing rules

$$B_0 = \sum_{i=1}^{c} N_i B_{0i} \qquad a = \left(\sum_{i=1}^{c} N_i a_i^{1/3}\right)^3 \qquad A_0 = \left(\sum_{i=1}^{c} N_i A_{0i}^{1/2}\right)^{1/2} \qquad b = \left(\sum_{i=1}^{c} N_i c_i^{1/3}\right)^3$$

$$C_0 = \left(\sum_{i=1}^{c} N_i C_{0i}^{1/2}\right)^2 \qquad \alpha = \left(\sum_{i=1}^{c} N_i \alpha_i^{1/3}\right)^3 \qquad b = \left(\sum_{i=1}^{c} N_i b_i^{1/3}\right)^3 \qquad \gamma = \left(\sum_{i=1}^{c} N_i \gamma_i^{1/2}\right)^2$$

3. Rate of reaction data§

For the reaction:

$$CH_3COOH + C_2H_5OH \underset{k'}{\overset{k}{\rightleftharpoons}} C_2H_5COOH + H_2O \quad \text{or} \quad A + B \underset{k'}{\overset{k}{\rightleftharpoons}} C + D$$

$$r_{jA} = k_{jA}C_A C_B - k_{jA}C_C C_D$$

$$k_{jA} = 29{,}000 \exp\left(-7150/T\right) \qquad \text{liters/g mol min}$$

$$k'_{jA} = 7{,}380 \exp\left(-7150/T\right) \qquad \text{liters/g mol min}$$

where T is in degrees Kelvin.

Table B-22 Enthalpies of perfect gases and latent heats of vaporization

1. Vapor enthalpies[†]

Component	$H_i^\circ = H_{f298, i}^\circ + \int_{298}^{T} c_p^\circ \, dT \; \text{cal/g mol}$

Component	
Monoethanol-amine	$H^\circ = -51,670 + 2.224T + 3.594 \times 10^{-2}T^2 - 1.447 \times 10^{-5}T^3 + 2.78 \times 10^{-9}T^4$
Water	$H^\circ = -60,135.8 + 7.701T + 2.2975 \times 10^{-4}T^2 + 8.4033 \times 10^{-7}T^3 \\ \quad - 2.1475 \times 10^{-10}T^4$
Carbon dioxide	$H^\circ = -96,130 + 4.728T + 8.77 \times 10^{-3}T^2 - 4.460 \times 10^{-6}T^3 + 1.024 \times 10^{-9}T^4$
Amine carbamate	$H^\circ = -229,500 + 20.26T + 6.14 \times 10^{-2}T^2 - 1.41 \times 10^{-5}T^3 - 1.29 \times 10^{-9}T^4$
Methane	$H^\circ = -19,840 + 4.598T + 6.225 \times 10^{-3}T^2 + 9.533 \times 10^{-7}T^3 - 6.758 \times 10^{-10}T^4$

2. Latent heats of vaporization[‡]

Component	λ_i at normal boiling point (cal/g mol)	T_{BP} (K)	T_c (K)
Monoethanolamine	12,000	443.5	614
Water	9,717	373.2	647.3
Carbon dioxide	4,100	194.7	304.2
Amine carbamate	17,162	350.13	417.3
Methane	1,955	111.7	190.6

where T_{BP} = normal boiling point temperature
T_c = critical temperature

At any temperature T, the latent heat λ_i may be approximate by use of

$$\lambda_{i, 2} = \lambda_{i, 1}\left(\frac{1 - T_{r, 2}}{1 - T_{r, 1}}\right)^{0.38}$$

where $T_r = T/T_c$

3. Liquid enthalpies

The liquid enthalpies were obtained by use of the vapor enthalpies and the latent heats of vaporization as follows

$$\hat{h}_i = \hat{H}_i - \lambda_i(T)$$

† R. C. Reid, J. M. Prausnitz, and T. K. Sherwood, *The Properties of Gases and Liquids*, McGraw-Hill Book Company, New York, 1977.
‡ R. C. Reid and T. K. Sherwood, *The Properties of Gases and Liquids*, McGraw-Hill Book Company, New York, 1966.

Table B-23 Vapor liquid equilibrium data a rate data

1. Vapor pressures†

Component	Vapor pressure P (mmHg)
Monoethanolamine	$P = \exp\left[17.8174 - 3988.33/(T - 86.93)\right]$
Water	$P = \exp\left[18.3036 - 3816.44/(T - 46.13)\right]$
Carbon dioxide	$P = \exp\left(11.1661 + 0.015359T\right)$
Amine carbamate	$P = \exp\left[17.766 - 3980/(T - 95)\right]$
Methane	$P = \exp\left(14.2005 + 0.010486T\right)$
	where T is in degrees Kelvin

2. Rate of reaction data‡

$$CO_2 + 2HOC_2H_4NH_2 \longrightarrow HOC_2H_4NH_3^+ + HOC_2H_4NHCOO^-$$

or

$$A + 2B \longrightarrow C + B$$

$$r_{jA} = k_{jA}C_A C_B$$

$$\log_{10} k_{jA} = 10.99 - 2152/T$$

where T is in degrees Kelvin and r_{jA} is in lb mol/(h ft^3) and k_{jA} is in ft^3/(lb-moles h).

† J. E. Davies and J. J. McKetta, "Solubility of Methane in Water," *Pet. Refiner*, **39**(3):205 (1960).

† G. Houghton, A. M. McLean, and P. D. Ritchie, "Compressibility, Fugacity, and Water-Solubility Carbon Dioxide in the Region 0–36 atm and 0–100°C," *Chem. Eng. Sci.*, **6**:132 (1957).

‡ H. Hikita, S. Asai, H. Ishikawa, and M. Honda, "The Kinetics of Reactions of Carbon Dioxide with Monethanolamine, Diethanolamine, and Triethanolamine by a Rapid Mixing Method," *Chem. Eng. J.* (printed in the Netherlands), **13**:7 (1977).

Table B-24 Equilibrium data†

$P = 400$ lb/in^2 abs; $(K_i/T)^{1/3} = a_{1i} + a_{2i}T + a_{3i}T^2 + a_{4i}T^3$ (T in °R)

Component	$a_1 \times 10$	$a_2 \times 10^3$	$a_3 \times 10^6$	$a_4 \times 10^9$
CH_4	-3.2551482	2.3553786	-3.1371170	1.3397973
C_2H_6	-2.7947232	1.4124232	-1.4582948	0.50974162
C_3H_8	-2.7980091	1.1811943	-1.0935041	0.35180421
i-C_4H_{10}	2.3209137	0.87122379	-0.66100972	0.1667774
n-C_4H_{10}	-2.3203344	0.83753226	-0.61774360	0.15243376
i-C_5H_{12}	-0.6981454	0.099962037	0.39689556	-0.29076073
n-C_5H_{12}	0.37103008	-0.36257004	0.99113800	-0.54441110
500	1.9642644	-0.81121972	1.0586630	-0.39478662

† Taken from *Equilibrium Ratio Data Book*, published by National Gasoline Association of American, 421 Kennedy Building, Tulsa, Oklahoma (1957).

AUTHOR INDEX

SUBJECT INDEX